PLANT CELL BIOL

PLANT CELL BIOLOGY

Editors

William V. Dashek
Retired from Adult Degree Program
Mary Baldwin College, Richmond/
Staunton, Virginia, USA

Marcia Harrison
Department of Biological Sciences
Marshall University, Huntington
West Virginia, USA

CRC Press
Taylor & Francis Group
Boca Raton London New York

CRC Press is an imprint of the
Taylor & Francis Group, an **informa** business

ISBN (Hardcover) 1-57808-408-3 [10 digits]
978-1-57808-408-1 [13 digits]
ISBN (Paperback) 1-57808-376-1 [10 digits]
978-1-57808-376-3 [13 digits]

Preface

Although there is a copious supply of cell biology textbooks, most are animal-oriented. The few plant cell biology textbooks are, in the main, not textbooks but expensive, methods-oriented, research volumes. Thus, there is need for a plant cell biology textbook for university undergraduates.

This textbook stresses concepts and is inquiry-oriented. To this end, there is extensive use of original research literature. As we live in an era of literature explosion, one must be selective. These judgements will naturally vary with the investigator. In establishing significance the input of colleagues was considered.

In addition to provision of select research literature, this volume presents citations and summaries of certain laboratory methods. In this connection, the textbook stresses quantitative data to enhance the student's analytical abilities. Thus, the volume contains computer-spread sheets and references to statistical packages, e.g. Harvard Graphics and Statistica.

In short, while the volume contains basic facts, the intent is to gain an appreciation for the scientific method and major research trends in plant cell biology.

William V. Dashek
Marcia Harrison

Dedication and Acknowledgment

Dr. Dashek dedicates this volume to his children, Kristin Ann Simpson and Karin Ann Bryant, who patiently dealt with his need for scholarship. He also thanks the following for his scientific development: Dr. W.G. Rosen, the late Dr. W.F. Millington, Dr. D.T.A. Lamport and the late Dr. J.E. Varner. Dashek extends his gratitude to Ms. Deanna Smith for her patience with clerical assitance.

Dr. Harrison is grateful to Susan Weinstein for careful reading and helpful advice concerning her manuscript. We are grateful to the late Ms. Margaret Majithia, copy-editor for Science Publishers, for her very thorough review of the manuscripts.

Contents

List of Contributors

Bowlby, Neil
Department of Biochemistry and Molecular Biology, Michigan State University, 212 Biochemistry, East Lansing, MI 48824, USA

Brewer, Karen J.
Department of Chemistry, Virginia Tech, Blacksburg, VA 24061-0212, USA

Dashek, William V.
Retired from Adult Degree Program, Mary Baldwin College, Staunton, Virginia, USA

Harrison, Marcia
Department of Biological Sciences, Marshall University, Huntington, West Virginia 25755, USA

Hoober, J. Kenneth
School of Life Sciences, Center for Photosynthesis, Arizona State University, P.O. Box 874501, Tempe, Arizona 85287, USA
E-mail: khoober@asu.edu

Kaneko, T.S.
Japan Women's University, Department of Biological and Chemical Sciences, Tokyo, Japan

Malmström, Susanna
Laboratoire de Biochimie et Phisiologie Moleculaire des Plantes, Agro-M/Université Montpelier II/INRA/CNRS UMR 5004, Place Viala, 34060 Montpellier cedex 1, France

Miglani, G.S.
Department of Genetics and Biotechnology, Punjab Agricultural University, Ludhiana, Punjab 141004, India. E-mail: gsmiglani45@yahoo.co.in

Shepherd, Virginia

University of New South Wales, Department of Physics, Sydney 2052, Australia

Śnieżko, Renata

Maria Curie-Sklodowska University, Institute of Cell Biology, Academicka 19, 20-033 Lublin, Poland

CHAPTER 1

Introduction

William V. Dashek

WHAT IS CELL BIOLOGY?

Cell biology is the study of cellular form and function at a microscopic and biochemical level. In contrast, molecular biology is concerned with investigating the structure and function of the biological macromolecules which comprise cells. Cell biology draws upon microscopy, biochemistry, immunology and to some extent molecular genetics. The topics that cell biology encompasses are: chemistry and function of biomolecules, cells and their organelles, movement of molecules across membranes, mitosis and meiosis, metabolism, photosynthesis, and cell signaling. Some cell biologists may have additions to this list.

RESEARCH METHODS OF CELL BIOLOGY

Table 1.1 presents the microscopical methods for investigating plant cells and their inclusions. It is apparent that immunology and microscopy have been wedded as immunocytochemistry and immunoelectron microscopy for the localization of cellular antigens. In addition, these immunomicroscopic methods have been employed to elucidate the "machinery" of mitosis and meiosis. Biochemical methods (Tables 1.2–1.9) have often been used by certain cell biologists to gain an understanding of the chemical composition and function of cells and their organelles. The reader is referred to Dashek (1997) for biochemical methods to isolate and characterize molecules other than macromolecules. There has been a growing trend to link cellular and molecular biology with a special interest in elucidating the genes regulating the biosynthetic pathway of cellular chemicals.

LITERATURE ON CELL BIOLOGY

The supply of cell biological research literature is copious, as evidenced by the bibliographies presented here. A useful Internet source book for cellular and molecular biologist is that of Cabibbo et al. (2004). With regard to cell biology facts, there is a variety of cell biology textbooks and monographs as well as cell and molecular images and videos on line (http://www.cellbio.com/images.html).

TABLE 1.1 Summary of light and electron microscopic techniques

Light microscopy		Electron microscopy and ancillary methods	
Technique	Application	Technique	Application
Bright field	Conventional microscopy	Atomic force	Mapping of surfaces to an atomic scale
Confocal scanning optical microscopy	Examination of cells in live tissue in bulk samples	Cryoelectron microscopy	Imaging of biological macromolecules in the absence of specimen dehydration and staining
Confocal fluorescence	DNA labeled with more than one fluorescent tag		
Dark field	Visualization technique for ashes produced by micro-incineration and fluorescence microscopy; useful for low-contrast subjects	Electron systems imaging EM shadowing	Detection, localization and quantitation of light elements Structural information from ordered arrays of macromolecules
Reflection contrast	Quantification in gap between light and EM microscopies	Immunoelectron	Localization of cellular antigens
Reflection-imaging microscopy	Useful for imaging highly reflective particles such as silver grains in autoradiographs		
Field ion microscopy	Atomic structure of crystals		
Nearfield scanning optical	Determination of single molecules on surface	Negative staining	Useful for detergent-extracted cytoskeletons, membrane fractions, organelles
Nuclear magnetic resonance microscopy	High-resolution 3-D imaging of living plants; forms images if H_2O in the body; water distribution and binding in transpiring plants and H_2O transport in plants with light-stressed foliage	Scanning electron microscopy	Surface topography
Nomarski differential interference contrast	Reveals edges in biological structures, e.g. organelle and nuclear boundaries, cell walls; also images fibrous subcellular components, e.g. microtubules	Scanning tunneling microscopy surface spectroscopy	Surface topography, image internal structure of macromolecules such as proteins, liquid crystals, and DNA
Phase contrast microscopy	Produces visible differences in retardation of light waves, useful for biological material which possesses limited inherent direct contrast	Transmission electron microscopy	Subcellular morphology
Polarization microscopy	Most useful for highly birefringent objects, e.g., cellulose microfibrils in cell walls and distinguishing crystalline and noncrystalline inclusions	X-ray microanalysis	Detection, localization and quantitation of elements
Raman microscopy	Analysis of bioaccumulations in plant vacuoles		

N.B. See microscopy reference at end of this chapter

TABLE 1.2 Summary of methods for separating and/or detecting sugars

Technique	Reference
Colorimetric detection of sugars	Dische (1962)
Chromatographic separations of sugars	
Column-carbon, celite, extrusion, gel permeation, ion exchange	Ares.umimet. edu.ve/quimica/ bpqi/ O2chromatog.pdf.
Gas liquid chromatography	Eklund et al. (1977)
High performance liquid chromatography	Rassi (1995)
Paper and thin layer chromatographies	Dashek (1997)

TABLE 1.3 Methods for the structural analysis of carbohydrates[a]

Technique	Reference
Chiral determination	James (1995)
Glycosidic link determination	Charlson et al. (1962)
Melting point determinations	Thompson and Wolfrom (1962)
C-Methyl determination	Maciak (1962)
Nuclear magnetic spectroscopy	Carpita et al. (1991) Vliegenthart et al. (1983)
Oligosaccharide sequencing	GlycoFace (1994)
Primary hydroxyl group determination	Lewis et al. (1962)
Structural determination of glycoprotein N-glucans	Schaumann et al. (1993)

[a]The reader is referred to Dashek, W.V. 1997 for methods pertaining to other molecules in plant cells and tissues.

TABLE 1.4 Summary of lipid separation techniques

Procedure	Reference
Solvent fractionation– Acetone precipitation	Kates (1982)
Column chromatography Adsorption Ion exchange Partition	Kates (1982)
Gas liquid chromatography	Shibamoto (1994)
High-pressure liquid chromatography	Kautsky (1981) Moreau (1990)
Paper and thin layer chromatography	Kates (1982)

TABLE 1.5 Summary of macromolecular lipid analysis

Method	Application
Acid analysis	Measures extent to which hydrolysis liberates fatty acids
Anisidine method	Measures of oxidation of secondary products
Gas chromatography/ mass spectrometry	Lipid structure analysis, e.g. sphingolipid profiling
Liquid chromatography	Separation of lipid classes
Nuclear magnetic resonance	Structural change of lipoprotein lipids
Saponification value	Mean molecular weight of the component fatty acids
Unsaponifiable matter content	Measure of proportion of lipid material other than fatty acids

TABLE 1.6 Summary of nucleic acid and separation techniques

Procedure	Reference
Gel electrophoresis	Allen and Budowle (1994)
High-pressure chromatography	Jones (1995) Lai and Birren (1991) Rickwood and Harris (1990) Tietz (1998) Brown (1984)

TABLE 1.7 Summary of nucleic acid structure research techniques

Procedure	Reference
DNA sequencing	Alphey (1997)
	Ball (1996)
	Brown (1984)
	Howe and Ward (1990)
Hybridization techniques	Hanes and Higgins (1985)
Nuclear magnetic resonance	Roberts (1993)
	Jones (1995)

TABLE 1.8 Methods for the separation of amino acids and peptides

Technique	Reference
Electrophoresis	Hedges et al. (1992)
	Rabilloud (2000)
Gas chromatography	Husek and Macek (1975)
	Kataoka et al. (2000)
	Zumwalt and Kuo (1987)
High performance liquid	Hill et al. (1979)
	Jen-Kin (1984)
	Hancock (1998)
	Wilkinson (1998)
	Cohen (2000)
	Kochhar et al. (2000)
Ion exchange chromatography	Jandik (2000)
Paper chromatography	Heilman (1992)
	Brenner and Niederwiser (1960)
Thin layer chromatography	Heilman (1992)

TABLE 1.9 Some relevant references for advanced methods for determining protein structure

Method	Reference
Gas chromatography/ mass spectrometry	McMaster and McMaster (1998)
High performance liquid chromatography	Tempst et al. (1987)
	Wakefield (1986)
	Hearn (1991)
	Mant and Hodges (1991)

Infrared spectroscopy	Singh (2000)
	Twardowski and Anzenbacker (1994)
Mass spectrometry	Chapman (1996)
	Johnstone and Rose (1996)
	Chapman (2000)
	Corthals et al. (2000)
Nuclear magnetic resonance	Cavanaugh et al. (1996)
Spectroscopy	Reid (1997)
Raman spectroscopy	Pelletier (1999)
	Twardowski and Anzenbacker (1994)
Sequence analysis	Inman and Apella (1986)
	Wittman-Liebold et al. (1986)
	Jornvall et al. (1991)
	Bryan and Smith (1996)
	Imahori and Sakiyama (1986)

These include: cells alive, common molecules page, the MIT hypertextbook, molecules and online service for biology. Table 1.10 displays certain biology online services. Of special is the online cell biology lab manual of W.H. Heidcamp. Finally cell biology practice problems have been published by MIT (http://www.cellbio.com/images.html). Karp and Pruitt (1999) have published problems in paperback form.

TABLE 1.10 Summary of certain cell biology online sources[a]

www.Cellbio.com
www.Nature.com/ncb/
cellbio.utmb.edu/cellbio/
www.ingenta.com/journals/browse/urban
www.cbc.umm.edu/nmwd/cell.html
www.mcb.harvard.edu/biolinks.html
users.vcn.com/jkimball.ma.ultranet/biologypages
www.trends.com/tcb/default/htm
www.campcell.appstate.edusun.science.wayne.edu/'cellbiol/

[a]There are 2,260,000 online cell biology sources. It is important that the students discriminate between websites by educators and those of noneducators. Those with edu in the address are usually prepared by educators.

References

MICROSCOPY

Burrells, W. 1977. *Microscope Technique. A Comprehensive Handbook for General and Applied Microscopy.* Halsted Press, Wiley, New York, NY.

Cherry, R.J. 1991. *New Techniques of Optical Microscopy and Microspectroscopy.* CRC Press, Boca Raton, FL, USA.

Gersh, I. 1973. *Submicroscopic Cytochemistry.* Academic Press, New York, NY.

Goldstein, J.L. 1981. *Scanning Electron Microscopy and X-ray Microanalysis: A Text for Biologists.* Material Scientists and Geologists, Plenum Press, New York, NY.

Hall, J.L. 1978. *Electron Microscopy and Cytochemistry of Plant Cells.* Elsevier North Holland Biomed. Press, Amsterdam.

Harder, D.P. 1992. *Image Analysis in Biology.* CRC Press, Boca Raton, FL, USA.

Harris, N. and Oparks, K.J. (eds.). 1994. *Plant Cell Biology.* IRL Press, Oxford, UK, pp. 156-157.

Hawes, C. 1994. Electron microscopy. In: *Plant Cell Biology.* Harris, N. and Oparka, K.J. (eds.). IRL Press, Oxford, UK, pp. 52-68.

Hayat, M.A. 1980. *X-ray Microanalysis in Biology.* Univ Park Press, Baltimore, MD, USA.

Herman, B. and LeMasters, J.J. 1993. *Optical Microscopy.* Academic Press, San Diego, CA, USA.

Jahne, B. 1997. *Practical Handbook on Image Processing for Scientific Applications.* CRC Press, Boca Raton, FL, USA.

Jones, C., Mulloy, B. and Thomas, H. 1994. *Optical Spectroscopy and Macroscopic Techniques.* Humana Press, Totowa, NJ, USA.

Juniper, B.C., Cox, C.C., Gilchrist, A.J. and Williams, P.R. 1970. *Techniques for Plant Electron Microscopy.* Blackwell Sci Publ, Oxford, UK.

Marmasse, C. 1980. *Microscopes and Their Uses.* Gordon and Breach, New York, NY.

Mohanty, S.B. 1982. *Electron Microscopy for Biologists.* Charles Thomas, Springfield, IL, USA.

Othmar, M. and Amrein, M. 1993. STM and SRM in Biology. Academic Press, San Diego, CA.

Posteck, M.A., Howard, K.S., Johnson, A.H. and McMichael, K.L. 1980. *Scanning Electron Microscopy: A Student's Handbook.* Ladd Res Indus, Inc., Burlington, VT, USA.

Reed, S.J.B. 1993. *Electron Microprobe Analysis.* Cambridge Univ Press, Cambridge, MA, USA.

Russ, J.C. 1995. *The Image Processing Handbook.* CRC Press, Boca Raton, FL, USA.

Shaw, P.J. and Rawlins, D.J. 1994. An introduction to optical microscopy for plant cell biology. In: *Plant Cell Biology, A Practical Approach,* pp. 1-26. Harris, N. and Oparka, K.J. (eds.). Oxford Univ Press, Oxford, UK.

Shotton, D. (ed.). 1993. *Electronic Light Microscopy. Modern Biomedical Microscopy.* Wiley-Liss, New York, NY.

Sigee, D.C. 1993. *X-ray Microanalysis in Biology: Experimental Techniques and Applications.* Cambridge Univ Press, Cambridge, UK.

Slayter, E.M. 1993. *Light and Electron Microscopy.* Cambridge Univ Press, Cambridge, UK.

Smith, R.F. 1994. *Microscopy and Photomicrography.* CRC Press, Boca Raton, FL, USA.

Swatland, H.J. 1997. *Computer Operation for Microscope Photometry.* CRC Press, Boca Raton, FL, USA.

Tribe, M.A., Evant, R.M. and Snook, R.K. 1975. *Electron Microscopy and Cell Structure.* Cambridge Univ Press, Cambridge, UK.

Turrell, G. and Corset, J. 1996. *Raman Microscopy Developments and Applications.* Academic Press, New York, NY, USA.

Williams, P.M., Cheema, M.S., Davies, M.C., Jackson, D.E. and Tedler, S.J.B. 1994. Methods in Molecular Biology, vol. 22, *Microscopy, Optical Spectroscopy and Microscopic Techniques.* Jones, C., Mulloy, B. and Thomas, A.H. (eds.). Humana Press, Totowa, NJ, USA.

BIOCHEMICAL METHODS

Carpita, N.C., Housley, T.L. and Hendrix, J.E. 1991. New features of plant fructan structure revealed by thylation analysis and carbon ^{13}R spectroscopy. Carbohyd. Res. 146: 129.

Charlson, A.J., Goin, P.A.J. and Perlin, A.S. 1962. Determination of the configuration of glycosidic linkages of oligosaccharides in: *Methods in Carbohydrate Chemistry.* Whistler, R.L. *et al.* (eds.). Academic Press, New York, NY.

Dashek, W.V. 1997. Carbohydrates. In: *Methods in Plant Biochemistry and Molecular Biology.* Dashek, W.V. (ed.). CRC Press, Boca Raton, FL, USA.

Dische, Z. 1962. Color reactions of carbohydrates: color reactions of pentoses. *Methods Carbohyd. Chem.* 1: 484-488.

Eklund, G.B., Jossefsson, B. and Roos, C. 1977. Gas liquid chromatography of monosaccharides at the picogram level using glass capillary column, trifluoroacetyl derivitization and electron capture detection. J. Chromatog. 142: 575-585.

GlycoFace. 1994. Carbohydrate Analysis. *Face Technology.* Glyco, Inc., Novato, CA.

James, T.D. 1995. Chiral discrimination of monosaccharides using a fluorescent molecular sensor. Nature 374: 345.

Lewis, B.A., Smith, F. and Stephen, M. 1962. Determination of primary hydroxyl groups. In: *Methods in Carbohydrate Chemistry.* Whistler, R.L. *et al.* (eds.) Academic Press, New York, NY.

Maciak, G. 1962. C-methyl determination chronic acid oxidation. In: *Methods in Carbohydrate Chemistry.* Whistler, R.L. *et al.* (eds.) Academic Press, New York, NY.

Rassi, Z. 1995. *Carbohydrate Analysis: High Performance Liquid Chromatography and Capillary Electrophoresis.* Elsevier, Amsterdam.

Schaumann, C., Oesch, F., Unger, K.K. and Wieser, R.C. 1993. Analytical technique for studying the structure of glycoprotein N-glycans. J. Chromatog. 646: 227-234.

Thompson, A. and Wolfrom, A.L. 1962. Melting points. In: *Methods in Carbohydrate Chemistry.* Whistler, R.L. *et al.* (eds.) Academic Press, New York, NY.

Vliegenthart, J.F.G., Darland, L. and Van Halbeek, H. 1983. High resolution H-nuclear magnetic resonance spectroscopy as a tool in the structural analysis of carbohydrates related to glycoproteins. Adv. Carbohydr. Biochem. 41: 290-374.

LIPIDS

Kates, M. 1982. *Techniques of Lipidology Isolation, Analysis and Identification of Lipids.* North Holland Publ. Co., Amsterdam, Netherlands.

Kautsky. 1981. *Steroid Analysis by HPLC: Recent Applications.* Marcel Dekker, New York, NY.

Moreau, R.A. 1990. Plant lipid class analysis by HPLC, pp. 20-22. In: *Plant Lipid Biochemistry, Structure and Utilization.* Quinn, P.J. and Harwood, J.L. (eds.). Portland Press, Ltd., London, UK.

Shibamoto, T. 1994. *Lipid Chromatographic Analysis.* Marcel Dekker, New York, NY.

NUCLEIC ACIDS

Allen, R.C. and Budowle, B. 1994. *Gel Electrophoresis of Proteins and Nucleic Acids: Selected Techniques.* De Gruyter, Berlin, Germany.

Alphey, L. 1997. *DNA Sequencing.* Springer-Verlag, Berlin, Germany.

Ball, J.R. 1996. *DNA Isolation and Sequencing.* John Wiley, New York, NY.

Brown, P.R. 1984. *HPLC in Nucleic Acid Research Methods and Applications. Chromatographic Science,* Series 28. Marcel Dekker, New York, NY.

Hanes, B.D. and Higgins, S.J. 1985. *Nucleic Acid Hybridization.* Oxford Univ. Press, Oxford, UK.

Howe, C.J. and Ward, E.S. 1990. *Nucleic Acids Sequencing: A Practical Approach.* Oxford Univ. Press, Oxford, UK.

James, T.L. 1995. *Nuclear Magnetic Resonance and Nucleic Acids.* Acad Press, San Diego, CA.

Jones, P. 1995. *Gel Electrophoresis: Nucleic Acids.* John Wiley, Chichester, UK.

Lai, E. and Birren, B.W. 1991. *Electrophoresis of Large DNA Molecules. Theory and Applications, Current Communications in Cell and Molecular Biology.* Cold Spring Harbor, NY.

Richwood, D. and Hames, B.D. 1990. *Gel Electrophoresis of Nucleic Acids: A Practical Approach.* IRL Press, Oxford Univ. Press, Oxford, UK.

Roberts, G.C.K. 1993. *NMR of Macromolecules: A Practical Approach.* IRL Press, Oxford Univ. Press, Oxford, UK.

Tietz, D. 1998. *Nucleic Acid Electrophoresis.* Springer-Verlag, Berlin, Germany.

PROTEINS

Brenner, M. and Niederwieser, A. 1960. Dunnschicht-Chromattographie von Aminosauren. Experientia 16: 378-383.

Bryan, J. and Smith, L.C. 1996. *Protein Sequencing Protocols.* Humana Press, Totowa, NJ, USA.

Cavanaugh, J., Fairbrother, W.J., Palmer, A.G. and Skeleton, N.J. 1996. *Protein NMR Spectroscopy Principles and Practice.* Academic Press, San Diego, CA.

Chapman, J. 1996. *Protein and Peptide Analysis by Mass Spectrometry.* Humana Press, Totowa, NJ, USA.

Chapman, J. 2000. *Mass Spectrometry of Protein and Peptides.* Humana Press, Totowa, NJ, USA.

Cohen, S.A. 2000. Amino acid analysis using pre-column derivatization with 6-aminoquinolyl-n-hydroxy-succinimidyl carbonate. In: *Amino Acid Analysis Protocols.* Cooper, C. (ed.). Humana Press, Totowa, NJ, USA.

Corthals, G., Gygi, S., Aeversold, R. and Patterson, D. 2000. Identification of proteins by mass spectrometry. In: *Proteome Research. Two-demensional Gel Electrophoresis and Identification Methods.* Springer-Verlag, Berlin, Germany.

Hancock, W. 1998. *CRC Handbook of HPLC for the Separation of Amino Acids, Peptides and Proteins.* CRC Press, Boca Raton, FL, USA.

Hearn, M.T.W. 1991. *HPLC of Proteins, Peptides and Polynucleotides: Contemporary Topics and Applications.* VCH, New York, NY.

Heilman, E. 1992. *Chromatography: Fundamentals and Applications of Chromatography and Related Differential Migration Methods.* Elsevier, New York, NY.

Hill, D.W., Watters, F.H., Wilson, T.D. and Stuart, J.D. 1979. High performance liquid chromatographic determination of amino acids in the picomole range. Anal. Chem. 51: 1338.

Husek, P. and Macek, K. 1975. *Gas chromatography of amino acids.* J. Chromatogr. 113: 139-230.

Imahori, K. and Sakiyama, F. 1993. *Methods in Protein Sequence Analysis.* Plenum Press, New York, NY.

Inman, J.K. and Apella, E. 1986. Methods of solid and liquid-phase sequence determination—personal views. pp. 449-471. In: *Practical Protein Chemistry – A Handbook.* Darbre, A. (ed.). John Wiley and Sons, Chichester, UK.

Jandik, P. 2000. Anion Exchange chromatography and integrated amperometric detection of amino acids. In: *Amino Acid Analysis and Protocols.* Cooper, C. (ed.). Humana Press, Totowa, NJ, USA.

Jen-kin, Lin. 1984. Dabsyl amino acids. In: *CRC Handbook of HPLC for the Separation of Amino Acids, Peptides and Proteins.* Hancock, W.S. (ed.). CRC Press, Boca Raton, FL, USA.

Johnstone, R.A.W. and Rose, M.E. 1996. *Mass Spectrometry for Chemists and Biochemists.* HD Science Limited, Notingham, 6 TP, UK (2nd ed.).

Jornvall, H., Hoog, J.D. and Gustavasson, A.M. 1991. *Methods in Protein Sequence Analysis.* Birkhauser-Verlag, Boston, MA, USA.

Kataoka, H., Matsumura, S., Yamamoto and Makita, M. 2000. Capillary Gas Chromatography Analysis of Protein and Non-Protein Amino Acids. In: *Biological Samples in Amino Acid Analysis and Protocols.* Cooper, E. (ed.). Humana Press, Totowa, NJ, USA.

Kataoka, H., Yukizo Uero, N., Nakai, K. and Makita, M. 2000. Analysis of o-phosphoamino acids in biological samples by gas chromatography with flame photometric detection. In: *Amino Acid Analysis Protocols.* Cooper, C. (ed.). Humana Press, Totowa, NJ, USA.

Kochhar, S., Mauratou, B. and Christen, P. 2000. Amino acid analysis by high performance liquid chromatography after derivatisation with 1-fkiyri-2, 4-dinitrophenyl-5-L-Alanine Amide (Marfey's reagent). In: *Amino Acid Analysis Protocols.* Cooper, C. (ed.). Humana Press, Totowa, NJ, USA.

Mant, C.T. and Hodges, R.S. 1991. *High Performance Liquid Chromatography of Peptides and Proteins. Separation, Analysis and Conformation.* CRC Press, Boca Raton, FL, USA.

Pelletier, M.J. 1999. *Analytical Application of Raman Spectroscopy.* Blackwell Science Ltd., Oxford, England.

Rabilloud, Th. 2000. *Proteome Research: Two-Dimensional Gel Electrophoresis and Identification Methods.* Springer-Verlag, Berlin, Germany.

Tempst, P., Hood, L.E. and Kent, S.B.H. 1987. Practical high performance liquid chromatography of proteins and peptides. pp. 179-208. In: *High Performance Liquid Chromatography in Plant Sciences.* Linskens, H.F. and Jackson, J.F. (eds.). Springer-Verlag, Berlin, Germany.

Twardowski, J. and Anzenbacher, P. 1994. Raman and IR spectroscopy. In: *Biology and Biochemistry.* Ellis Horwood, New York, NY, USA.

Wakefield, M.D. 1986. Separation of mixtures of proteins and peptides by high performance liquid chromatography. pp. 181-205. In: *Practical Protein Chemistry—A Handbook.* Darbre, A. (ed.). John Wiley and Sons, Chichester, UK.

Wilkinson, J.M. 1998. Dansyl amino acids. In: *CRC Handbook of HPLC for the Separation of Amino Acids, Peptides and Proteins*. Hancock, W.S. (ed.). CRC Press, Boca Raton, FL, USA.

Wittman-Liebold, B., Salnikow, J. and Erdmann, V.A. 1986. *Advanced Methods in Protein Microsequence Analysis*. Springer-Verlag, Berlin, Germany.

Zumwalt, R.W. and Kuo, K.C. 1987. *Amino Acids by Gas Chromatography*. Franklin, Elkins Park, PA, USA.

CELL BIOLOGY JOURNALS

Acta Histochemica

Annual Review of Cell and Developmental Biology

Applied Immunohistochemistry and Molecular Morphology

Biochemistry and Cell Biology

Biology of the Cell

Cell

Cell and Tissue Research

Cell Biochemistry and Function

Cell Biology and Toxicology

Cell Biology International

Cell Communication and Adhesion

Cell Death and Differentiation

Cell Motility and the Cytoskeleton

Cell Stress and Chaperones

Cell Tissue Organs

Cellular and Molecular Neurobiology

Cellular Physiology and Biochemistry

Cellular Signaling

Chromosome Research

Current Advances in Cell and Developmental Biology

Current Opinion in Cell Biology

DNA and Cell Biology

European Journal of Cell Biology

Experimental Cell Research

Histochemical Journal

Histochemistry and Cell Biology

International Journal of Biochemistry and Cell Biology

Journal of Cell Biology

Journal of Cell Science

Journal of Cellular Biochemistry

Journal of Cellular Physiology

Journal of Histochemistry and Cytochemistry

Journal of Membrane Biology

Methods in Cell Science

Methods in Molecular and Cellular Biology

Mitochondrion

Molecular and Cell Biology of Lipids

Molecular and Cellular Biochemistry

Molecular and Cellular Biology

Molecular Biology of the Cell

Molecular Cell

Molecular Cell Research

Molecular Membrane Biology

Molecules and Cells

Plant Cell

Plant Cell Reports

Plant, Cell and Environment

Protoplasma

Tissue and Cell

Trends in Cell Biology

CELL BIOLOGY TEXTBOOKS

Alberts, B. 1998. *Essential Cell Biology: An Introduction to the Molecular Biology of the Cell*. Garland Publ., New York, NY.

Altman, P.L. and Katz, D.D. 1976. *Cell Biology*. Fed. Amer. Soc. Exper. Biol. Bethesda, MD, USA.

Ambrose, E.J. and Easty, D.M. 1978. *Cell Biology*. Univ. Park Press, Baltimore, MD, USA.

Avers, C.J. 1976. *Cell Biology*. Van Nostrand, New York, NY.

Avers, C.J. 1978. *Basic Cell Biology*. Van Nostrand, New York, NY.

Avers, C.J. 1986. *Molecular Cell Biology*. Benjamin Cummings, Menlo Park, CA.

Bregman, A.A. 1986. *Laboratory Investigations in Cell Biology*. John Wiley, New York, NY, USA.

Burke, J.D. 1970. *Cell Biology*. Williams and Wilkins, Baltimore, MD, USA.

Cabibbo, A., Grant, R. and Helmoer-Citterich. 2004. *The Internet for Cell and Molecular Biologists*. Horizon Bioscience. Norfolk, UK.

Celis, J.E. 1997. *Cell Biology: A Laboratory Handbook*, vols. I, II, and III. Acad. Press, New York, NY.

Choinski, J.S. 1997. *Experimental Cell and Molecular Biology*. McGraw Hill – WCB, Dubuque, IA, USA.

Cooper, G.M. 2000. *The Cell: A Molecular Approach*. Sinauer, Sunderland, MA, USA (2nd ed.).

Darnell, C. 1990. *Cell Biology*. W.H. Freeman, East Lansing, MI, USA.

De Robertis, E.D.P., Saez, F.A. and DeRoberts, E.M.F. 1975. *Cell Biology*. Saunders, Philadelphia, PA, USA.

Dealtry, G.B. and Rickwood, D. 1992. *Cell Biology Labfax*. Oxford: BIOS Sci. Publ., New York, NY.

Dyson, R.D. 1975. *Essentials of Cell Biology*. Allyn and Bacon, Boston, MA, USA.

Dyson, R.D. 1978. *Cell Biology: A Molecular Approach*. Allyn and Bacon, Boston, MA, USA.

Garrett, R.H. and Grisham, C.M. 1995. *Molecular Aspects of Cell Biology*. Harcourt College Publishers, New York, NY, USA.

Gunning, B.C.S. and Steer, M.W. 1975. *Plant Cell Biology an Ultrastructural Approach*. Crane Publ., Russak, NY.

Harris, N. and Oparka, K.J. 1994. *Plant Cell Biology*. IRL Press, Oxford, UK.

Hedges, S.B., Hass, C.A. and Maxson, L.R. 1992. *Carribean biogeography: Molecular evidence for dispersal in West Indian terrestrial vertebrates*. Proc. Natl. Acad. Sci., USA 89: 1909-1913.

Heidcamp, W.H. 1995. Cell Biology Laboratory Manual. www.gov.edu/rcellab/

Howland, J.L. 1975. *Environmental Cell Biology*. Harwal Publ. Co., Media, PA, USA.

Inman, J.K. and Apella, E. 1986. Newer methods of solid and liquid-phase sequence determination-personal views. In: *Practical Protein Chemistry-A Handbook*. Darbre, A. (ed.). John Wiley and Sons, Chichester, England, pp. 449-471.

Imahori, K. and Sakiyama, F. 1993. *Methods in Protein Sequence Analysis*. Plenum Press, New York, NY, USA.

Johnson, K.E. 1991. *Histology and Cell Biology*. Harwal Publ. Co., Media, PA, USA.

Karp, G. 1979. *Cell Biology*. McGraw Hill, New York, NY.

Karp, G. 1996. *Cell and Molecular Biology: Concepts and Experiments*. John Wiley and Sons, New York, NY.

Karp, G. and Pruitt, N.L. 1999. *Problems Book and Study Guide: Cell and Molecular Biology: Concepts and Experiments*. John Wiley, New York, NY.

Kimball, J.W. 1984. *Cell Biology*. Addison-Wesley, Reading, MA, USA.

King, B. 1986. *Cell Biology*. Allen and Unwin, Boston, MA, USA.

Lackie, J.M. and Dow, J.A. 1999. *The Dictionary of Cell and Molecular Biology*. Academic Press, London, England.

Lodish, H., Berk, A., Zipursky, S.L., Matsudaria, P., Baltimore, D. and Dannel, J. 2000. *Molecular Cell Biology*. W.H. Freeman, New York, NY.

McElroy, W.D. and Swanson, C.P. 1976. *Modern Cell Biology*. Prentice Hall, Englewood Cliffs, NJ, USA.

McMaster, M.C. and McMaster, C. 1998. '*GC/MS: A Practical User's Guide*.' John Wiley and Sons, New York, NY, USA.

Negrutiu, I. and Gharti-Chhetri, G.B. 1991. *A Laboratory Guide for Cellular and Molecular Biology*. Birkhauser, Boston, MA, USA.

Pain, R.H. and Smith, B.J. 1973. *New Techniques in Biophysics and Cell Biology*. John Wiley, New York, NY.

Parsons, J.A. 1975. *Exercises in Cell Biology*. McGraw Hill, New York, NY.

Reid, D.G. 1997. *Protein NMR Protocols*. Humana Press, Totowa, NJ, USA.

Rickwood, D. and Harris, J.R. 1996. *Cell Biology Essential Techniques*. John Wiley, New York, NY.

Ross, R. 1997. *Foundation of Allied Health Sciences. An Introduction to Chemistry and Cell Biology*. W.C. Brown Publ., Dubuque, IA, USA.

Sadava, D. 1993. *Cell Biology Organelle Structure*. Jones and Bartlett, Sudbury, MA, USA.

Sheeler, P. and Bianchi, D.E. 1980. *Cell Biology Structure, Biochemistry and Function*. John Wiley, New York, NY.

Singh, B.R. 2000. *Infrared Analysis of Peptides and Proteins. Principles and Applications.* American Chemical Society, Washington DC, USA.

Smith, C.A. and Wood, E.J. 1996. *Cell Biology.* Chapman and Hall, New York, NY, USA.

Staw. 1998. *Cell Biology and Genetics.* Wadsworth Publ., Belmont, CA.

Stephenson, W.K. 1978. *Concepts in Cell Biology.* John Wiley, New York, NY, USA.

Thorpe, N.O. 1984. *Cell Biology.* John Wiley, New York, NY.

Verner, B. 2000. *Cell Biology.* Saunders, Philadelphia, PA, USA.

Virgil, E.L. and Hawes, C.R. 1989. *Cytochemical and Immunological Approaches to Plant and Cell Biology.* Academic Press, San Diego, CA.

Widnell, C.C. and Pfenniger, K.H. 1990. *Essential Cell Biology.* Williams and Wilkins, Baltimore, MD, USA.

Wolfe, S.L. 1983. *Introduction to Cell Biology.* Wadsworth Publ. Co., Belmont, CA.

Wolfe, S.L. 1985. *Cell Ultrastructure.* Wadsworth Publ. Co., Belmont, CA.

CHAPTER 2

Scientific Method

William V. Dashek

INTRODUCTION

The scientific method is a process that attempts to generate a reliable, consistent, and nonarbitrary view of the world (Grinnell, 1992; Carey, 1994; Giere, 1997). Thus, this method endeavors to minimize a researcher's personal/cultural bias (Rusbult, 2004) when testing a hypothesis or theory. To overcome bias, standardized procedures (Fig. 2.1) have been employed. Rusbult has proposed an integrated scientific method (ISM). He views creativity and critical thinking as "mutually supportive skills". He prefers to think of ISM as a "roadmap for creative wandering" rather than as a "flowchart for describing a predictable sequence".

COMPONENTS OF SCIENTIFIC METHOD

There appears to be a general consensus that the method involves at least four major steps: 1) observation and description of a phenomenon, 2) formulation of a testable hypothesis to explain the phenomenon, 3) employment of the hypothesis to predict the existence of other phenomena or to predict (quantitatively) the results of new observations, and 4) performance of controlled experiments by several experimenters (http://teacher.nsrl.rochester.edu/phy-labs).

NOMENCLATURE

The differences among hypotheses, models, theories, and laws are summarized in Table 2.1.

EXPERIMENTAL DESIGN

Experiments are designed (Quinn and Keough, 2002; Cobb, 2003; Myers and Well, 2003; Maxwell and Delaney, 2004) to test the null, denoted Ho, and an alternative hypothesis, designated Hi. The null hypothesis represents a theory believed to be true but yet not proven.

The final conclusion from an experiment is given in terms of the null hypothesis, i.e. reject Ho in favor of Hi, or do not reject Ho. Hypothesis testing can involve two types of error, type I and type II errors (Table 2.2). The former occurs when

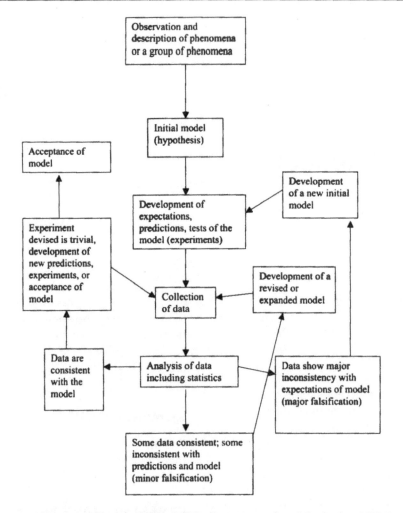

Fig. 2.1 Summary of the scientific method. Adapted http://maps.unomaha.edu/maher/geo117/scimethod.html.

TABLE 2.1 Summary of important terms*

Terms	Meaning
Hypothesis	A logical but unproven explanation for a set of facts; refers to knowledge state prior to experimentation
Model	When a hypothesis possesses at least limited validity
Theory	A hypothesis or a group of related hypotheses which has (have) been confirmed through repeated experimentation
Law	Laws are theories that have been repeatedly tested and not disproved

*Adapted from http://teacher.nsrl.rochester.edu/phy-labs...
and http://pasadena.wr.usgs.gov/office/garderson...

TABLE 2.2 Summary of type I and type II errors[a, b]

Statistical choice	Null Hypothesis Positive Ho	(Ho) Decision Negative Ho
Negate Ho	Type I error[c] P (type I error) = α More serious that a type II error Error of the first kind	Affirm Decision
Ho not negated, Hi	Affirm Decision	Type II error[d] P (type II error) = β Frequently due to a small sample size Error of the second kind

[a]Type I and II errors are inversely related.
[b]The student should consult statistics texts (see References for discussions of p values and confidence levels (Rohlf, 1994; Zav, 1998; Rosner, 1999; Looney, 2002).
[c]Type I errors occur when the null hypothesis is true but rejected.
[d]Type II errors result when the null hypothesis is false but not rejected. Power of a hypothesis test = 1 − P (Type II error) = 1 − B.
Note: This table is a composite from multiple readings, e.g.,
http://www.cas.lancs.ac.uk/glossary-/vbl/hyptest.html and other websites.

the null hypothesis is rejected when it is true, i.e. Ho is wrongly rejected. A type II error occurs when the null hypothesis is not rejected but is actually false. This type of error frequently results from the sample size being too small. Field and Hole (2003) discuss how to report experiments.

STATISTICAL ANALYSES

Quantitatively oriented cell biologists are often confronted with determining whether there is a statistically meaningful difference between experimental and control groups with a sufficient number of individuals in each group (Hampton, 1993; Cobb, 2003; Ruxton and Colegrave, 2003; Maxwell and Delaney, 2004). Indeed, the very design of such experiments is dictated by statistics. Table 2.3 provides a summary of the statistical methods available to cell biologists. Statistical software packages are presented in Table 2.4.

TABLE 2.3 Summary of statistical test choices [a, b]

Variables	Other parameters	Recommended choice	Description
One	One group	$\bar{x} \pm$ SD	$\bar{x}$ = sum of all the numbers of items in the set, SD = Square root of the variance
	Two groups	T-test[c]	A parametric statistical test requiring normality assumptions
	Three groups	ANOVA	An acronym for a category of tests termed analysis of variance
Two	Both continuous	Correlation	Strength of the relationship between variables
	One continuous One discrete	ANOVA	See above
	Both categorical	Chi-square	Nonparametric alternative to the t-test

(Contd. Table 2.3)

(Contd. Table 2.3)

Three or More	One group	Multiple regression	Serves to make predictions regarding multiple values
		Factor analysis	Used to discover simple patterns in the pattern of relationships of any variable
		ANOVA repeated measures	Used for repeated measures of variables
		Analysis of covariance	Combines features of simple linear regression with one-way analysis of variance
		Multivariate ANOVA	A set of tests used when there are three or more independent variables with two or more treatments per variable
		Discriminant function	Employed to determine which variable discriminates between two or more naturally occurring groups

[a]Modified from T. Lee Willoughby – http://researchmed.unkc.edu/ttwbiostats/

[b]Descriptions of these tests can be found in Fry (1993 a,b), Iles (1993 a,b), Dytham (1999), Wardlaw (2000), Myers and Well (2003), and Maxwell and Delaney (2004).

[c]See http://www.cas.lancs.ACUK/glossary

Note: Experiments may be those that lead to predictions or physical experiments with observations. The latter require controls, statistically significant sample sizes (N numbers) and repetition. The significance of the scientific method is its predictive power.

Adapted from: http://maps.unomaha.edu/Maher/geo117/scimethod.html

TABLE 2.4 Statistical software packages[e]

Package	Usefulness
GLIM Version 3.77[a]	ANOVA and linear and multiple regression
Minitab release 7.1 and 8.2[b]	Comprehensive statistical package ANOVA, linear bivariate and multiple regression
SAS release 6.03[a]	Comprehensive statistical package, ANOVA, linear and multiple regression, cluster analysis, and others
SAS, Sigma plot and Deltagraph[a]	Nonlinear curve fitting
SPSS/PC + version 3.0[c]	Comprehensive statistical package, ANOVA, linear and multiple regression, cluster analysis, and others
Stratagraphics version 2.1[a]	Comprehensive statistical package, ANOVA, bivariate linear and nonlinear regression, multiple regression, cluster analysis, and others
Delta Graph, Table Curve, Sigma Plot, Origin, Igor and Prism[d]	Curve fitting of complex functions
Matlab, Lab View[d]	Data analysis programs

[a]Adapted from: Appendix A, Fry (1993b)

[b]Wardlaw (2000). Provides extensive laboratory examples which use Minitab.

[c]SPSS is highlighted by Campbell (1989).

[d]Young (2001).

[e]Harvard Graphics and Statistica are well established software packages with spreadsheets.

References

Campbell, R.C. 1989. *Statistics for Biologists*. Cambridge Univ. Press, Cambridge, UK.

Carey, S.S. 1994. *A Beginner's Guide to Scientific Method*. Wadsworth, Belmont, CA.

Cobb, G.W. 2003. *Introduction to Design and Analysis of Experiments*. Key College Pub., Emeryville, CA.

Dytham, C. 1999. *Choosing and Using Statistics: A Biologist's Guide*. Blackwell Science, Oxford, Malden, MA, USA.

Field, A. and Hole, G. 2003. *How to Design and Report Experiments*. Sage Publ., Thousand Oaks, CA.

Fry, J.C. 1993a. One way analysis of variance. In: *Biological Data Analysis. A Practical Approach*, pp. 1-39. Fry, J.C. (ed.). IRL press, Oxford, UK.

Fry, J.C. 1993b. Bivariate regression. In: *Biological Data Analysis. A Practical Approach*, pp. 81-125. Fry, J.C. (ed.). IRL Press, Oxford, UK.

Giere R. 1997. *Understanding Scientific Reasoning*. Harcourt Brace, Fort Worth, TX, USA (4th ed.).

Grinnell, F. 1992. *The Scientific Attitude*. Guilford Press, New York, NY (2nd ed.). http://teacher.NSRL.rochesteredu/phy_labs/appendixE/appendixE.html

Hampton, R. 1993. *Introductory Biological Statistics*. McGraw-Hill, New York, NY.

Iles, T.C. 1993a. Crossed and hierarchical analysis of variance. In: *Biological Data Analysis. A Practical Approach*, pp. 41-80. Fry, J.C. (ed.). IRL Press, Oxford, UK.

Iles, T.C. 1993b. Multiple regression. In: *Biological Data Analysis. A Practical Approach*, pp. 127-172. Fry, J.C. (ed.). IRL Press, Oxford, UK.

Looney, S.W. 2002. *Biostatistical Methods*. Humana Press, Totowa, NJ, USA.

Maxwell, S.E. and Delaney, H.D. 2004. *Designing Experiments and Analysis Data: A Model Comparison Perspective*. Lawrence Erlbaum Assoc., Mahwok, NJ.

Myers, J. and Well, A.D. 2003. *Research Design and Statistical Analysis*. Lawrence Erlbaum Assoc., Mohwok, NJ, USA.

Quinn, G.P. and Keough, M.J. 2002. *Experimental Design and Data Analysis for Biologists*. Cambridge Univ. Press, Cambridge, UK.

Rohlf, F.J. 1994. *Biometry: The Principles and Practice of Statistics in Biological Research*. W.H. Freeman, New York, NY.

Rosner, B. 1999. *Fundamentals of Biostatistics*. Duxbury Press, Pacific Grove, CA.

Rusbult, C. 2004. http://www.asa3.org/ASA/education/think/science.html

Ruxton, G. and Colegrave, N. 2003. *Experimental Design for the Life Sciences*. Oxford Univ. Press, New York, NY.

Wardlaw, A.C. 2000. *Practical Statistics for Biologists*. John Wiley and Sons, New York, NY.

Young, S.S. 2001. *Computerized Data Acquisition and Analysis for the Life Sciences*. Cambridge Univ. Press, Cambridge, UK.

Zav, J.H. 1998. *Biostatistical Analysis*. Prentice Hall, Upper Saddle River, NJ, USA.

CHAPTER 3

Basic Chemical Principles

Karen J. Brewer

INTRODUCTION

Many basic chemical principles are important to the understanding and study of plant cell biology. Chemistry is the study of **matter**. Anything you can touch, see or smell is a form of matter. Matter can be thought of as anything that has mass and volume. Matter is typically classified by the state of the matter: solid, liquid, or gas. All matter is composed of chemicals and the study of chemicals is the field of chemistry. This chapter provides an overview of some of the key topics in chemistry as they relate to plant cell biology. More detailed discussions of these topics can be found in chemistry texts. Topics to be discussed include basic concepts of elements and molecules with a focus on water and its properties. The environment in which plants grow and the large amount of water present in plants dictates a need for a clear understanding of water and some of its unique properties.

ATOMS, ATOMIC STRUCTURE, ISOTOPES, AND PERIODIC TABLE OF ELEMENTS

Atoms, Elements, and the Mole

A chemical view of the world involves the description of matter as composed of atoms. An **atom** is the smallest unit of an element that will still have the properties of that element. Atoms are composed of nuclei that contain the subatomic particles protons and neutrons and a surrounding electron cloud (Fig. 3.1). It is the number of protons in a nucleus of an atom that defines the identity of that atom. Chemical symbols and names have been given to all the known atoms based on the number of protons contained in the nucleus.

All the currently known elements have been collected according to their properties into a table called the **Periodic Table** of Elements (Fig. 3.2). The periodic table provides much information about the elements.

The arrangement of the elements is such that periodic trends in the properties are apparent based on the location of an element in the periodic table. Important information about each element is summarized in the periodic table, Fig. 3.2, enlarged for hydrogen in Fig. 3.3.

The number at the top of each box represents the **atomic number** of that element. This number (1 for H or hydrogen) is the number of protons in the nucleus of a single atom of that element. Therefore, each hydrogen atom contains one proton. It is this characteristic of the atom that makes it a hydrogen atom. The **chemical symbol** for each element is given below the atomic number, H for hydrogen. This is the chemical shorthand method used to represent atoms of this element. You see these chemical symbols in formulas for molecules such as water, H_2O. Below the chemical symbol in this periodic table is given the name for each element, e.g. hydrogen. Finally, the number below the name is the standard **atomic weight** of the element. This is the average mass of one atom of that element, 1.00794 **amu (atomic mass units)** for hydrogen. An amu is equivalent to 1.66×10^{-24} grams (g). This is a very small mass, so chemists developed a method to convert this unit of measure into something easier to measure, the mole (or mol). A mole of atoms of any element has a mass of the atomic weight of that element in grams. For example, a mole of hydrogen atoms has a

TABLE 3.1 Properties of the subatomic particles

	Symbol	Charge	Mass
Proton	p	1+	1.672×10^{-24} g
Neutron	n	0	1.675×10^{-24} g
Electron	e⁻	1–	9.109×10^{-27} g

mass of 1.00794 g. A **mole** of anything is composed of **Avogadro's number** of those things, 6.022×10^{23}. Therefore, 6.022×10^{23} hydrogen atoms, or one mole of hydrogen atoms, weighs 1.00794 g.

Atomic Structure

The subatomic particles of interest are the **proton, neutron,** and **electron,** Table 3.1.

The number of protons is what determines the identity of an atom and is given by the atom's **atomic number. Protons** have a 1+ charge and add significant mass to atoms since a proton's mass is 1.672×10^{-24} g. The number of electrons that a neutral atom possesses is equal to the number of protons. An **electron** carries a 1- charge and has a much smaller mass than a proton; the mass of an electron is 9.109×10^{-27} g. The number of electrons an atom has can vary in charged species, or ions, of that atom. A positively charged hydrogen atom, or proton (H^+), generates the positive charge by loss of an electron. The **charge on an ion** (Z) is equal to the number of protons (p) minus the number of electrons (e⁻).

$$Z = \#p - \#e^-$$

Positive ions have more protons than electrons and negative ions have more electrons than protons. An atom cannot change the number of protons it possesses and remain the same element, i.e. all hydrogen atoms have one proton. Most elements have several stable isotopes that vary by the number of neutrons in the atom's nucleus.

Electron cloud

Nucleus that contains protons and neutrons

Fig. 3.1 Representation of an atom

IUPAC Periodic Table of Elements

Key

atomic number
symbol
name
standard atomic weight

Notes

- Aluminium and caesium are commonly used English-language spellings for aluminum and cesium.
- IUPAC 2001 standard atomic weights (mean relative atomic masses) are listed with uncertainties in the last figure in parentheses [R. D. Loss, *Pure Appl. Chem.* 78, 1107-1122 (2003)].
- These values correspond to current best knowledge of the elements in natural terrestrial sources. For elements with no IUPAC assigned standard value, the atomic mass (in unified atomic mass units) or the mass number of the nuclide with the longest known half-life is listed between square brackets.
- Element with atomic number 111 has not yet been named. The IUPAC provisional name is shown.
- Elements with atomic numbers 112, 114, and 116 have been reported but not fully authenticated.

Copyright © 2003 IUPAC, the International Union of Pure and Applied Chemistry. For updates to this table, see http://www.iupac.org/reports/periodic table. This version is dated 7 November 2003.

Fig. 3.2 IUPAC Periodic Table of Elements. (Copied with permission from the IUPAC web site http://www.iupac.org/reports/periodic_table that is copyrighted by the IUPAC.)

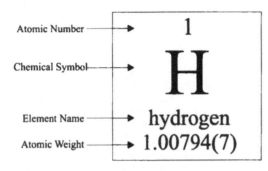

Atomic Number

Chemical Symbol

Element Name

Atomic Weight

1

H

hydrogen

1.00794(7)

Fig. 3.3

For example ^1H (or hydrogen one) is the most abundant isotope of hydrogen and possesses zero neutrons. The number of neutrons an isotope possesses is the isotope number minus the number of protons that element possesses (in this case: 1-1 = 0). ^2H or hydrogen two (commonly known as deuterium) has 2-1 or 1 neutron. **Neutrons** possess no charge but contribute significantly to the mass of an atom. A neutron's mass is 1.675×10^{-24} g. Many elements have more than one naturally occurring **isotope**, some of which are radioactive. Another common isotope of hydrogen is tritium, or hydrogen three, ^3H, which contains 2 neutrons, 1 proton, and 1 electron, Table 3.2.

Periodic Table of Elements

The Periodic Table is organized in a manner that groups atoms together that have similar properties (Fig. 3.2). This is achieved by the unusual shape of the Table and arrangement of the atoms by atomic number. A copy of the periodic table is shown in Figure 3.4 that highlights the classifications of the elements in it.

TABLE 3.2 Common isotopes of hydrogen

Symbol	Name	Number of Protons	Number of Neutrons	Number of Electrons
^1H	Hydrogen	1	0	1
^2H or D	Deuterium	1	1	1
^3H or T	Tritium	1	2	1

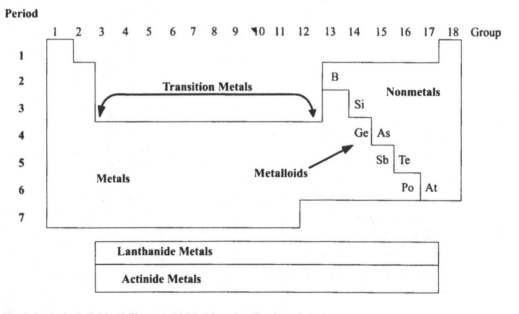

Fig. 3.4 Periodic Table of Elements highlighting classification of elements

The vertical rows of elements are called groups and are numbered 1-18. Groups of elements often possess similar properties and form molecules with similar stoichiometry and properties. Some of the groups are given special names that are often used in place of the group number. The most common examples are the alkali metals (group 1), the alkaline earth metals (group 2), the halogens (group 17) and the noble gases (group 18). The horizontal rows of elements are called **periods.** The elements on the left side of the periodic table are **metals** with groups 1-2 being the **main group metals** and groups 3-12 the **transition metals.** The two horizontal rows of elements at the bottom of the table are the **lanthanide metals** and **actinide metals.** Metals are typically good conductors of electricity and form sheets and wires. The elements on the upper righthand side of the periodic table are the **nonmetals.** Nonmetals are typically poor conductors of electricity, are often gases, and form hard, brittle solids. The metals and nonmetals are separated by a staircase arrangement of elements known as the **metalloids** that possess properties of the metals and nonmetals, B, Si, Ge, As, Sb, Te, and Po.

One of the properties of elements that varies in a predicable manner throughout the periodic table is electronegativity. Electronegativity is the ability of an atom to attract shared electrons in a chemical bond. Electronegativity tends to increase to the right and upward in the Periodic Table, excluding group 18. Many scales are used to measure **electronegativity** but the most common is the Pauling scale. The Pauling values for electronegativity of some common elements are given in Table 3.3.

Occurrence of Elements in Plants

Many of the elements in the periodic table occur in plants. The most common are the typical components of organic materials, C, H, N, O and P. Many metals and transition metals also commonly occur and play important biological roles. Table 3.4 lists common elements found in plants, their use in plants, and signs of deficiency. The elements contained in ≥ 1000 ppm are considered major elements in plant tissue and those < 1000 ppm minor elements.

MOLECULES AND CHEMICAL BONDING

The combination of atoms into molecules is a fundamental aspect of chemistry. This combination of atoms occurs through interactions of the atoms in a defined manner leading to substances that have unique properties relative to the elements from which they are composed. When atoms combine to form new substances they do so through chemical bonding. **Chemical bonding** is the force that attracts atoms to each other. When chemical bonding exists, the arrangement of the atoms together is lower in energy than the discrete atoms. Two common types of chemical bonding are ionic and covalent.

Ionic Bonding

Ionic bonding is the force of attraction between oppositely charged ions. This type of bonding predominates in salts, e.g. sodium chloride (NaCl), the table salt. Ionic bonding

TABLE 3.3 Pauling Electronegativities (Atkins and Jones, 2002)

			H 2.2			
Li 1.0	Be 1.6	B 2.0	C 2.6	N 3.0	O 3.4	F 4.0
Na 0.93	Mg 1.3	Al 1.6	Si 1.9	P 2.2	S 2.6	Cl 3.2
K 0.82	Ca 1.3	Ga 1.6	Ge 2.0	As 2.2	Se 2.6	Br 3.0

TABLE 3.4 Occurrence of select elements in plants and their significance in plant growth and development*

Concentration in dry tissue (ppm)	Element	Importance	Deficiency symptoms
15,000	Nitrogen N	Amino acids, peptides, proteins, nucleotides, chlorophyll structure	Reduced growth, yellowing (chlorosis), reduced lateral bud breaks; symptoms appear first on older growth
10,000	Potassium K	Cofactor of many enzymes, necessary for guard cell movements and syntheses, for example, protein biosynthesis	Reduced growth, shortened internodes, marginal brown leaf edges, and necrosis
5,000	Calcium Ca	Regulatory functions, cell wall structure, stabilizes membranes, controls movement	Inhibition of bud growth, death of shoot tips, blossom endrot of many fruits, pits on root vegetables
2,000	Phosphorus P	Energetic bonds (ATP), component of nucleic acids, important in phosphorylations of sugars and proteins	Reduced growth, thin stems, reduced lateral bud breaks, loss of lower leaves, reduced flowering
2,000	Magnesium Mg	Chlorophyll components, counter ion of ATP, important for protein biosynthesis	Reduction in growth, marginal chlorosis, reduction in seed production
1,000	Sulfur S	Amino acid and protein component, coenzyme A	General yellowing of affected leaves or the entire plant
100	Iron Fe	Necessary for chlorophyll synthesis, component of cytochromes, and ferredoxin	Intervenal chlorosis on younger leaves with white necrotic lesions
100	Chlorine Cl	Takes part in osmotic processes	Wilting at leaf tips, general chlorosis and necrosis, bronzing, stunted
6	Copper Cu	Cofactor of some enzymes	Young leaves wilted, wilted terminal bud, dark green leaves with necrosis
50	Manganese Mn	Like copper, component of protein biosynthesis	Chlorosis (intervenal), necrosis
20	Zinc Zn	Like copper (for example carboxypeptidase, DNA-dependent RNA polymerase)	Rosette growth, small leaves
0.1	Molybdenum Mo	Controls nitrogen metabolism	Intervenal chlorosis, necrosis, poor flowering, can cause N deficiency
20	Boron B	Influences use of Ca	Young leaves of the terminal bud light green, leaves twisted, stalk dies at bud

*Signs of excess and deficiency can be found in most plant physiology texts or http://extemsopm.oregonstate.edu/mg/botany/table4-html from which a portion of this table was adapted.

or salt formation is common between elements that are metals (e.g. Na) and elements that are nonmetals (e.g. Cl). Ionic bonding tends to occur between elements with a large difference in electronegativity. Ionic bonding can be quite strong and is the predominate force that holds salt lattices together. For NaCl the lattice energy is 769 kJ mol^{-1} (*CRC Handbook of Chemistry and Physics*). The strength of an ionic bond typically increases with increasing charge on the ions and decreasing distance between the ions. Typically, when a salt dissolves in a solvent, the ionic bonds are broken. This occurs through a process called solvation, in which the solvent interaction with the ions becomes strong enough to overcome the ionic attraction of the oppositely charged ions. Solvation can be particularly important in aqueous systems and thus has a major impact on the properties of substances in plants.

Covalent Bonding

Covalent bonding is that between atoms in which the atoms share bonding electrons. This type of bonding leads to formation of molecules. Covalent bonding is the predominant form of bonding between elements of nonmetals. It is also very common between transition metals and nonmetals. Covalent bonding is common between elements with similar electronegativity. Covalent bonds are named according to the number of pairs of electrons shared between two adjacent atoms. If the two atoms share one pair of electrons, this is called a **single bond** and is commonly represented with a single solid line. A single bond exists between the two hydrogen atoms in a hydrogen molecule, H-H or H_2. If two atoms share two pairs of electrons this is called a **double bond**. Double bonds are typically stronger than single bonds but are not typically twice the strength of a single bond. A double bond is represented as two solid lines. A double bond exists between the two oxygen atoms in an oxygen molecule, O = O or O_2. **Triple bonds** exist for the sharing of three pairs of electrons and are represented with three solid lines. A triple bond exists between the two nitrogen atoms in a nitrogen molecule, $N \equiv N$ or N_2. Covalent bonds vary in strength as a function of the atoms involved in the particular bond and the bond order (single, double, etc.) of the bond. A summary of typical covalent bond strengths are given in Table 3.5.

TABLE 3.5 Typical covalent bond strengths in kJ mol^{-1} (*CRC Handbook of Chemistry and Physics*). Bond type, molecule for which bond strength is given, bond strength in kJ mol^{-1}.

H-H (H_2) 436	C-H (CH_4) 438	O-H (H_2O) 497	N-H (NH_3) 453
S-H (SiH_4) 384	C-O $(H_3COCH_2C_6H_5)$ 280	C=O (CO_2) 532	Ca≡O (CO) 1077
C-C (H_3CCH_3) 376	C=C (H_2CCH_2) 728	Ca≡C (HCCH) 965	C-N $(C_6H_5CH_2NH_2)$ 298
C-S (H_3CSH) 312	C-Cl (CH_2Cl_2) 350	Na≡N (N_2) 945	O=O (O_2) 498

Covalent bonding is also possible in situations wherein more than two atoms formally share a pair of electrons. This type of bonding is known as **delocalized bonding** and is common among aromatic organic systems such as benzene, C_6H_6 (Fig. 3.5).

It is often represented with curved or dashed lines. Delocalized bonding frequently leads to bond orders of noninteger values such as the 1.5 C-C bond order in benzene.

Covalent bonds can be **nonpolar** or **polar** depending on the identity of the two atoms involved in a particular covalent

Fig. 3.5 Representation of benzene, C_6H_6

bond. When two atoms share a pair of electrons equally, this results in the formation of a nonpolar covalent bond. **Nonpolar covalent bonds** occur most commonly between two atoms of the same element. **Polar covalent bonds** occur when two atoms share pairs of electron unevenly. The degree to which the electron pairs are polarized toward one atom determines the extent of polarity of the covalent bond. Polar covalent bonds occur commonly when the atoms sharing the electrons have differing electronegativity. The atom with the higher electronegativity will typically have a higher electron density and thus possess a partial negative charge, δ^-. The atom with the lower electronegativity will typically have a lower electron density and thus possess a partial positive charge, δ^+. The magnitude of these partial charges is typically proportional to the difference in electronegativity of the two atoms involved in the bond.

Dipole-Dipole and Ion-Dipole Attractions

The presence of polar bonds leads to partial charges on molecules, δ^+ and δ^-. These partial charges play an important role in intermolecular interactions. In ionic bonding we described the attractive force between ions of different charges. A similar attractive force is possible in molecular systems in which the molecule is polar and thus parts of the molecule possess partial positive and partial negative charges. The

electrostatic attraction between the oppositely charged parts of two molecules is called a **dipole-dipole attraction** (Fig. 3.6).

Dipole-dipole attractions can be quite large and the strength of this attraction is dictated by the magnitude of the partial charges on the interacting molecules and the distance between these interacting charges. Larger partial charges and smaller distances between the interacting charges increase the magnitude of a dipole-dipole attraction.

Ion-dipole attractions are also possible and are the primary force that leads to dissolution of salts in polar solvents, such as water. Ion-dipole attractions are electrostatic, attractive forces between oppositely charged species. For example, an ion-dipole attraction can occur between a positively charged sodium ion, Na^+, and the negative side of the dipole on a water molecule, the oxygen atom in H_2O. Ion-dipole attractions tend to be stronger than dipole-dipole attractions, largely due to the increased charge on one species (the ion) involved in the electrostatic attraction. Like all electrostatic attractions, the magnitude of an ion-dipole attraction increases with increasing charge and decreasing distance between the interacting charges.

Van der Waals' Attractions

All atoms can have attractive forces between them, even in the absence of covalent or ionic bonding. Attractive forces between molecules are also possible. One

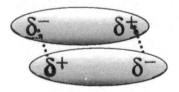 

Fig. 3.6 Two representations of a dipole-dipole attraction between two polar molecules

type of attractive force between atoms is Van der Waals' attractions, also called **London forces**. These forces are present in all materials and are typically quite weak, ca 2 kJ mol^{-1}. The commonly accepted explanation of **Van der Waals' attractive forces** is that the electron cloud around any atom or molecule is in motion and thereby can be asymmetrically distributed at any given point in time. This asymmetrical distribution of electron density leads to formation of an instantaneous dipole in that substance. This instantaneous dipole in one molecule can then impact the electron density in an adjacent molecule, leading to an induced dipole in it. These two dipoles would then have a dipole-dipole attraction, which is referred to as a Van der Waals' attraction. The strength of the Van der Waals' attraction depends on the strength of the induced dipole which is related to the **polarizability** of a molecule of atom. Van der Waals' attractions will increase with increasing polarizability and decreasing distance between the interacting species. Typically larger species with larger numbers of electrons are going to be more polarizable than smaller species with fewer numbers of electrons. Van der Waals' attractions are the primary attractive force in large, saturated organic molecules (hydrocarbons). These forces can be significant in large molecules. These types of forces are important in the large organic molecules that make up cell membranes and proteins.

Hydrogen Bonding

Hydrogen atoms bonded to very electronegative atoms often have significant partial positive charges due to a very polar covalent bond. This makes it possible for the hydrogen atom to be attracted to atoms with partial negative charges. It has been observed that this attraction is particularly strong when it occurs between hydrogen atoms with partial positive charges and other small, more electronegative, nonmetal atoms with partial negative charges and nonbonding, or lone pairs of electrons. This type of attractive force is called a **hydrogen bond**. Although this bond is significantly weaker than ionic or covalent bonds, it is the strongest type of intermolecular force with typical hydrogen bond strengths of ca 20 kJ mol^{-1}. Hydrogen bonds are very common between hydrogen atoms and oxygen, nitrogen and fluorine atoms. It is the hydrogen bond that is responsible for many of the unique and useful properties of water.

The hydrogen bond between two water molecules is quite strong and typically denoted by a dotted line (Fig. 3.7). A hydrogen atom involved in a hydrogen bond is bonded to an electronegative atom, oxygen in the case of water. As a result, the covalent bond connecting the atoms is polar and the hydrogen atom possesses a partial positive charge. Since the electronegativity difference between hydrogen and the attached atom is often large, the hydrogen atom often possesses a significant partial positive charge. This positively charged hydrogen atom is then attracted to the

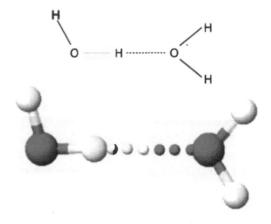

Fig. 3.7 Representation of a hydrogen bond

negatively charged atom in an adjacent molecule, oxygen in the case of two water molecules. The oxygen atom has a lone pair of electrons that is localized near the site of attraction of the hydrogen atom. As a result the hydrogen atom can interact with this pair of electrons to form a hydrogen bond.

WATER AND PROPERTIES OF AQUEOUS SYSTEMS

Water has many interesting and relatively unique properties that impart some of the common properties to aqueous systems. These properties are critical to plants and plant cell biology as plants contain relatively large amounts of water and dissolve species in this aqueous environment. Water is a highly polar molecule that has strong dipole-dipole interactions with many molecules and is capable of hydrogen bonding. Water also has a high freezing point and boiling point. Further, water possesses a high heat capacity and heat of vaporization and fusion. All these properties together make water a rather unique molecule.

Structure of Water and Surface Tension

Many of the properties of water result from the structure of the H_2O molecule. An H_2O molecule has two hydrogen atoms bonded to a central oxygen atom, each through a single covalent bond (Fig. 3.8).

The water molecule displays a bent arrangement of the two hydrogen atoms around the central oxygen atom, with an H-O-H **bond angle** of 104.5°. This nonlinear arrangement of the hydrogen atoms around the central oxygen atom results from the oxygen atom having 6 valence electrons, two of which are involved in covalent bonds to the hydrogen atoms and four that form two nonbonding pairs (or lone pairs) of electrons. The two atoms involved in each

covalent bond in water, H and O, have very different electronegativities, 2.2 and 3.4 respectively. Whenever two atoms with differing electronegativity form a covalent bond, that covalent bond is polar. In the case of the O-H bond in water, the electron pair is polarized toward the more electronegative oxygen atom, generating a δ^- charge on the oxygen atom and a δ^+ charge on the hydrogen atom. Since each oxygen atom participates in two such polar covalent bonds, the partial negative charge on the oxygen atom is twice the magnitude of the partial positive charge on each hydrogen atom. If these two O-H bonds were arranged at a bond angle of 180°, then these two bond dipoles would cancel and the water molecule would be nonpolar. Since the bond angle between these two bond dipoles is 104.5°, the individual bond dipoles do not cancel and this results in an overall molecular dipole and a polar water molecule. The molecular dipole in water has the negative end on the oxygen atom and the positive end between the two hydrogen atoms.

Water is a molecule ideally suited for hydrogen bonding. The presence of the hydrogen atoms bound to the small, electronegative oxygen atom with two lone pairs of electrons allows water to form strong hydrogen bonds. Each water

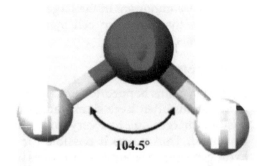

Fig. 3.8 Representation of a water molecule showing the direction of the molecular dipole

molecule possesses two hydrogen atoms and two lone pairs of electrons, therefore each water molecule can form two hydrogen bonds. This results in water forming a network in which at any given time, many water molecules are held together into a large array via hydrogen bonding.

The large degree of hydrogen bonding between water molecules leads to strong cohesion in water and high surface tension. **Cohesion** is the attractive force between objects and on the molecular scale is the attraction between molecules. Water is very cohesive because of the strong hydrogen bonding network. This leads to a very strong surface tension in water. **Surface tension** results from the forces of attraction on molecules at the surface of a liquid exerted by adjacent molecules. Since surface molecules only have other like molecules to exert this attraction in the directions parallel to and away from the surface, this leads to a net inward force (Fig. 3.9). Water has one of the highest surface tensions of all liquids. Surface tension is responsible for the fact that water forms spherical drops with what appears to be a skin that is hard to penetrate on the surface.

Freezing and Melting Point of Water

The presence of many strong hydrogen bonds in a sample of water is responsible for many of the properties of water that one might not predict based on its molecular weight. The **molecular weight** of a substance is the sum of atomic weights of the atoms in that molecule. For water the molecular weight is 18.02 g mol^{-1}. Water has a **freezing point** of 0°C (32°F) and a **boiling point** of 100°C (212°F). The conversion between **Celsius** and **Fahrenheit** temperature scales is given by:

$$\text{Temperature in Fahrenheit} = \{9/5 \times \text{Temperature in Celsius}\} + 32$$

Given the small molecular weight of water, you would predict a much lower freezing point and boiling point. For example, molecular nitrogen, N$_2$, has a molecular weight of 28.02 g mol^{-1} and a freezing point of –210°C and a melting point of –196°C. Hydrogen sulfide, H$_2$S, has a molecular weight of 34.09 g mol^{-1} and a freezing point of –86°C and a melting point of –60°C. The presence of an extensive strong hydrogen bonding network is responsible for water being a liquid at room temperature (RT) and atmospheric pressure.

Density and Specific Heat of Water

Water is a liquid that maintains a very constant volume as pressure is applied. In general, liquids are far less compressible than gases due to the close contact of atoms in the liquid phase. Water functions well as a hydraulic liquid and the hydraulic properties of water are very evident in plants. When a plant loses much of its water volume, it wilts. This wilting is the result of a decrease in the hydraulic pressure associ-

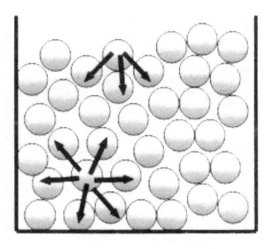

Fig. 3.9 Relative attractive forces of a surface molecule and a molecule in the bulk

ated with the water that helped the plant maintain its shape in its hydrated form. The **density of water** in its liquid form is 1.00 g cm^{-3} at 0°C and varies little with temperature. One interesting and very unique property of water is that its density decreases upon going from the liquid to the solid form. The density of ice or frozen water is 0.92 g cm^{-3}, significantly less than liquid water. This is why ice floats on liquid water (Fig. 3.10).

The decrease in density of water upon freezing is a direct result of the hydrogen bonding network in water. When a substance changes from liquid to solid, it does so by becoming more organized with a regular arrangement of the molecules in the solid phase. In water, this is accomplished by forming a very regular, open network. The openness of the network leads to decrease of density in solid water. The fact that water decreases its density upon freezing leads to extensive damage to plant and animal cells when they freeze. The decreased density upon freezing means that the same amount of solid water occupies more space than that amount of liquid

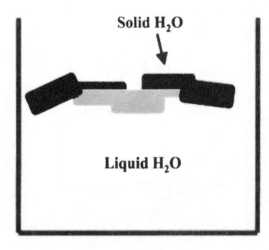

Fig. 3.10 Ice is less dense than water so ice floats on top of a sample of liquid water

water. As a result water expands upon freezing, typically leading to rupture of cell membranes.

Water has a relatively high **specific heat**, 1 cal g^{-1} °C. This means it takes one calorie to raise the temperature of 1 gram of water by 1 degree Celsius. The calorie, cal, is a small unit of energy. The commonly used nutritional Calorie, Cal or C, is equal to 1,000 calories, cal. One calorie is exactly equal to 4.184 J. The specific heat of water is among the highest of all liquids. This allows aqueous systems such as plants and animals to absorb large amounts of heat without changing temperature dramatically. The high specific heat of water imparts high temperature stability to plants and animals.

Heat of Fusion and Vaporization of Water

Water also requires significant amounts of energy to change phases, i.e. to change between a liquid and a gas (**vaporization**) or a liquid and a solid (**fusion**). The **heat of fusion** of water is 6.0 kJ mol^{-1} or 333 J g^{-1}. This means that it requires 333 joules to change 1 gram of ice at 0°C to water at °C. The **heat of vaporization** of water is 40.7 kJ mol^{-1} or 2260 J g^{-1}. This means that it requires 2,260 J to change 1 gram of water at 100°C to steam at 100°C. Both the large heat of fusion and the large heat of vaporization of water are a direct result of the extensive hydrogen bonding network in water. This hydrogen bonding network has to be disrupted in order to change the phase of water from either a solid to a liquid (melting) or a liquid to a gas (vaporization).

Water as a Solvent and Measures of Concentrations of Solutions

Water is a very good solvent for a number of materials. A **solution** is defined as a mixture of materials in which the materials are well mixed on the molecular level. In a solution, the material that is predominant is called the **solvent**. The material mixed on a molecular level is smaller in amount and called the **solute** (Fig. 3.11).

It is possible to make aqueous solutions (wherein water acts as the solvent) of many solutes. Water is able to dissolve, or make a true solution, of more solutes than almost any solvent. This is the result of the polar nature of water and its ability to form strong hydrogen bonds. The polar nature of water allows it to form strong dipole-dipole and dipole-ion attractions that allow it to solvate polar molecules and salts. The hydrogen bonding ability of water allows water to solvate materials capable of forming hydrogen bonds. This ability of solvents to dissolve similar species is commonly expressed as "like dissolves like." When water solvates a solute it forms a type of organized cage around the solute known as the **solvent cage**, which can extend through many layers of water molecules. This solvent cage is important in many of the properties of dissolved species in aqueous solution. Pure water is not a very good conductor of electricity. When water solutions are made of

salts, the solutions become good conductors of electricity.

The composition of a solution is commonly described in terms of the amount of solute in the solvent. Concentration can be expressed in a number of ways. The most common chemical method to express concentration is **molarity** or M. This unit expresses the amount of solute in moles per liter of solution. A solution of 58.44 g of NaCl in 1.0 liter of water is a 1.0 M or **molar** solution of NaCl. **Normality** (N) is another common unit of concentration used with acids or bases and represents the concentration in moles of the acidic or basic unit per liter of solution. The concentration unit, **molality**, expresses the amount of solute (in moles) per kg of solution. A 1.0 **molal** or m solution of NaCl would be 58.44 g of NaCl in 1.0 kg solution. Other common units to measure concentration include ratios. The volume/volume (v/v) ratio is commonly used for liquid/liquid solutions. This gives the ratio of volume of one liquid (the solute) to the volume of the other liquid (the solvent). Mass/volume ratios are also commonly used for solids dissolved in liquids, i.e. g L^{-1}. The 1.0 M solution of NaCl described above is a 58.44 g L^{-1} solution of NaCl. The unit of **ppm** is commonly used for very dilute solutions and represents the parts per million of the solute in the solvent.

Ionization of Water, pH, Acids, and Bases

Water undergoes an ionization equilibrium to form its conjugate acid and base. This ionization equilibrium is illustrated below:

$$2H_2O \rightleftharpoons H_3O^+ + OH^-$$

This equilibrium, like all equilibria, is governed by an equilibrium constant, K. The equilibrium constant for water ionization is called K_w and has a value of 1.0×10^{-14} at 25°C. K_w is the product of the concentration

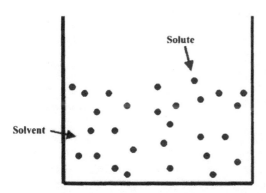

Fig. 3.11 Solution of a solute (dark blue) in a solvent (light blue) showing that the solvent is the bulk material and the solute is mixed on the molecular level with the solvent

of H_3O^+ and OH^-. Therefore in pure water at 25°C the $[H_3O^+]$ or hydronium ion concentration and the $[OH^-]$ or hydroxide concentration are both equal to 1.0×10^{-7} M. The hydronium ion, H_3O^+, is also commonly referred to as the proton, H^+, since an aqueous proton is a solvated hydronium ion. The hydronium ion is the conjugate acid of water and the hydroxide ion is the conjugate base. The **Bronsted-Lowry acid base theory** is commonly used to describe aqueous acids and bases. According to this theory, acids and bases and their equilibria are described using conjugate acid base pairs. A **conjugate acid** of any species is that species plus a proton, H^+. Therefore the conjugate acid of H_2O is H_3O^+. The **conjugate base** of any species is that species minus a proton. Since the product of the H_3O^+ and OH^- concentrations is a constant, K_w, any increase in H_3O^+ concentration will lead to a concurrent decrease in OH^- concentration.

The concentration of hydronium ions and hydroxide ions in an aqueous solution is a very important property and therefore a shorthand method has been developed to express their concentrations, the pH system. Since the concentration of protons is related by K_w to the concentration of hydroxide ions in aqueous solution, these concentrations can be expressed with a single number. The **pH** scale expresses the proton concentration using the following relationship:

$$pH = -\log [H^+]$$

Therefore in pure water the pH is 7.00. This is refereed to as a **neutral** solution. If the $[H^+] > [OH^-]$ then the pH < 7.00 and the solution is **acidic**. If the $[H^+] < [OH^-]$ then the pH > 7.00 and the solution is **basic**. A material that when added to neutral water leads to a solution with pH < 7.00 is an **acid**. A material that when added to neutral water leads to a solution with pH > 7.00 is a **base**.

It is often important to make an aqueous solution with a pH that does not vary dramatically as solutes are added. This is accomplished by the use of a buffer. A **buffer** is a solution that resists changes in pH upon addition of acids or bases. This is accomplished by adding large amounts of a weak acid and its conjugate base or a weak base and its conjugate acid to water. The pH in a buffer solution is related to the concentration of the conjugate acid base pair and the acid dissociation equilibrium constant for the buffer:

$$pH = pK_a - \log ([HA]/[A])$$

In this expression pK_a is minus the log of the acid dissociation equilibrium constant (K_a) for the acid form of the buffer. [HA] is the concentration of the acid form of the buffer and [A] is the concentration of the conjugate base of the buffer. Since large amounts of HA and A are added to a buffer solution, concentrations of HA and A change very little upon addition of acid or base to the buffer solution. This is the property that allows buffer solutions to maintain relatively constant pH upon acid or base addition.

DIFFUSION AND OSMOTIC PRESSURE

In liquids and gases the molecules in the system are in constant motion. In a gas, the distance between molecules is large and much of the space is empty. Gas molecules are not in direct contact with other gas molecules and are able to move reasonable distances without contact with another gas molecule. Molecules in a liquid are in direct contact with other molecules and therefore move only very small distances before their motion is impended by another molecule. As a result, motion in liquids over a large, macroscopic distance is slow and macroscopic motion of gaseous molecules much more rapid. The overall, net movement of molecules from one location to another as a result of their random kinetic movement is a process known as **diffusion**. Diffusion is slow in liquids and much more rapid in gases. Diffusion can lead to the net transfer of materials from one location to another. If a system has a concentration gradient, i.e. the concentration of any species is less in one location than in another, diffusion will take place in the direction necessary to equalize this concentration gradient. Biological systems often have rather large concentration gradients and diffusion therefore occurs as an attempt at equalizing these gradients.

Differences in pressure, temperature and solute concentration can all lead to diffusion of solvent or solute to equalize gradients. Increasing pressure on part of a sample will lead to diffusion of the molecules from the area of higher pressure to the area of lower pressure. Increasing the temperature on one part of a sample relative to another location will lead to diffusion of the molecules toward the area of lower temperature.

A system with a concentration gradient of a solute (for example NaCl) in a solvent (for example H_2O) can equalize the concentration gradient by two different types of diffusion. Diffusion of the solute, NaCl, can occur from the location of higher concentration to the location of lower concentration to equalize the concentration gradient. Alternatively, diffusion of the solvent, H_2O, can occur from the location of a lower solute concentration to the location of a higher solute concentration. By moving the solvent to the more concentrated solute area, dilution of the solute occurs, thereby equalizing the concentration gradient.

In biological systems the areas of differing concentration are often separated by semipermeable membranes. A **semipermeable membrane** is a membrane that allows movement of some species from one side to the other but does not allow movement of all species through the membrane. Common semipermeable membranes allow movement of water molecules through the membrane but do not allow movement of solute species through the membrane. When two solutions of differing concentration of solute are separated by a semipermeable membrane that only allows movement of water, only diffusion of water is possible. As a result, diffusion of water takes place from the side with a lower solute concentration to the side with the higher solute concentration to equalize the concentration gradient. This process of solvent movement to equalize a concentration gradient of the solute is known as **osmosis**. Osmosis is a very important process in biological systems. Osmosis leads to generation of a pressure build-up in an enclosed system. **Osmotic pressure, π,** is the pressure differential necessary to stop the flow of solvent across a **semipermeable membrane** against a concentration gradient and is related to solute concentration (π is osmotic pressure, i the i factor (explained below), R the gas constant $0.0821 \text{ L} \cdot \text{atm} \cdot \text{mol}^{-1} \cdot \text{K}^{-1}$, T the temperature in Kelvin, and c the molarity of the solute):

$$\pi = iRTc$$

The Van't hoff i factor is typically determined experimentally. For dilute solutions of molecular species (nonelectrolytes) i = 1. For salts the i factor is equal to the number of ions generated when the salt dissolves, i.e. 2 for NaCl or 3 for $CaCl_2$.

THERMODYNAMICS

The study of the energy of a system is called **thermodynamics**. Thermodynamics are important to the study of plant systems. In thermodynamics it is important to clearly identify the system under study. The

system is defined as the part of space or the amount of matter under scrutiny. Everything else in a thermodynamic sense is known as the **surroundings**. In the description of thermodynamic properties of a system, we often focus on interactions and heat transfer between the system and its surroundings.

Thermodynamics is governed by several important concepts known as the first, second and third law of thermodynamics. The **first law of thermodynamics** simply states that energy is neither created nor destroyed, i.e. the conservation of energy. The **second law of thermodynamics** dictates that when all things are considered, the system and surroundings, the sum of the changes always leads to an increase in disorder. This has an interesting ramification that no process can be 100 percent efficient. For example, when converting heat into work, less than 100 percent conversion is always achieved. The **third law of thermodynamics** states that the randomness of anything at absolute zero is zero.

Thermodynamics uses several measures of a system. **Entropy**, S, is a measure of the randomness of a system. The second law of thermodynamics dictates that the change in entropy, ΔS, of the system and its surroundings for any process is always positive for a **spontaneous** process, i.e. one that occurs on its own with no input of energy. When the entropy of a system and its surroundings has reached maximum, the system has reached equilibrium. **Equilibrium** is a dynamic state in which motions or reactions are still taking place

but the rate in one direction equals the rate in the opposite direction so that no net change occurs. **Enthalpy**, H, is the state function that relates to heat transferred and allows for quantification of energy changes at constant pressure. For any process, the change in enthalpy, ΔH, is a measure of the heat change at constant pressure. If ΔH is positive, the process is **endothermic** and heat is supplied to the system from the surroundings. If ΔH is negative, the process if **exothermic** and heat is given off from the system to the surroundings. The **Gibbs free energy** of a system is represented by G and related to H and S:

$$G = H - TS \text{ and } \Delta G = \Delta H - T\,\Delta S$$

Any process with ΔG < 0 for the system is a **spontaneous** process and can occur automatically with no input of energy. Any process with ΔG > 0 for the system is a **nonspontaneous** process and cannot occur without an input of energy to the system. For a system with ΔG = 0, the system is at equilibrium. The free energy, G, is related to the equilibrium constant for a reaction:

$$\Delta G = -RT \ln K_{eq}$$

For a spontaneous process, K_{eq} is greater than one leading to the expected negative value of ΔG.

References

Atkins, P. and Jones, L. 2002. *Chemical Principles: The Quest for Insight*. W.H. Freeman and Company, New York, NY, p. 95 (2nd Ed.).

IUPAC, 2004. *IUPAC Periodic Table*, http://www.iupac.org/reports/periodic table.

CRC Handbook of Chemistry and Physics, (84th Ed.). 2004. CRC Press, Baca Raton, FL, USA, pp. 12-24, 9-53, 9-65.

CHAPTER 4

Biomolecules I: Carbohydrates, Lipids, Proteins, and Nucleic Acids

William V. Dashek

CARBOHYDRATE CHEMISTRY AND FUNCTION

Monosaccharides and Their Derivatives

Monosaccharides are simple sugars (Sturgeon, 1990) consisting of two families, the aldoses (Fig. 4.1) and the ketoses (Fig. 4.2) distinguished by the position of carbonyl oxygen (Preiss, 1980; Churms, 1982; Loewus and Tanner, 1982; Collins, 1987; Binkley, 1988; El Kadem, 1988; Kennedy, 1988; Thiem, 1990; Bols, 1996; Dashek, 1997; Lehman, 1998; Robyt, 1998). Various monosaccharide derivatives (Table 4.1) possess marked biological significances (Cordy, 1980; Lehman, 1998).

Oligosaccharides

Oligosaccharides (Kandler and Hopf, 1980; Avigad, 1982, 1990) are carbohydrates composed of two to ten monosaccharides joined in glycosidic linkage (Fig. 4.3). Table 4.2 presents some oligosaccharides commonly found in plants.

Polysaccharides

In contrast to monosaccharides and oligosaccharides, polysaccharides are higher molecular weight polymers (Meuser et al., 1993). Polysaccharides are classified as storage, e.g. starch (Fig. 4.4, Morrison and Karkalas, 1990) and fructans (Table 4.3) or structural cell wall (cellulose, Franz and Blaschek, 1990), hemicellulose, pectin (Jarvis, 1984; Carpita, 1987; O'Neill et al., 1990; Carpita et al., 1996). With regard to integration of the latter to yield a primary cell wall (Aspinall, 1980), Carpita and Gibeaut (1993) have published molecular models of the primary cell wall (Fig. 4.5) and Bacic et al. (1988) discussed the structure and function of plant cell walls. Selvendran and Ryder (1990) reported their isolation.

For those students interested in computer modeling of carbohydrates, the book by French (1990) is worthwhile. In addition, students considering careers in bioengineering should consult Hwa-Kwan et al. (1996) and Mombarg (1997) regarding carbohydrate engineering.

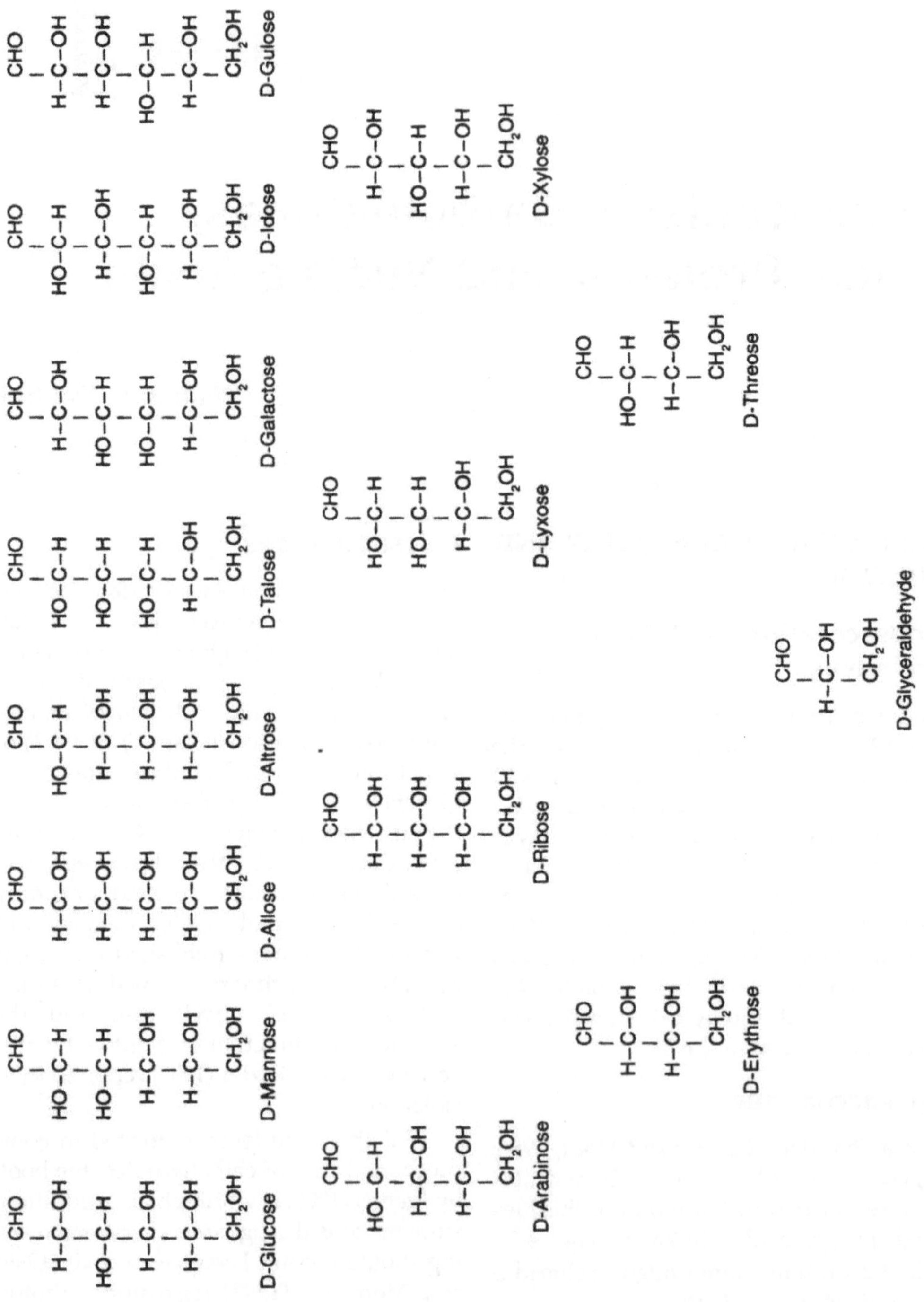

Fig. 4.1 Structure of aldoses

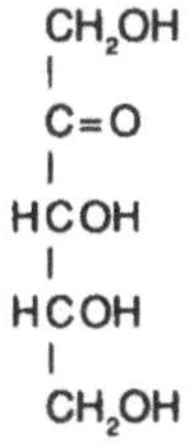

CH_2OH
|
C=O
|
HCOH
|
HCOH
|
CH_2OH

D-ribulose

(D-*erythro*-pentulose)

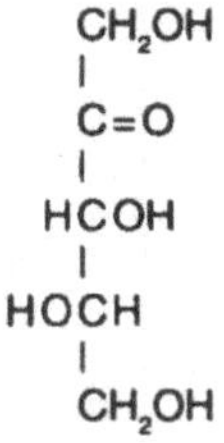

CH_2OH
|
C=O
|
HCOH
|
HOCH
|
CH_2OH

L-xylulose

(L-*thrco*-pentulose)

CH_2OH
|
C=O
|
HOCH
|
HCOH
|
HCOH
|
CH_2OH

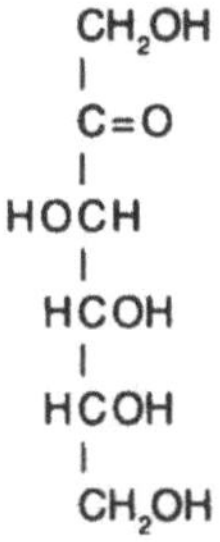

keto-D-fructose

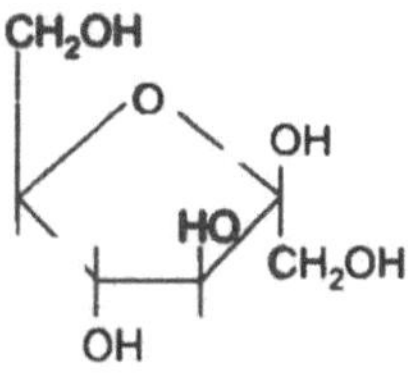

β-D-fructofuranose

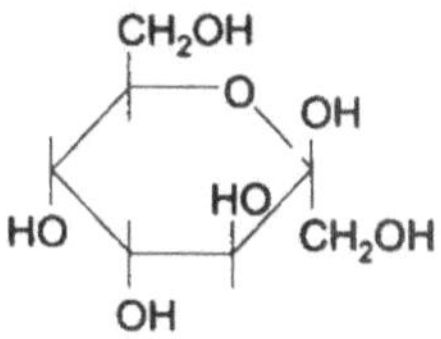

sedoheptulose

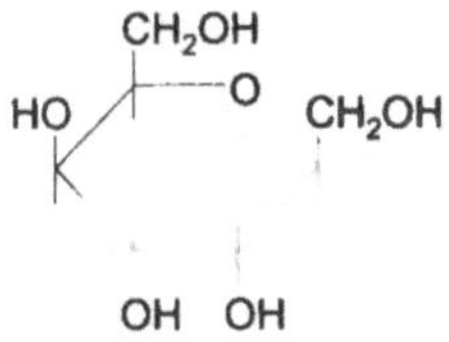

α-D-*manno*-heptulose

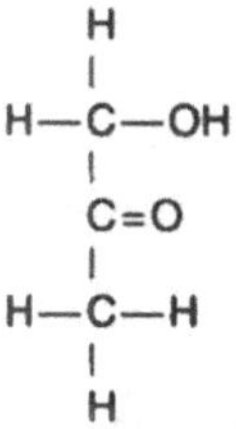

H
|
H—C—OH
|
C=O
|
H—C—H
|
H

Dihydroxyacetone

Fig. 4.2 Chemistry of ketoses

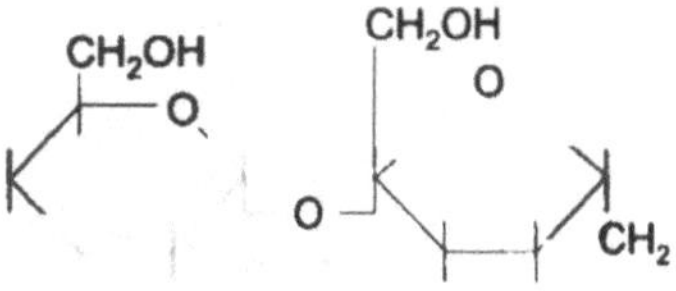

Sucrose

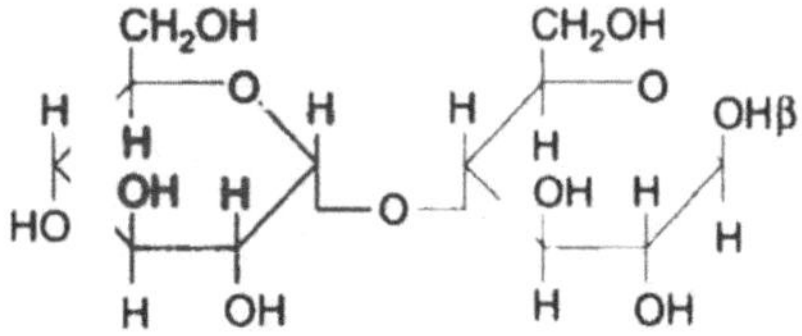

β-Maltose

Fig. 4.3 Examples of a glycosidic linkage

Kaushal *et al.*, 1988; Raju *et al.*, 1996), research on these macromolecules has focused on those associated with the cell wall (Varner and Linn, 1989; Cassab, 1998).

Certain plants contain glycolipids (Kates, 1990), especially those associated with photosynthetic tissues (Table 4.4). In addition, lipid-linked saccharides (Fig. 4.6) involved in the synthesis of complex carbohydrates are known to occur in certain plants (Elbein, 1982).

Synthesis of Carbohydrates

While Bednarski and Simon (1991) detailed the enzymes involved in carbohydrate synthesis, Dey and Dixon (1985) reviewed biosynthesis of storage carbohydrates. The biosynthesis of cell walls (Delmer and Stone, 1988), which proceeds from nucleotide sugars (Fig. 4.7), has been reviewed a number of times (Carpita and Gibeaut, 1988; Lewis and Paice, 1989). Feingold and Avigad (1990) presented the reactions which form nucleotide sugars in plants. In addition, they have provided a list of nucleotide sugars which occur in plants (Table 4.5). Their list contains the plant sources of these nucleotides.

Glycoproteins and Glycolipids

While a variety of glycoproteins (Table 4.4) occurs in plant cells/tissues (Lamport, 1980; Bowles, 1982; Selvendran and O'Neill, 1982;

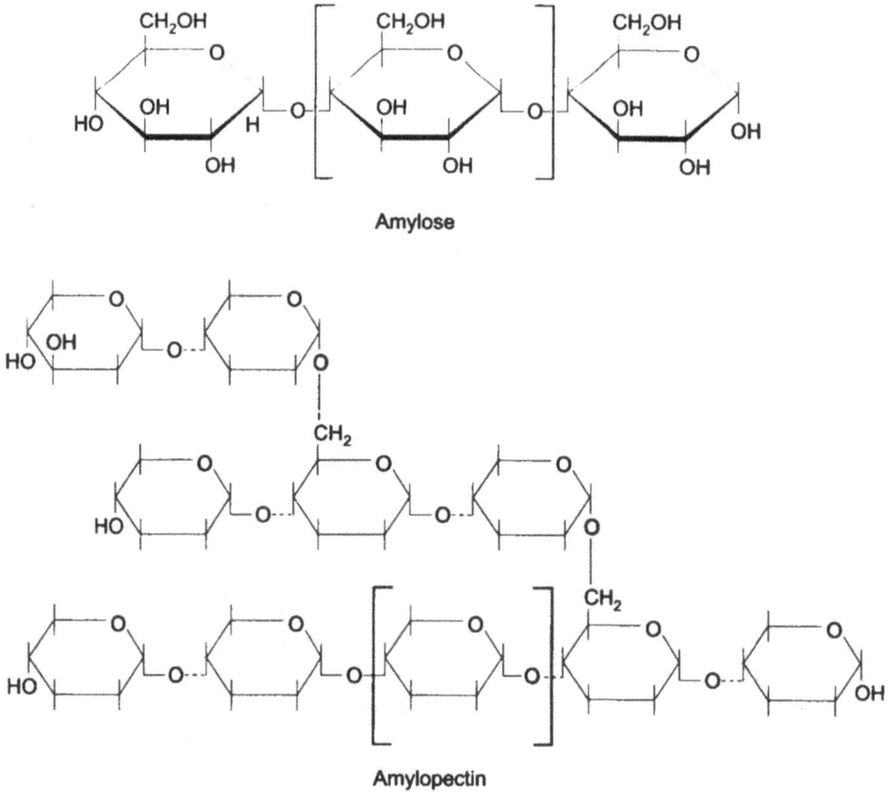

Amylose

Amylopectin

Fig. 4.4 Starch molecule

The student is referred to the section on respiration for a discussion of carbohydrate metabolism. In this connection, the monographs of Beitner (1985), Duffus and Duffus (1984), and Roehrig (1984) are useful.

Immunology of Carbohydrates

Much of this topic centers about plant lectins and their carbohydrate-binding specificity (Van Damme et al., 1998). Lectins are widespread natural products which can promote symbiosis between plants and specific nitrogen-fixing bacteria. Lectins can also serve as natural defense molecules.

During the past few years, the immu-

nology of carbohydrates has gained importance. This complex area of carbohydrates is beyond the scope of this chapter and the interested reader is referred to Wu (1988) and Axford (1998).

LIPID CHEMISTRY

Lipids are water insoluble, oily, or greasy organic substances extractable from cells and tissues by nonpolar solvents, e.g. chloroform or ether (Mead et al., 1980; Stumpf, 1980; Siegenthaler and Eichenberger, 1984; Stumpf et al., 1987; Moreau, 1990; Quinn and Harwood, 1990; Vance and Vance, 1991).

TABLE 4.1   Some biologically significant monosaccharide derivatives

Compound	Chemical structure	Biological function
Sugar acids		Cell wall constituents

Uronic acids
 (carbonyl oxidation and carbon at the
 other end of chain)

Glucuronic

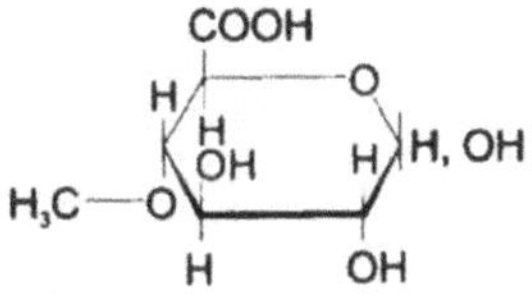

Galacturonic

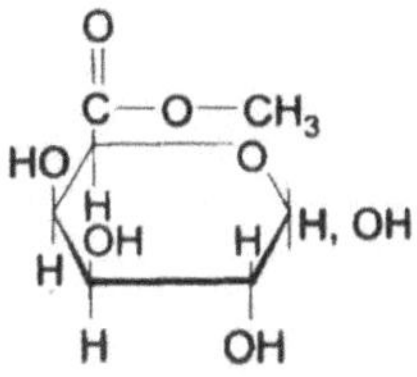

Amino sugars
 (Replacement of a sugar OH with an
 amino group)

Components of
glycoproteins

β-D-mannosamine

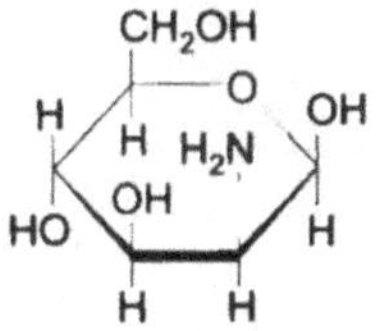

β-D-galactosamine

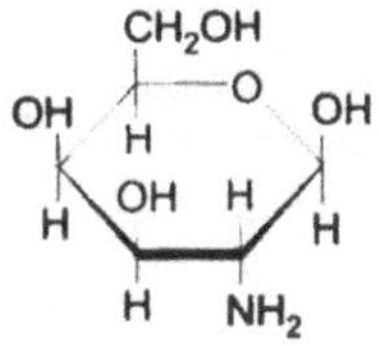

β-D-glucosamine

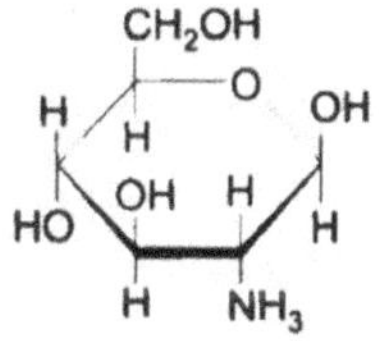

Sugar phosphates (Condensation of phosphoric
 acid with one of the hydroxyl groups of a sugar
 to form a phosphate ester)

Intermediates in
metabolism

glucose-6-phosphate

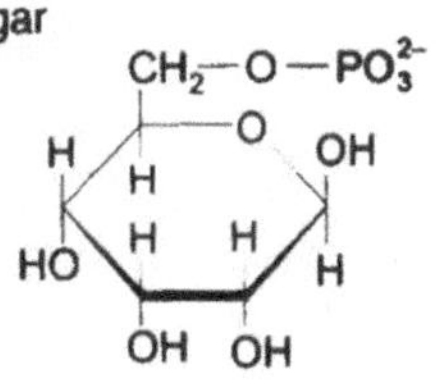

TABLE 4.2	Some oligosaccharides commonly found in plants*

Oligosaccharide

Trisaccharides
Umbelliforose
Raffinose
Planteose
Tetrasaccharides
Stachyose
Lychnose
Isolychnose
Sesamose

*Higher homologues are based on tetrasaccharides. A thorough discussion of oligosaccharides can be found in Dey (1990).

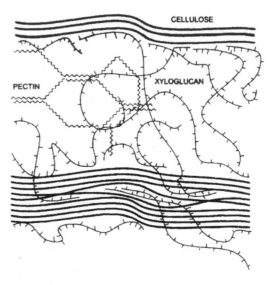

Fig. 4.5 Model of primary cell wall. From: Cosgrove (1998).

Types and Composition of Lipids

Fatty acids (Table 4.6, Fig. 4.8) are basic units of most lipids (Stumpf *et al.*, 1987; Mostofsky and Yehuda, 1996). They are long-chain organic acids possessing from 4-24 C atoms and responsible for the H_2O-insoluble nature of lipids. Fatty acids do not occur free or uncombined in cells of tissues but rather covalently bound in lipids. Fatty acids differ in chain length and presence, and number and position of double bonds. Most fatty acids possess an even number of Cs with fatty acids of 16 and 18 Cs most prevalent.

The long hydrocarbon chain can be fully saturated (only single bonds) or it may be unsaturated (one or more double bonds). The latter fatty acids are more prevalent than the former in plant lipids. When fatty acids contain two or more double bonds, the double bonds are separated by a methylene group, e.g. CH=CH-CH=CH. The double bonds are in the cis geometrical configuration. Whereas *saturated fatty acids* are solids possessing a waxy consistency, *unsaturated fatty acids* are liquids.

TABLE 4.3	Summary of the biology of fructans*

Fructans	Occurrence
Inulin 2, 1 linked	Tubers of dahlia and Jerusalem artichoke
Levans 2, 6 linked	Abundant in leaves and stems of certain plants
Unnamed highly branched 2, 1 and 2, 6 linkages	Common in leaves, stems and inflorescences
Unnamed small molecules with no more than 5 fructose units	Onion roots and asparagus

*Extensive discussions of fructans occur in Pollock and Chatterton (1988), Carpita *et al.* (1991), Suzuki and Chatterton (1993), and Farrar and Pollock (1995).

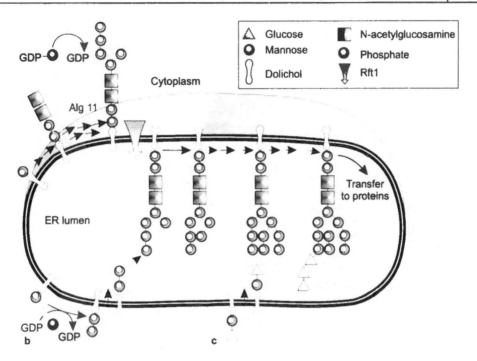

Fig. 4.6 Lipid-linked saccharides for synthesis of carbohydrates. Source: Helenius *et al.* Nature, Vol 415, 24 January 2002

TABLE 4.4 Examples of plant glycolipids and glycoproteins*

Glycoconjugate	Examples
Glycolipids	From photosynthetic tissues
	Monogalactosyl diglyceride
	Digalactosyl diglyceride
	Sulfoquinovosyl diglyceride
	Steryl glucoside
	Cerebrosides
	Glucocerebroside
	Phytoglycolipid
	Lipid-linked saccharides
	N-acetyl glucosaminyl-pyrophosphyoxyl-dolichol
Glycoproteins	Lectins
	Enzymes
	Bromelain
	Ficin
	Horseradish peroxidases
	Invertases
	Storage proteins
	7S soybean protein
	Kidney bean glycoprotein
	Vicilin and legumin
	Cell wall proteins
	Arabinogalactan proteins (AGPs)
	Hydroxyproline-rich glycoproteins or extensins (HRGPs)
	Proline-rich proteins (PRPs)
	Toxins
	Ricin

*Expansin proteins associated with cell wall loosening do not appear to be glycoproteins (Cosgrove, 1998, 1999).

Table Contd.

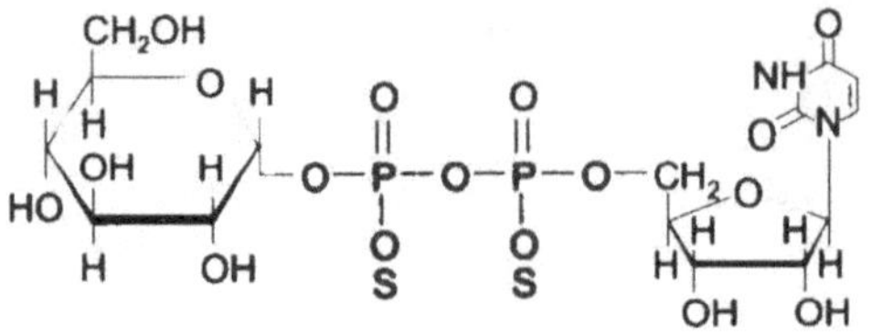

Fig. 4.7 Example of nucleotide sugar

Triacylglycerols are esters of glycerol with three fatty acids (Fig. 4.9; compare monoaclyglycerols, diaclyglycerols, and triacylglycerols). These are the simplest and most abundant lipids (Harwood, 1980), containing fatty acids and known as *fats*, *neutral fats*, or *triglycerides* (Gurr et al., 2002). Triaclyglycerols can be degraded by lipases (Wooley and Peterson, 1994). Triacylglycerols are storage lipids and often occur as oily droplets within the cytosol.

Waxes are fatty acid esters of long-chain alcohols possessing 14 – 36 Cs with long chain alcohols of 16 – 22 Cs (Table 4.7). Waxes are derived from C^{16} and C^{18} fatty acids that are elongated to form oxyl chains up to 5 Cs (Kolattukudy, 1980; Hamilton, 1995).

TABLE 4.5 Some nucleotide sugars occuring in plants*

Sugar	Nucleotide
L-arabinose	UDP
	ADP
	GDP
D-ribose	ADP
D-xylose	UDP
D-galactose	UDP
	ADP
	GDP
D-glucose	UDP
	GDP
	TDP
	ADP
D-mannose	GDP
	ADP
L-galactose	GDP
D-fructose	UDP
	ADP
L-fucose	GDP
L-rhamnose	UDP
D-apiose	UDP

*In addition to the above, Feingold and Barber (1990) have included some uronic acid nucleotide associations.

TABLE 4.6 Some naturally occurring fatty acids

Carbon structure	Common name
Atoms	Saturated fatty acids
$^{12}CH^3(CH^2)^{10}COOH$	Lauric acid
$^{14}CH^3(CH^2)^{12}COOH$	Myristic
$^{16}CH^3(CH^2)^{14}COOH$	Palmitic
$^{18}CH^3(CH^2)^{16}COOH$	Stearic
$^{20}CH^3(CH^2)^{18}COOH$	Arachidic
$^{24}CH^3(CH^2)^{22}COOH$	Lignoceric
	Unsaturated fatty acids
$^{16}CH^3(CH^2)^5CH{=}CH(CH^2)^7COOH$	Palmitoleic
$^{18}CH^3(CH^2)^7CH{=}CH(CH^2)^7COOH$	Oleic
$^{18}CH^3(CH^2)^4CH{=}CHCH^2CH{=}CH(CH^2)^7COOH$	Linoleic
$^{18}CH^3(CH^2)\ CH{=}CHCH^\bullet CH{=}CHCH^2CH{=}CH(CH^2)^7COOH$	Linolenic

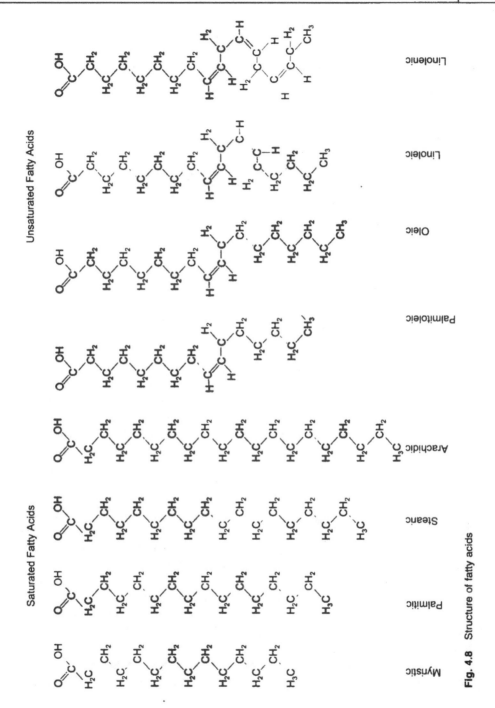

Fig. 4.8   Structure of fatty acids

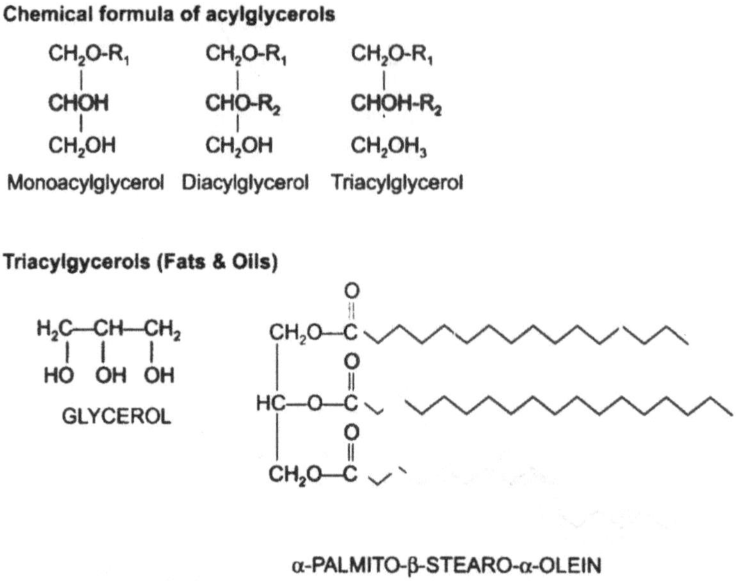

Chemical formula of acylglycerols

$$
\begin{array}{lll}
CH_2O\text{-}R_1 & CH_2O\text{-}R_1 & CH_2O\text{-}R_1 \\
| & | & | \\
CHOH & CHO\text{-}R_2 & CHOH\text{-}R_2 \\
| & | & | \\
CH_2OH & CH_2OH & CH_2OH_3 \\
\text{Monoacylglycerol} & \text{Diacylglycerol} & \text{Triacylglycerol}
\end{array}
$$

Triacylgycerols (Fats & Oils)

α-PALMITO-β-STEARO-α-OLEIN

Fig. 4.9 Chemistry of acylglycerols

TABLE 4.7 Components of plant waxes

Compound	Structure
n-alkanes	$CH^3(CH^2)_nCH^3$
Iso-alkanes	$CH^3CH(CH^3)R$
Ketones	$R^1CO\ R^2$
Secondary	$R^1CH(OH)\ R^2$
Alcohols	$CH^3CH(OH)\ R$
β-diketones	$R^1COCH^2CO\ R^2$
Primary alcohols	$R\ CH^2OH$
Acids	$R\ COOH$

Phospholipids are the major structural components of membrane lipids (Kates, 1990; Dennis and Vance, 1992; Cevc, 1993; Cevc and Paltauf, 1995). These are polar lipids that possess one or more polar heads in addition to their hydrocarbon tails (Fig. 4.10). Phospholipids contain phosphorus in the form of phosphoric acid with an ester linkage involving the third hydroxyl of glycerol and phosphoric acid. The first and second hydroxyls are esterified with fatty acids. Phospholipids can be degraded by phospholipases (Wiate, 1987; Dennis, 1991).

Sphingolipids are the second large class of membrane lipids consisting of a polar head and two nonpolar tails but lacking glycerol. These lipids are composed of one molecule of long-chain fatty acid, one molecule of long-chain alcohol, sphingosine, or derivatives, and a polar head alcohol (Fig. 4.11). The various types of sphingolipids are ceramides, cerebrosides, and sphingomyelin.

Steroids are the nonsaponifiable fat soluble lipids (Law and Rolling, 1985; Good et al., 1987; Lukas, 1994). The most abundant steroids are sterols consisting of polar head and a remaining nonpolar moiety (Fig. 4.12).

While cholesterol is the major sterol of animal membranes, stigmasterol is the chief sterol of plant membranes. Glycolipids are

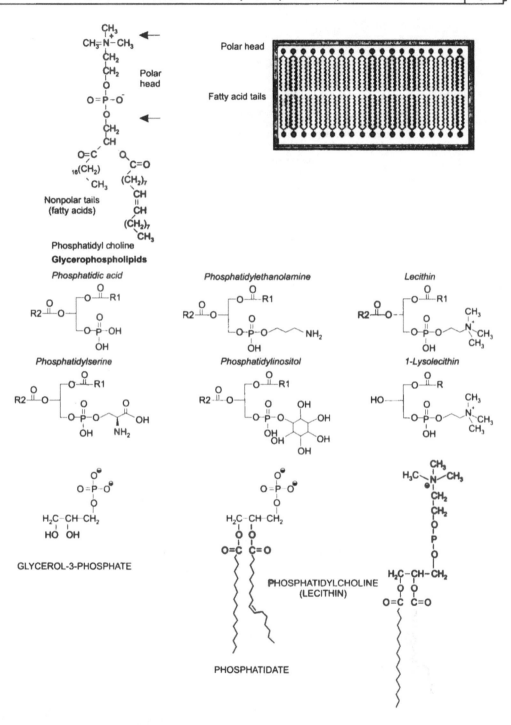

Phosphatidyl choline

Glycerophospholipids

Phosphatidic acid   *Phosphatidylethanolamine*   *Lecithin*

Phosphatidylserine   *Phosphatidylinositol*   *1-Lysolecithin*

GLYCEROL-3-PHOSPHATE

PHOSPHATIDYLCHOLINE
(LECITHIN)

PHOSPHATIDATE

Fig. 4.10 Phospholipid composition

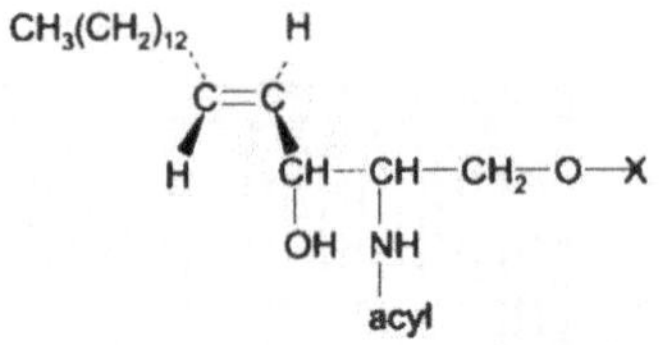

X = H   ceramide (acylated sphingosine)
X = sugar   cerebroside
X = phosphocholine   sphingomyelin
X = phosphoinositol   ceramide phosphoinositol (present in yeasts)

Sphingomyelin

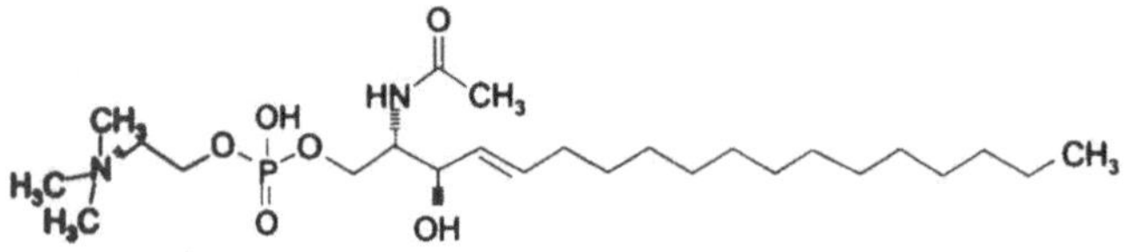

Fig. 4.11   Sphingomyelins

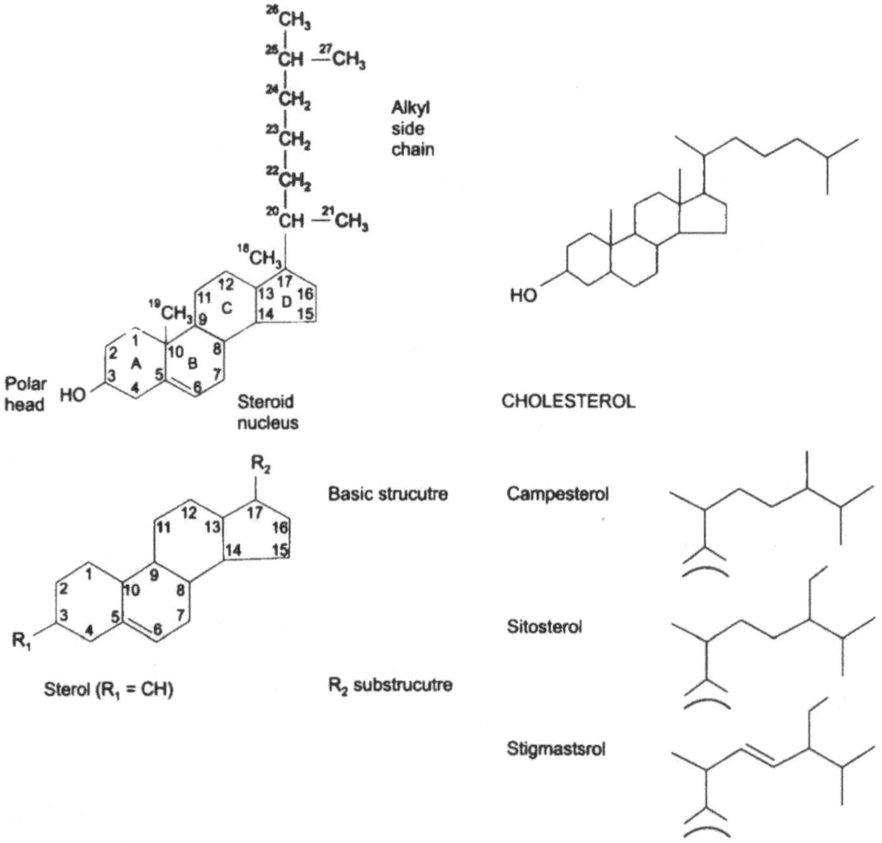

Fig. 4.12   Example of steroids

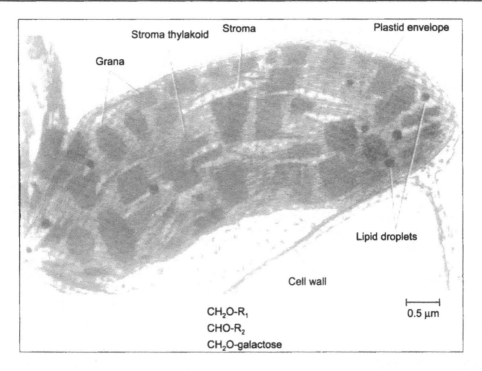

Fig. 4.13 Chloroplast lipids. Electron Micrograph from Evert, R. Botany Dept., Univ. of Wisconsin, Madison, WI

chloroplast lipids such as mono- and digalactosyldiacylglyerols and sulphoquinovosyldiacylglyerol. They are important structural components of thylakoid membranes (Fig. 4.13) (Horowitz, 1982; Fukuda and Hindsgaul, 1994; Heinz, 1996; Benning, 1998). Mudd and Kleppinger-Sparace (1987) have discussed plant sulfolipids in detail.

Lipoproteins are globular, biologically active proteins that adhere to the lipid bilayer of membranes (Schumaker, 1994). These lipoproteins act as mediators of specific and active transport (Jost and Griffith, 1982; Watts, 1993; Schumaker, 1994). In addition, lipid microdomains in the plasma membrane can contain signaling proteins (Van Meer, 2002). Finally, Biocs *et al*. (1989) and Larssan (1994) have discussed the biological roles of plant lipids in detail.

AMINO ACIDS, PEPTIDES, AND PROTEINS

Amino acids are organic compounds composed of both amino (NH_2) and carboxyl (COOH) groups. The two groups are attached to a single carbon atom, the alpha carbon (Fig. 4.14). In addition, R groups are attached to the atom and are used to classify amino acids (Table 4.8). The

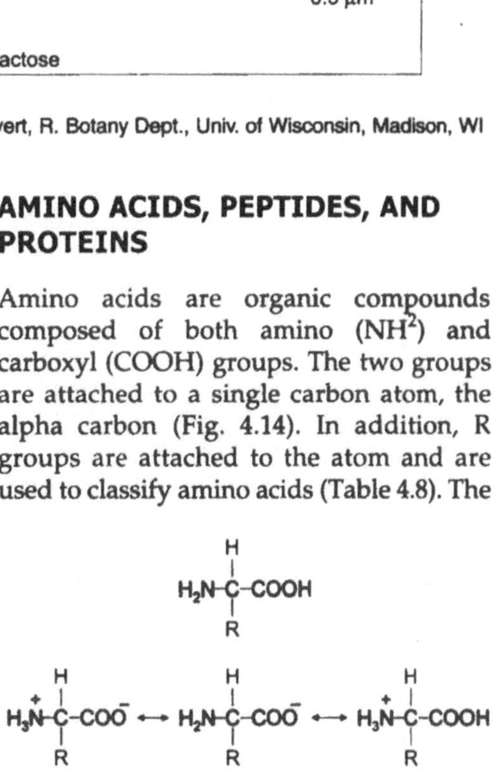

Fig. 4.14 Structure of an amino acid

TABLE 4.8 Classification of amino acids based on their R groups*

Classification	Amino acids
Nonpolar amino acids whose R groups contain CH^2 and CH^2 groups	Alanine, isoleucine, leucine, methionine, phenylalanine, proline, tryptophan, and valine
Polar uncharged amino acids whose R groups contain O^2	Asparagine, cysteine, glutamine, glycine, serine, threonine, and tyrosine
Negatively charged R groups	Glutamic acid, aspartic acid
Positively charged R groups	Arginine, histidine, and lysine

*Some amino acids possess specialized functions, e.g., cysteine (links chains together), methionine (initiation of polypeptide chains), and proline (promotes 'kinks' in polypeptides and proteins).

characteristics of amino acids, summarized in Table 4.9, have been thoroughly discussed in biochemistry textbooks as well as the monographs of Boulton *et al.* (1985), Davies (1985, 1993), Blackburn (1986) and Barrett and Elmore (1998). The structures of plant protein amino acids (Singh, 1999) and certain of their derivatives are presented in Table 4.10a. Rosenthal (1982) published an extensive list of amino acids and imino acid derivatives. Amino acids, except for hydroxyproline (hyp), can occur in the free amino acid pool in various concentrations. Nonprotein amino acids (Table 4.10b) are structural analogues of the protein amino acids (Miflin, 1980; Wallsgrove, 1995) and appear to be of restricted occurrence and unclear roles. Some investigators believe they serve as a nitrogen pool while others suggest that they are defensive agents and/or intermediates in metabolism.

In addition to amino acids, plants contain amines and amides. Table 4.11 displays some amides which can occur in plants. These are hydroxycinnamic amide conjugates of polyamines (Galston and Sawhney, 1990; Cohen, 1997) which can accumulate within the floral apex during flower development (Rhodes, 2000).

The amino, carboxyl, and R groups can undergo a variety of reactions (Table 4.12) that are useful in the chromatographic separation of amino acids as well as determining protein structure.

TABLE 4.9 Characteristics of amino acids

Zwitterions or dipolar ions

Water soluble, degree of solubility varies

Form hydrochlorides with concentrated HCl

Can be esterified

Exist as L-stereoisomers except for glycine (see biochemistry text for discussion of stereoisomerism)

Possess characteristic titration curves with pKas

Exhibit characteristic chemical reactions

Proteins are a major group of macro-molecules (Howard and Brown, 2002) which are comprise of amino acids (Blackburn, 1986; White and White, 2002) in peptide linkage (Fig. 4.15). The *primary structure* of proteins consists of the linear sequence of amino acids with amino and carboxyl termini (Fig. 4.16). The number of amino acids can vary from two (dipeptide) to many (polypeptides). The R groups of amino acids contribute to the structural and functional properties of proteins. Hydrogen bonds can be formed between polar side chains and within the peptide bonds (Bodansky, 1993). Thus polypeptides (Wielan and Bodansky, 1991) tend to form folds resulting in either sheets or helices. The α-helix and β-pleated sheets are two folding patterns which contribute to the *secondary structure* (Fig. 4.16) of proteins. The α-helices are rodlike structures in which peptide bonds are regularly bonded to those of nearby amino acids. With regard

TABLE 4.10A Structures of protein amino acids

Name	Abbr	Linear structure formula
Alanine	ala e	CH_3-$CH(NH_2)$-COCH
Arginine	arg r	HN-$C(NH_2)$0NH-$(CH_2)_3$-$CH(NH_2)$-COOH
Asparagine	asn n	H_2N-CO-CH_2-$CH(NH_2)$-COOH
Aspartic acid	asp d	HOOC-CH_2-$CH(NH_2)$-COOH
Cysteine	cys c	HS-CH_2-$CH(NH_2)$-COOH
Glutamine	ln q	H_2N-CO-$(CH_2)_2$-$CH(NH_2)$-COOH
Glutamic acid	glu e	HOOC-$(CH_2)_2$-$CH(NH_2)$-COOH
Glycine	gly g	NH_2-CH_2-COOH
Histidine	his h	NH-CH=N-CH=C-CH_2-CH-$CH(NH_2)$-COOH
Isoleucine	ile i	CH_3-CH_2-$CH(CH_3)$-$CH(NH_2)$-COOH
Leucine	leu l	$(CH_3)_2$-CH-CH_2-$CH(NH_2)$-COOH
Lysine	lys k	H_2N-$(CH_2)_4$-$CH(NH_2)$-COOH
Methionine	met m	CH_3-S-$(CH_2)_2$-$CH(NH_2)$-COOH
Phenylalanine	phe t	Ph-CH_2-$CH(NH_2)$-COOH
Proline	pro p	NH-$(CH_2)_3$-CH-COOH
Serine	ser s	HO-CH_2-$CH(NH_2)$-COOH
Threonine	thr t	CH_3-CH(OH)-$CH(NH_2)$-COOH
Tryptophan	trp w	Ph-NH-CH=C-CH_2-$CH(NH_2)$-COOH
Tyrosine	tyr y	PH-p-Ph-CH_2-$CH(NH_2)$-COOH
Valine	val v	$(CH_3)_2$-CH-$CH(NH_2)$-COOH

gly g
ala a
arg r 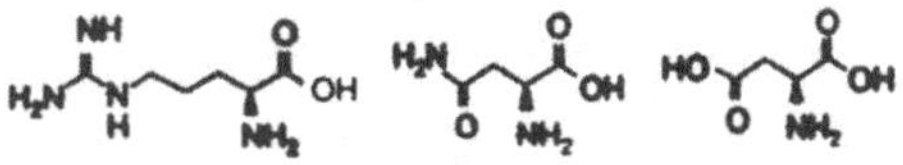
asn n
asp d
cys c 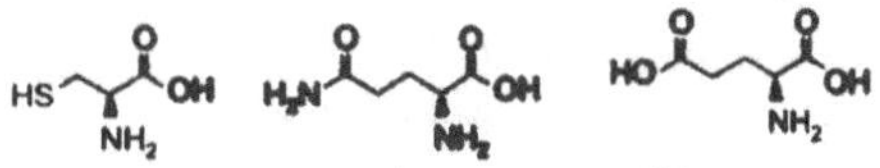
gln q
glu e
his b 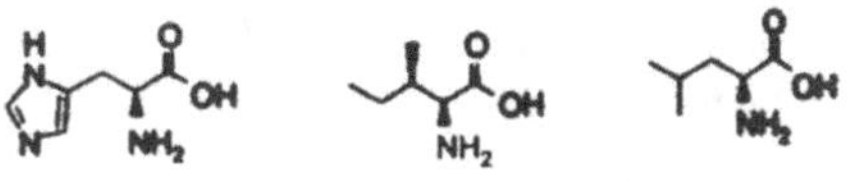
ile i
leu l
lys k 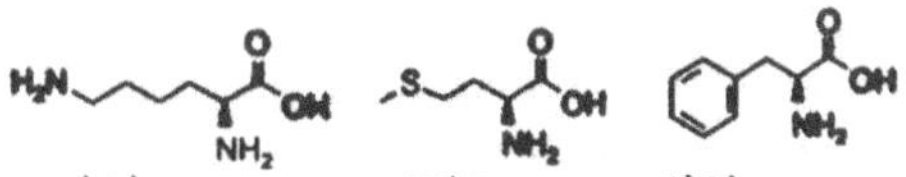
met m
phe t
pro p 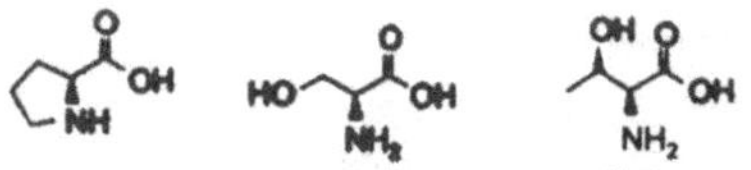
ser s
thr t
trp w 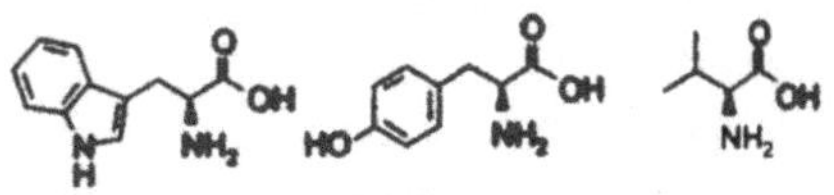
tyr y
val v

Hydroxyproline is similar to proline except that there is a hydroxyl on carbon 4

TABLE 4.10B Structure of non-protein amino acids

B-alanine (3-alanine)
4-aminobutyrate (GABA)
3-cyanolalanine (B-cyanolanine)
2-aminobutyric acid
2-methylene-4-aminobutyric acid
3-methylene-4-aminobutyric acid
2-aminoisobutyric acid
5-aminolevulinic acid
2-amino-4-methylhexanoic acid (homoisoleucine)
2-amino-4-methylhex-4-enoic acid
2-amino-4-methylhex-4-ynoic acid
2-amino-3-methylpentanoic acid
2-aminoadipic acid
4-ethylideneglutamic acid
3-aminoglutaric acid
2-aminopimelic acid
N^4-ethylasparagine
N^4-methylasparagine
erythro-4-methylglutamic acid
4-methyleneglutamic acid
4-methyleneglutamine
N^5-methylglutamine
N^5-isopropylglutamine
2-amino-4(aminoxy)butyric acid canaline
2,4-diaminobutyrate
N^4-acetyl-2,4-diaminobutyrate
N^4-lactyl-2,4-diaminobutyrate
N^4-ocalyl-2,3-diaminopropionic acid
N^6-acetyllysine
N^6-methylysine
N^6-trimethyllysine (lamine)
Ornithine (2,5-diaminopentanoic acid)
saccharopine (N6-2'glutamyl)lysine
2,6-diaminopimelic acid
N^4-(2-hydroxylethyl)asparagine
erythro-3-hydroxyaspartic acid
4-hydroxyarginine
4-hydroxycitrulline
threo-4-hydroxyglutamic acid
3,4-dihydroxyglutamic acid
3-hydroxy-4-methylglutamic acid
3-hydro-4-methyleneglutamic acid
4-hydroxy-4-methylglutamic acid

Table Contd.

4-hydroxy-glutamine
N5-(2-hydroxyethyl)glutamine
5-hydroxynorleucine
threo-4-hydroxynorleucine
threo-4-hydroxyhomoarginine
homoserine
O-acetylhomoserine
O-oxlylhomoserine
O-phosphohomoserine
4-hydroxyisoleucine
5-hydroxymethylhomocysteine
threo-3-hydroxyleucine
5-hydroxyleucine
2-hydroxylysine
4-hydroxylysine
5-hydroxylysine
N^6-acetyl-5-hydroxylysine
N^6-trimethyl-5-hydroxylysine
4-hydroxyornithine
mimosine
4-hydroxynorvaline
5-hydroxynorvaline
2-amino-4,5-dihydroxypentanoic acid
2-amino-4-hydroxypimelic acid
4-hydroxyvaline
O-acetylserine
O-phosphoserine
Pipecolic acid (piperidine-2-carboxylic acid)
3-hydroxypipecolic acid
trans-3-hydroxyproline
trans-4-hydroxyproline
trans-4-hydroxymethylproline
azetidine-2-carboxylic acid
N-(3-amino-3-carboxypropyl)azetidine-2-carboxylic acid
4,5-dehydropipecolic acid (baikiain)
3-amino-3-carboxypyrrolidone (cucurbitine)
2-(cyclopent02'-enyl) glycine
5-hydroxytryptophan
albizzine (2-amino-3-ureidopropionic acid)
arginosuccinic acid
canavinosuccinic acid
citrulline
homoarginine
homocitrulline

Table Contd.

Table Contd.

indospicine

O-ureidohomoserine

6-hydroxykynurenine

3-(4-aminophenyl)alanine

3-(3-aminomethylphenyl)alanine

3-(3-carboxyphenyl)alanine

3-carboxytyrosine

3-(3-hydroxymethylphenyl)alanine

3-(3-hydroxyphenyl)alanine

3-(3,4-dihydroxyphenyl)alanine (L-DOPA)

2-(phenyl)glycine

2-(3-carboxyphenyl)glycine

2-(3-carboxy-4-hydroxyphenyl)glycine

2-(3-hydroxyphenyl)glycine

4-aminopipecolic acid

guvacine

2-amino-4-(isxazolin-5-one)-2-yl)butyric acid

lathyrine

tetrahydrolathyrine

Source: Rhodes (2000). Structures of some of the known nonprotein plants amino acids. http://newcrop.hort.purdue.edu

to β-pleated sheets, there appear to be two types—antiparallel form and parallel. Whereas the polypeptide chain is folded back and forth upon itself in the antiparallel form, the parallel sheet consists of polypeptide sections arranged in the same direction. Lastly, a secondary structure can be stabilized via formation of disulfide (S-S) linkages between the SH groups of cysteines. The *tertiary structure* of polypeptides results from the folding of the secondary structure in H_2O producing a complex globular shape (Fig. 4.16). Domains are compact local units resulting from folding of portions of the polypeptide chain. Finally, a protein's *quaternary structure* reflects its subunit structure. Subunits arise when two or more polypeptide chains associate, yielding thereby a functional aggregate. Elaboration of protein structure has been facilitated by x-ray crystallography and

TABLE 4.11 Some examples of hydroxycinnamic acid conjugates of polyamines

Polyamines

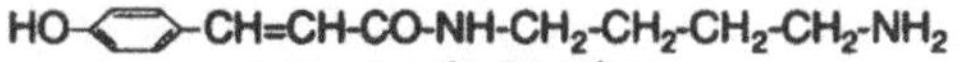

HO-⟨⟩-CH=CH-CO-NH-CH₂-CH₂-CH₂-CH₂-NH₂
  p-coumarylputrescine

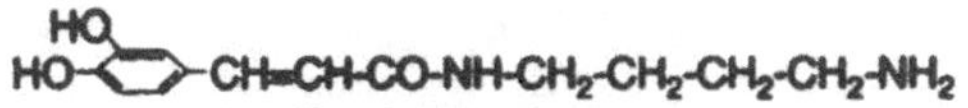

HO / HO-⟨⟩-CH=CH-CO-NH-CH₂-CH₂-CH₂-CH₂-NH₂
  caffeoylputrescine

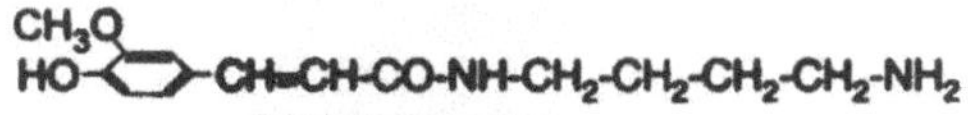

CH₃O / HO-⟨⟩-CH=CH-CO-NH-CH₂-CH₂-CH₂-CH₂-NH₂
  ferulylputrescine

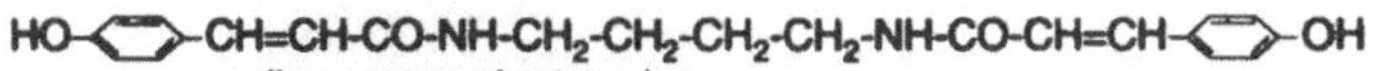

HO-⟨⟩-CH=CH-CO-NH-CH₂-CH₂-CH₂-CH₂-NH-CO-CH=CH-⟨⟩-OH
  di-p-coumarylputrescine

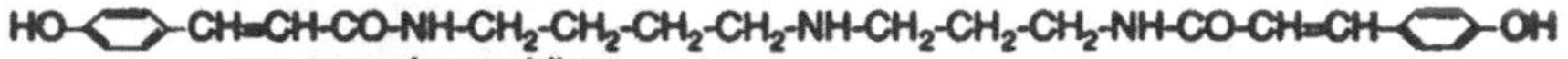

HO-⟨⟩-CH=CH-CO-NH-CH₂-CH₂-CH₂-CH₂-NH-CH₂-CH₂-CH₂-NH-CO-CH=CH-⟨⟩-OH
  p-coumarylspermidine

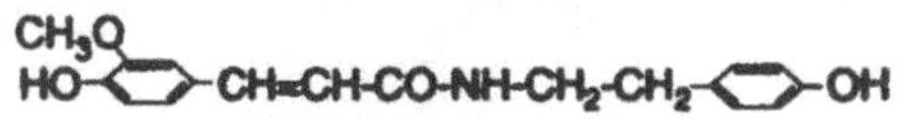

CH₃O / HO-⟨⟩-CH=CH-CO-NH-CH₂-CH₂-⟨⟩-OH
  ferulyltyramine

HO-⟨⟩-CH=CO-NH-CH₂-CH₂-⟨⟩-OH
  p-coumaryltyramine

Source: Rhodes (2000).
http://newcrop.hort.purdue.edu/rhodcv/hort640c/polyam/po00005.htm

TABLE 4.12 Summary of reactions of the carboxyl group and amino group

Amino (NH²)	Acetylation
	Acylation
	Dabsylation[a]
	Dansylation[b]
	Guanidination
	N-benzylation
	Schiff-base
	Sulfonylation
	Thioacylation
	Thiocarbamylation
Carboxyl (COOH)	Acylhalide formation
	Esterification
	Oxidative
	Decarboxylation

[a]Dabsylation: 4-4-dimethylaminazobenzene-4¹-sulfonylchloride in alkaline medium binds ovalently to primary amino, secondary amino, phenolic, imidazole, and sulfhydryl groups, which are useful for HPLC separation of amino acids.

[b]Dansylation: Determination of N-terminal amino acids in peptides and proteins; reagent for precolumn derivatization and HPLC analysis of amino acids.

nuclear magnetic, infrared, and Raman spectroscopies. Nuclear magnetic resonance is a technique which yields data regarding the molecular structure of a substance. For NMR, a substance is positioned in a strong magnetic field, which affects the spin of the atomic nuclei of certain isotopes of common elements (Cavanaugh et al., 1996; Reid, 1997). These nuclei are reoriented by passing a radiowave through the substance. The nuclei release a pulse of energy when the wave is terminated, providing information regarding molecular structure. In contrast to NMR, x-ray diffraction crystals diffract x-rays and create a diffraction pattern. The pattern can be interpreted mathematically by a computer providing information about the structure of a molecule. These techniques require very complex, expensive instrumentation and considerable training.

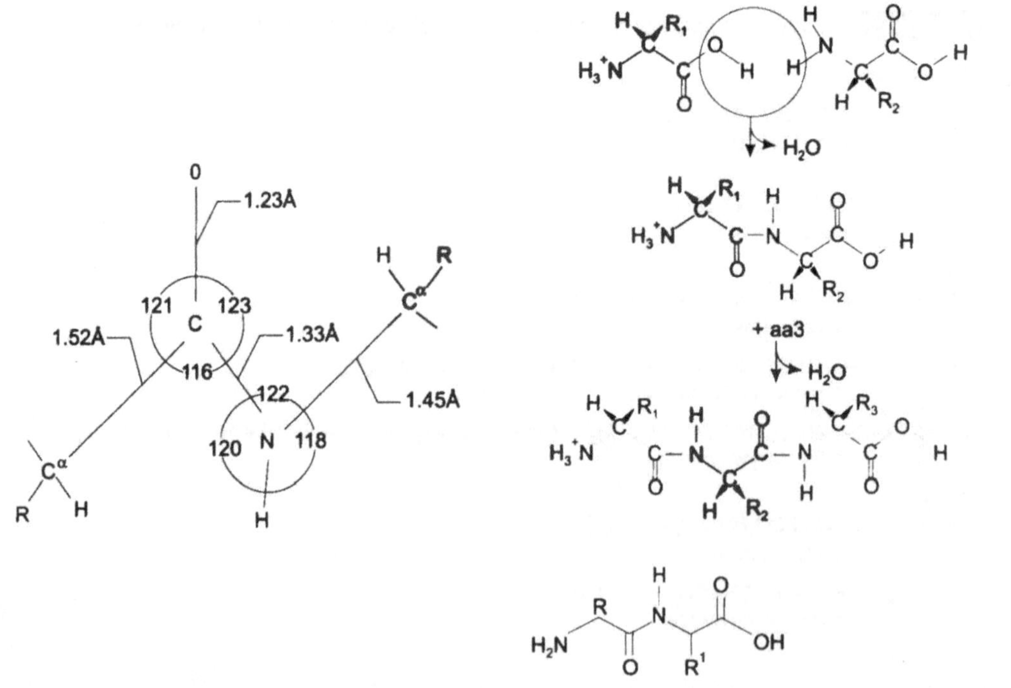

Fig. 4.15 Peptide linkage

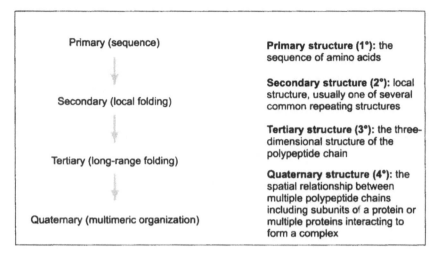

Fig. 4.16 Hierarchies of protein structure

Infrared spectroscopy probes are universally available and ascertain peptide bonds which display distinct IR signals for differentially folded peptides and proteins. Fourier transformation infrared spectroscopy is widely employed to investigate the secondary structure of proteins distinguishing between α-helices and β-pleated sheets. Lastly, Raman spectroscopy provides valuable information regarding amino acid side chains that complements IR data.

The student is referred to the following for in-depth discussions of protein structure and/or methods for protein structure analysis (Austin, 1987; L'Italien, 1987; Vilafranca, 1990; Allen, 1994; Dare and Creighton, 1994; Atassi, 1995; Creighton, 1995; Kyte, 1995; Havel, 1996; Kamp *et al.*, 1997, 2004; Angeletti, 1998). Lastly, the synthesis of peptides (Jones, 1992) and proteins is discussed in Chapter 11.

CLASSIFICATION OF PROTEINS

Proteins are divided into two main groups, fibrous and globular. The former are highly elongated structures existing as fibers or sheets. In contrast, globular proteins are more complex consisting of one or more polypeptide chains folded back on themselves many times. This results in a spherical shape. In globular proteins, some of the amino acid side chains are buried while fibrous proteins possess amino acids whose side chains are exposed to H_2O. Plant proteins have been discussed by Orly (1986), while Schumaker (1994) reviewed lipoproteins and Allen (1995) described cell surface proteins.

Protein Purification

The purification of proteins from cellular homogenates, cell walls, and/or growth media is discussed in the section regarding structure and function of enzymes below. In addition, this topic has been the subject of many monographs (see References under enzymes given at the end of this chapter) and thus has received thorough treatment.

Factors Affecting Enzymatic Activity

Enzymes (E) and substrate (S) form a complex (Fig. 4.17) in an enzyme-catalyzed

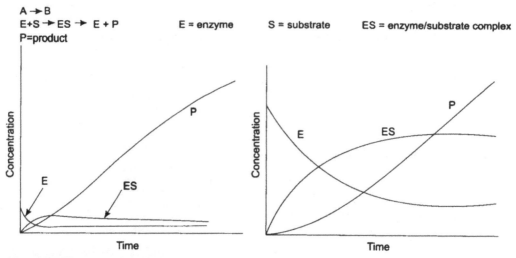

Fig. 4.17 ES complex formation

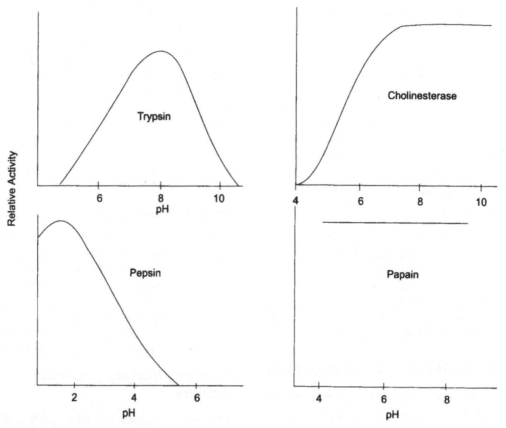

Fig. 4.18 pH optima of enzymes

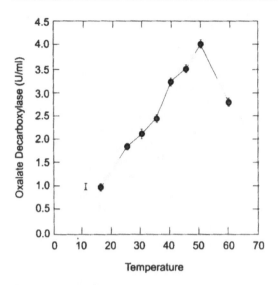

Fig. 4.19 Temperature optimum of enzymes. From: Dashek and Michaels (1997)

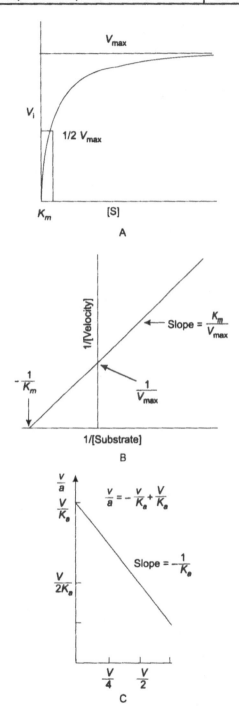

Fig. 4.20 Contd.

reaction (Cooper *et al.*, 1988; Kolby, 1990). These reactions are affected by pH, temperature, enzyme amount, substrate specificity, inhibitors, modulators, Allosteric enzymes, and product amounts (feedback mechanisms). The pH optima for various enzymes are depicted in Fig. 4.18. While the optima are often quite narrow, for certain enzymes they can be substantially broader. On the other hand, the temperature optimum (Fig. 4.19) is usually fairly narrow. Increasing the enzyme amount results in an initial linear acceleration in enzyme activity which subsequently "levels off". In contrast, raising the substrate concentration results in a hyperbolic curve (Fig. 4.20). This curve represents Michaelis-Menten kinetics in which velocity (Y axis) is plotted against substrate concentration (X-axis). The Michaelis-Menten constant

$$k_m \left(V_o = \frac{V_{max}[S]}{K_m + [S]} \right)$$ equals the substrate

concentration at half maximum velocity ($1/2\ V_{max}$). The significance of k_m is that it

Fig. 4.20 Contd.

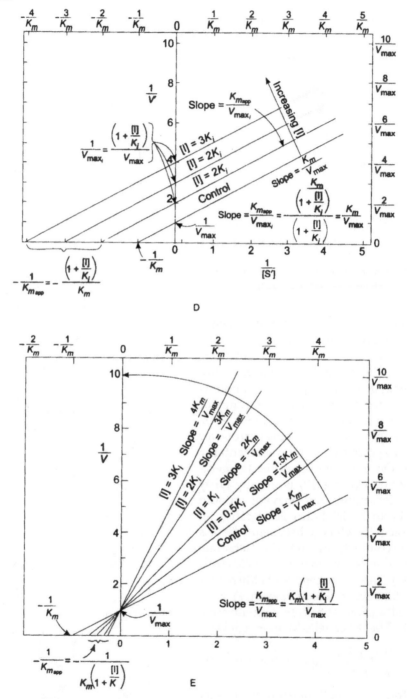

D

E

Fig. 4.20 Enzyme kinetics (from Segel, 1993)

TABLE 4.13 Summary or the effects of inhibitors on Lineweaver-Burk plots 1/V vs 1/[S]

	Slope	Intercept on ordinate
No inhibitor	$\dfrac{K_m}{V\text{max}}$	$\dfrac{1}{V\text{max}}$
Competitive	$\dfrac{K_m}{V\text{max}}\left(1+\dfrac{[I]}{K_1}\right)$	$\dfrac{1}{V\text{max}}$
Uncompetitive	$\dfrac{K_m}{V\text{max}}$	$\dfrac{1}{V\text{max}}\left(1+\dfrac{[I]}{K_1}\right)$
Noncompetitive	$\dfrac{K_m}{V\text{max}}\left(1+\dfrac{[I]}{K_1}\right)$	$\dfrac{1}{V\text{max}}\left(1+\dfrac{[I]}{K_1}\right)$

indicates how tightly a particular substance binds to the active site of an enzyme (Friedrich, 1984). However, for a multistep reaction the relationship between k_m and rate constants is not so simple (Zubay, 1998). It is possible to graph the Michaelis-Menten plot as a double reciprocal, i.e. $1/v$ vs $1/[S]$, thereby yielding a Lineweaver-Burk plot (Fig. 4.20) reflecting the equation $1/vo = \dfrac{k_m}{V_{max}} 1/[S] = \dfrac{1}{V_{max}}$. The negative reciprocal of k_m, $-1/k_m$ results.

An Eadie-Hofste plot can be derived by rearranging the Michaelis-Menten equation to a linear form, $1/vo = -k_m \dfrac{Vo}{[S]} + V_{max}$. The value of a Lineweaver-Burk plot is its importance in investigating the effects of inhibitors on enzyme-catalyzed reactions. Table 4.13 presents a summary of the effects of competitive, uncompetitive, and noncompetitive inhibitors on Lineweaver-Burk plots, which enable discernment of the type of enzyme inhibitor. The student is referred to Purich (1993) and Segel (1993) for additional treatments of enzyme kinetics.

Enzyme Classification

Enzymes are proteins (Dressler, 1995; Copeland, 1996) that catalyze a variety of chemical reactions occurring in microbial, plant, and animal systems (Coolbear, 1992; Tamanoi and Sigman, 2000). Table 4.14 presents classification of the enzymes together with their EC numbers and the reactions they catalyze. While White and White (1997) have published a source book of enzymes, Burrell (1993) and McDonald (1996) have discussed the enzymes of molecular biology.

Thermodynamics and Enzyme Catalysis

Zubay (1998) discussed thermodynamics and Cooper et al. (1988), Schultz (1988) and Sigman and Boyer (1992) described enzyme catalysis in depth. The laws of thermodynamics as well as important terms and concepts relating to the laws are presented in Table 4.15. Enzyme structure serves two purposes: 1) To absorb the particular substrate molecule onto the complementary surface (active site) of the enzyme, and 2) To catalyze conversion of this substrate to a

TABLE 4.14 Classification of enzymes

Enzyme structures are classified by their E.C. number (using the May 2000 v. 26.0 release of the ENZYME Data Bank)

Enzyme classification (E.C.)	Reaction catalyzed
E.C.1.–.–.–Oxidoreductases [1064 PBD entries] $A-B+C=A=B-C$	Electron transfer
E.C.2.–.–.–Transferases [1435 PDB entries] $A-B+C = A+B-C$	Group transfer
E.C.3.–.–.–Hydrolases [3173 PDB entries] $A-B+H_2O \rightleftarrows A-H+B-OH$	
E.C.4.–.–.–Lyases [419 PDB entries $\frac{XY}{AB}=A=B+X-Y$	Group additional removal to/from double bond
E.C.5.–.–.–Isomerases. [309 PDB entries] $\frac{XY}{AB}=\frac{YY}{AB}$	Transfer of groups in molecules resulting in isomeric forms
E.C.6.–.–.–Ligases [162 PDB entries] $A-B = A-B$	Generation of C–C, S–S, G–O and C–N bonds condensation and ATP cleavage

The enzyme structures can also be accessed via the PDBsum database:
• PDBsum – Summaries and structural analyses of PDB data files
[a]The student is referred to Nomenclature Committee of the International Union of Biochemistry and Molecular Biology for discussion of enzyme nomenclature.
[b]From: Dashek and Michaels (1997) and R. Laskowski and A. Wallace: http://www.biochem.ucl.oc.uk/bsm/enzymes/

TABLE 4.15 Summary of the laws of thermodynamics and related terms and concepts

Laws of Thermodynamics

    First Law – Total energy of the universe remains constant

    Second Law – Universe constantly changes and becomes more disordered or systems change spontaneously from states of lower probability (more ordered) to states of higher probability.

Terms and Concepts

    Free Energy (G) – Criterion for an increase in disorder

      ΔG Change in free energy, i.e. a measure of the extent of disorder

      ΔG Release of free energy, reactions occur spontaneously, creation of disorder, energetically favorable

      $+\Delta G$ Occurs only if coupled to a 2nd reaction with a $-\Delta G$; creates order but energetically unfavorable

Entropy = Degree of disorder of a state

(S)

    $\Delta S = R\ln pB/pA$

Change in free energy that occurs when A → B, i.e., one mole of A is converted into one mole of B. pA and pB = probabilities of two states, Note: Reaction with a large increase in S occurs spontaneously

R = gas constant

Enthalpy (H) or E + PV Change in free energy ΔH = heat absorbed

Free energy change is a direct measure of entropy change in the universe

H = Enthalpy	V = Volume
G = Gibbs free energy	E = Energy
T = Absolute temperature	S = Entropy
P = Pressure	

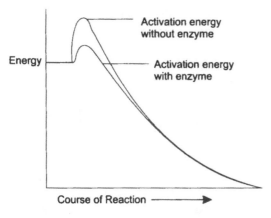

Fig. 4.21 Mechanism of enzyme catalysis

product which is released back to the solution, leaving the enzyme catalyst unaltered (Rosevear *et al.*, 1987). To understand the mechanism of enzyme catalysis, the reader should refer to Fig. 4.21. A compound A can achieve a lower and more favorable energy state by being converted to compound Y. However, conversion cannot occur unless A can acquire sufficient activation energy from its surroundings to undergo the reaction. Enzymes, organic catalysts, lower the activation energy (Fersht, 1985; Cooper *et al.*, 1988; Sigman and Boyer, 1992).

TABLE 4.16	Some coenzymes and the groups they transfer

Coenzyme	Group transferred
ATP	Phosphate
Biotin	Carboxyl
Coenzyme A	Acetyl
NADH, NADPH	Hydrogen and electron
S-adenosylmethionine	Methyl
Thiamine pyrophosphate	Aldehyde

Cofactors

Coenzymes are small, nonpolypeptide molecules tightly bound to enzyme surfaces and essential for enzyme activity. The function of coenzymes is group transfer during certain biochemical reactions (Table 4.16). In addition to coenzymes, certain elements are required for the activity of metalloenzymes.

Multienzyme Complexes

These are large protein assemblies of various enzymes involved in a reaction sequence. These assemblies can increase reaction rates without raising substrate concentrations. Two well-known multienzyme complexes are the pyruvate dehydrogenase complex and enzymes of electron transport chain. The process of protein targeting to a proteasome for the former degradation appears to involve three enzymes: E^1 binds to and activates a ubiquitin molecule and then transfers the latter to E^2, which joins to E^3. As E^3 binds to protein, the ubiquitin molecule carried by E^2 is detached and transferred to the protein. This cascade of events repeats until the protein is "tagged" with a chain of ubiquitins. Subsequently, the chain binds to the proteasome, whereupon the protein is unfolded for protease degradation in the proteasome's interior (Goldberg, 2000).

Immobilized Enzymes

Immobilization refers to retention of a biologically active catalyst (Woodward, 1985) within either a reactor or analytical system (Rosevear *et al.*, 1987). Soluble protein catalyst or neutral buoyancy cells (White, 1985; Laskin, 1985; Mosbach, 1988) are confined within or on a supporting material in a reactor. There are two types of reactors, continuous and batch; in the former, the immobilized material may be retained in a column reactor through which fluid is continuously passed. The result is conversion of the substrate to a catalyst-free product. In contrast, the batch reactor

TABLE 4.17 Benefits and limitations of the immobilized state*

Benefits	Limitations
Catalyst retention in the reactor	Loss of catalyzed activity during immobilization
High concentration of catalyst in the reactor	Mass transfer problems
Control of catalyst microenvironment	Physical determination between catalyst and fluid
Quantitative and rapid removal of catalyst	Prolonged operation
Separation of catalyst	Empirical nature of immobilization
Production from its use	Technology

*Adapted from Rosevear *et al.* (1987) who discuss each of the above in detail.

involves repeated movement of the immobilized catalyst from one fluid to another. The substrate is converted to product on a batch principle. Table 4.17 presents the benefits and limitations of the immobilized state. Immobilized enzymes are beneficial for ELISAS, wastewater treatment and a variety of industrial purposes (Guibault, 1984; Giubault and Mascini, 1993; Uhlig, 1998).

Immobilization of enzymes can occur by absorption and covalent coupling, metal link/chelation process, gel entrapment and microencapsulation. The student can find details of these methods in Woodward (1985), Rosevear (1987), and Bickerstaff (1997).

Enzyme and Total Protein Assays

Enzyme assays have been reviewed by Eisenthal and Danson (1992). The choice of enzyme assay depends on whether disappearance of the substrate is being followed or the appearance of the product (Eisenthal and Danson, 1992). The choices are several. Spectrophotometic methods measure the absorption of the substrate or the product at a given wavelength, i.e. absorption maximum. A specific example is the polyphenol oxidase catalyzed conversion of catechol to benzoquinone (Taylor, 1987; Moore *et al.*, 1989). The latter absorbs maximally at 440 nm. Radioactive assays are sensitive but present many difficulties to the investigator who must assess benefits versus difficulties. These assays utilize radioactive substrates, thereby mandating that the investigator separate and purify the radioactive product from unreacted radioactive substrate in a reaction mixture. Thus, considerable biochemical techniques such as time- consuming chromatography and electrophoresis may be involved. This is complicated by institutional regulations regarding the use and disposal of radionuclides.

Once isolated the radioactive product can be monitored by gas-flow counting, radiochromatogram strips scanning, or liquid scintillation counting (Burns and Steiner, 1991; Peng *et al.*, 1980). Each of these techniques requires considerable training.

TABLE 4.18 Summary of colorimetric procedures for proteins in solution

Procedure	Sensitivity	Reference(s)*
Protein-coomassie Blue binding	Microgram	Bradford (1976)
Gold reagents	Nanogram	Burton *et al.* (1989)
Colloidal gold	Low nanogram	Cheley and Bayley (1991)
		Ciesiolka and Gobinis (1988)
Stain auro dye	Subnanogram	Li *et al.* (1989)
Auro dye, Ferri dye, and India ink	Nanogram	
Folin phenol reagent	Microgram	Lowry *et al.* (1956)

*Citations of these references occur in Dashek (1997).

TABLE 4.19 Some enzyme purification techniques

Technique	Principle	Reference
Ammonium sulfate Fractionation	"Salting-out"	Scopes (1987)
Dialysis	Removal of unwanted low molecular weight substances via dialysis tubing; diffusion through dialysis tubing	Scopes (1987)
Chromatographies Affinity	Adsorption of specific proteins onto a ligand covalently attached to a matrix	Lowe (1979) Scopes (1987) Bailon et al. (2000)
Dye ligand	Pseudoaffinity adsorbents, i.e. dyes as affinity adsorbents	Amicon Corporation (1980) Scopes (1987)
Gel filtration	Molecular sieving	Fischer (1980)
High pressure liquid	Differential migration rates between a stationary column and a mobile phase	Mant and Hodges (1991)
Hydrophobic interaction	Hydrophobic interaction between hydrocarbon side chains	Hodgkinson and Lowry (1981)
Hydroxyapatite and calcium phosphate gels	Selective adsorption of proteins and release by higher salt concentrations	Gorbunoff (1985)
Immunoadsorbents	Use of immobilized antibodies	Livingstone (1974)
Ion exchange	Simple displacement of counterions ionically bound to a matrix followed by displacements of sample eluting ions	Pharmacia (1967)
Other techniques Chromatofocusing	Isoelectric focusing on ion-exchange columns	Sluyterman and Elgersma (1978)
Isoelectric focusing	Separation of proteins by their isoelectric points in a pH gradient	Rigletti (1983) Goodman and Baptist (1979) Scopes (1987)
Isotachophoresis	Separation of proteins on the basis of mobility (mobility per unit electric field)	
Liquid-liquid chromatography		Scopes (1980)

The pH Stat monitoring of enzyme-catalyzed reactions is dependent on the knowledge that enzyme-catalyzed reactions are accompanied by a release or consumption of protons. Because the above assays are time consuming, automated analyses for numerous samples have been generated (Uhlig, 1998).

The amount of protein in the enzyme preparation must be quantified in order for the specific activity (μmol product formed or substrate disappeared min^{-1} mg $protein^{-1}$

or $\dfrac{\Delta A\ min^{-1} \times 100}{mg\ protein}$) to be calculated.

Protein amount can be quantified by either UV-spectroscopy (Demchenko, 1986) or colorimetric (Table 4.18) assays.

Enzyme Purification

Because protein purification has been reviewed many times (Doonan, 1989;

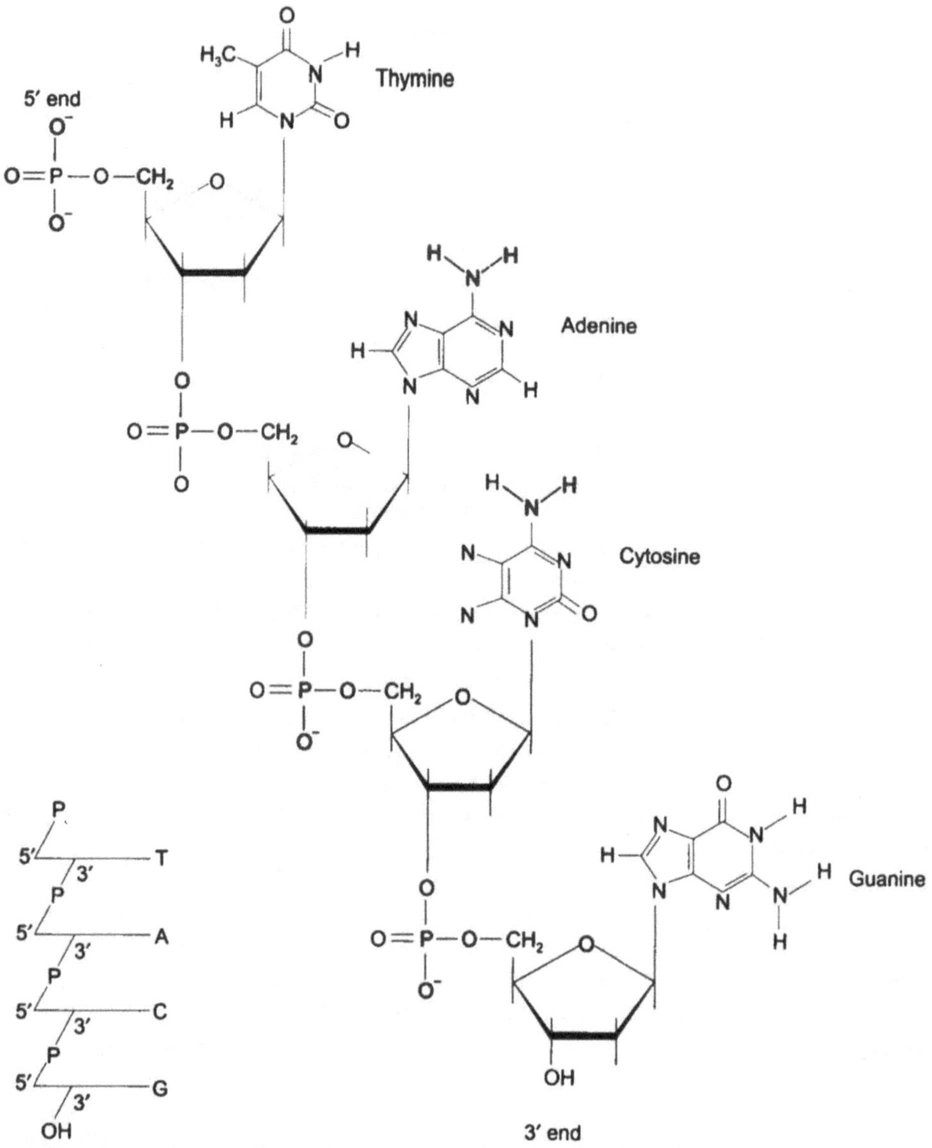

Fig. 4.22 Covalently-linked nucleotides

Deutscher, 1990; Englard and Seifter, 1990; Harris and Angral, 1990), the topic is not discussed here. Table 4.19 presents a list of techniques commonly employed in designing a protein purification protocol.

Nucleic Acids

Nucleic acids (DNA and RNA) consist of covalently linked nucleotides (Fig. 4.22) which are composed of a pentose (deoxyri-

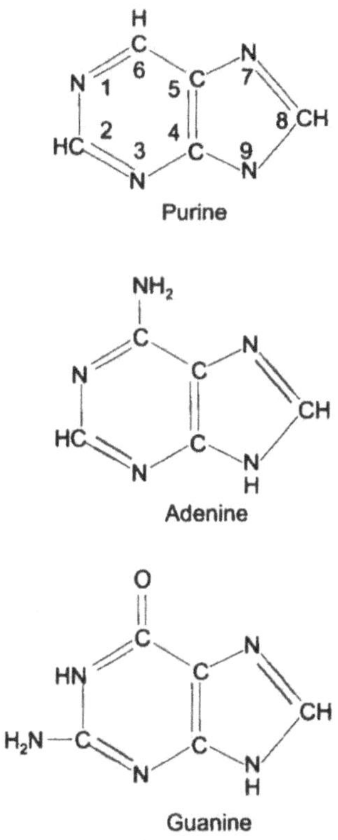

Fig. 4.23 Purine chemistry

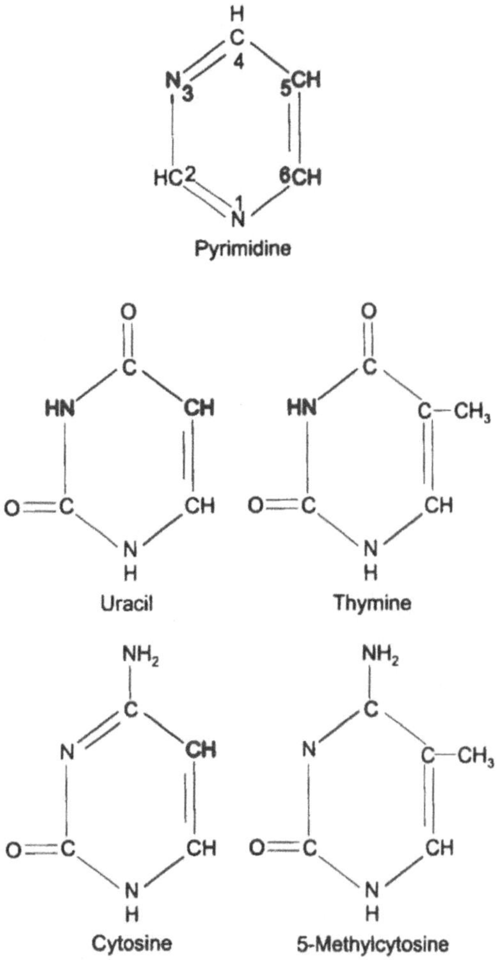

Fig. 4.24 Pyrimidine chemistry

bose or ribose), a phosphate group, and nitrogenous bases (purines and pyrimidines). The student is referred to the following works for in-depth discussions (Hall and Davis, 1979; Marcus, 1981; Boulter, 1982 a, b; Blackburn and Gait, 1990; Adams *et al.*, 1992; Brickell and Darling, 1995) of nucleic acid chemistry.

The phosphate groups are linked to carbons 5 and 3 of the monosaccharides. The nitrogenous bases are joined to deoxyribose in DNA and ribose in RNA through carbon 1 of the monosaccharides. The purine (Fig. 4.23) and pyrimidine (Fig. 4.24) bases of both DNA and RNA are depicted in Table 4.20.

A nitrogenous base on one strand pairs with a base on the other strand of DNA via hydrogen bonding, e.g. A with T and G with C (Brielman, 1999). Nucleosides (Townsend, 1988) consist of a nitrogenous base and sugar (Fig. 4.25). In contrast, a nitrogenous base, sugar, and phosphate form a nucleotide (Fig. 4.26). Table 4.21 illustrates nucleosides and nucleotides.

DNA (Fig. 4.27) is a duplex (double-stranded molecule) but is single-stranded in

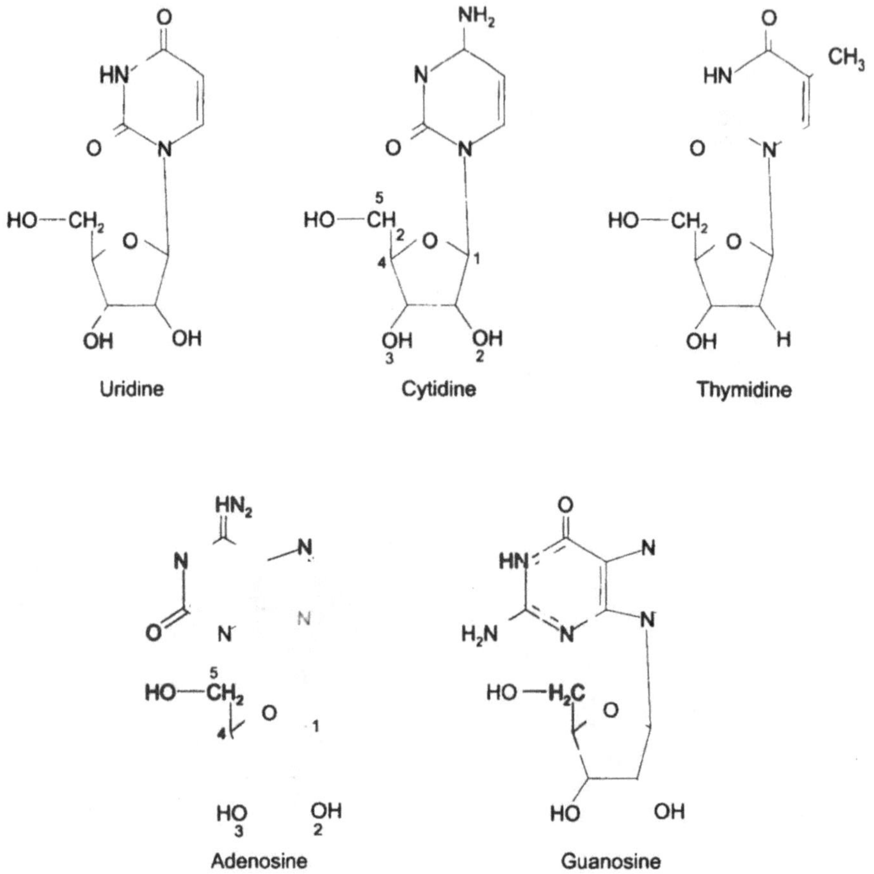

Fig. 4.25 Composition of nucleosides

some phages and viruses (Saluz and Wiebauer, 1995). The backbone of DNA is maintained by repeating 3'5' phosphodiester linkage (Fig. 4.28). The side chains are purines and pyrimidines. Conformation of DNA is β-DNA (Neidle, 1999) but other conformations are possible (Fig. 4.29). The latter are contingent upon nucleotide sequences, degree of hydration and interactions with proteins. Circular DNA molecules (Fig. 4.30) can occur. Finally the structure of DNA has been investigated by nuclear magnetic resonance (Roberts, 1993; James, 1995).

RNA is a lengthy, single strand of nucleotides (Townsend and Tripsin, 1991). The types of RNA are m-RNA (messenger), r-RNA (ribosomal) and t-RNA (transfer). Messenger is transcribed from a single strand of DNA. This RNA is heterogeneous in both size and sequence and contains a 5' cap consisting of 5' to 5' triphosphate linkages between two nucleotides, which are 7-methyl-guanosine and another containing 2' o-methyl purine. Most m-RNAs possess a polyadenosine tail at the 3' terminus. Both the 5' cap and 3' tail (Fig. 4.31) are additions to transcribed RNA.

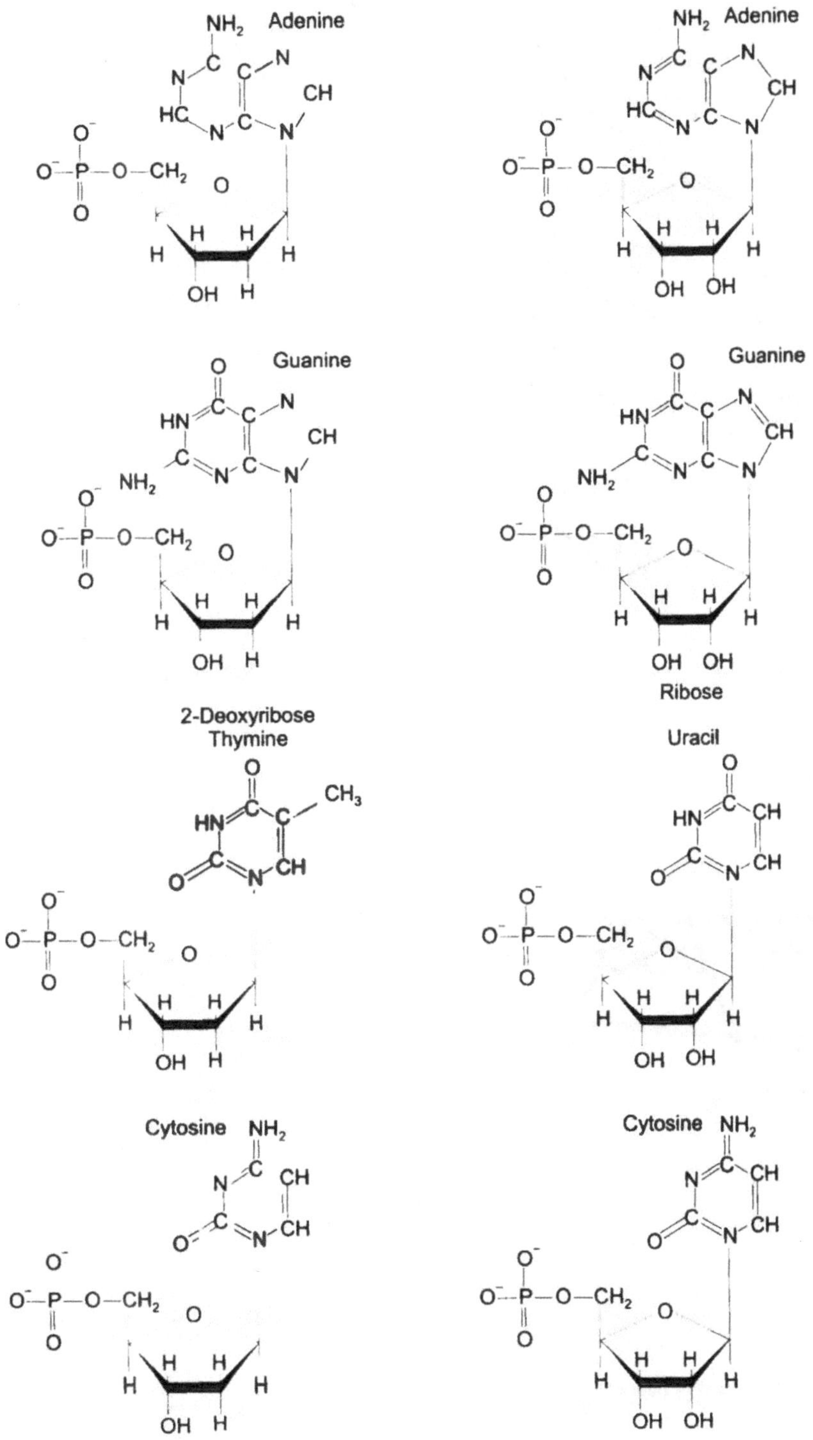

Fig. 4.26 Nucleotide structure

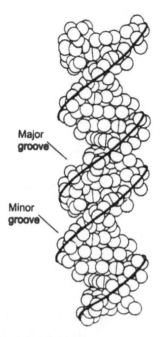

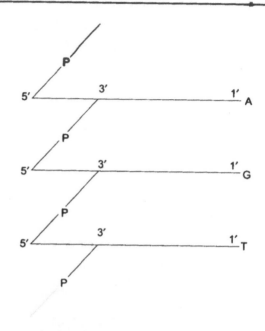

Fig. 4.27 The DNA molecule

Fig. 4.28 DNA backbone

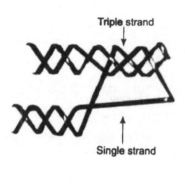

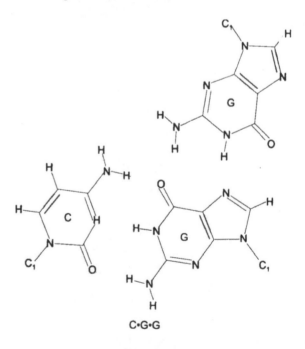

C•G•G

Fig. 4.29 β-DNA conformations

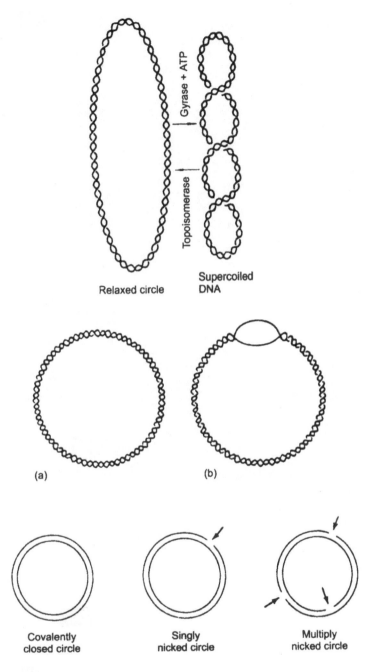

Fig. 4.30 Other DNA conformations

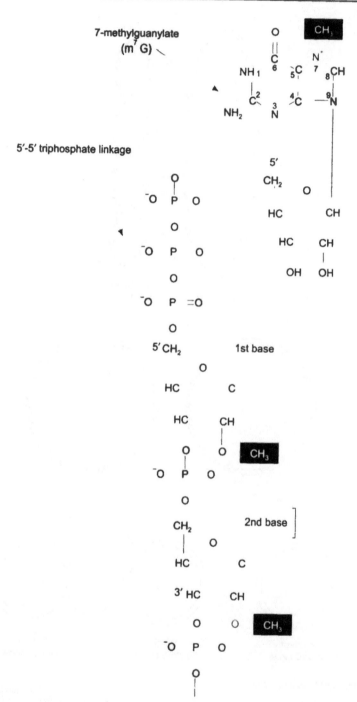

Fig. 4.31 5' cap and 3' tail of transcribed RNA

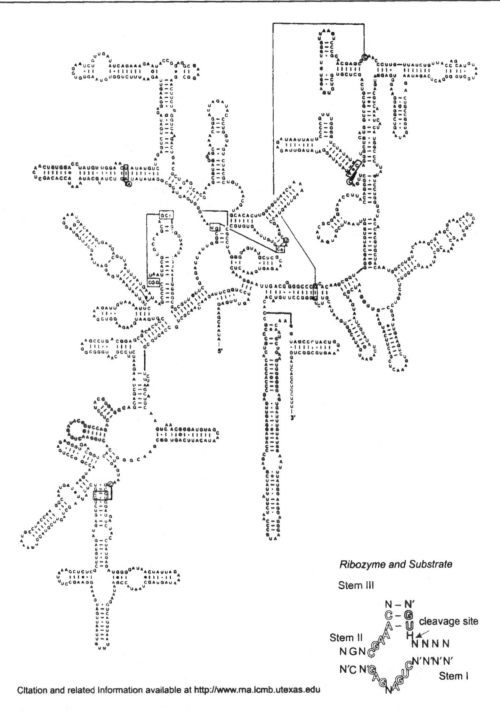

Ribozyme and Substrate

Citation and related information available at http://www.rna.lcmb.utexas.edu

Fig. 4.32 Example of a ribozyme

TABLE 4.20 Summary of purines and pyrimidines

Purines (Lister, 1996)	Pyrimidines (Brown, 1994)
Adenine (A)	Uracil (U) in RNA
Guanine (G)	Cytosine (C)
	Thymine (T) in DNA

TABLE 4.21 Nucleosides and nucleotides

Nucleosides	Nucleotides
Adenosine	Adenylate
Deoxyadenosine	Deoxyadenylate
Guanosine	Guanylate
Deoxyguanosine	Deoxyguanylate
Cytidine	Cytidylate
Deoxycytidine	Deoxycytidylate
Thymidine	Thymidylate
Deoxythymidine	Deoxythymidylate
Uridine	Uridylate

TABLE 4.22 Comparison of eukaryotic and prokaryotic ribosomes

	Eukaryotic	Prokaryotic
Sedimentation coefficient	80S	5S, 16S, 23S
Subunits	5S, 5.8S, 18S, 28S	5S, 16S, 23S
Number of nucleotides	5S, 5.8S, 18S, 28S	5S, 16S, 23S
	120 156 1900 4700	120 1542 2904

Messenger – RNA carries the genetic information from the nucleus to cytoplasmic ribosomes.

Ribosomes are nearly spherical and about 20-30 mm in diameter. Ribosomes can exist free in the cytoplasm or attached to the endoplasmic reticulum. Polysomes are clusters of ribosomes. Table 4.22 presents a comparison of eukaryotic and prokaryotic ribosomes.

Transfer-RNA (t-RNA) contains a loop and occurs as an RNA strand folds back on itself yielding a secondary structure. A tertiary structure likewise results from such folds. Transfer-RNA is a major player in protein synthesis as this RNA brings the appropriate amino acid to the ribosome. Protein synthesis has been reviewed many times and the student is referred to any college level introductory biology or botany text.

Ribozymes (Fig. 4.32) are recently discovered RNA molecules which catalyze biochemical reactions (Rossi, 1999). Contemporary reviews of the molecular biology of nucleic acids have been given by Eckstein and Lilley (1996), Leontis and Santalucia (1998), and Walker (1984).

References

CARBOHYDRATES

Aspinall, G.O. 1980. Chemistry of cell wall polysaccharides. In: *Carbohydrates. The Biochemistry of Plants. A Comprehensive Treatise*, vol. 3, pp. 463-500. Preiss, J. (ed.). Academic Press, New York, NY.

Avigad, G. 1982. Sucrose and other disaccharides. In: *Encyclopedia of Plant Physiology, New Series. Plant Carbohydrates I. Intracellular Carbohydrates*. vol. 13A, pp. 216-247. Loewus, F.A., and Tanner, W. (eds.). Springer-Verlag, Berlin, Germany.

Avigad, G. 1990. Dissacharrides. In: *Carbohydrates. The Biochemistry of Plants. A Comprehensive Treatise*, vol. 14, pp. 111-188. Preiss, J. (ed.). Academic Press, San Diego, CA.

Axford, J.S. 1998. *Glycoimmunology* 2. Plenum Press, New York, NY.

Bacic, A., Harris, P.J., and Stone, B.A. 1988. Structure and function of plant cell walls. In: *Carbohydrates. Structure and Function. The Biochemistry of Plants. A Comprehensive Treatise*, vol. 14. Preiss, J. (ed.). Academic Press, San Diego, CA.

Bednarski, M.D. and Simon, E.W. 1991. *Enzymes in Carbohydrates Synthesis*. Amer. Chem. Soc., Washington, DC.

Beitner, R. 1985. *Regulation of Carbohydrate Metabolism*. CRC Press, Boca Raton, FL, USA.

Binkley, R.W. 1988. *Modern Carbohydrate Chemistry*. Marcel Dekker, New York, NY.

Bols, M. 1996. *Carbohydrate Building Blocks*. John Wiley and Sons, Inc., New York, NY.

Bowles, D. 1982. Membrane glycoproteins. In: *Plant Carbohydrates I. Intracellular Carbohydrates*. pp. 584-601. Loewus, F.A. and Tanner, W. (eds.). Springer-Verlag, Berlin, Germany.

Burns, D.P. and Steiner, R. 1981. Advanced Technology Guide for LS 6000 Series Scintillation Counters. Bechman Instruments Palo Alto, CA, USA. Pp. 3-12.

Carpita, N.C. 1987. The biochemistry of the 'growing' plant cell wall. In: *Physiology of Cell Expansion during Plant Growth*, pp. 28-45. Cosgrove, D.J. and Knievel, D.P. (eds.). Amer. Soc. Plant Physiol., Rockville, MD, USA.

Carpita, N.C. and Gibeaut, D.M. 1988. Biosynthesis and secretion of plant cell wall polysaccharides. In: *Current Topics in Plant Biochemistry and Physiology*, vol. 7, pp. 112-133. Randall, D.D., Blevins, D.B. and Campbell, W.H. (eds.). Univ. Missouri, Columbia, MO, USA.

Carpita, N.C. and Gibeaut, D.M. 1993. Structural models of primary cell walls in flowering plants: Consistency of molecular structure with physical properties of the walls during growth. Plant J 3:1-30.

Carpita, N.C., Housley, T.L. and Hendris, J.E. 1991. New features of plant fructan structure revealed by methylation analysis and carbon [13]NMR spectroscopy. Carbohy. Res. 146:129.

Carpita, N.C., McCarr, M. and Giffing, L.R. 1996. The plant extracellular matrix. News from the cell's frontier. Plant Cell 8:1451-1463.

Cassab, G.I. 1998. Cell wall proteins. Ann. Rev. Pl., Physiol., Pl. Molec. Biol. 49:281-309.

Churms, S.C. 1982. *Carbohydrates*. CRC Press, Boca Raton, FL, USA.

Collins, P.M. 1987. *Carbohydrates*. Chapman and Hall, New York, NY.

Cordy, D.J. 1980. *Biological Functions of Carbohydrates*. John Wiley and Sons, Inc., New York, NY.

Cosgrove, D.J. 1998. Cell wall loosening by expansin. Pl. Physiol. 118:333-339.

Cosgrove, D.J. 1999. Enzymes and other agents that enhance cell wall extensibility. Ann. Rev. Pl. Physiol. Pl. Molec. Biol. 50:391-419.

Dashek, W.V. 1997. Carbohydrates. In: *Methods in Plant Biochemistry and Molecular Biology*, pp. 29-46. Dashek, W.V. (ed.). CRC Press, Boca Raton, FL, USA.

Delmer, D.P. and Stone, B.A. 1988. Biosynthesis of plant cell walls. In: *Carbohydrates. The Biochemistry of Plants. A Comprehensive Treatise*, vol. 14, pp. 373-420. Preiss, J. (ed.). Academic Press, San Diego, CA.

Dey, P.M. 1990. Oligosaccharides. In: *Carbohydrate Methods in Plant Biochemistry*, vol. 2, pp. 189-218. Dey, P.M. (ed.). Academic Press, San Diego, CA.

Dey, P.M. and Dixon, R.A. (eds.). 1985. *Biochemistry of Storage Carbohydrates in Green Plants*. Academic Press, New York, NY.

Duffus, D.M. and Duffus, J.H. 1984. *Carbohydrate Metabolism in Plants*. Longman, London, England.

El Khadem, H. 1988. *Carbohydrate Chemistry: Monosaccharides and Their Oligomers*. Academic Press, New York, NY.

Elbein, A.D. 1982. Glycolipids and other glycosides. In: *Plant Carbohydrates*, pp. 601-612. Loweus, F.A. and Tanner, W. (eds.). Springer-Verlag, Berlin, Germany.

Farrar, J.F. and Pollock, C.J. 1995. *The Biology of Fructans*. Cambridge Univ. Press, New York, NY.

Feingold, D.S. and Avigad, G. 1980. Sugar nucleotide transformation in plants. In: *The Biochemistry of Plants. A Comprehensive Treatise. Carbohydrate Structure and Function*. Preiss, J. (ed.). Academic Press, NY, USA, Vol. 3, pp. 221-270.

Feingold, D.S. and Barber, G.A. 1990. Nucleotide sugars. In: *Methods in Plant Biochemistry*. Vol. 2, pp. 39-78. Dey, P.M. (ed.). Academic Press, London, England.

Franz, G. and Blaschek, W. 1990. Cellulose. In: *Carbohydrates. The Biochemistry of Plants. A Comprehensive Treatise*, vol. 14, pp. 291-322. Preiss, J. (ed.). Academic Press, San Diego, CA.

French, A.D. 1990. *Computer Modeling of Carbohydrate Molecules.* Amer. Chem. Soc., Washington, DC.

Hwa-Kwan, Park, Robyt, J.F. and Yang-Do Choi. 1996. *Enzymes for Carbohydrate Engineering.* Elsevier, Amsterdam, The Netherlands.

Jarvis, M.C. 1984. Structure and properties of pectin gels in plant cell walls, Plant Cell Envir. 7:153-164.

Kandler, O. and Hopf, H. 1980. Occurrence, metabolism and function of oligosaccharides. In: *Carbohydrates. The Biochemistry of Plants. A Comprehensive Treatise.* Preiss, J. (ed.). Academic Press, New York, NY.

Kates, M. 1990. *Glycolipids, Phospholipids and Sulfolipids.* Plenum Press, New York, NY.

Kaushal, G.P., Szumilo, T. and Elbein, A.D. 1988. Structure and biosynthesis of plant n-linked glycoproteins. In: *Carbohydrates. The Biochemistry of Plants. A Comprehensive Treatise*, vol. 14, pp. 421-463. Stumpf, P.K. and Conn, E.E. (eds.). Academic Press, New York, NY.

Kennedy, J.F. 1988. *Carbohydrate Chemistry.* Oxford University Press, NY.

Lamport, D.T.A. 1980. Structure and function of plant glycoproteins. In: *The Biochemistry of Plants* pp. 501-541. Stumpf, P.K. and Conn, E.E. (eds.). Academic Press New York, NY.

Lehman, J. 1998. *Chemistry of Carbohydrates. Structure and Biology.* Kohlenhydrate Chemis and Biologie, Stuttard, Germany.

Lewis, H.G. and Paice, M.G. 1989. *Plant Cell Wall Polymers. Biogenesis and Biodegradation.* Amer. Chem. Soc., Washington, DC.

Loewus, F.A. and Tanner, W. 1982. *Plant Carbohydrates.* Springer-Verlag, Berlin, Germany.

Meuser, F., Manner, D.J. and Seibel, W. 1993. *Plant Polymeric Carbohydrates.* Roy. Soc. Chem., Cambridge, England.

Mombarg, E. 1997. *Catalytic Modifications of Carbohydrates.* Delft Univ. Press, Delft, The Netherlands.

Morrison, W.R. and Karkalas, J. 1990. Starch. In: *Methods in Plant Biochemistry*, pp. 323-352. Dey, P.M. (ed.). Academic Press, New York, NY.

O'Neill, M., Albershiem, P. and Darvill, A. 1990. The pectin polysaccharides of primary cell wall. In: *Carbohydrates. The Biochemistry of Plants. A Comprehensive Treatise*, vol. 17, pp.

415-441. Preiss, J. (ed.). Academic Press, San Diego, CA.

Pollock, C.J. and Chatterton, J.J. 1988. Fructans in Carbohydrates. Preiss, J. (ed.), vol. 14. In: *Carbohydrates. The Biochemistry of Plants. A Comprehensive Treatise.* Stumpf, P.K. and Conn, E.E. (eds.). Academic Press, Inc., New York, NY.

Preiss, J. 1980. Carbohydrates: Structure and Function. In: *Carbohydrates. The Biochemistry of Plants. A Comprehensive Treatise*, vol. 3. Stumpf, P.K. and Conn, E.E. (eds.). Academic Press, Inc., New York, NY.

Raju, T.S., Lerner, T. and O'Connor, J.V. 1996. Glycoprotein-biological significance and methods for the analysis of complex carbohydrates of recombinant glycoproteins. Biotechnology and Applied Biochemistry. 24:191-194.

Robyt, J.F. 1998. *Essentials of Carbohydrate Chemistry.* Springer, New York, NY.

Roehrig, K.L. 1984. *Carbohydrate Biochemistry and Metabolism.* AVI Publ. Co., Westport, CT, USA.

Selvendran, R.R. and O'Neill, M.A. 1982. *Plant Carbohydrates, I. Intracellular Carbohydrates.* Loewus, F.A. and Tanner, W. (eds.). Springer-Verlag, Berlin, Germany, pp. 515-583.

Selvendran, R.R. and Ryder, P. 1990. Isolation and analysis of plant cell walls: In *Carbohydrates. The Biochemistry of Plants and a Comprehensive Treatise*, vol. 14, pp. 549-579. Preiss, J. (ed.). Academic Press, San Diego, CA.

Sturgeon, R.J. 1990. Monosaccharides. In: *Carbohydrate Methods in Plant Biochemistry*, vol. 2, pp. 1-37. Dey, P.M. (ed.). Academic Press, San Diego, CA.

Suzuki, M. and Chatterton, N.J. 1993. *Science and Technology of Fructans.* CRC Press, Boca Raton, FL, USA.

Theim, J. 1990. *Carbohydrate Chemistry.* Springer-Verlag, New York, NY.

Van Damme, J.M., Peumans, W.J., Puztai, A. and Bardocz, S. 1998. *Handbook of Plant Lectins: Properties and Biomedical Applications.* John Wiley, New York, NY.

Varner, J.E. and Lin, L.S. 1989. Plant cell wall architecture. Cell 56: 231-239.

Wu A.M. 1988. *The Molecular Immunology of Complex Carbohydrates.* Plenum Press, New York, NY, USA.

LIPIDS

Benning, C. 1998. Biosynthesis and function of the sulfolipids sulfoquinovosyldiacylglycerol. Ann. Rev. Plant Physiol. Molec. Biol. 49: 53-75.

Biocs, P.A., Cruiz, K. and Krenner, T. 1989. *Biological Role of Plant Lipids.* Plenum Press, New York, NY, USA.

Cevc, G. 1993. *Phospholipids Handbook.* Marcel Dekker, New York, NY.

Cevc, G. and Paltauf, F. 1995. *Phospholipids Characterization, Metabolism and Novel Biological Applications.* AOCS Pr., Champaign, IL, USA.

Dennis, E.A. 1991. *Phospholipases.* Academic Press, San Diego, CA.

Dennis, E.A. and Vance, D.E. 1992. *Phospholipid Biosynthesis.* Academic Press, San Diego, CA.

Fukuda, M. and Hindsgaul, O. 1994. *Glycobiology. A Practical Approach.* IRL Press, New York, NY.

Good, L.J., Zimowski, J., Evershed, R.P. and Male, V.L. 1987. The sterol esters of higher plants. In: *The Metabolism Structure and Function of Plant Lipids,* pp. 95-102. Stumpf, P.K., Mudd, J.B. and Ness, W.D. (eds.). Plenum Press, New York, NY.

Gurr, M.I., Harwood, J.L. and Frayn, K.N. 2002. *Lipid Biochemistry.* Blackwell Science, Malden, MA, USA.

Hamilton, R.J. 1995. *Waxes, Chemistry, Moleuclar Biology and Function.* Oily Press, Dundee, Scotland.

Harwood, J.L. 1980. Plant acyl lipids: structure, distribution and analysis. In: *The Biochemistry of Plants.* Stumpf, P.K. and Conn, E.E. (eds.). Academic Press, New York, NY.

Heinz, E. 1996. Plant glycolipids: structure, isolation and analysis. In: *Advances in Lipid Methodology,* pp. 211-332. Christie, W.W. (ed.). Oily Press, Dundee, Scotland.

Horowitz, M.I. 1982. *Glycoproteins, Glycolipids and Proteoglycans.* Academic Press, New York, NY.

Jost, P.C. and Griffith, O.H. 1982. *Molecular Biology of Lipid-Protein Interactions.* John Wiley, New York, NY.

Kates, M. 1990. *Glycolipids, Phosphoglycolipids and Sulfoglycolipids.* Plenum Press, New York, NY.

Kolattukudy, P.E. 1980. Cutin, suberin and waxes and their role in plant microbe interactions. In: *The Metabolism, Structure and Function of Plant Lipids,* pp. 291-314. Stumpf, P.K. (ed.). Plenum Press, New York, NY.

Larssan, K. 1994. *Lipids: Molecular Organization Physical Functions and Technical Applications.* Oily Press, Dundee, Scotland.

Law, J.H. and Rolling, H.C. 1985. *Steroids and Isoprenoids.* Acad. Press, Orlando, FL, USA.

Lukas, S.E. 1994. *Steroids.* Enslow Publs. Springfield, NJ, USA.

Mead, J.F., Alfin-Slater, R., Howtan, D.R. and Pojjack, G. 1980. *Lipids: Chemistry, Biochemistry and Nutrition.* Plenum Press, New York, NY.

Moreau, R.A. 1990. Plant lipid class analysis by HPLC. In: *Plant Lipid Biochemistry, Structure and Utilization,* pp. 20-22. Quinn, P.J. and Harwood, J.L. (eds.). Portland Press Ltd., London, UK.

Mostofsky, D.I. and Yehuda, S. 1996. *Fatty Acids Biochemistry and Behavior.* Humana Press, Totowa, NJ, USA.

Mudd, J.B. and Kleppinger-Sparace, K. 1987. Sulfolipid, vol. 9, pp. 275-289. In: *The Biochemistry of Plants.* Stumpf, P.K. (ed.). Academic Press, New York, NY.

Quinn, P.J. and Harwood, J.L. 1990. *Plant Lipid Biochemistry, Structure and Utilization.* Portland Press Ltd., London, UK.

Schumaker, V.N. 1994. *Lipoproteins, Apolipoproteins and Lipases.* Academic Press, San Diego, CA.

Siegenthaler, P.A. and Eichenberger, W. 1984. *Structure, Function and Metabolism of Plant Lipids.* Elsevier, Amsterdam, Holland.

Stumpf, P.K. 1980. *Lipids, Structures and Functions. The Biochemistry of Plants,* vol. 4. Academic Press, New York, NY.

Stumpf, P.K., Mudd, J.B. and Ness, W.D. (eds.). 1987. *The Metabolism, Structure and Function of Plant Lipids.* Plenum Press, New York, NY.

Van Meer, G. 2002. The different hues of lipid rafts. Science 216: 815-867.

Vance, D.E. and Vance, J.E. 1991. *Biochemistry of Lipids, Lipoproteins and Membranes.* Elsevier, New York, NY.

Watts, A. 1993. *Protein-Lipids Interactions.* Elsevier, Amsterdam, Holland.

Wiate, M. 1987. *The Phospholipases.* Plenum Press, New York, NY.

Wooley, P. and Petersen, S.B. 1994. *Lipases: Their Structure, Biochemistry and Application.* Cambridge Univ. Press, New York, NY.

AMINO ACIDS, PEPTIDES, AND PROTEINS

Allen, G. 1994. *Proteins: Physical and Chemical Properties of.* Jai Press, Greenwich, CT, USA.

Allen, G. 1995. *Proteins: Cell Surface Proteins.* Jai Press, Greenwich, CT, USA.

Angeletti, R.H. 1998. *Protein Analysis and Design.* Academic Press, San Diego, CA.

Atassi, M.Z. 1995. *Methods in Protein Structure Analysis.* Kluwer Acad. Publ., Dordrecht, Netherlands.

Austin, R. 1987. *Protein Structure.* Springer-Verlag, Berlin, Germany.

Barrett, G.C. and Elmore, D.T. 1998. *Amino Acids and Peptides.* Cambridge Univ. Press, Cambridge, England.

Blackburn, S 1986. *Peptides.* CRC Press, Boca Raton, FL, USA.

Bodansky, M. 1993. *Peptide Chemistry: A Practical Textbook.* Springer-Verlag, Berlin, Germany.

Boulton, A.A., Baker, G.B. and Wood, J. 1985. *Amino Acids.* Humana Press, Totowa, NJ, USA.

Cavanaugh, J., Fairbrother, J., Palmer, A.G. and Skeleton, N.J. 1996. *Protein NMR Spectroscopy Principles and Practice.* Academic Press, San Diego, CA.

Cohen, S. 1997. *A Guide to the Polyamines.* Oxford Univ. Press, Oxford, England.

Creighton, T.E. 1995. *Proteins: Structures and Molecular Properties.* W.H. Freeman, New York, NY.

Dare, N.J. and Creighton, T.E. 1994. *Protein Structure.* Oxford Univ. Press, Oxford, England.

Davies, J.S. 1985. *Amino Acids and Peptides.* Chapman and Hall, London, England.

Davies, J.S. 1993. *Amino Acids and Peptides.* CRC Press, Boca Raton, FL, USA.

Galston, A.W. and Sawhney, R.K. 1990. Polyamines in plant physiology. Pl. Physiol. 94: 406-410.

Gottschalk, W. and Muller, H.P. 1983. *Seed Proteins: Biochemistry, Genetics, Nutritive Value.* M. Nijhoff, W. Junk Publ., Hingham, MA, USA.

Havel, H. 1996. *Spectroscopic Methods for Determining Protein Structure in Solution.* John Wiley and Sons, New York, NY.

Howard, G.C. and Brown, E. 2002. *Modern Protein Chemistry. Practical Aspects.* CRC Press, Boca Raton, FL, USA.

Jones, J.H. 1992. *Amino Acids and Peptide Synthesis.* Oxford Univ. Press, Oxford, England.

Kamp, R.M., Calvette, J.J. and Choli-papadopulou, T. 2004. *Methods in Protease and Protein Analysis.* Springer-Verlag, Berlin, Germany.

Kamp, R.M., Choli-Papadopoulou, T., Wittmann-Liebold, B. 1997. *Protein Structure Analysis Preparation, Characterization and Microsequencing.* Springer-Verlag, Berlin, Germany.

Kyte, J. 1995. *Structure in Protein Chemistry.* Garland Publ., New York, NY.

L'Italien, 1987. *Proteins: Structure and Function.* Plenum Press, New York, NY.

Miflin, B.J. 1980. Amino acids and derivatives. In: *The Biochemistry of Plants. A Comprehensive Treatise.* Stumpf, R.K. and Conn, E. (eds.). Academic Press, New York, NY.

Orly, R.L. 1986. Plant Proteins: *Applications, Biological Effects and Chemistry.* Amer. Chem. Soc., Washington, DC.

Reid, D.G. 1997. *'Protein NMR Protocols.'* Humana Press, Totowa, NJ, USA.

Rhodes, D. and Nadolskav, A. 2001. *Plant Stress Physiology.* Inc. of Life Sciences, Nature Publishing Group. www.eds.net.

Rosenthal, G.A. 1982. *Plant Nonprotein Amino and Imino Acids: Biological, Biochemical and Toxicological Properties.* Academic Press, New York, NY.

Schumaker, V.N. 1994. *Lipoproteins, Apolipoproteins and Lipases.* Acad. Press, San Diego, CA.

Singh, B.K. 1999. *Plant Amino Acids: Biochemistry and Biotechnology.* Dekker, New York, NY.

Sluyterman, L.A. and Elgersma, O. 1978. Chromatofocusing: Isoelectric focusing on ion-exchange columns I. General Principles. J. Chromatogr. 150-17-30.

Vilafranca, J.J. 1990. *Current Research in Protein Chemistry: Techniques, Structure, and Function.* Academic Press, San Diego, CA.

Wallsgrove, R.M. 1995. *Amino Acids and Their Derivatives in Higher Plants.* Cambridge Univ. Press, Cambridge, England.

White, J.S. and White, D.C. 2002. *Proteins, Peptides and Amino Acids Sourcebook.* Humana Press, Totowa, NJ, USA.

Wielan, T. and Bodansky, M. 1991. *The World of Peptides. A Brief History of Peptide Chemistry.* Springer-Verlag, Berlin, Germany.

ENZYMATIC STRUCTURE AND FUNCTION

Bailon, P., Ehrlich, G.K., Fung, W-J. and Berthold, W. 2000. *Affinity Chromatography Methods and Protocols.* Humana Press, Totowa, NJ, USA.

Bickerstaff, G.F. 1997. *Immobilization of Enzymes and Cells.* Humana Press, Totowa, NJ, USA.

Bradford, M.M. 1976. A rapid and sensitive method for the quantitation of microgram quantities of protein utilizing the principle of protein dye binding. Anal. Biochem. 72:248-254.

Burns, D.P. and Steiner, R. 1991. *Advanced Technology. Guide for LS6000 Series Scintillation Counters.* Beckman Instruments, Palo Alto, CA.

Burrell, M. 1993. *Enzymes of Molecular Biology.* Humana Press, Totowa, NJ, USA.

Burton, M.L., Onstott, L.T. and Polars, A.S. 1989. The use of gold reagents to quantitate antibodies eluted from nitrocellulose blot applications to electron microscopic immunocytochemistry. Anal. Biochem. 183:225.

Cheley, S and Bayley, H. 1991. Assaying nanogram amounts of dilute proteins. Biofeedback, 10:2.

Ciesiolka, T. and Gobruis, H. 1988. A 1 to 10-fold enhancement of sensitivity for quantitation of proteins by modified application of colloidal gold. Anal. Biochem. 16:280-283.

Coolbear, T. 1992. *Enzymes and Products from Bacteria, Fungi and Plant Cells.* Springer-Verlag, Berlin, Germany.

Cooper, A., Houber, J.L. and Chein, L.C. 1988. *The Enzyme Catalysis Process. Energetics, Mechanism and Dynamics.* Plenum Press, New York, NY.

Copeland, R.A. 1996. *Enzymes: A Practical Introduction to Structure, Mechanism and Data Analysis.* VCH Publ., New York, NY.

Dashek, W.V. 1997. *Methods in Plant Biochemistry.* CRC Press, Boca Raton, FL, USA.

Dashek, W.V. and Michaels, J.A. 1997. Assay and purification of enzymes—oxalate decarboxylase. In: *Methods in Plant Biochemistry and Molecular Biology*, pp. 49-71. W.V. Dashek (ed.). CRC Press, Boca Raton, FL, USA.

Demchenko, A.P. 1986. *Ultraviolet Spectroscopy of Proteins.* Springer-Verlag, Berlin, Germany.

Deutscher, M.P. (ed.) 1990. *Guide to Protein Purification, Method in Enzymology.* Academic Press, New York, NY, vol. 182.

Doonan, S. 1989. *Protein Purification Protocols. Methods in Molecular Biology Series.* Humana Press, Totowa, NJ, USA.

Dressler, D. 1995. *Enzymes.* W.H. Freeman, New York, NY.

Eisenthal, R. and Danson, M.J. 1992. *Enzyme Assays. A Practical Approach.* Oxford Univ. Press, Oxford, England.

Englard, S. and Seifter, S. 1990. In: *Precipitation Techniques. Guide to Protein Purification, Methods in Enzymology.* Deutscher, M.P. (ed.). Academic Press, New York, NY, vol. 182.

Fersht, A. 1985. *Enzyme Structure and Mechanism.* W.H. Freeman, San Franciso, CA, USA.

Fischer, L. 1980. *Gel Filtration Chromatography.* Elsevier, New York, NY.

Friedrich, P. 1984. *Supramolecular Enzyme Organization.* Pergawan Press, Oxford, England.

Goldberg, A.L. 2000. Probing the proteosome pathway. Nature Biotechnology 18:494-496.

Goodman, W.F. and Baptist, J.N. 1979. Isoelectric point electrophoresis: a new technique for protein purification. J. Chromatog. 179: 330-332.

Gorbunoff, M.J. 1985. Protein chromatography on hydroxyapatite columns. Methods Enzym. 117: 370-381.

Guilbault, G.C. 1984. *Analytical Uses of Immobilized Enzymes.* M. Dekker, New York, NY.

Guilbault, G.C. and M. Mascini. 1993. *Uses of Immobilized Biological Compounds.* Kluwer Acad. Publ., Dordrecht, Netherlands.

Harris, E.L. and Angral, S. 1990. *Protein Purification Methods: A Practical Approach.* Oxford Univ. Press, Oxford, England.

Hodgkinson, S. and Lowry, P.J. 1981. Hydrophobic-interaction chromatography and anion-exchange chromatography in the presence of acetonitrile. Biochem. J. 199:430-433.

Kolby, J. 1990. *Enzymes: A Comprehensive Study,* vols. I and II. CRC Press, Boca Raton, FL, USA.

Laskin, A.I. 1985. *Enzymes and Immobilized Cells in Biotechnology.* Butterworth-Heinemann, Woodburn, MA, USA.

Li, K., Geraerts, W., Van Elk, P.M.R. and Joosee, J. 1989. Quantification of proteins in the subnanogram and nanogram range. Comparison of Auro dye, Ferri dye and India ink staining methods. Anal. Biochem. 182:44-47.

Livingstone, D.M. 1974. Immunoaffinity chromatography of proteins. Methods Enzymol. 34: 723-731.

Lowe, C.R. 1979. *An Invitation to Affinity Chromatography*. Elsevier, New York, NY, USA.

Lowry, O.H., Rosenbrough, N.J., Fair, A.L. and Randall, R.J. 1956. Protein measurement with the Folin phenol reagent. J. Biol. Chem 193:265-275.

Maggio, E.T. 1980. *Enzyme-immunoassay*. CRC Press, Boca Raton, FL, USA.

Mant, C.T. and Hodges, R.S. 1991. *High Performance Liquid Chromatography of Peptides and Proteins. Separation, Analysis and Conformation.* CRC Press, Boca Raton, FL, USA.

Martonosi, A.N. 1985. *The Enzymes of Biological Membranes.* Plenum Press, New York, NY.

McDonald, C.J. 1996. *Enzymes in Molecular Biology.* J. Wiley and Sons, Chichester, England.

Mosbach, K. 1988. *Immobilized Enzymes and Cells,* Part D. Academic Press, San Diego, CA.

Moore, N.L., Mariam, D.H., Williams, A.L. and Dashek, W.V. 1989. Substrate specificity, de novo synthesis and partial purification of polyphenol oxidase from the wood decay fungus, *Coriolus versicolor.* J. Indust. Microbiol. 4:349-363.

Peng, C.T., Horrocks, D. and Alphen, E.L. 1980. *Liquid Scintillation Counting: Recent Applications and Development.* Academic Press, New York, NY.

Pharmacia Laboratory Separation Division. 1967. *Ion Exchange Chromatography. Principles and Methods.* Pharmacia, Uppsala, Sweden.

Purich, D.L. 1993. *Contemporary Enzyme Kinetics and Mechanisms.* Academic Press, New York, NY.

Rigletti, P.G. 1983. *Isoelectric Focusing Theory, Methodology and Applications.* Elsevier, New York, NY.

Rosevear, A., Kennedy, J.F. and Cabral, J.M. 1987. *Immobilized Enzymes and Cells.* IOP Publishing, Bristol, England.

Schultz, P.G. 1988. The interplay between chemistry and biology in the design of enzymatic catalysts. Science 240:426.

Scopes, R.K. 1987. *Protein Purification. Principles and Practice.* Springer-Verlag, New York, NY, USA.

Segel, I.H. 1993. Enzyme kinetics. *Behavior and Analysis of Rapid Equilibrium and Steady State Enzyme Systems.* John Wiley, New York, NY.

Sigman, D. and Boyer, P.D. 1992. *Enzyme Mechanisms of Catalysis.* Acad. Press, New York, NY.

Tamanoi, F. and Sigman, D.S. 2000. *Enzymes.* Academic Press, New York, NY.

Taylor, R., Mayfield, J.E., Shortle, W.C., Llewellyn, G.C. and Dashek, W.V. 1981. Attempts to determine whether the products of extracellular polyphenol oxidase modulate the catechol-induced bimodal growth response of 'Coriolus versicolor' pp. 43-62. Biodeterioration Research I, Plenum Press, New York, USA.

Uhlig, H. 1998. *Industrial Enzymes and Their Applications.* John Wiley and Sons, Inc., New York, NY.

White, J. 1985. *Immobilized Cells and Enzymes.* IRL Press, Oxford, England.

White, J.S. and White, D.C. 1997. *Source Book of Enzymes.* Boca Raton, FL, USA.

Woodward, J. 1985. *Immobilized Cells and Enzymes.* IRL Press, Oxford, England.

Zubay, G. 1998. *Biochemistry.* McGraw Hill, New York, NY, 4th ed.

NUCLEIC ACIDS

Adams, R.L.P., Knowler, J.T. and Leader, V.P. 1992. *The Biochemistry of the Nucleic Acids.* Chapman and Hall, London, England.

Blackburn, G.M. and Gait, M.J. 1990. *Nucleic Acids in Chemistry and Biology.* Oxford Univ. Press, Oxford, England.

Boulter, D. 1982a. *Nucleic Acids and Proteins in Plants I. Structure, Biochemistry and Physiology of Proteins,* vol. I, Springer-Verlag, Berlin, Germany.

Boulter, D. 1982b. *Nucleic Acids and Proteins in Plants I. Structure, Biochemistry and Physiology of Proteins,* vol. II, Springer-Verlag, Berlin, Germany.

Brickell, P. and Darling, D. 1995. *Nucleic Acid XXX: The Basics.* Oxford Univ. Press, Oxford, England.

Brielman, H.L. 1999. *Phytochemicals: The Chemical Components of Plants in Natural Products from Plants.* Kaufman, P.P., Cseke, L.J., Warber, S., Duke, J.A. and Brielman, H.L. (eds.). CRC Press, Boca Raton, FL, USA.

Brown, D.J. 1994. *The Pyrimidines.* John Wiley, New York, NY.

Eckstein, F.D. and Lilley, D.M. 1996. *Nucleic Acid and Molecular Biology.* Springer-Verlag, Berlir, Germany.

Hall, T.C. and Davis, J.W. 1979. *Nucleic Acids in Plants.* CRC Press, Boca Raton, FL, USA.

James, T.L. 1995. *Nuclear Magnetic Resonance and Nucleic Acids.* Academic Press, San Diego, CA.

Lister, J.H. 1996. The Purines supplement 1. John Wiley, New York, NY.

Leontis, N.B. and Santalucia, J. 1998. *Molecular Modeling of Nucleic Acids.* Academic Press, San Diego, CA.

Marcus, A. 1981. *Proteins and Nucleic Acids, vol. 6. The Biochemistry of Plants. A Comprehensive Treatise.* P.K. Stumpf and E.E. Conn (eds.). Academic Press, New York, NY, USA.

Neidle, S. 1999. *Oxford Handbook of Nucleic Acid Structure.* Oxford Univ. Press, Oxford, England.

Roberts, G.C.K. 1993. *NMR of Macromolecules: A Practical Approach.* IRL Press at Oxford Univ. Press, Oxford, England.

Rossi, J.J. 1999. Ribozymes in the nucleolus. Science 285:1685.

Saluz, H.P. and Wiebauer, K. 1995. *DNA and Nucleoprotein Structures in Vivo.* R.G. Landes Co., Austin, TX, USA.

Townsend, L.B. 1988. *Chemistry of Nucleosides and Nucleotides.* Plenum Press, New York, NY.

Townsend, L.B. and Tripsin, R.S. 1991. *Nucleic Acid Chemistry.* John Wiley, New York, NY.

Walker, J.M. 1984. *Nucleic Acids. Methods in Molecular Biology Series.* Humana Press, Totowa, NJ, USA.

CHAPTER 5

Biomolecules II: Biologically Important Molecules Other than Carbohydrates, Lipids, Proteins, and Nucleic Acids

William V. Dashek

This chapter is concerned with organic constituents in plant systems other than the big four discussed in Chapter 4. Table 5.1 lists these constituents and some of their functional significances. The chemistry, metabolism, and function of these compounds are detailed herein.

ALKALOIDS

Alkaloids are basic substances which contain one or more nitrogen atoms, often as part of a cyclic system (Southan and Cordell, 1989). Certain alkaloids are terpenoids (Rahman, 1990) while some are aromatic compounds (Fig. 5.1). Table 5.2 presents the chemical characteristics and possible functions of some alkaloids. A summary of the distribution (Raffauf, 1996), chemistry and pharmacological (Roberts and Wink, 1998) effects of plant alkaloids is shown in Table 5.3. Detailed schematic pathways for alkaloid biosynthesis occur on kegg pathway database (2004).

ORGANIC ACIDS

Organic acids (Fig. 5.2) are colorless, weak acids which are insoluble in nonpolar solvents, i.e. benzene or pet ether (Hanai, 1982). They are soluble in water, ethanol and ether. These acids are weakly acidic (Duke, 1992; Fox and Powell, 2001) and can accumulate within vacuoles of certain plants. Organic acids have multiple roles.

PHENOLIC COMPOUNDS

A classification of phenolic compounds is depicted in Table 5.4. Phenolic compounds (Fig. 5.3), widely distributed in the plant kingdom, are characterized by a hydroxylated benzene ring (Gross *et al.*, 1999). Smith (1997) stated that in trees, phenolics frequently occur as polymeric acids or glycosylated esters and perform diverse function. For example, induction of phenolics is a response to injury and infection.

TABLE 5.1 Plant chemicals other than carbohydrates, proteins, lipids, and nucleic acids

Compound	Chemistry	Examples of classes
Alkaloids	Natural products that contain nitrogen usually as part of a cyclic system	Chelidonine, lycodopine, senicionine, intermedine, caffeine, hygrine, scopolamine, cocaine, platynecine, nicotine, morphine, lupine, berberine, papaverine, psilocybin, corynantheine, ajmaline, ellipticine
Flavonoids	Compounds which have two benzene rings separated by a propane unit	Classes—catechins, leucoanthocyanidins, flavanones, flavanonols, chalcones, aurones, isoflavones, flavanols
Nonphenolic aromatics	Tetrapyrroles Cyclical porphyins Linear pyrroles	Chlorophylls and cytochromes Phytochrome, phycocyanin and phycoerythrin
Organic acids	Weakly acidic acids	Respiratory acids such as citric acid
Phenol ethers	Methyl esters of phenol	Khellin, visnagin, trans-anethole and apiol
Phenolic simple phenols	Monomeric components of polymeric polyphenols and acids	Hydroquinone resorcinol, catechol, salicylic acid, thymol
Phenylpropanoids	Phenol with an attached 3C side chain	Coumarin, umbelliferone, psoralen, pinoresinol, eugenol
Plant hormones	Diverse chemical structures	Absicsic acid Auxins Brassinosteroids Cytokinins Ethylene Gibberellins Jasmonates
Quinones	Strongly colored pigments	Benzoquinone, napathoquinone, anthroquinone
Tannins	Various phenolics	Classes—condensed (formed by condensation of catechins) hydrolysable (derived from gallic acid)
Terpenes	Widespread and chemically diverse groups of natural products	Hemiterpenes—isoamyl alcohol, isovaleric acid, senecioic acid, tiglic acid, angelic acid, β-furoic acid
	Derived from isoprene yielding isopentane units	Monoterpenes—myrcene, geranoil, menthol, carvone, linalool, safranal, eucalyptol, camphor, aucubin
	Hemiterpene C_5 Monoterpene C_{10} Sesquiterpene C_{15}	Sesquiterpenes—farnesol, cadinenes, carophyllene, helanalin, psilostaychin, acorone
	Diterpene C_{20}	Diterpenes—camphorene, abietic acid, marrubin, phytol, zoapatanol, clerodin, gibberrellic acid, taxol
	Triterpene C_{30}	Triterpenes—squalene, limonin, cucurbitacin, azadirachtin, polygallic acid
	Tetraterpenes C_{40}	Tetraterpenes—lycopene, β-carotene, α-carotene, lutein, and rhodoxanthin
Vitamins	Organic compounds which are either water or fat soluble	B vitamins, vitamins A, C, D, E, K

Adapted from: Brielmann, 1999; Dey and Harborne, 1999.

Chemical characteristics and possible functions of alkaloids in plants	
Chemical characteristics	*Possible functions*
Basic substances	Growth regulators
Contain one or more nitrogen atoms	Insect repellents or attractants
Usually colorless	Maintain ionic balance
Often optically active	Nitrogen storage reservoirs
Bitter taste	Nitrogen waste products
Many are terpenoids	

TABLE 5.2

Other possible functions of phenolics include: (Table 5.5) deterring herbivores from feeding on leaves and limiting the spread of pathogens.

Lignin (Fig. 5.4) is a polymer of phenylpropanoid units linked together in a complex and irregular pattern which varies from species to species, tissue to tissue, and cell to cell (Dean, 1997).

In the past some phenolics and some esterified phenolics were termed tannins (Haslam, 1989) because of their ability to tan

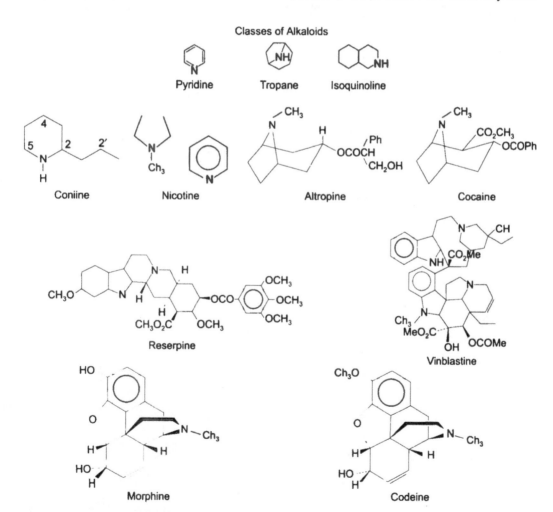

Fig. 5.1 Classes of alkaloids

TABLE 5.3 Summary of the distribution, chemistry and possible functions of plant alkaloids*

Distribution	Organism	Alkaloid	Chemistry	Pharmacological effect
Liliaceae (monocot)	Colchium autumnole	Colchicine	Tropolone	Inhibitor of mitosis
Loganiaceae (dicot)	Strychnos nux-vomica	Strychnine	Pentocyclic	Neural stimulant
Papaveraceae (dicot)	Papaver somniferum	Morphine	Morphine	Narcotic
Rubiaceae (dicot)	Cinchora officinalis	Quinine	Quinoline	Cardiac depressant
	Uragoya ipecucuanba	Emetine	Emetine-type	Emetic
Solanaceae (dicot)	Atropa belladonna	Atropine	Tropane	Antispasmodic
	Nicotaina tabacum	Nicotine	Pyridine	

*Other alkaloids with physiological actions in man are camptothecin from *Camptotheca acuminata* = antitumor, cocaine for coca = local anesthetic: pelletierine from pomegranate = vernifuge; pilocarpine form *Pilocarpus pennatifolius* = diaphoretic; reserpine from *Rauwolfia serpentina* = tranquilizer; tubocurarine from *Chonodedron* = muscle paralyzer (taken in part from Robinson, 1980). Rahman (1990) discussed the diterpenoid and steroidal alkaloids and in (1994) reviewed the isoquinoline alkaloids. Rizk (1990) described the pyrrolizidine alkaloids and Zenk and Phillipson (1992) elaborated the indole alkaloids.

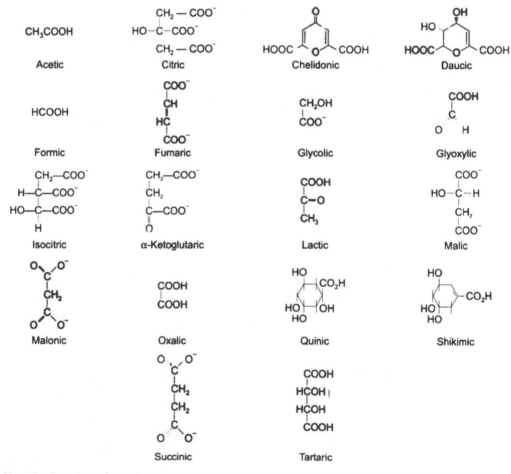

Fig. 5.2 Chemistry of organic acids

TABLE 5.4	Summary of phenolic compounds in plants

Phenolic class	Examples
Flavonoids	Anthocyanins
	Aurones
	Flavonol
	Catechins
	Chalcones
	Isoflavonoids
Hydroxycinnamic Acid derivatives	Caffeic acid
	Chlorogenic acid
	Coumarins
	Ferrulic acid
Pheobic acids	Ellagic acid
	Tannic acid

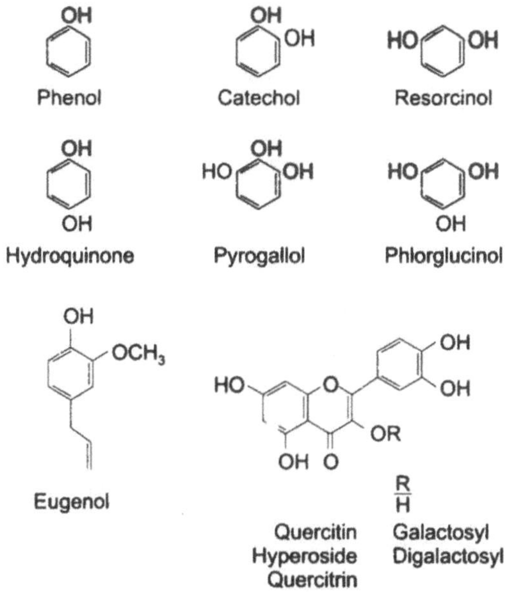

TABLE 5.5	Biological effects of phenolic compounds

Phenolic group	Function
Quinones	Allelopathic compounds
Phenols	Allelopathic compounds
Phenolcarboxylic acids	Allelopathic compounds
Isoflavones	Fungicide
Dihydrochalcones	Fungicide
Anthocyanins	Floral pigments
Chalcones	Floral pigments
Aurones	Floral pigments
Yellow flavonoids	Floral pigments
Flavones	Floral pigments
Anthocyanins	Fruit pigments
Isoflavones	Fruit pigments
Chalcones	Fruit pigments
Stilbenes	Phytoalexins
Phenylanthrenes	Phytoalexins
Isoflavones	Phytoalexins
Phenylpropanoids	Phytoalexins
Quinones	Protection against pests
Tannins	Protection against pests
Flavonols	Protection against pests

Source: Harborne (1986)

Fig. 5.3 Structure of some phenolic compounds

FLAVONOIDS

Flavonoids (Fig. 5.6) are plant secondary compounds (Mantley and Busling, 1998; Packer, 2001). Some are phenolic pigment compounds. Much of flavonoid research (Busling and Mantley, 2002) has centered on the importance of these compounds in chemotaxonomy and phylogeny. The synthesis of flavonoids (Shirley, 1996) is illustrated in Fig. 5.7. Current research involves the genetics of flower pigments (anthocyanins), the quantitative evaluation of medically important compounds (Harborne, 1986; Middleton, 2000; Havsteen, 2002), and the secondary responses of CO_2 enriched air (Clark and Titus, 1997).

or preserve leather. Today, tannins (Fig. 5.5) occur in two major groups, the proanthocyanins (formerly the condensed tannins (Hemingway and Karchesy, 1988) and gallic acid derivatives (formerly the hydrolyzable tannins) that hydrolyze to yield phenolic acids and sugars.

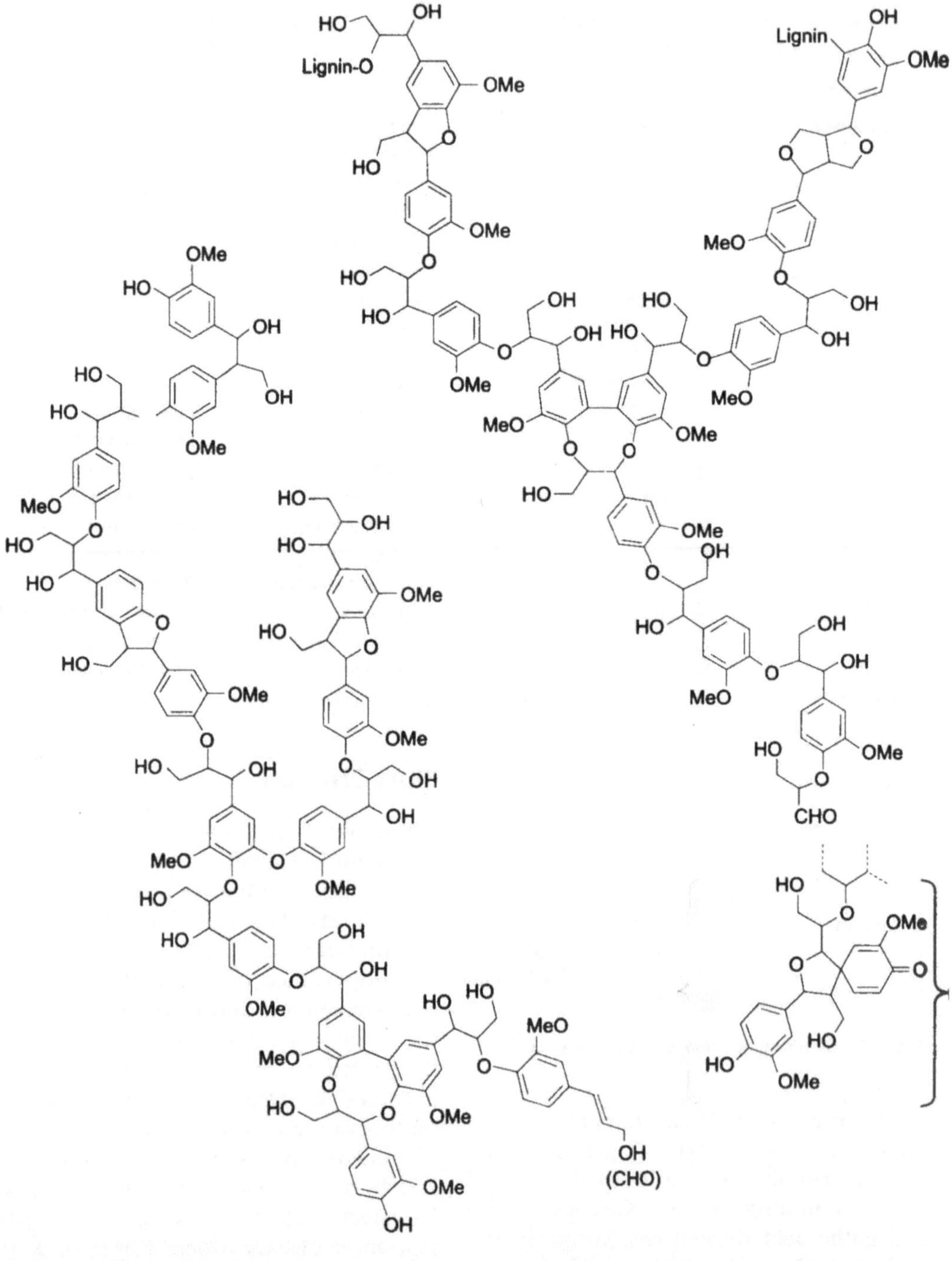

Fig. 5.4 Partial structure of softwood lignin. From http://www.helsinki.fir/~orgkm-wwlignin-structure.html

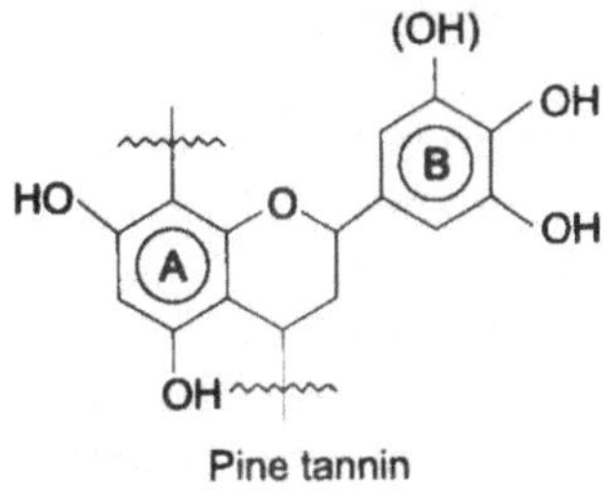
(OH)
OH
OH
HO
O
B
A
OH
Pine tannin

Fig. 5.5 Chemistry of pine tannin (From: Kim, S. *et al.* 2003)

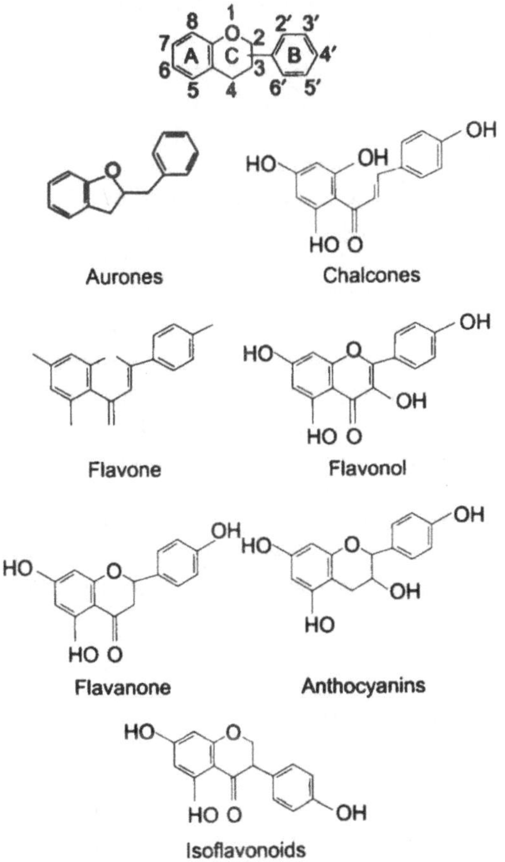
Aurones
Chalcones
Flavone
Flavonol
Flavanone
Anthocyanins
Isoflavonoids

Fig. 5.6 Types of flavonoids

Phenylpropanoids (Fig. 5.8) are plant compounds with an attached 3C side chain (Seigler, 1998). Some are derived from medicinal plants (Kurkin, 2003).

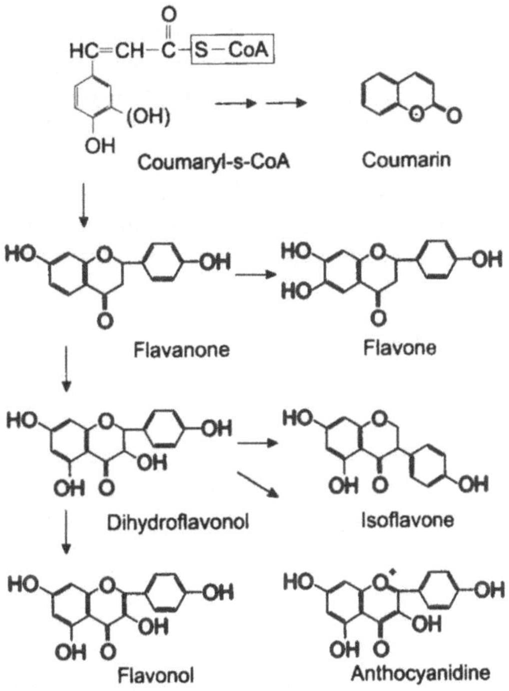
HC=CH–C–S–CoA
(OH)
OH
Coumaryl-s-CoA
Coumarin
Flavanone
Flavone
Dihydroflavonol
Isoflavone
Flavonol
Anthocyanidine

Fig. 5.7 Synthesis of flavonoids

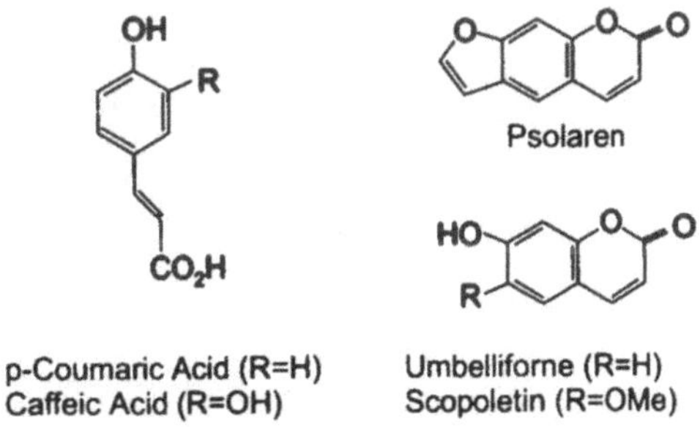
Psolaren
p-Coumaric Acid (R=H)
Caffeic Acid (R=OH)
Umbelliforne (R=H)
Scopoletin (R=OMe)

Fig. 5.8 Chemical composition of some phenylpropanoids

PHYTOALEXINS

Phytoalexins are antimicrobial substances (Daniel and Purkayastha, 1995) synthesized by a plant in response to infection (Kuc, 1994). Phytoalexins are diverse in chemical composition (Fig. 5.9). Small proteins and polysaccharides (elicitors) produced by pathogens prompt certain plant tissues to synthesize phytoalexins (Dixon, 1986), components of the hypersensitive response (a defensive reaction). Elicitin is a 10.2 to 10.4 kd extracellular protein which is produced by *Phytophthora* (see literature in Yu, 1997).

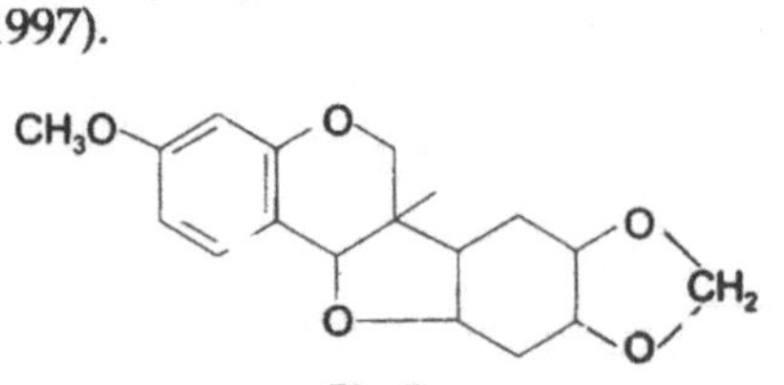

Pisatin

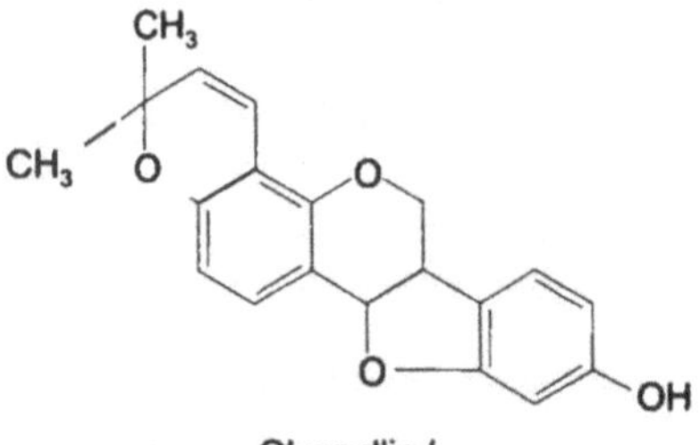

Glyceollin I

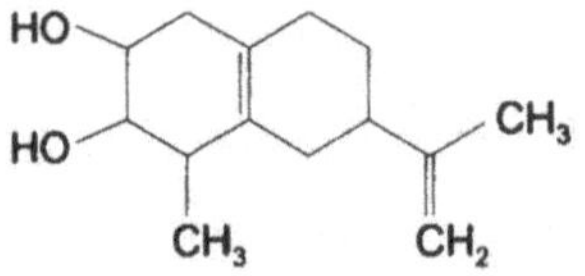

Rishitin

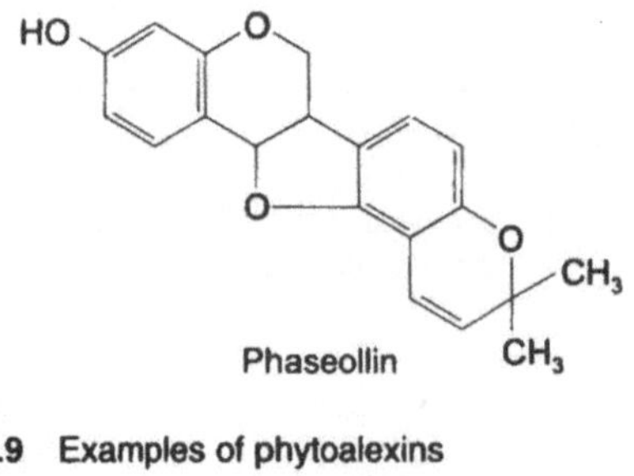

Phaseollin

Fig. 5.9 Examples of phytoalexins

PLANT HORMONES

Plant hormones are organic compounds produced in one part of the plant and transported to another part where they influence physiological and biochemical events (see References). These hormones are operative at very low concentrations, e.g. $\angle 1 \times 10^{-7}$ M. Another widely used term for plant hormones is phytohormones. Some horticulturists define plant hormones as biostimulants assisting plants to develop higher plant antioxidant activity. The term plant growth regulator is used by some plant biologists to mean phytohormone. However, regulators produce their effects when they are synthesized. The main classes of plant hormones are: abscisic acid, auxins, brassinolides, cytokinins, ethylene, gibberellins and jasmonates. Some plant physiologists also consider salicylic acid as a plant hormone.

Abscisic acid (Fig. 5.10.) differs from other classes of plant hormones in that it is a single compound (Walton and Li, 1995). Originally, it was thought to be involved in the abscission of fruits (abscisin II) and, subsequently, to play a major role in bud dormancy (dormin). Abscisin II and dormin are now recognized as the same compound, called abscisic acid or ABA (Arteca, 1996). It

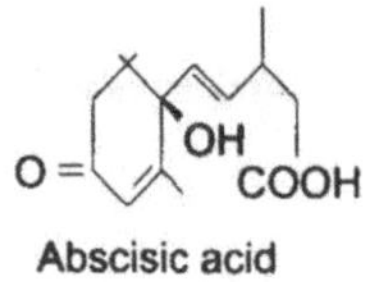

Abscisic acid 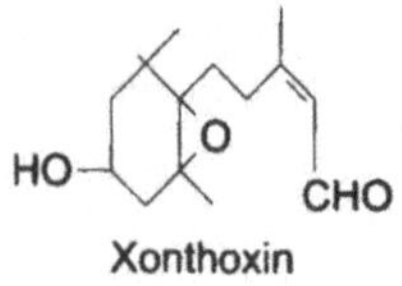   Xonthoxin

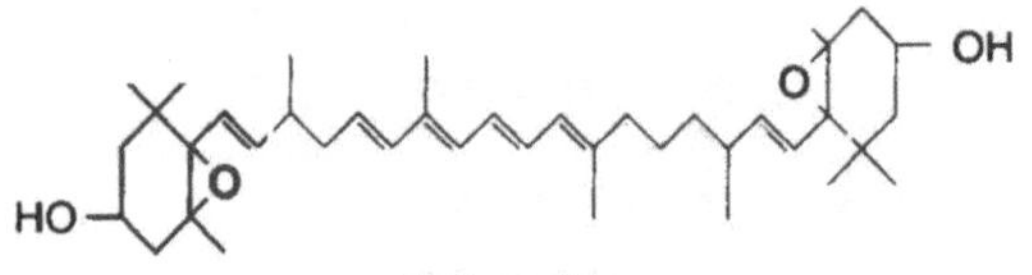

Violaxanthin

Fig. 5.10 Structure of abscisic acid and related compounds

occurs naturally in plants and is a sesquiterpenoid which is partially synthesized via the mevalonic pathway (Walton and Li, 1995) in chloroplasts and other plastids (Fig. 5.11). Synthesis of ABA is stimulated by water loss and freezing temperatures (see References in Arteca, 1996). Table 5.6 displays the functions of ABA in plant growth and development.

Auxins were the first plant hormones to be discovered. Their history, including discovery, isolation and identification, has been reviewed numerous times. While certain auxins are indole derivatives, others are not (Fig. 5.12). Considerable attention has been paid to structure – activity relationships and the minimal requirements for auxin activity (see literature in Steward and Krikorian, 1971). Table 5.7 presents the known functions of auxin in plant growth and development. With regard to auxin biosynthesis, indole-3-acetic acid, the naturally occurring auxin, appears to be derived from tryptophan (Bartel, 1997). The biosynthetic pathway (Bandurski *et al.*, 1995) presented in Fig. 5.13 is based on a pathway proposed for decades. In more

recent times, Juyang *et al.* (2000) studied the IAA biosynthetic pathway in *Arabidopsis thaliana* (Fig. 5.13). They suggested that IGP is an important branch point in Trp-independent IAA biosynthesis in *A. thaliana*.

Indole-3-acetic acid   2,4 dichlorophenoxyactic acid

Naphthalene-1-acetic acid

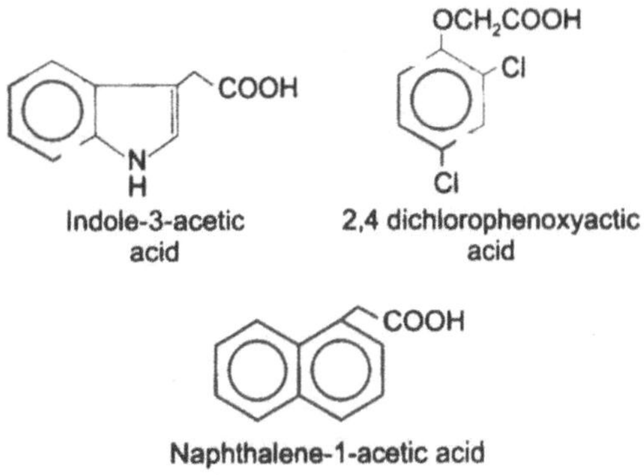

Fig. 5.12 Chemistry of some auxins

TABLE 5.7 Functions of auxin

Auxin from the apical bud suppresses growth of lateral buds.

Can inhibit or promote (via ethylene stimulation) leaf and fruit abscission.

Delays leaf senescence.

Delays fruit ripening.

Enhances production of ethylene at high concentrations.

Induces cell division in the cambium and in combination with cytokinins in tissue culture.

May induce fruit set and growth in some plants.

May influence phloem transport.

Mediates tropistic responses in response to gravity and light.

Promotes flowering in bromeliads.

Promotes (via ethylene production) femaleness in dioecious flowers.

Stimulates cell elongation.

Stimulates growth of floral parts.

Stimulates phloem and xylem differentiation.

Stimulates root initiation on stem cuttings and lateral root development in tissue culture.

Adapted from: http://www.plant-hormones.info/auxins.htm. Zizimalova and Napier, 2003 discuss points of regulation of auxin action.

Neoxanthin ⟶ Xanthoxin ⟶ Abscisic ⟶ ABA
acid aldelhyde

Fig. 5.11 Abscisic acid biosynthetic pathway

TABLE 5.6 Functions of ABA

Affects induction and maintenance of dormancy.

Induces gene transcription for proteinase inhibitors in response to wounding which may explain an apparent role in pathogen defense.

Induces seeds to synthesize storage proteins.

Inhibits shoot growth.

Inhibits the effect of gibberellins on stimulating *de novo* synthesis of α-amylase.

Stimulates the closure of stomata (water stress increases ABA synthesis).

Adapted From: http://www.plant-hormones.info/abscisicacid.htm

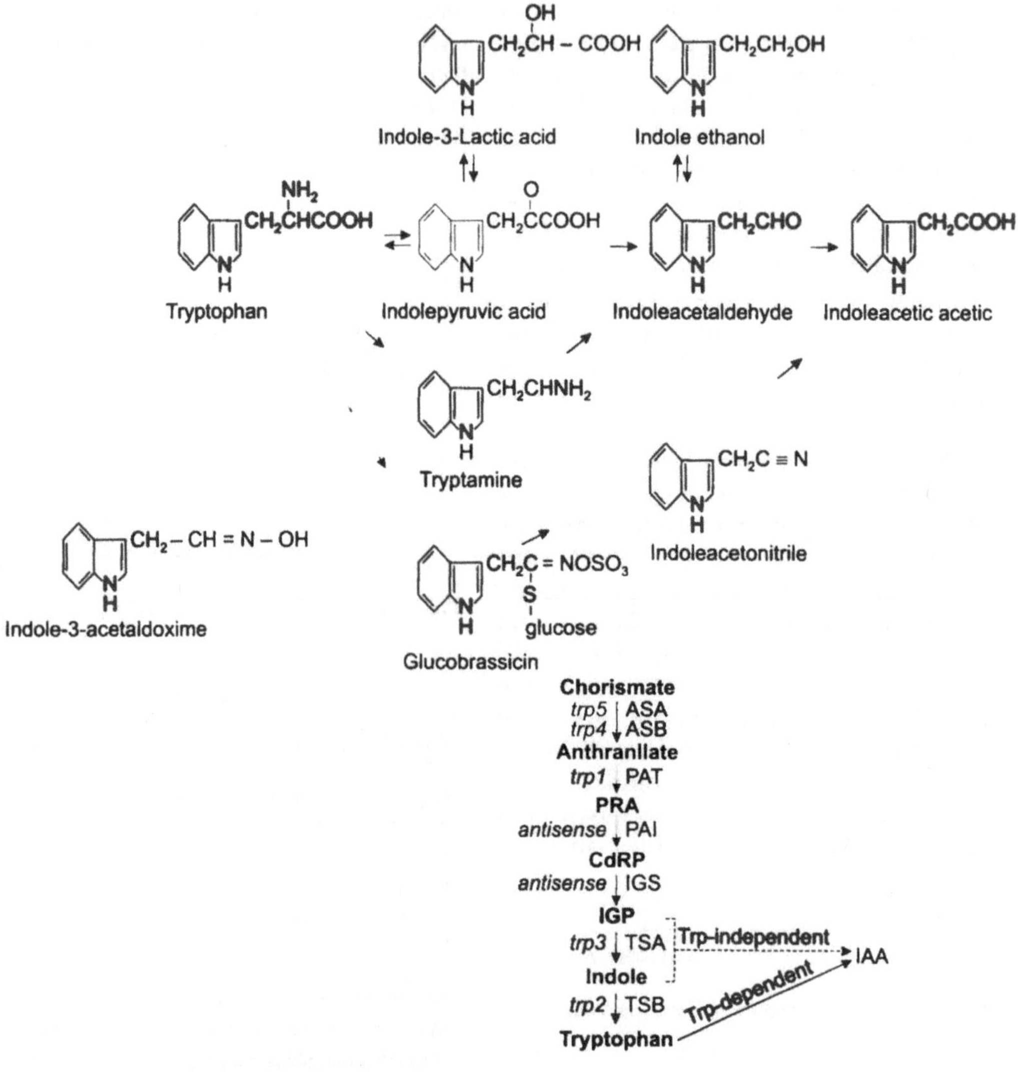

Fig. 5.13 Proposed auxin biosynthetic pathways

Top: conventional pathways; indole-3-acetaldoxime may be a precursor to indoleacetonitrile

Bottom: pathway in *Arabidopsis thaliana*.

The IAA biosynthetic pathway in *Arabidopsis thaliana*.

IAA can be synthesized through the Trp-dependent and/or Trp-independent pathway in plants. Trp mutants or antisense RNA transgenic plants are indicated in the Trp biosynthetic pathway starting from chorismate to tryptophan in the gene encoding each corresponding enzyme is also given at each step. Dashed arrow indicates possible Trp-independent IAA biosynthesis. ASA, anthranilate synthase α subunit; ASB, anthranilate synthase β subunit; CdRP, 1-(0-carb-oxyphebylamino)-1-deoxyribulose-5-phosphate; IGP, indole-3-glycerol phosphate; IGS, indole-3-glycerol phosphate synthase; PAI, phosphor-ribosylanthranilate isomerase; PAT, phosphoribosylanthranilate trans-ferase; PRA, 5-phosphoribosylanthranilate; TSA, Tryptophan synthase α subunit; TSB, tryptophan synthase β subunit (from Juyang *et al.*, 2000)

Lastly, much current auxin research centers on binding sites (Jones and Lomax, 1997).

Brassinolides (Fig. 5.14) are steroidal compounds (Khripach *et al.*, 1999; Yokota *et al.*, 1999) which promote plant growth when administered at $\angle 1 \times 10^{-4}$ µg ml^{-1} (Culter *et al.*, 1991). They were first isolated, but not characterized from Isonuki, a Japanese evergreen (see References in Arteca, 1995). Later, it was reported that an oil fraction from *Brassica majus* (L.), rape pollen, contained compounds which affect plant growth similar to the evergreen-derived compound. Subsequently, a steroidal lactone, isolated from bee pollen of *B. majus*, was characterized and identified. Brassinolides appear to arise from squalene. The biological effects of brassinosteroids are illustrated in Table 5.8.

Cytokinins are purines related to adenine and stimulate cell division in conjunction with auxin (Mok and Mok, 1994; Haberer and Kieber, 2002). The first discovered cytokinin, kinetin, does not occur naturally. However, dimethylallyadenine, zeatin, and zeatin riboside (Fig. 5.15) do. The activity of cytokinins is influenced by side groups on N_6. The synthetic compound, 6 benzylamino purine, possesses marked cytokinin activity. Other synthetic cytokinins include benzyladenine and ethoxyethyladenine. The functions of cytokinins in plant growth and development are displayed in Table 5.9 and the proposed biosynthetic pathway (McGaw and Burch, 1995; Kakimoto, 2003) for cytokinins illustrated in Fig. 5.16. In addition tRNAs appear to be a source of cytokinins.

Ethylene (Fig. 5.17) is a gaseous hormone (Abeles *et al.*, 1992) and is the only member of its class. The functions of ethylene (Chang *et al.*, 1997) are presented

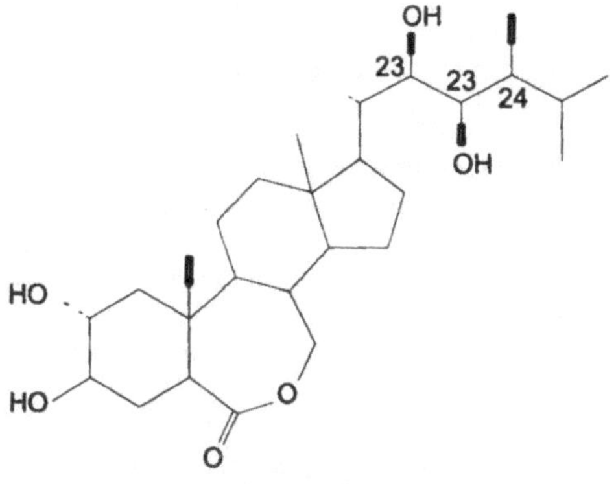

TABLE 5.8 Biological effects of brassinosteroids

Effects

1. Affect plasmalemma energization and transport
2. Decrease fruit abortion and drop
3. Enhance resistance to chilling diseases, herbicide and, salt stress
4. Enhance xylem differentiation
5. Increase RNA and DNA polymerase activities and synthesis of RNA, DNA, and protein
6. Inhibit root growth and development
7. Promote ethylene biosynthesis and epinasty
8. Promote germination
9. Promote shoot elongation at low concentrations

Adapted from http://www.hos.ufl.edu/mooreweb

TABLE 5.9 Cytokinin functions

Enhances leaf expansion resulting from cell enlargement

May enhance stomatal opening in some species

Promote cell division

Promotes conversion of etioplasts into chloroplasts via stimulation of chlorophyll synthesis

Stimulates shoot initiation/bud formation in tissue culture

Stimulates growth of lateral bud release of apical dominance.

Adapted from: http://www.plant-hormones.info/cytkinins.htm

Brassinolide

Fig. 5.14 A brassinolide

in Table 5.10 and the suggested biosynthetic pathway (Kende, 1993) is shown in Fig. 5.17. A role for ethylene as a gaseous signal molecule has been reviewed by Bleecher and Kende (2000).

Gibberellins (Fig. 5.18) are acidic compounds classified on the basis of structure and function (Takahashi *et al.*, 1991). These diterpenes are derived from the gibberellin skeleton. The most prominent of the gibberellins is GA$_3$, gibberellic acid. There are over 100 known gibberellins and they are distinguishable chemically by subscripts. Gibberellins are widespread in

both flowering and nonflowering plants. They also occur in some ferns, mosses and algae. The biosynthetic pathway from mevalonic acid (Sponsel, 1995; Hedden, 1999) is depicted in Fig. 5.19. A number of growth retardants (Fig. 5.20) can influence gibberellin biosynthesis (Rademacher, 2000). Gibberellin functions are shown in Table 5.11.

Jasmonates (Fig. 5.21) were isolated from plants (Demole *et al.*, 1962; see Reference in Arteca, 1995). Jasmonic acid, synthesized from the fatty acid linolinelic acid (Fig. 5.22), occurs in many plant species. The biological effects of jasmonates are summarized in Table 5.12 and have been

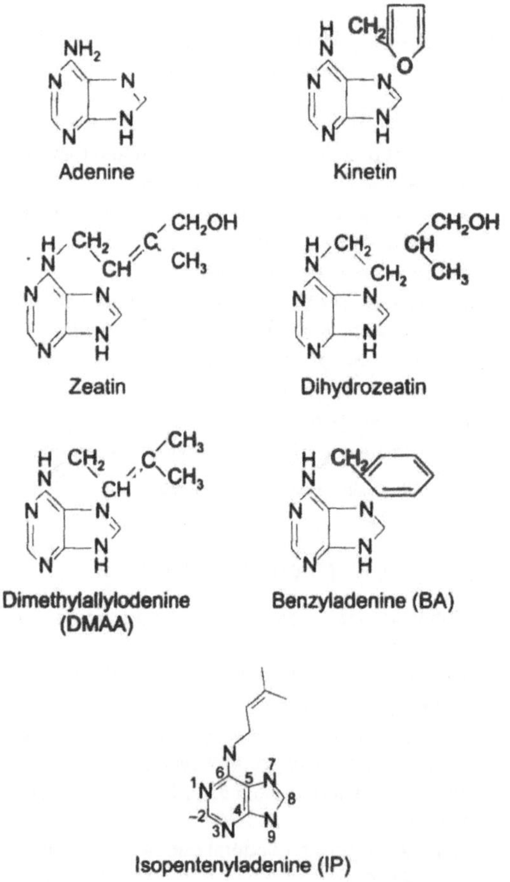

Fig. 5.15 Cytokinins

TABLE 5.10 Functions of ethylene

Affects release of dormancy
Enhances leaf and fruit abscission
Induces fruit ripening
Induction of femaleness in dioecious flowers
May have a role in adventitious root formation
Promotes flower opening
Stimulates bromeliad flower induction
Stimulates flower and leaf senescence
Stimulates shoot and root growth and differentiation (triple response)

Adapted from: http://www.plant-hormones.info/ethylene.htm

TABLE 5.11 Gibberellin actions

Breaks seed dormancy in some plants which require stratification or light to induce germination
Can cause parthenocarpic (seedless) fruit development
Enhances stem elongation by stimulating cell division and elongation
Induces maleness in dioecious flowers (sex expression)
May delay senescence in leaves and citrus fruits
Stimulates α-amylase production in germinating cereal grains for mobilization of seed reserves
Stimulates bolting/flowering in response to long days

Adapted from: http://www.plant-hormones.info/gibberellins.htm

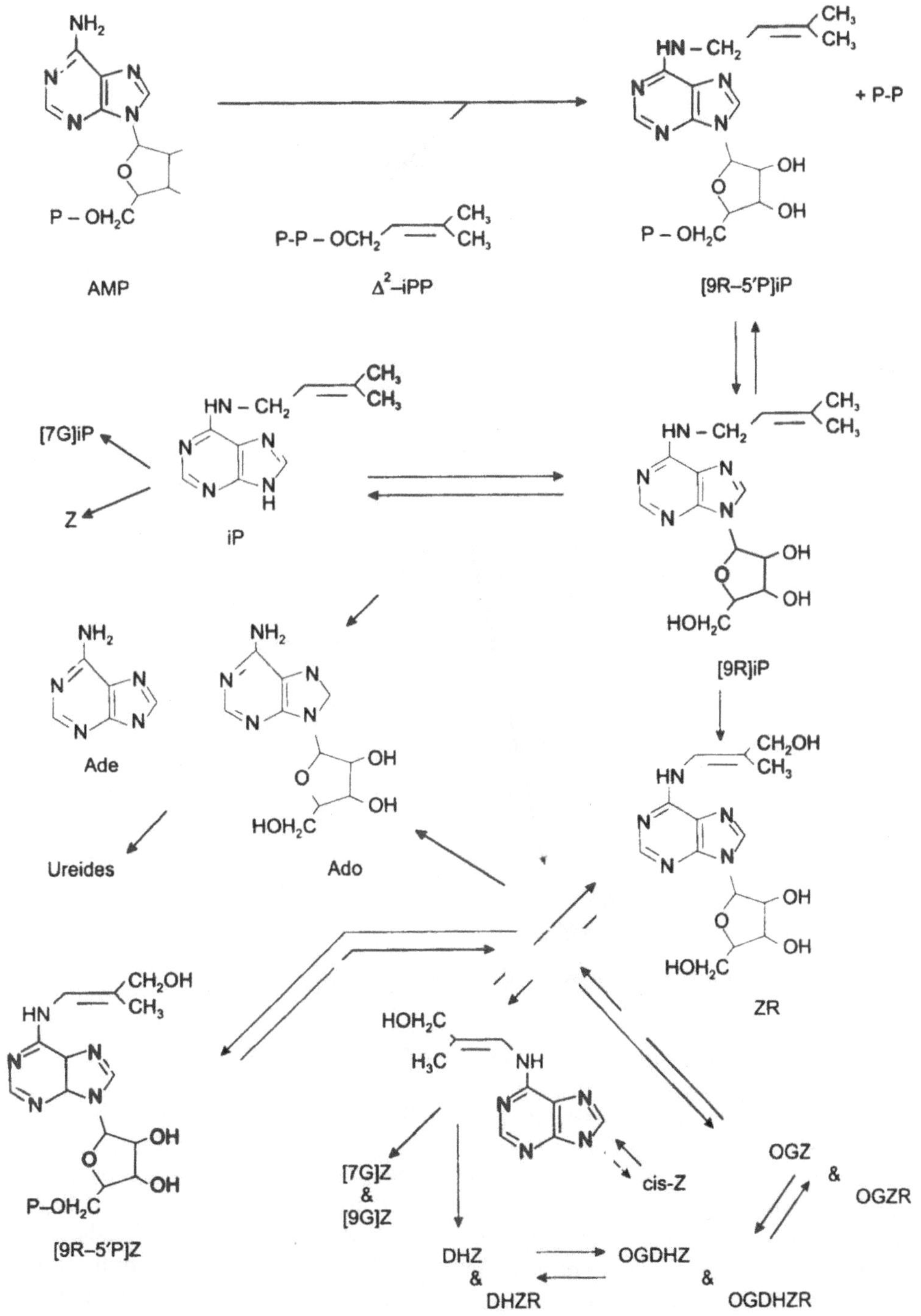

Fig. 5.16 Proposed pathway for cytokinins (from Lightfoot *et al.*, 1997)

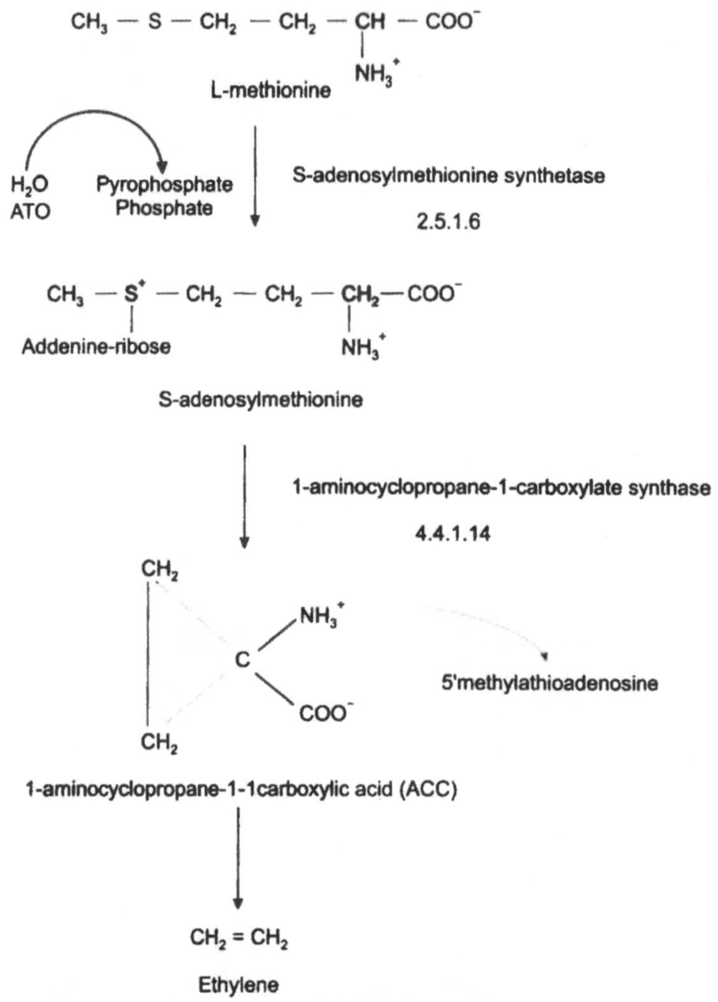

$$CH_3 - S - CH_2 - CH_2 - \underset{\underset{NH_3^+}{|}}{CH} - COO^-$$

L-methionine

H₂O → Pyrophosphate
ATO   Phosphate

S-adenosylmethionine synthetase

2.5.1.6

$$CH_3 - \overset{+}{\underset{\underset{\text{Addenine-ribose}}{|}}{S}} - CH_2 - CH_2 - \underset{\underset{NH_3^+}{|}}{CH_2} - COO^-$$

S-adenosylmethionine

1-aminocyclopropane-1-carboxylate synthase

4.4.1.14

$$\underset{\underset{CH_2}{|}}{\overset{\overset{CH_2}{|}}{C}} \overset{NH_3^+}{\underset{COO^-}{\diagdown}}$$

5'methylathioadenosine

1-aminocyclopropane-1-1carboxylic acid (ACC)

$$CH_2 = CH_2$$

Ethylene

Fig. 5.17 Structure of ethylene and its possible biosynthetic pathway

reviewed by Creelman and Millet (1997). Salicylic acid (Fig. 5.23), which occurs in many plant tissues (see References in Arteca, 1995) is synthesized from phenylalanine. The acid promotes flowering, stimulates plant pathogenesis, protein production, can enhance flower longevity, and may inhibit ethylene biosynthesis. In addition, salicylic acid can inhibit seed germination, interfere with wound responses, and reverse certain ABA effects (http://www.hos.ufl.edu/mooreweb).

PORPHYRINS

Chlorophylls (Scheer, 1991) belong to a class of compounds known as porphyrins (Lavallee, 1987), i.e., four pyrrole rings (Kadish *et al.*, 1999) joined in a cyclic manner (Fig. 5.24). Magnesium is chelated in the center of the chlorophyll molecule (Milgram, 1997). Chlorophylls are the major pigments localized in chloroplast thylakoids (see Chapter 13). Because chlorophylls possess a phytol chain, they

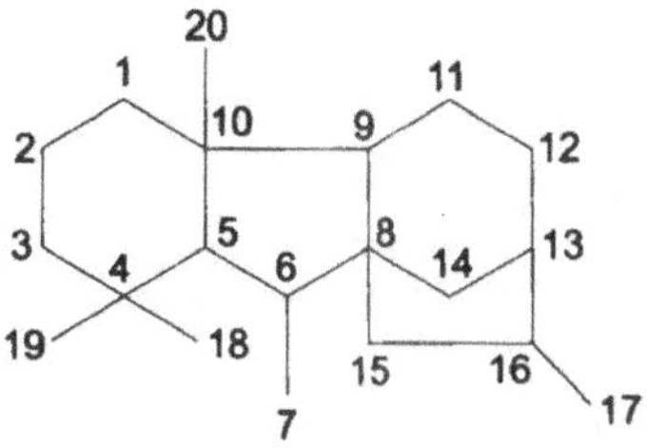

Gibberellane

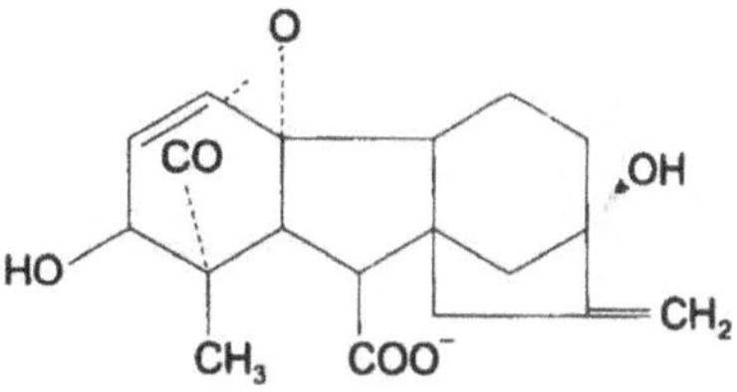

Gibberellic Acid

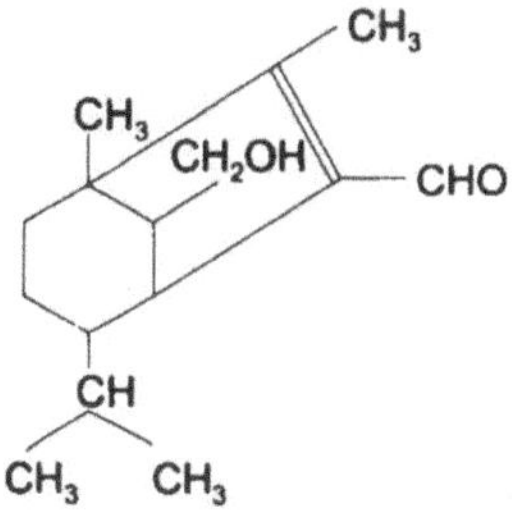

Helminthosporal

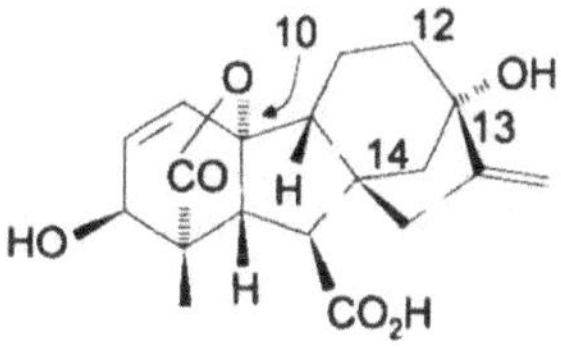

1

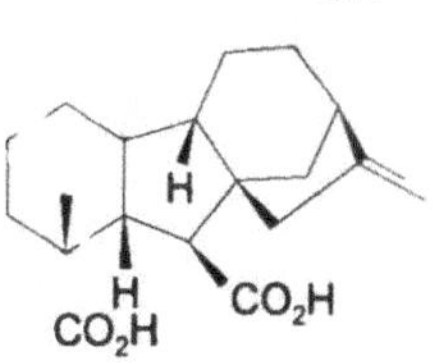

2 (GA111)

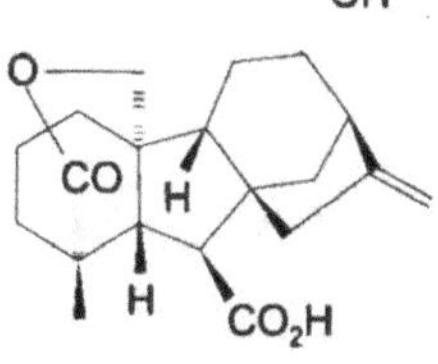

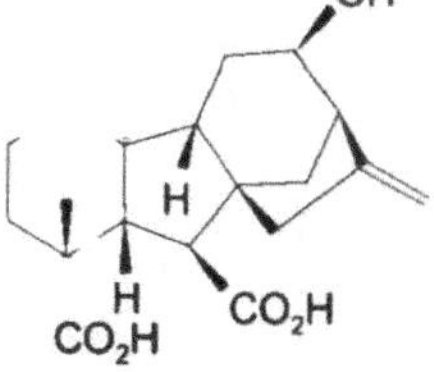

3 (GA112)

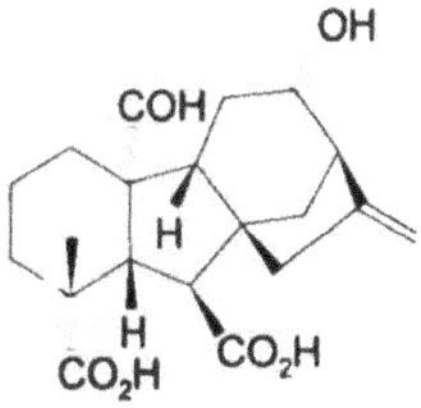

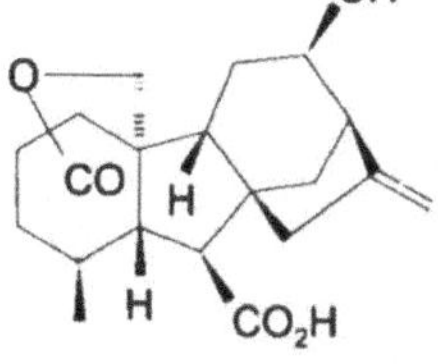

4 (GA113)

5 (GA114)

6 (GA115)

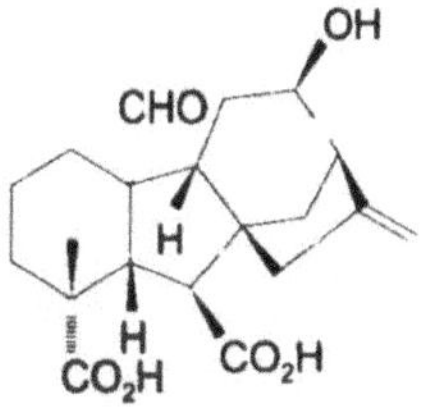

7 (GA116)

Fig. 5.18   Examples of gibberellins

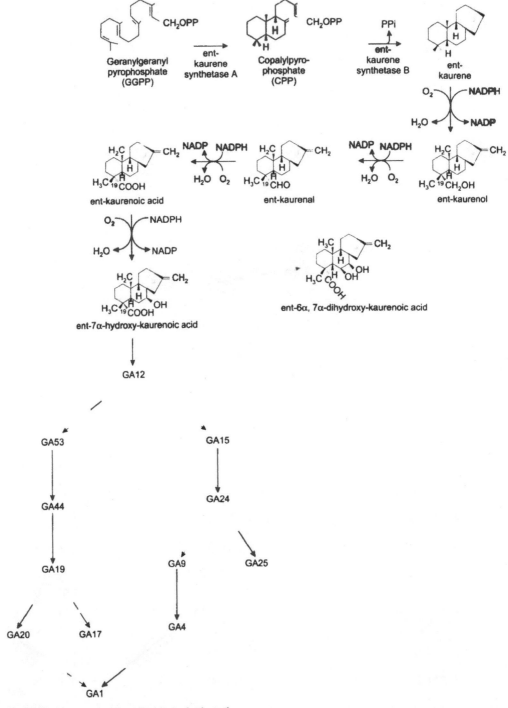

Fig. 5.19 Proposed gibberellin biosynthetic pathway

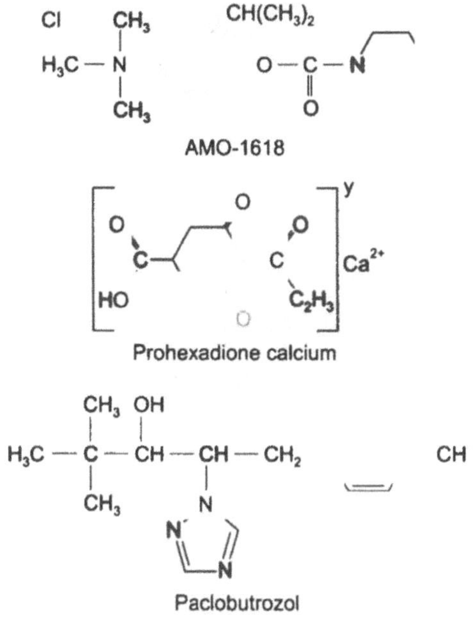

AMO-1618

Prohexadione calcium

Paclobutrozol

Fig. 5.20 Some growth retardants

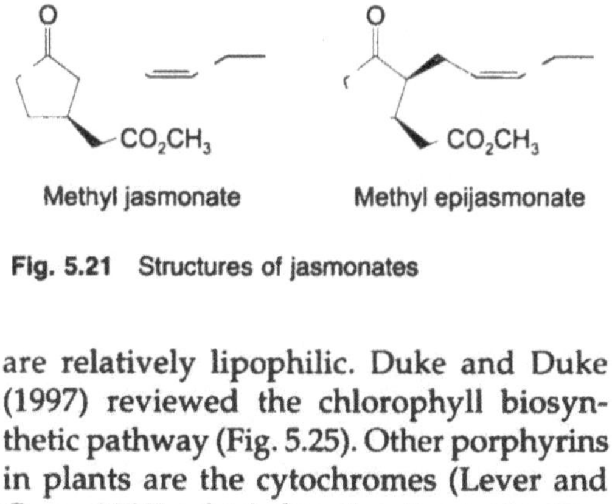

Methyl jasmonate   Methyl epijasmonate

Fig. 5.21 Structures of jasmonates

TABLE 5.12 Biological effects of jasmonate

Effects	
Promotion	Inhibition
Senscence	Carotenoid biosynthesis
Abscission	Chlorophyll formation
Adventitious root formation	Embryogenesis
Breaking of seed dormancy	Floral bud formation
Chlorophyll degradation	Photosynthetic activities
Differentiation in plant tissue culture	Pollen germination
Ethylene biosynthesis	Root length growth
Microtubule disruption	Seedling longitudinal growth
Pigment formation	Tissue culture growth
Protein synthesis	
Respiration	
Stomatal closure	
Tendril coiling	
Tuber formation	

Adapted from: http://www.hos.uf.edu/mooreweb.

TERPENES

Terpenes (Fig. 5.26) are organic compounds derived from isoprene units and are widely distributed (Connolly, 1991; Harborne and Tomos-Barberan, 1991). They originate via the union of two or more isoprene units (Fig. 5.27). Terpenoids are usually lipid-soluble and are often located in the cytoplasm of the plant cell. However, some of them are preferentially distributed. Table 5.13 presents a classification of the terpenoids.

Plants contain compounds which are terpenoid-like, e.g. sterols and others are the mixed terpenoids—chlorophylls, plasto-quinones, phylloquinones, tocopherol quinines, and ubiquinone. Quinones are cyclic ketones (Fig. 5.28). Coenzyme Q is a quinone derivative containing a 10 isoprene unit carbon chain. Plastoquinone is a com-

are relatively lipophilic. Duke and Duke (1997) reviewed the chlorophyll biosynthetic pathway (Fig. 5.25). Other porphyrins in plants are the cytochromes (Lever and Gray, 1983) which function in respiration. Phytochrome, the red-far red reversible photoreceptor (Johnson, 1991) for photomorphogenetic events, is not a porphyrin but contains pyrrole rings in linear array. This is also true of phycoerythrin, a photosynthetic algal pigment (Alison, 2001).

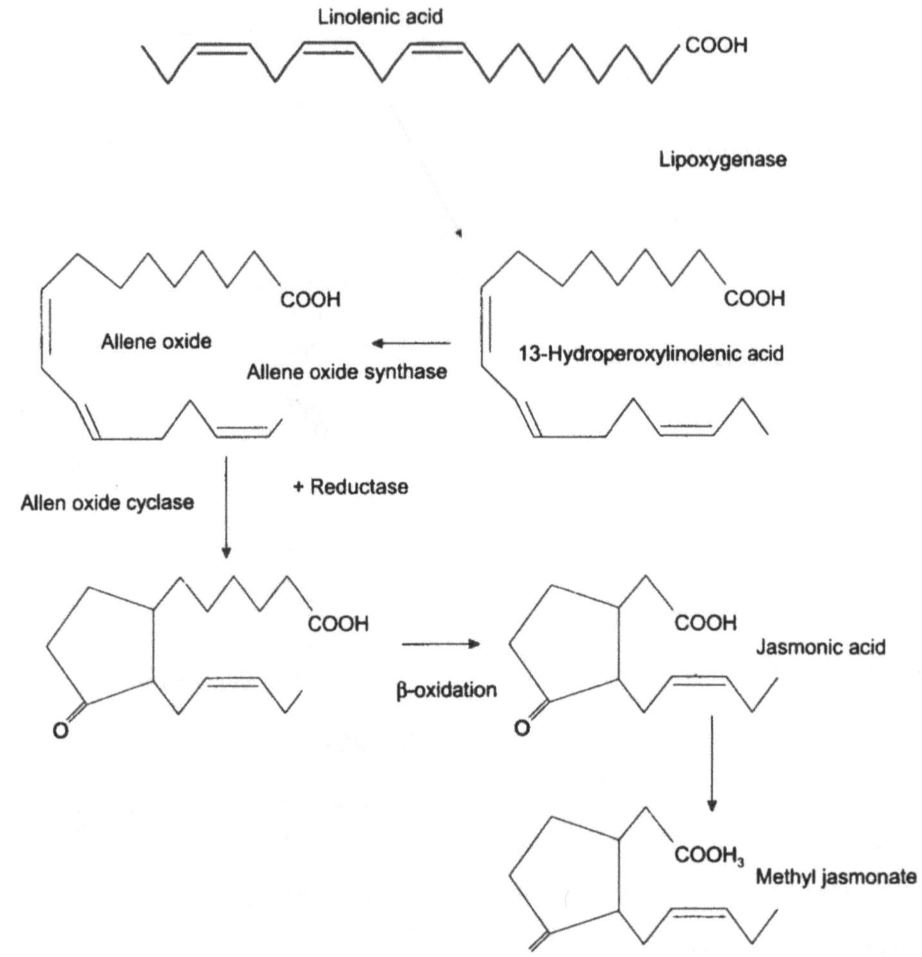
Linolenic acid
COOH
Lipoxygenase
COOH
Allene oxide
Allene oxide synthase
COOH
13-Hydroperoxylinolenic acid
Allen oxide cyclase
+ Reductase
COOH
O
β-oxidation
COOH
Jasmonic acid
O
COOH₃
Methyl jasmonate

Fig. 5.22 Possible jasmonic acid biosynthetic pathway

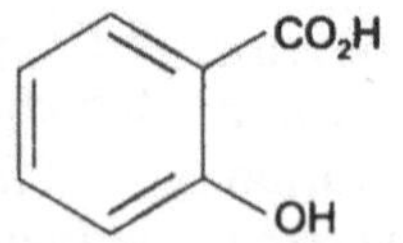
CO₂H
OH
2-hydroxybenzoic acid (salicylic acid)

Fig. 5.23 Salicylic acid structure

pound which exists in the lipid of chloroplast membranes. It transports electrons over short distances within the membranes. As for quinone synthesis, plastoquinone and ubiquinone are benzoquinone, isoprenoid derivatives. Hydroquinones are formed by oxidative decarboxylation of hydroxybenzoate derivatives (Fig. 5.29).

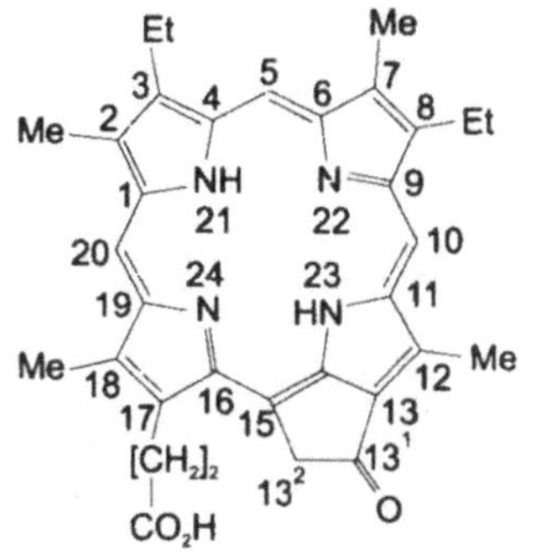

Phytoporphyrin

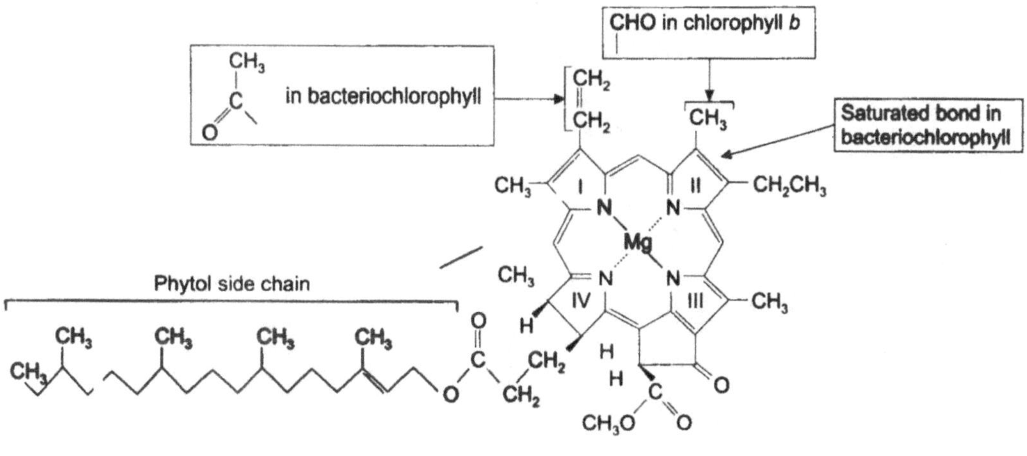

Chlorophyll a

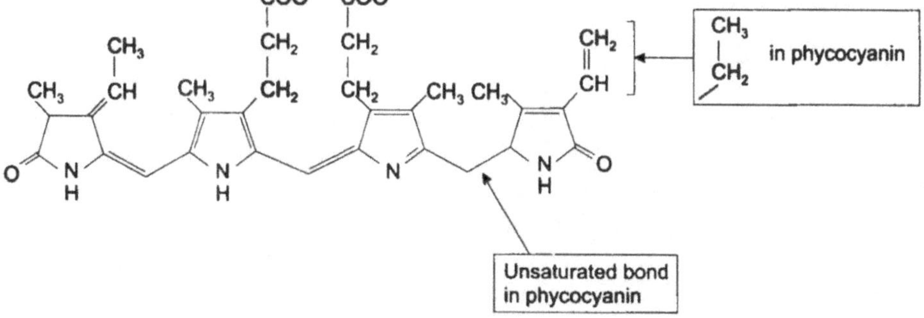

Phycoerythrin

Fig. 5.24 Chemical composition of chlorophylls and linear pyrroles

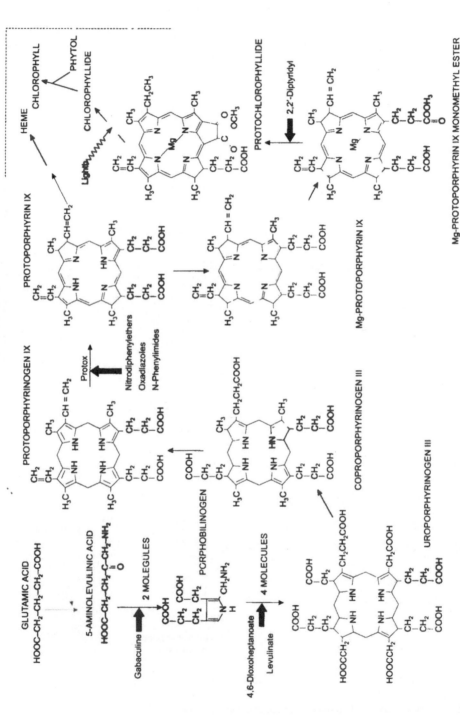

Fig. 5.25 Chlorophyll biosynthetic pathway (from Duke and Duke, 1997)

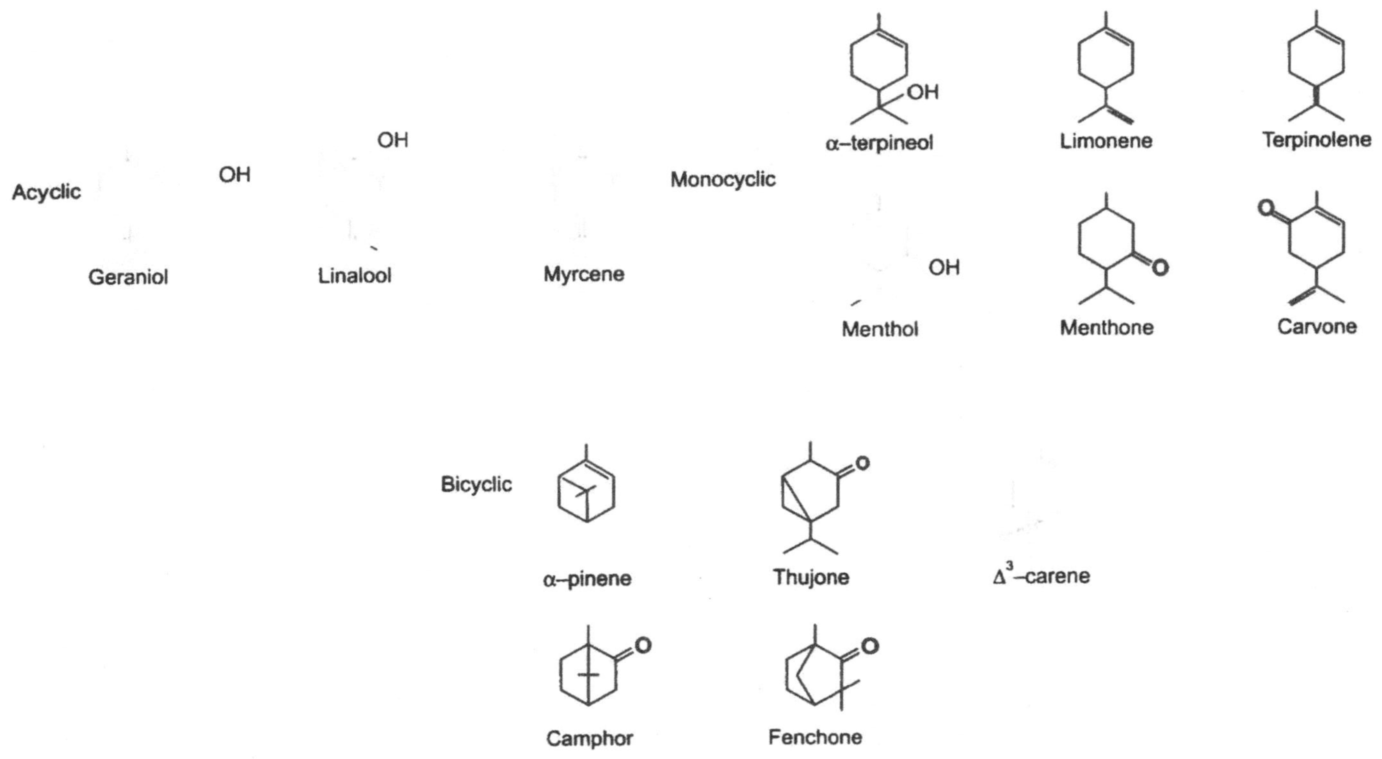

Fig. 5.26   Terpene chemistry

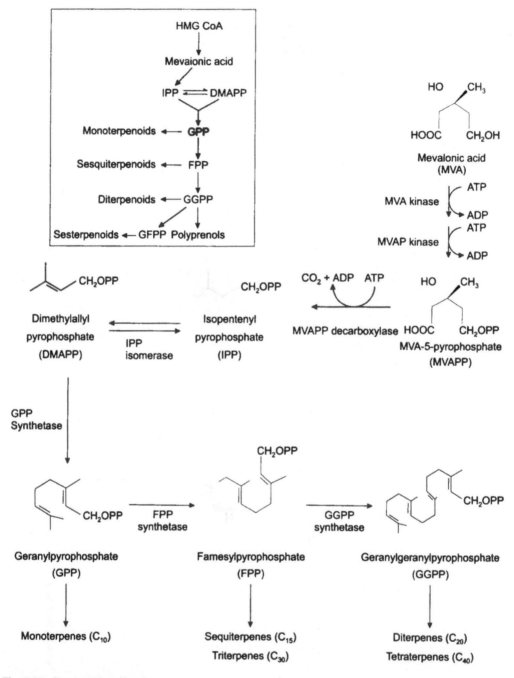

Fig. 5.27 Terpene biosynthesis

VITAMINS

Vitamins (Fig. 5.30, incorporated in Table 5.14) are organic compounds (Robinson, 1973; Friedrich, 1988; Rucker *et al.*, 2001) comprising both water- and lipid-soluble (Diplock, 1985) forms (Table 5.14). These molecules are essential for human health (Mozafar, 1994). Many vitamins are modified as coenzymes, which act together with enzymes to catalyze biochemical reactions. A coenzyme usually functions as the major component of the active site of an enzyme. Prosthetic groups are tightly bound coenzymes. Table 5.14 presents the chemistry, occurrence, and function of plant vitamins.

TABLE 5.13 Summary of isoprenoid forms[a]

Types of terpenoid and examples	Empirical formula and composition	Distribution[b]
Hemiterpenes	C_5H_8	Do not accumulate in plants by themselves, but rather combine with another type of compound
Monoterpenes (volatile)	$C_{10}H_{16}$	Higher plants (see reference for distribution of acyclic, monocyclic, and bicyclic monoterpenoid distribution)
Essential oils, iridoids, and tropolones	Combination of two isoprene units[a]	
Sesquiterpenes (volatile)	Combination of three isoprene units	Certain higher plants and fungi; plant hormone ABA controls dormancy in seeds of herbaceous plants and buds of wood plants
Essential oils, abscisic acid (ABA)	$C_{15}H_{24}$	
Diterpenes (less volatile) acids in plant resins gibberellins	Combination of four isoprene units	Gibberellins universally distributed in fungi and green plants
Triterpenes (nonvolatile) sterols, triterpenes, saponins, cardiac glycosides	$C_{30}H_{48}$ Combination of six isoprene units ·	Distributed in the entire plant kingdom
Tetraterpenes Carotenoids	$C_{40}H_{64}$ Combination of eight isoprene units	Carotenoids widely distributed
Polyterpenes (rubber)	$(C_5H_8)^n$ Any member of isoprene units over 8	Rubber in *Hevea brasiliensis*

[a]These types are actually formed by isopentylpyrophosphate—$CH_2 = {}_c(CH_3CH_2CH_2OPP)$—rather than isoprene itself.
[b]The reader is referred to Dev and Misra (1985), Harrenijn *et al.* (2001) and Seaman *et al.* (1990) for extensive discussions of terpenoids.

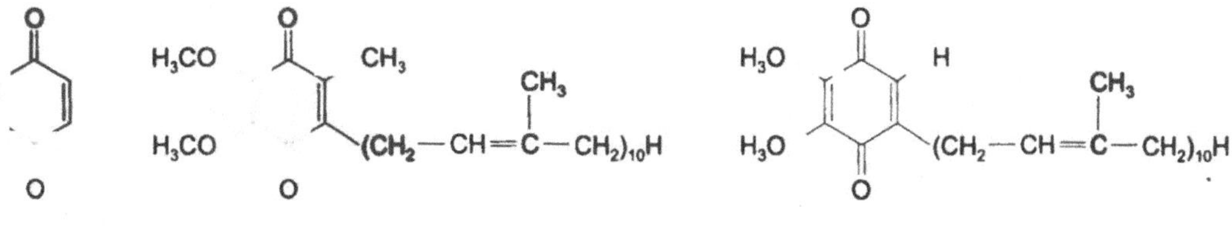

Fig. 5.28 Examples of plant quinones

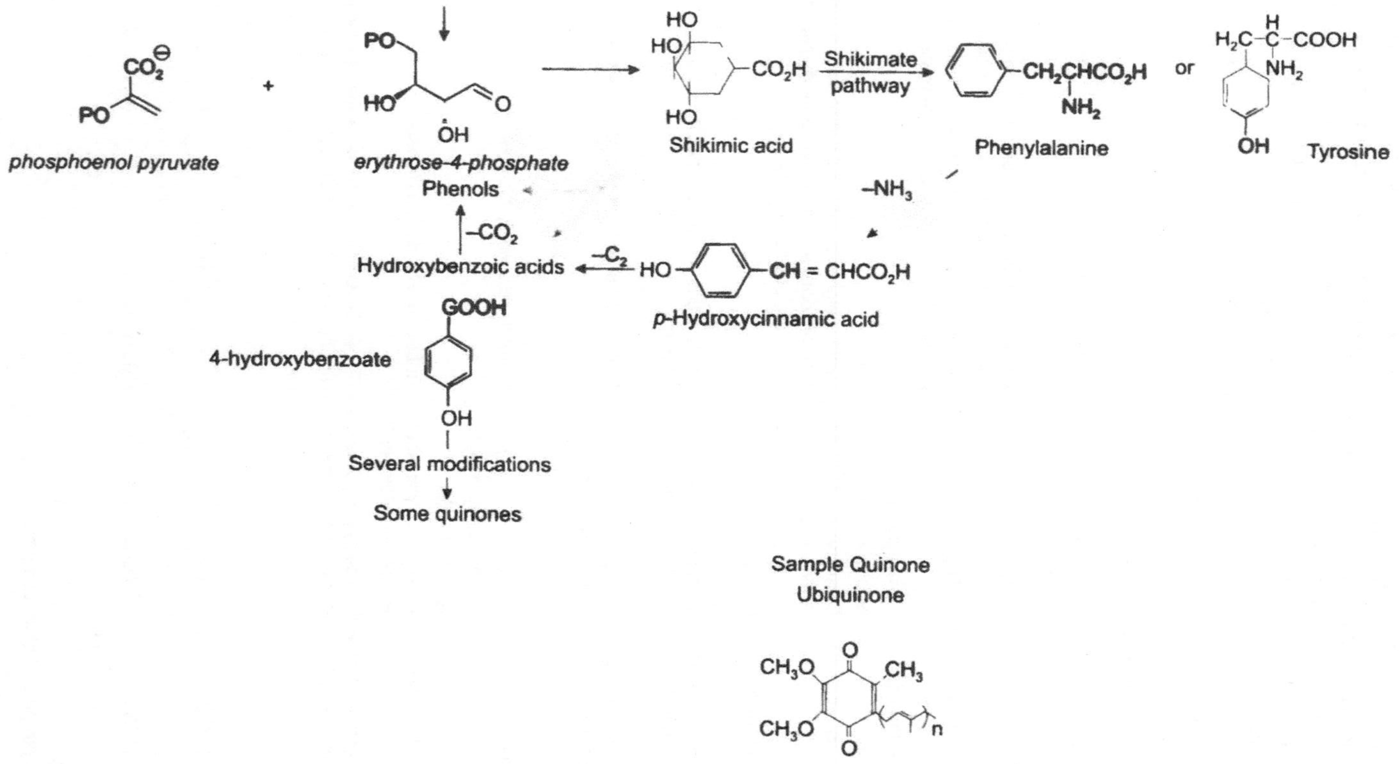

Fig. 5.29 Pathway leading to quinones. From: Dubey, V. 2002. Current Science 83: 685.

TABLE 5.14 Summary of the occurrence and function of vitamins in plants

Vitamin	Structure	Plant source	Function
Thiamine	Most of thiamine in plants present in free thiamine with small amounts of disphosphothiamine and a very small amount of triphosphothiamine; active form is thiamine pyrophospate	Rice grain and most plants; green vegetables = 0.2 mg/100 g; seeds of leguminous plants = 0.8 mg/100 g	Anthineuritic properties
Nicotinic acid and nicotinic amide	Constituent of two coenzymes, NAD, and NADP	Fruits contain 1.0 mg/100 g and some < 0.5 mg/10 g; nuts = 6 to 13 mg/100 g; vegetables contain from 0.5 to 1.5 mg/100 g	Carbohydrate metabolism
Riboflavin		Alfalfa and dandelion; fruits contain < 0.05 mg/100 g; nuts = up to creals (0.02 to 0.3 mg/100 g); vegetable juices: 0.03 to 0.2 mg/100 g	Carbohydrate metabolism
Pyridoxine, pyridoxal, and pyridoxamine	Pyridoxine, Pyridoxal-5'-phosphate, Pyridoxamine	Occur in foodstuffs in varying proportions present in fruits at 1.5 mg/100 g; vegetables contain 0.04–0.1 mg/100 g	Antidermatitis activity in animals; decarboxylation of various amino acids; transamination; photosynthetic phophosphorylation
Panthotenic acid		Widespread occurrence, fruits = 0.5 mg/100 g; vegetables = 0.1 mg to 0.5 mg/100 g; legumes = 2 to 5 mg/100 g	Biological acetylations

(Contd. Table 5.14)

(Contd. Table 5.14)

Biotin and biotin adenylate	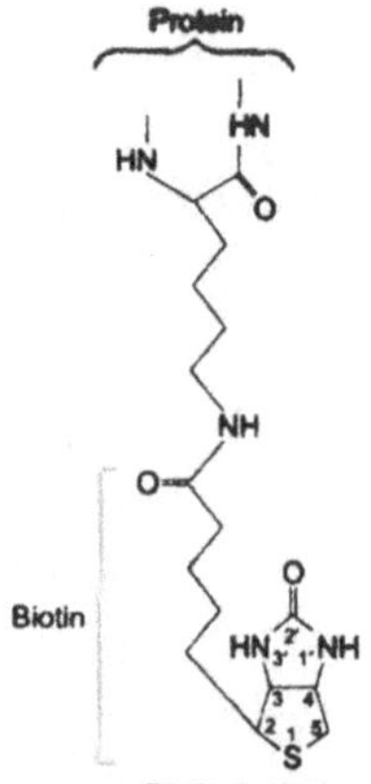 o-Biotinyl-protein 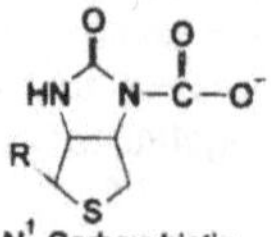 N¹-Carboxybiotin	Present in many plant tissues; nuts = 16 µg/10 g	Growth factor for microorganisms, carbon dioxide fixation, and decarboxylation
Folic acid	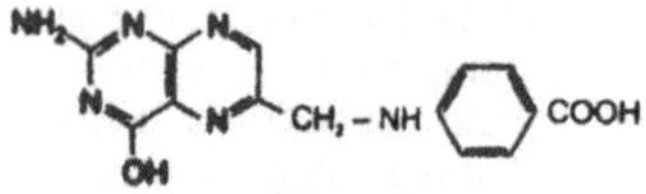 IX	Spinach leaves; fruits = 0.5 to 5 µg/10 g; vegetables = 3 to 26 µg/10 g	Biological methylations
Vitamin B12	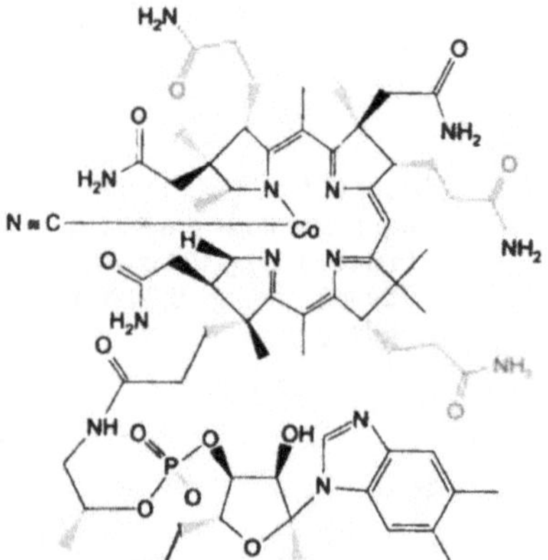	Soybean meal, yellow corn, and wheat = 1 µg/100 g; present in larger amounts in root modules	Synthesis of choline in animals; function in plants uncertain
Vitamin C ascorbic acid and dehydroascorbic acid	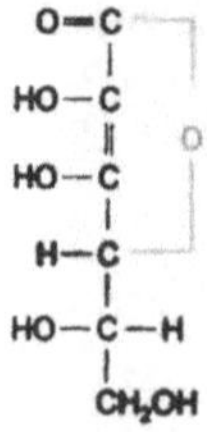 (a) Ascorbic acid 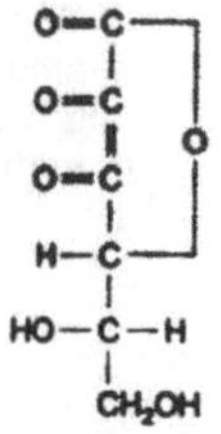 (b) Dehydroascorbic acid	Orange juice, cabbage juice, and paprika; fruits = 3–10 mg/100 g; green vegetables = 60–140 mg/100g	Prevents scurvy in animals; electron donor and acceptor in various plant processes

(Contd. Table 5.14)

(Contd. Table 5.14)

Vitamin D	XIII	Does not appear to occur in plants	Prevents rickets in animals
Vitamin E		Wheat germ; vegetable oils = 0.1–0.3%; most fruits and vegetables = 0.1–0.75 mg/100 g	Nonspecific anti-oxidant
Vitamin K		Leaves contain 5–20 mg/100 g vitamin K	Constituent of chloroplasts
Ubiquinone		Alfalfa	**Photosynthesis and oxidative phosphorylation**
Plasto-quinone		Alfalfa	**Photophosphory-lation and photochemical evolution of oxygen by the chloroplast**

References

ALKALOIDS

Breilmann, H.L. 1999. *Phytochemicals: The Chemical Components of Plants in Natural Products from Plants.* Kaufman, P.B., Ceske, L.J., Warber, S., Dike, J.A. and Brielmann, H.L. (eds.). CRC Press, Boca Raton, FL, USA.

Dey, P.M. and Harborne, J.B. 1999. *Methods in Plant Biochemistry*, vol. 1 Plant Phenolics. Academic Press, San Diego, CA.

Kegg Pathway Database 2004. html and www.genome.od.jp.

Raffauf, R.F. 1996. *Plant Alkaloids A Guide to Their Discovery and Distribution.* Forest Products Press, Press World, New York, NY.

Rahman, A. 1990. *Diterpenoid and Steroidal Alkaloids.* Elsevier Health Sciences, Amsterdam, The Netherlands.

Rahman, A. 1994. *Isoquinoline Alkaloids.* Elsevier Health Sciences, Amsterdam, The Netherlands.

Rizk, A.M. 1990. *Naturally Occurring Pyrrolizidine Alkaloids.* CRC Press, Boca Raton, FL, USA.

Roberts, M.F. and Wink, M. 1998. *Alkaloids: Biochemistry Ecology and Medicine Applications.* Plenum Press, New York, NY.

Robinson, T. 1980. *The Organic Constituents of Higher Plants.* Cordus Press, North Amherst, MA, USA.

Southan, T.W. and Cordell, G.A. 1989. *Dictionary of Alkaloids.* CRC Press, Boca Raton, FL, USA.

Zenk, M.H. and Phillipson, J.D. 1992. *Indole and Biogenetically Related Alkaloids.* Academic Press, New York, NY.

ORGANIC ACIDS

Duke, J.A. 1992. *Handbook of Biologically Active Phytochemicals and Their Activities.* CRC Press, Boca Raton, FL, USA,

Fox, R.B. and Powell, W.H. 2001. *Nomenclature of Organic Compounds: Principles and Practice.* Oxford Univ. Press. Oxford, England.

Hanai, T. 1982. *Phenols and Organic Acids.* CRC Press, Boca Raton, FL, USA.

Robinson, T. 1980. *The Organic Constituents of Higher Plants, Their Chemistry and Interrelationships.* Cordus Press, North Amherst, MA, USA.

PHENOLS

Gross, G.C., Hemingway, R.W., and Yoshida, J. 1999. *Plant Polyphenols 2. Chemistry, Biology, Pharmacology, Ecology.* Kluwer Acad. Publ., New York, NY.

Haslam, E. 1989. *Plant Polyphenols Vegetable Tannins Revisited.* Cambridge Univ. Press, New York, NY.

FLAVONOIDS

Busling, B.S. and Mantley, J.A. 2002. *Flavonoids in Cell Function.* Plenum Publ., Kluwer Acad., New York, NY.

Clark, W.D. and Titus, G.P. 1997. Applications in flavonoid research. In: *Methods in Plant Biochemistry and Molecular Biology,* pp. 217-227. Dashek, W.V. (ed.). CRC Press, Boca Raton, FL, USA.

Harborne, J. 1986. *Plant Flavonoids in Biology and Medicine. Biochemical, Pharmacological, and Structure Activity Relationships.* Liss, New York, NY.

Havsteen, B.H. 2002. The biochemistry and medical significance of the flavonoids. Pharm. Ther. 96: 67-202.

Mantley, J.A. and Busling, B.S. 1998. *Flavonoids in the Living System.* Plenum Press, New York, NY.

Middleton, E. 2000. The effects of plant flavonoids on mammalian cells. Implications for inflammation, heart disease and cancer. Pharm. Res. 52: 673-751.

Packer, L. 2001. *Flavonoids and Other Polyphenols.* Acad. Press, San Diego, CA.

Shirley, B.W. 1996. Flavonoid biosynthesis new functions for an old pathway. Trends Plant Sci. 1: 377-387.

PHENYLPROPANOIDS

Dean, J.F. 1997. Lignin analysis. In: *Methods in Plant Biochemistry and Molecular Biology,* pp. 199-215. Dashek, W.V. (ed.). CRC Press, New York, NY.

Kurkin, V.A. 2003. Phenylpropanoids from medicinal plants: Distribution, classification, structural analysis and biological activity. Chem. Natural Comp 39: 123-153.

Seigler, D.S. 1998. *Plant Secondary Metabolism.* Kluwer Acad. Publ., Dordrecht, The Netherlands.

PHYTOALEXINS

Daniel, M. and Purkayastha, R.P. 1995. *Handbook of Phytoalexin Metabolism and Action.* Marcel Dekker, New York, NY.

Dixon, R.A. 1986. The phytoalexin response elucidation, signaling and control of host gene expression. Biol Rev. 61: 239-291.

Kuc, J. 1994. Relevance of phytoalexins—a critical review. Acta Hort. 381: 529-539.

Yu, L. 1997. The isolation and assay of elicitins. In: *Methods in Plant Biochemistry and Molecular Biology,* pp. 265-279. Dashek, W.V. (ed.). CRC Press, Boca Raton, FL, USA.

GENERAL PLANT HORMONES

American Society of Plant Physiologists. 1984. *The Molecular Biology of Plant Hormone Action.* Rockville, MD., USA.

Chailakhyan, M.Kh. and Khrianin, V.N. 1987. *Sexuality in Plants and Its Hormonal Regulation.* Springer-Verlag, New York, NY.

Crozier, A. and Hillman, J.R. 1984. *The Biosynthesis and Metabolism of Plant Hormones.* Cambridge Univ. Press, New York, NY.

Davies, P.J. 1987. *Plant Hormones and Their Role in Plant Growth and Development.* M. Nijhoff, Hingham, MA, USA.

Hooykaas, P.J.J., Hall, M.A. and Libbenga, K.R. 1999. *Biochemistry of Plant Hormones.* Elsevier, Amsterdam, The Netherlands.

Klambt, D. 1987. *Hormone Receptors.* Springer-Verlag, Berlin, Germany.

Moore, T.C. 1989. *Biochemistry and Physiology of Plant Hormones.* Springer-Verlag, New York, NY.

Rivier, L. and Crozier, A. 1987. *Principles and Practice of Plant Hormone Analysis.* Academic Press, New York, NY.

Takahaski, N. 1996. *Chemistry of Plant Hormones.* CRC Press, Boca Raton, FL, USA.

Tucker, G.A. and Roberts, J.A. 2000. *Plant Hormone Protocols.* Humana, Totowa, NJ, USA.

Venis, M. 1985. *Hormone Binding Sites in Plants.* Longman, New York, NY.

ABSCISIC ACIDS

Arteca, R.N. 1996. *Plant Growth Substances: Principles and Applications.* Kluwer Acad. Publ., Dordrecht, The Netherlands.

Walton, D.C. and Li, Y. 1995. Abscisic acid biosynthesis and metabolism. In: *Plant Hormones Physiology. Biochemistry and Molecular Biology.* Davies, D.J. (ed.). Kluwer Acad. Publ., Higham, MA, USA.

AUXINS

Bandurski, R.S., Cohen, J.D., Slovin, J.P. and Renecke, M. 1995. Auxin biosynthesis and metabolism. In: *Plant Hormones Physiology Biochemistry and Molecular Biology*, pp. 39-65. Davies, P.J. (ed.). Kluwer Acad., Amsterdam, The Netherlands.

Bartel, B. 1997. Auxin biosynthesis. Ann. Rev. Pl. Physiol. Pl. Molec. Biol. 48: 51-66.

Jones, A.M. and Lomax, T.L. 1997. Photo affinity labeling with 5- A zido indole–3-acetic acid. In: *Methods in Plant Biochemistry and Molecular Biology*, pp. 115-132. Dashek, W.V. (ed.). CRC Press, Boca Raton, FL, USA.

Juyang, J., Shao, X. and Li, J. 2000. Indole -3- glycerol phosphate, a branchpoint of indole -3- acetic acid biosynthesis from the tryptophan biosynthetic pathway in *Arabidopsis thaliana.* Plant J. 24: 327-333.

Steward, F.C. and Krikorian, A.D. 1971. *Plants, Chemicals and Growth.* Academic Press, New York, NY.

Zizimalova, E. and Napier, R.M. 2003. Points of regulation for auxin action. Plant Cell Rept. 21: 625-634.

BRASSINOSTEROIDS

Arteca, R.N. 1995. *Plant Growth Substances: Principles and Applications.* Kluwer Acad. Publ., Amsterdam, The Netherlands.

Culte, H.G., Yokota, T. and Adam, G. 1991. *Brassinosteroids: Chemistry, Bioactivity and Applications.* American Chemical Society, Washington, DC, USA.

Khripach, VA, Zhabinski, V.N. and de Grout, A.E. 1999. *Brassinosteroids A New Class of Plant Hormones.* Academic Press, San Diego, CA.

Yokota, T., Sokurai, A. and Clouse, S.D. 1999. *Brassinosteroids Steroidal Plant Hormones.* Springer-Verlag, New York, NY.

CYTOKININS

Haberer, G. and J.J. Kieber. 2002. Cytokinin: new insights into a class of phytohormone. Pl. Physiol. 128: 354-362.

Kakimoto, T. 2003. Biosynthesis of cytokinins. J. Plant Res. 116: 223-239.

Lightfoot, D.A., McDaniel, K.L., Ellis, J.K., Hammerton, R.H. and Nicander, B. 1997. Methods for analysis of cytokinin content, metabolism and response. In: *Methods in Plant Biochemistry and Molecular Biology.* CRC Press, Boca Raton, FL, USA.

McGaw, B.A. and Burch, L.R. 1995. Cytokinin biosynthesis and metabolism. In: *Plant Hormones Physiology, Biochemistry and Molecular Biology*, pp. 98-117. Davies, P.J. (ed.). Kluwer Acad., Amsterdam, The Netherlands.

Mok, D.W.S. and Mok, M.C. 1994. *Cytokinins.* CRC Press, Boca Raton, FL, USA.

ETHYLENE

Abeles, F.B., Morgan, P.W. and Saltveit, M.E. 1992. *Ethylene in Plant Biology.* Academic Press, San Diego, CA.

Bleecher, A.B. and Kende, H. 2000. Ethylene: A gaseous signal molecule in plants. Annu. Rev. Cell Dev. Biol. 16: 1-18.

Chang, C. Grievson, D., Kanellis, A.K. and Kende, H. 1997. *Biology and Biochemistry of the Plant Hormone Ethylene.* Kluwer Acad. Publ., Boston, MA, USA.

Kende, J. 1993. Ethylene biosynthesis and its regulation in higher plants Annu. Rev. Pl. Pysiol. Pl. Molec. Biol. 44: 283-307.

GIBBERELLINS

Hedden, P. 1999. Recent advances in gibberellin biosynthesis. J. Exper. Bot. 50: 553-563.

Rademacher, W. 2000. Growth retardant effects on gibberellin biosynthesis and other metabolic pathways. Ann. Rev. Pl. Physiol. Pl. Molec. Biol. 51: 501-531.

Sponsel, V.M. 1995. The biosynthesis and metabolism of gibberellins in higher plants. In: *Plant Hormones Physiology, Biochemistry and*

Molecular Biology, pp. 66-97. Davies, P.J. (ed.). Kluwer Acad., Amsterdam, The Netherlands.

Takahashi, N. Phinney, B.O. and Macmillan, J. 1991. *Gibberellins*. Springer-Verlag, New York, NY.

JASMONATES

Arteca, R.N. 1995. *Plant Growth Substances: Principles and Applications*. Kluwer Acad. Publ., Amsterdam, The Netherlands.

Creelman, R.A. and Millet, J.E. 1997. Biosynthesis and action of jasmonates in plants. Ann Review Pl. Physiol. Pl. Molec. Biol. 48: 355 – 381.

PORPHYRIN AND LINEAR PYRROLES

Alison, S. 2001. *Heme, Chlorophyll and Bilins Methods and Protocols*. Humana Press, Totowa, NJ, USA.

Duke, M. and Duke, S.O. 1997. Analysis and manipulation of the chlorophyll pathway in higher plants. In: *Methods in Plant Biochemistry and Molecular Biology*. Dashek, W.V. (ed.). CRC Press, Boca Raton, FL, USA.

Johnson, C.B. 1991. *Phytochrome Properties and Biological Action*. NATO Asi Services H: *Cell Biology*, vol. 50. Springer-Verlag, New York, NY.

Kadish, K.M., Smith, K.M. and Guilard, R. 1999. *The Porphyrin Handbook*. Academic Press, New York, NY.

Lavallee, D.K. 1987. *The Chemistry and Biochemistry of N-Substituted Porphyrins*. VCH Publ., New York, NY.

Lever, A.B.P. and Gray, H.B. 1983. *Iron Porphyrins*. Addison Wesley, Reading, MA, USA.

Milgram, L.R. 1997. *The Colours of Life: an Introduction to the Chemistry of Porphyrins and Related Compounds*. Oxford Univ. Press, Oxford, England.

Scheer, H. 1991. *Chlorophylls*. CRC Press, Boca Raton, FL, USA.

TANNINS

Haslam, E. 1989. *Plant Polyphenols. Vegetable Tannins Revisited*. Cambridge Univ. Press, New York, NY.

Hemingway, R.W. and Karchesy, J.J. 1988. *Chemistry and Significance of Condensed Tannins*. Plenum Press, New York, NY.

Kim, S., Lee, Y.K., Kim, H.J. and Lee, H.H. 2003. Physico-mechanical properties of particleboards bonded with white pine and wattle tannin-based adhesives. J. Adhes. Sci. Tech. 17: 1863-1875.

Smith, K.T. 1997. Phenolics and compartmentalization in the sapwood of broad leaves, pp. 189-198. In: *Methods in Plant Biochemistry and Molecular Biology*. Daskek, W.V. (ed.). CRC Press, Boca Raton, FL, USA.

TERPENOIDS

Connolly, J.D. 1991. *Dictionary of Terpenoids*. Chapman Hall, London, England.

Dev, S. and Misra, R. 1985. *CRC Handbook of Terpenoids: Diterpenoids*. CRC Press, Boca Raton, FL, USA.

Harborne, J.B. and Tomos-Barberan, F.A. 1991. *Ecological Chemistry and Biochemistry of Plant Terpenoids*. Clarendon Press, Oxford, England.

Harrenijn, P., Van Osten, A.M. and Piron, P.G.M. 2001. *Natural Terpenoids as Messengers. A Multidisciplinary Study of Their Production, Biological Functions and Practical Application*. Kluwer Acad. Publ., Boston, MA, USA.

Seaman, F., Bohmann, F., Zdero, C. and Mabry, T.J. 1990. *Diterpenes of Flowering Plants. Compositae Asteraceae*, Springer-Verlag, New York, NY.

VITAMINS

Diplock, A.T. 1985. *Fat Soluble Vitamins: Their Biochemistry and Application*. Heinemann, London, England.

Friedrich, W. 1988. *Vitamins*. deGruyter, New York, NY.

Mozafar, A. 1994. *Plant Vitamins Agronomic, Physiological and Nutritional Aspects*. CRC Press, Boca Raton, FL, USA.

Robinson, F.A. 1973. Vitamins (plants). Phytochemistry 3:195.

Rucker, R.B., Suttie, J.W., McCormick, D.D. and Machlin, L.J. 2001. *Handbook of Vitamins*. Marcel Dekker, New York, NY.

CHAPTER 6

Subcellular Organelles: Structure and Function

W.V. Dashek and T.S. Kaneko

STRUCTURE OF CELLULAR MEMBRANES

Transmission electron microscopy (TEM) together with biochemical methods has revealed a diversity of functional organelles comprising most higher plant cells. In addition, coupling of these techniques has provided significant knowledge regarding organelle structure-function relationships. Furthermore, applications of these techniques as well as other technologies such as freeze-fracturing and freeze-etching have resulted in a critical reevaluation of the Davson-Danelli model of cellular membranes. Currently, most students of the life sciences learn the fluid-mosaic model (Singer and Nicolson, 1972) of the membrane (Fig. 6.1). According to this widely accepted model, cellular membranes are composed of a lipid bilayer in which globular proteins are embedded. Integral proteins can traverse the bilayer and protrude on either side of the membrane. While the embedded protein is hydrophobic (Engelman, 1996), the exposed portion is hydrophilic. Short-chain carbohydrates attached to the protruding proteins are

thought to function in cell-to-cell adhesion. The chemical composition of plant membranes is presented in Table 6.1. Like most areas of contemporary life science research, both the structure and function of cellular membranes are constantly being reevaluated (Jacobson *et al.*, 1995; Smallwood *et al.*, 1996).

ENDOMEMBRANE CONCEPT

One striking advance in contemporary subcellular biology is a more thorough understanding of the interrelationships of certain plant cell organelles (Fig. 6.2). Rather than existing as discrete entities in physiological and biochemical isolation of one another, it is now generally agreed that the rough endoplasmic reticulum (RER), Golgi apparatus, and cell surface comprise a functional continuum, i.e., the endomembrane system (Morré, 1990). Recently, Morré and Keenan (1997) reviewed the current opinions of the system which, in their opinion, is "still a valid explanation for the transport of proteins and lipids from the ER to the plasma membrane." The hypothesis underlying the endomembrane concept

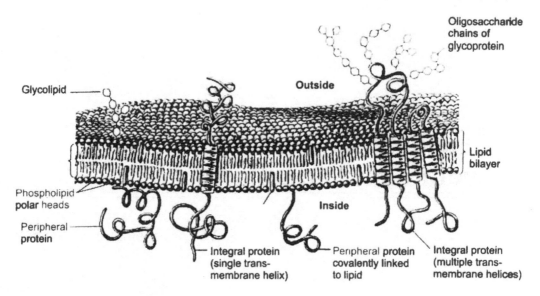

Fig. 6.1 Fluid-mosaic model of the membrane
(http://www.chem.umd.edu/courses/chem243davis/weds-4-28-notes.pdf)

TABLE 6.1 Composition of certain cellular membranes

Chemical Composition

Fatty acyl groups in membrane lipids
16:0, 16:1, t-16:1, 16:3, 18:0, 18:1, 18:2, á 18:3, ã 18:3, 18:4, 22:0, 22:1, 24:0, 24:1

Electroneutral phospholipids	Phosphatidylcholine, phosphatidylethanol, phosphatidylethanolamine
Anionic phospholipids	Phosphatidylserine, phosphatidylglycerol, phosphatidylinositides
Lyo-phopholipids	
Sphingolipids	Cerebrosides
Chloroplast-specific glycerolipids	Galactolipids, sulpholipids
Mitochondrial phospholipids	Diphosphatidylglycerol and monophosphatidylglycerol
Sterols	
	Sitosterol
	Campersterol
	Stigmasterol
	Unusual sterols
	Cycloartenol
	Cholesterol, minute quantities
Sterolglycosides	
Lanosterol	Pathogenic fungal membranes
Water	
Extramembrane water	Membrane is a bilayer sandwiched between two layers of water
	Water is located within the bilayer which is attached to or in approximate contact with the expanses of membrane constituents

(Contd. Table 6.1)

(Contd. Table 6.1)

Proteins	
Integral	May cross the membrane once or several times and are linked either electrostatically or by means of biophysical lipophilicity to the inner domains of the bilayer
Simple integral proteins	Classic α-helical structures that traverse the membrane only once
Complex integral proteins	Globular: comprises several α-helical loops that may span the membrane several times
Peripheral proteins	Associated with only leaflet: easily isolated by altering ionic strength of pH of the encasing medium
Transport proteins	Pumps, carrier, and channels

Summarized from Leshem *et al.* (1991)

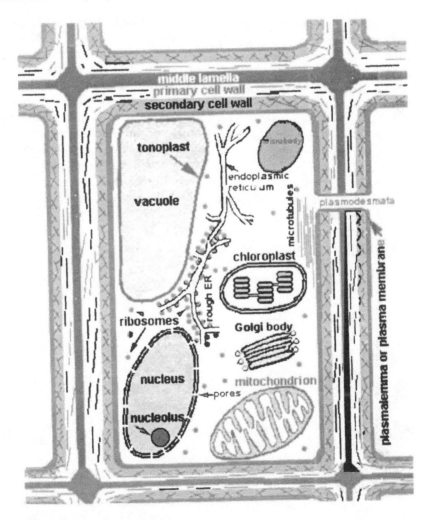

Fig. 6.2   Diagram of plant cell structure
(http://www.generalhorticulture.tanner.edu/lectsupl/print/page06.html)

is that membranes flow through an interconnected endoplasmic reticulum, Golgi apparatus, and plasma membrane (Fig. 6.3). Both ultrastructural and biochemical evidence suggest that interconnections of these elements involve "vesicular and transitional membrane forms". Morré and Keenan (1997) thoroughly described the nature of these forms. In particular, they evaluated ER to Golgi apparatus transfer and Golgi apparatus to plasma membrane transfer as well as models for conveying "materials" through the Golgi apparatus. Morré and Keenan (1997) further critically examined the evidence for the vesicle shuttle model (Fig. 6.4a) and the flow differentiation model (Fig. 6.4b). They concluded that the latter model of Golgi apparatus functioning remains as relevant as when it was first proposed. The student is encouraged to read the comprehensive review of Morré and Keenan (1997) to gain a historical perspective for the evolution of thinking concerning mechanisms of plant cell secretion. A more recent account of the endomembrane system's role in secretion can be found in Vitale and Galili (2001).

PRIMARY CELL WALL

Components of the plant cell wall (Bacic *et al.*, 1988; Carpita and Gibeaut, 1993; Carpita *et al.*, 1996) are the middle lamella (intercellular substance), primary wall, and secondary wall. The middle lamella comprises the pectic layer between cells and holds adjoining cells together. The primary wall is thin (1-3 μm) and elastic, containing cellulose, hemicelluloses, pectins, and glycoproteins. This wall provides mechanical strength, maintains cell shape, regulates cell expansion (Cosgrove, 2001) and intercellular transport, protects against certain organisms and can function in cell signaling. Cellulose is a β 1,4 D-glucan and the pectic substances consist of galacturonans, rhamnogalacturonans, arabinans, galactans, and arabinogalactans I. In contrast, hemicelluloses comprise xylans, glucomannans, galactoglucomannans, xyloglucans, β-D-glucans, and other polysaccharides. The latter include: β, 1-3-linked D-glucans

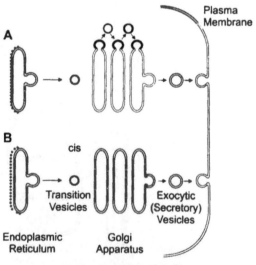

Fig. 6.4 Comparison of the vesicle shuttle and (A) flow differentiation (B) models of intra-Golgi transport. (Morré and Keenan, 1997; as cited by Dashek, 2000)

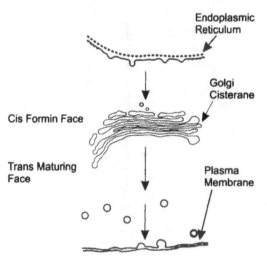

Fig. 6.3 Endomembrane concept

(callose), arabinogalactans and glucurono-mannans. In addition to polysaccharides, primary walls 'house' proteins (Cassab, 1998), e.g., hyp-containing glycoproteins and expansin (Fleming *et al.*, 1997). Expansin may enhance plant cell wall plasticity *in vitro* and possibly tissue expansion *in vivo*. A model of the primary wall is presented in Figs. 6.5.

SECONDARY CELL WALL

The secondary wall is located internal to the primary wall and can possess pits and sometimes three distinct layers, S_1, S_2, and S_3. The secondary wall can contain 25% lignin (a noncarbohydrate) and cellulose is more abundant than in the primary walls. Support and resistance to decay (Highley and Dashek, 1998) are the chief functions of the secondary wall.

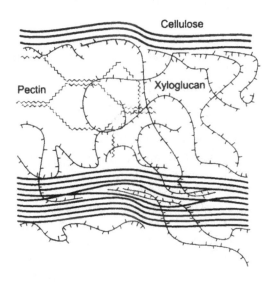

Fig. 6.5 Model of the primary cell wall. From: Cosgrove (see chapter 4).

PLASMODESMATA

Plasmodesmata (Robards and Lucas, 1990; Lucas and Wolf, 1993; McLean *et al.*, 1997) may occur throughout the plant cell wall providing a pathway for transport of certain substances between cells. Plasmodesmata may be aggregated in primary pit fields or in the pit membranes between pit pairs. Viewed with the TEM, plasmodesmata appear as narrow canals (2 µm) lined by a plasma membrane and traversed by a desmotubule, a tubule of endoplasmic reticulum. The functionally diverse plasmodesmata are dynamic, altering their dimensions (Kragler *et al.*, 1998). For example, while some transport endogenous plant transcription factors, others transport numerous proteins from companion cells to sieve elements (see Table 6.1).

PLASMALEMMA

The plasmalemma (Marmé *et al.*, 1982; Sussman and Harper, 1989; Laisson and Moller, 1990) is the outer limiting membrane of the plant cell (Fig. 6.6). An intervening space, the periplasmic space, can occur between the plasmalemma and the cell wall. Paramural bodies, plasmalemmasomes and multivesicular bodies are plasmalemma derivatives and presumably function in solute transport, especially in transfer cells. Contacts between the plasmalemma and cell wall have been reviewed by Kohorn (2000). Functions of the plasmalemma include: mediating transport of substances, coordinating synthesis of cell wall cellulosic microfilaments, and translating hormonal and environmental signals for control of cell growth and differentiation.

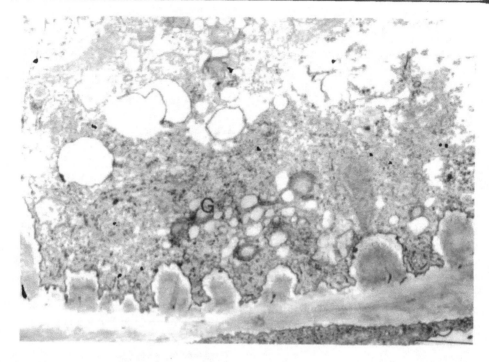

Fig. 6.6 Plasmalemma as illustrated by electron microscopy. (Mayfield and Dashek, 1997). Bar = 1 μm

CYTOPLASM

Plastids

Plant cells abound with various types of plastids (Bogorad and Vasil, 1991), each delimited by an envelope consisting of two membranes. Plastids are often classified on the basis of developmental stage and the pigments that they contain. Chloroplasts (Edelman *et al.*, 1982; Hoober, 1984) are the sites of photosynthesis and their ribosomes can perform protein synthesis (Halliwell, 1984). Chloroplasts, which contain chlorophylls and carotenoids, are disk-shaped and 4-6 μm in diameter (Fig. 6.7). These plastids comprise a ground substance (stroma) and traversed by thylakoids (flattened membranous sacs) which are usually stacked as grana. In addition, the chloroplasts of green algae and plants contain starch grains, small oil droplets and DNA.

Chromoplasts are pigmented plastids which contain carotenoids but lack chlorophyll. These plastids are attractants to insects and animals.

Leucoplasts are nonpigmented plastids which synthesize starch. Some leucoplasts, proteinoplasts, contain protein.

Proplastids are small, colorless to pale green, undifferentiated plastids which occur in the meristematic cells of roots and stems. Proplastids are believed to be precursors of more highly differentiated plastids. Etioplasts are proplastids containing prolamellar bodies and precursors of chloroplasts developmentally arrested by low light levels (Pyke, 1999).

Nucleus

The nucleus (Fig. 6.8) is the most prominent structure within the cytoplasm (Hadjiolov, 1985). It is bounded by a nuclear envelope perforated with circular pores of 30-100 nm in diameter. The outer nuclear envelope may be continuous with the ER. The nucleus contains chromatin (Adolph, 1991) within the nucleoplasm. The chromatin is comprised in part of DNA, the genetic information which regulates cellular activities by determining which proteins are produced and when.

MITOCHONDRIA

Mitochondria (Moore and Beechey, 1987; Levings and Vasil, 1995, Science, Special Issue, 1999; Bremmer and Krohner, 2000)

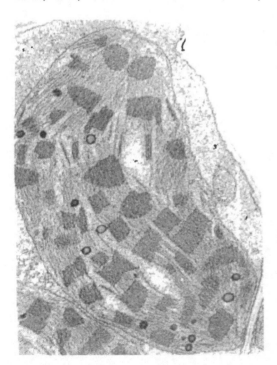

Fig. 6.7 Chloroplast ultrastructure. (Evert, 2005, www.botany.org/bsa/millen) Magnification not known. Chloroplasts 1-10 µm in size

are organelles possessing a double membrane, the inner of which is invaginated as cristae (Fig. 6.9). An intermembranel space exists between the inner and outer membranes. The inner membrane consists of an unusually high amount of protein and possesses spherically shaped particles, ~9 nm in diameter. These particles seem to be equivalent to F_0, F_1 ATPase. In contrast to the inner membrane, the outer membrane is smooth and appears to be attached to the smoother er. This outer membrane is permeable to all molecules of 10,000 MW or less. A mitochondrial matrix is enclosed by the inner membrane and consists of a ground substance of particles, nucleoids, ribosomes and electron transparent regions containing DNA (Mackenzie and McIntosh, 1999).

PROTEIN BODIES

These organelles reside in the endosperm of cereal grains and their structures are tissue specific. They are about 2-5 µm in diameter and often "house" globoid and occasionally crystalloid inclusions. Prolamin accumulates in either small or large spherical bodies. Crystalline protein bodies are the sites of accumulation of nonprolamin storage proteins (Herman and Larkins, 1999).

ENDOPLASMIC RETICULUM (ER)

The ER (Lee and Chen, 1988; Sitia and Meldolesi, 1992; Etkin, 1997) is a complex three-dimensional membrane system (Fig. 6.10). As visualized in the TEM, there are two parallel membranes with an intervening electron transparent space, the lumen. The form and abundance of the ER vary (Staehlein, 1997). The rough ER (RER) appears as flattened sacs with numerous attached ribosomes (15-20 nm). In contrast, the smooth ER (SER) lacks ribosomes. The

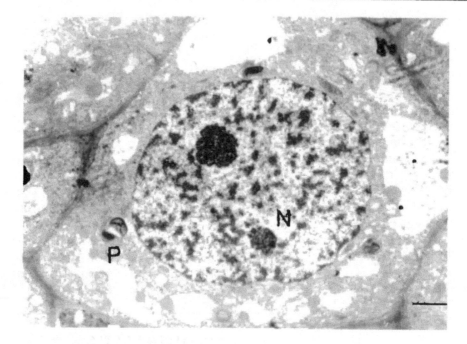

Fig. 6.8 Electron micrograph depicting a nucleus (N). (Mayfield and Dashek, 1997). Bar = 2 μm

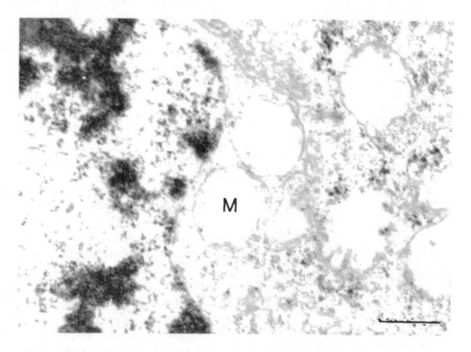

Fig. 6.9 Subcellular morphology of a mitochondrion (M). Mayfield and Dashek, 1997). Bar = 0.5 μm

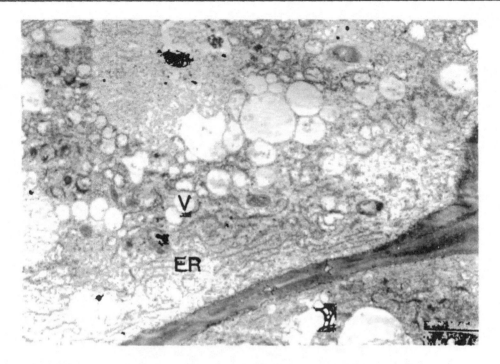

Fig. 6.10 Electron micrograph depicting endoplasmic reticulum (ER). (Mayfield and Dashek, 1997). Bar = 2 μm

ER seems to function as a communication system (Vitale and Derecke, 1999) within cells and can be continuous with the outer nuclear envelope. While the RER is involved in protein synthesis, the SER functions in glycosylation and other protein modifications; e.g. oxidative protein folding (Tu *et al.*, 2000).

GOLGI APPARATUS

Morphology

The Golgi apparatus is basically composed of the Golgi stack (Staehelin and Moore, 1995; Berger and Roth, 1997). The Golgi stacks usually consist of 4–6 flattened single-membrane-enclosed sacs (cisternae), which are parallel to each other (Fig. 6.11a, Kaneko *et al.*, 1994). The number of Golgi

stacks per cell varies greatly among plant cells. These parallely lined cisternae are about 0.2 μm thick in total and the cisternae are about 0.7 μm long as shown in Fig. 6.11b (higher magnifications of G1 and G2 shown in Fig. 6.11a). The Golgi apparatus appears to possess opposite poles, the cis- (forming) and trans- (maturing) faces (Mazzarello and Bentivoglio, 1998). Whereas the cis-membranes are similar to the endoplasmic reticulum (ER), the transmembranes resemble the plasmalemma. In plant cells, the Golgi apparatus is divided into individual Golgi stacks, which are dispersed throughout the cytoplasm, however, in animal cells it is localized in the perinuclear position.

In animal cells, rather than existing as discrete entities in physiological and biochemical isolation from one another, it is

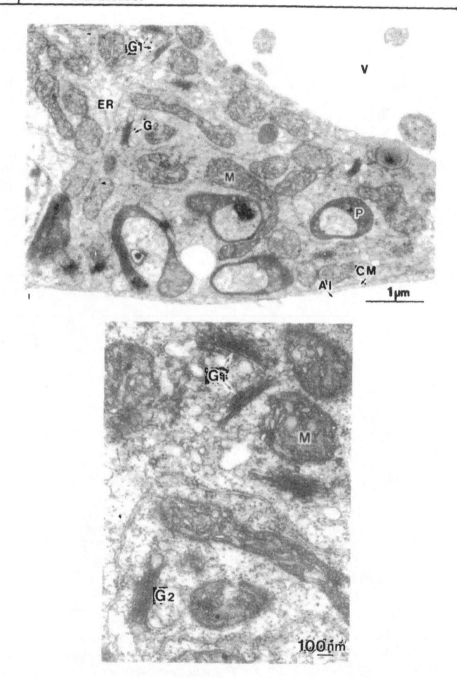

Fig. 6.11a, b (a) Electron micrograph of Golgi apparatus in a tobacco protoplast. The Golgi cistenae are parallel to each other. (b) Higher magnification of Golgi apparatus indicated by G_1 and G_2 in Figure 6.11a. The Golgi stack consists of 4 to 6 flattened cisternae. Al, alginate bead; CM, cell membrane; ER, endoplasmic reticulum; G, Golgi apparatus; M, mitochondrion; P, proplastid; V, vacuole (Kaneko et al. 1994)

now generally accepted that the XER, Golgi apparatus, and cell surface exist as a functional continuum, that is, the endomembrane system. To date, the hypothesis at the heart of the endomembrane concept is that the membranes flow through an interconnected system of the ER, Golgi apparatus, and plasma membrane. Ultrastructural and biochemical evidence exists that the interconnections of these structures with one another involves "vesicular and transitional membrane forms" (see earlier discussion).

However, in plant cells, the continuous close association of the Golgi apparatus with the ER was observed by fluorescence microscopy using green fluorescent protein (GFP)-based techniques and by electron microscopy using zinc iodide/osmium tetroxide fixation to selectively stain the ER and the Golgi apparatus. Between the Golgi apparatus and ER no vesicular membrane forms were observed on zinc iodide-and osmium tetroxide-impregnated cells (Fig. 6.12; Kaneko *et al.*, 2000). In conventional cross-sectional observations, direct connections between the ER and Golgi cisternae toward the cis face of the stack were observed, which may be permanent or transient.

Functions

Functions of the Golgi apparatus are secretion, cell wall synthesis, transport of glycoproteins, and transformation of ER-like membranes into plasma membrane-like membranes.

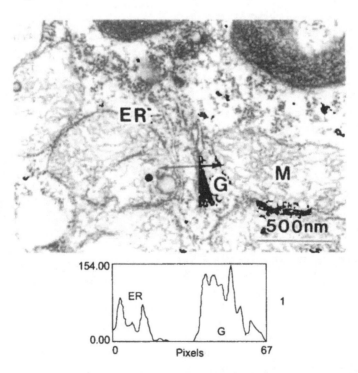

Fig. 6.12 Electron micrograph of the ER and the Golgi apparatus stained by the ZIO method. Line profile through the Golgi cisternae and ER. ER, endoplasmic reticulum; G, Golgi apparatus; M, mitochondrion (Kaneko *et al.* 2000)

Transport of Proteins from ER to Golgi Apparatus

In animal cells, proteins destined for secretion from the ER to the Golgi apparatus are transported as follows: Proteins are first packaged into coat protein (COP) II-coated transport vesicles. A protein coat called COPII on the membrane of ER pinches off to form the vesicles. These vesicles contain a separate group of proteins collectively called SNAREs[a] (vesicle-associated SNAREs), which target and fuse with the target membrane (the membrane of the Golgi apparatus). Syntaxin 5, membrin, and rbet1 are member proteins of SNAREs of these vesicles (Fig. 6.13). Rab proteins, members of the Ras-like guanosine triphosphatase (GTPase), such as Rab1 and tethers, a set of nonconserved proteins such as p115, work together during docking of transport vesicles into target membrane. One possibility for explaining the contribution of the membrane of the Golgi apparatus to the targeting/fusion process is that the vesicle-associated SNAREs and the SNAREs in the target membrane assemble into a four-α-helix bundle, which either directly or indirectly results in membrane fusion (Fig. 6.13).

In plant cells, a model has been proposed for the transport of proteins from the ER to the Golgi apparatus, which differs from that in animal cells on the basis of recent findings obtained using the GFP-expression technique and transmission electron microscopy (TEM). The reason for this difference is that in plant cells, the ER-to-Golgi protein transport may follow a simpler route than in animal cells (Zhang and Staehlein, 1992). In a plant cell, no structures equivalent to the vesicular-tubular clusters (VTCs) or the ER-Golgi intermediate compartment (ERGIC) of animal cells have been identified. It has been suggested that in plant cells direct connections in the form of tubular extensions between the ER and the cis-face of Golgi stacks may be substituted for the vesicles that mediate protein transport between the two organelles in animal cells. An *in vivo* imaging of leaf tissue revealed an intimate relationship between the ER and the Golgi stacks in these cells. Golgi stacks were observed to move rapidly and extensively along the ER, with an actomyosin system moving toward the Golgi apparatus stacks along the ER network. Direct connections between ER and Golgi cisternae toward the cis-face of the stacks observed by electron microscopy suggest that protein traffic between these two organelles may not require vesicle carriers. Although the components of COPII-dependent traffic machinery were identified in plants by EST database searches and by biochemical research, it is likely that COPII components simply determine the site of formation of ER-to-cis-Golgi connections, and that vesicle vectors are not required for transport of proteins to the Golgi apparatus in plant cells.

Screening of the *Arabidopsis* genome for sequences related to key players of intracellular trafficking in nonplant organisms suggests conservation of key regulators of vesicle trafficking. However, in the case of protein families, various expressions of subfamilies over the course of plant evolution point to possibly divergent roles of individual members.

Transport of Proteins out of Golgi Apparatus

In nondividing plant cells, the destination for proteins existing in the Golgi apparatus is either the plasma membrane/cell wall

---

[a]SNAREs: SNAP receptors: soluble NSF attachment protein NSF: N-[ethylmaleimide]-sensitive factor.

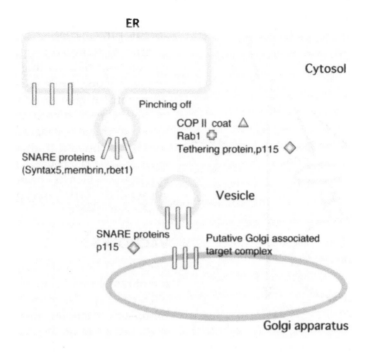

Fig. 6.13 Transport of proteins from the ER to the Golgi apparatus in an animal cell. Proteins are transported from the ER to the Golgi apparatus in vesicles which bud from the membrane of the ER. ER, endoplasmic reticulum, Syntax 5; orange, membrin; green, rbet1; purple green

(secretion) or the vacuolar system. Here the mechanism of sorting proteins to the plasma membrane/cell wall (secretion) or the vacuolar system (Hofte and Chrispeels, 1992) is the focus of discussion. In mammalian cells, the trans Golgi network (TGN) is the place of sorting of proteins destined for the lysosome, the plasmamembrane or the extracellular space. (Fig. 6.14). Whether or not the term "trans Golgi network" (TGN) from the mammalian "endomembrane nomenclature" commonly used in plant literature is adequate for the place of sorting has been discussed. In plant cells, no discrete protein-sorting compartments with a characteristic set of proteins downstream from the transface of the Golgi apparatus has yet been described. Whether the transface of the Golgi apparatus in plant cells is

homologous to the mammalian TGN must be determined.

Secretion: transport towards plasma membrane and cell wall. Soluble secretory proteins carry no positive targeting information in plant cells. It has been suggested that the default destination of soluble proteins is the plasma membrane.

Transport toward the vacuolar system: various types of vacuole, vesicle, sorting signal, receptor and sorting site.

In plants, various peptides as sorting signals are required for transport of proteins to either of the two main vacuolar systems. These systems are distinguishable by their different sets of tonoplast intrinsic proteins (TIPs) and lumenal contents. Sorting toward either of the two relies on different

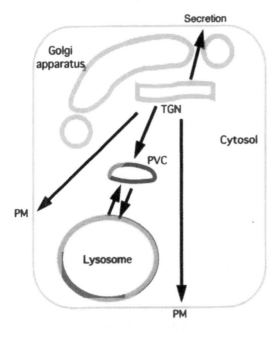

Fig. 6.14 Transport of proteins from the Golgi apparatus in an animal cell. Proteins are sorted in the trans-Golgi network and transported in vesicles to their final destinations; the plasmamembrane, the lysosome or the extracellular space. TGN, trans Golgi network; PVC, prevacuolar compartment; PM, plasma membrane

peptide targeting signals and is mediated by different sets of transport vesicles.

The lytic vacuole is characterized by the presence of γ-TIPs. It was reported that proteins destined to the lytic vacuole are transported via an intermediate compartment called the prevacuolar compartment (PVC) and that the transport of proteins depends on amino-terminal sequence-specific propeptides (NPIR or equivalent) recognized by vacuolar sorting receptors (VSRs). The first VSR identified was BP8 in *Pisum sativum*. It is likely that packaging of cargo destined to the lytic vacuole in transface of the Golgi apparatus-located clathrin-coated vesicles. Proteins destined to the protein storage vacuole (α-TIP vacu-

oles) are delivered in vesicles without any specific protein coating. Sorting depends on a carboxyl-terminal propeptide and an as yet unidentified sorting receptor. Other storage vacuole proteins as well as integral membrane proteins are sorted even earlier, at the ER level, where they are packaged into so-called precursor-accumulating (PAC) vesicles and transported to the storage vacuole via a route bypassing the Golgi apparatus, which has been characterized by a vacuolar sorting receptor (Barieu and Chrispeels, 1999).

VACUOLES

Vacuoles (Wink, 1993; Leigh *et al.*, 1997) are membrane-bound regions of the cell filled with cell sap. They are surrounded by a tonoplast (vacuolar membrane) (Marmé *et al.*, 1982) and can be diverse with distinct functions. Currently, most investigators believe that lysosomes and plant vacuoles are identical. Vacuoles can develop turgor pressure and maintain tissue rigidity. They are storage components for various metabolites such as reserve proteins in seeds and malic acid in CAM plants. Concomitantly, vacuoles can remove toxic secondary products and are sites of pigment deposition (Marty, 1999). A recently published review (Deepesh, 2000) provides a contemporary introduction to vacuoles.

RIBOSOMES

Ribosomes (Spirin, 1986; Zimmerman, 1995) are small organelles, 17-23 nm in diameter (see Fig. 6.9). They can exist either as clusters known as polysomes or attached to the ER where they bind to pores in the ER membrane. A major constituent of the ER pore is the translocon complex, a protein complex which binds to the 80s ribosomes. Ribosomes consist of subunits, a 30s subunit (16s rRNA and 21 proteins) and a 50s subunit

(23s and 5s RNAs, proteins and the catalytic site of peptidyl transferase). Ribosomes are the sites of protein synthesis. Very recently, the ribosome has been proposed to be a ribozyme, an RNA molecule (Cech, 2000)

CYTOSKELETON

The cytoskeleton is a complex network of protein filaments. Microtubules (Saifer, 1986; Hyans and Lyod, 1994) are long cylindrical structures of ~24 nm in diameter and varying lengths. Each microtubule (0.5-1.0 μm in diameter and 25 μm in internal diameter) is composed of tubulin subunits. The assembly of microtubules occurs at nucleating sites or a microtubule organizing center. During polymerization of tubulin, guanosine triphosphate (GTP) is hydrolyzed to GDP.

Microfilaments (5-7 nm in diameter) are composed of filamentous actin (Volkman and Baluska, 1999). In contrast, microtubules structures are formed by α, β-tubulin heterodimer (Fosket and Morejohn, 1992). The wall is composed of 13 parallel protofilaments. Various microtubule-associated proteins and motor proteins (kinesin and dynein) are bound to the wall (Vallee, 1991). The microtubule is a polar structure, i.e., plus and minus ends.

The functions of microtubules are: moving vesicles toward the developing wall, comprising spindle fibers and cell plate formation. In addition, microtubules may be involved in cell wall deposition, tip growth (Heath, 1990; Geitmann et al., 2000), nuclear migration, and cytoplasmic streaming. Quader (1998) has updated our current knowledge of microtubules as components of the cytoskeleton. Although the volume written by Gavin (2000) draws upon protozoan and animal systems, it contains molecular, microscopic, and genetic methods and protocols which may be applicable to investigating plant, cytoskeletal proteins.

MICROBODIES

These are spherical organelles (0.1 – 2.0 μm in diameter) bounded by a single membrane. They possess a granular interior and sometimes a crystalline protein body. A specialized type of microbody is the glyoxysome (0.5-1.5 μm) containing enzymes of the glyoxylate cycle. Glyoxysomes (Kindle and Lazarow, 1982) are found in either the endosperm or cotyledons of oily or fatty seeds. Plant peroxisomes are often found in association with chloroplasts and may detoxify certain photosynthetic by-products (Maseuth, 1991). Molecular regulation of their functions has been discussed by Hayaski (2000). Electron micrographs of plant organelles have appeared in a number of textbooks (Maseuth, 1991; Raven et al., 1992; Moore and Clark, 1995) and the original research literature (see references in Plant Cell, Special Issue, 1999). Lastly, marked advances concerning how water and mineral elements are transported across cell membranes have been made (Chrispeels et al., 1999) These investigators discovered that aquaporins transport water and mineral elements (Maurel and Chrispeels, 2001).

Plant Cells, Tissues, and Organs

Several compilations of structure-function relationships for plant cells, tissue and organs have been published. The student is referred to Steeves and Sussex (1989), Fahn (1990), Maseuth (1991), Raven et al., (1992), Fosket (1994), Moore and Clark (1995) for in-depth discussions of the morphology and physiology of plant cells (Table 6.2), tissues (Table 6.3) and organs. Finally, Table 6.4 in Dashek (2000) presents a summary of contemporary light and electron microscopies available for characterizing plant cells, tissues and organs. These should be learned if

TABLE 6.2 Structure and function of plant cells[a, b]

Cell type	Structure	Function
Parenchyma	Isodiametric, thin-walled primary cell wall; in some instances can have secondary cell walls; not highly differentiated	Photosynthesis, secretion, organic nutrient, and water storage, regeneration as in wound healing
Transfer cells	Specialized parenchyma cells, plasmalemma greatly extended, irregular extensions of cell wall into protoplasm	Transfer dissolved substances between adjacent cells, presence correlated with internal solute flux
Chlorenchyma	Parenchyma cells containing chloroplasts	Photosynthesis
Collenchyma	Rectangular in longitudinal section, retain a protoplast at maturing, primary wall unevenly thickened, nonlignified, highly hydrated	Provide support for growing and mature organs, aerial portions of the plant only
Sclerenchyma	Thick cell walls containing lignin-nonextensible secondary walls, lack protoplasts at maturity, living or dead at functional maturity	Strength and supporting elements of plant parts, scattered throughout plant (also see conducting tissue)
Epidermal cells	Tabular; are layered sheets on surface of leaves and young roots, stems, flowers, fruits, seeds, ovules	Secrete a fatty substance, cutin, which forms a protective layer, the cuticle; cuticle covered by an epicuticular wax
Guard cells	Specialized epidermal cells, crescent-shaped, contain chloroplasts, form defines stomatal pore	Regulate stomatal aperture/pore for gas exchange
Subsidiary cells	Surround guard cells of stomata	Reservoirs for water and ions
Trichomes	Single-celled or multicellular outgrowths of epidermal cells	Produce volatile oils for glandular trichomes; reflect bright light in some desert plants; increase boundary layer; mechanically discourage predators; salt secretion in some halophytes; digestion in sundews
Sieve tube elements (sieve tube is a vertical row of sieve tube elements)	Vertical row of elongated, specialized cells; possess multiperforated end walls or sieve plates; possess living protoplasts at maturity but lack nuclei; young sieve element contains one or more vacuoles separated by a tonoplast; at maturity tonoplast disappears; protoplasts of sieve-tube members of dicots and some monocots contain P-protein	
Albuminous cells	Specialized parenchyma cells associated with sieve cells in gymnosperms	May function like companion cells
Companion cells	Specialized parenchyma cells; possess numerous plasmodesmatal connections with sieve-tube members	May play a role in delivery of substances to sieve-tube members

[a]Summarized from: Maseuth (1991) and Moore and Clark (1995).
[b]Tables 6.2 and 6.3 were taken from Dashek (2000).

TABLE 6.3 Summary of plant tissues

Tissue type	Composition	Function
Meristematic	Meristematic cells	Cell division and growth
		Increase in length
		Increase in girth
Apical or primary	Meristematic cells	Produce primary body of root-and-shoot system
Lateral or secondary	Fusiform, ray initials	Produce secondary xylem and secondary phloem in woody plants
Protective or dermal		Mechanical protection
		Restriction of transpiration aeration
Epidermal tissue of leaves	Epidermal cells with and without modifications	Mechanical protection
		Restriction of transpiration aeration
Peridermal tissue of stems and leaves	Cork cells and cork cambium	Mechanical protection
		Restriction of transpiration aeration
Fundamental or ground		Production and storage of food; wound healing
Parenchyma	Parenchyma cells	
Chlorenchyma	Chlorenchyma cells	Photosynthesis
Collenchyma	Collenchyma cells	Flexible/extensible support for growing primary tissues
Sclerenchyma	Sclerenchyma cells	Mechanical support
Conductive tissue[a]		
Xylem	Tracheids, vessel elements,	Conduction of water and minerals;
Phloem	sieve cells, sieve tubes	translocation of organic nutrients

See Table 6.2 for references.
[a]See Figs. 6.15 and 6.17

the student is considering a career in plant cell biology.

SHOOT DEVELOPMENT

The shoot apex of the young angiosperm stem consists of an apical meristem (Lyndon, 1998), i.e., a hemispherical or conical-shaped tissue comprising meristamatic cells (Fig. 6.15). Both the size and shape of the meristem vary from species to species. The stem apex gives rise to differentiated stem tissue as well as leaf and lateral bud primordia. The tunica-corpus of the apex (Fig. 6.16) is composed of several discrete cell layers. The outer tunica contains one to four cell layers and covers the corpus, an unlayered core of cells. The corpus and the inner tunica layer provide the majority of the stem tissues.

ROOT DEVELOPMENT

Roots increase their length solely at their tips via the addition of new cells by the apical meristem and subsequent cell elongation and differentiation. In contrast to shoot apical meristems, root apical meristems do not produce leaf primordia. However, they manufacture a rootcap (Fig. 6.17). The root is often divided into zones, i.e. rootcap, meristem, elongation region (1,500 to 5,200 μm behind the apex), and zone of differentiation. Three concentric tissues—the protoderm, ground meristem and procambium give rise to the root structure.

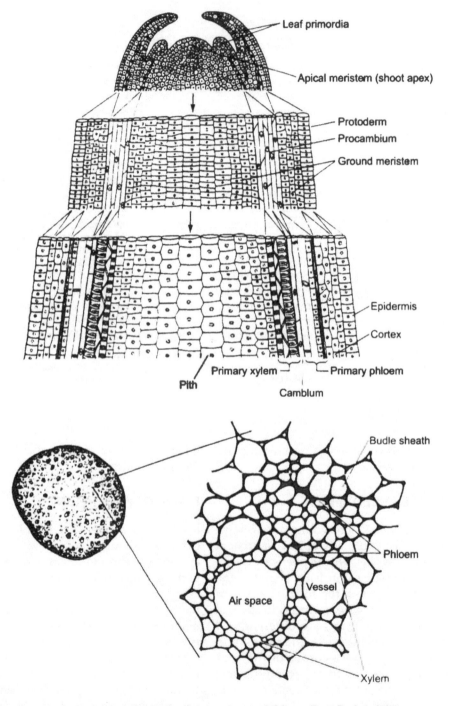

Fig. 6.15 Diagram of shoot. Note: Conductive tissue, xylem and phloem, (from Fosket, 1994)

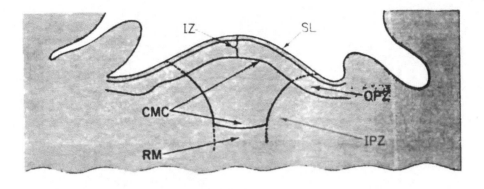

Fig. 6.16 Shoot apex organization, SL = tunica (Steves and Sussex, 1989)

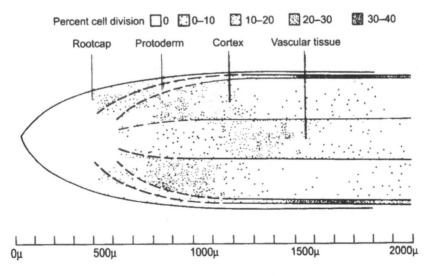

Fig. 6.17 Diagram of root organization (Steves and Sussex, 1989)

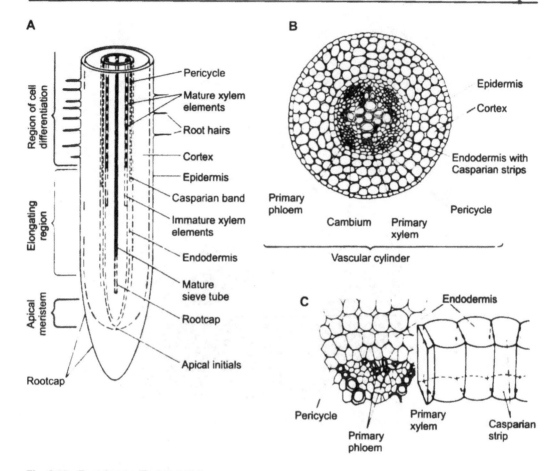

Fig. 6.18 Root tissues (Fosket, 1994)

Whereas the protoderm differentiates into the epidermis and lateral rootcap, the procambium and ground meristem yield the vascular and cortical tissues (Howell, 1998). The vascular tissue forms a cylindrical central core (the stele) and is surrounded by the pericycle, endodermis, cortex and epidermis (Fig. 6.18).

Altman and Waisel (1997) can be consulted for hormonal and molecular aspects of root development. Finally, the student is referred to Fosket (1994) and Howell (1998) for the molecular aspects of shoot and root development.

References

Adolph, K.W. 1991. *Advanced Technique in Chromosome Research*. Marcel Dekker, New York, NY.

Ahmed, S.U., Rojo, E., Kovaleva, V., Venkataraman, S., Dombrowski, J.E., Matsuoka, K. and Reikhel, N.V. 2000. The plant vacuolar sorting receptor AtELP is involved in transport of NH(2)-terminal propeptide-containing vacuolar proteins in *Arabidopsis thaliana*. JCB 149: 1335-1344.

Alberts, B., Johnson, A., Lewis, J., Raff, M., Roberts, K. and Walter, P. 2002. *Intracellular Vesicular Traffic, Molecular Biology of the Cell*, pp. 711-766. Garland Science, Taylor & Francis Group, New York, NY (4th ed).

Altman, A. and Waisal, Y. 1997. *Biology of Root Formation and Development*. Plenum Press, New York, NY.

Bacic, A., Harris, P.J. and Stone, B.A. 1988. Structure and functions of plant cell walls. In: *The Biochemistry of Plants*, vol. 14. J. Preiss (ed.). Academic Press, New York, NY.

Barieu, F. and Chrispeels, M.J. 1999. Delivery of a secreted soluble protein to the vacuole via a membrane anchor. Plant Physiol. 120: 961-968.

Berger, E.G. and Roth, J. (eds.) 1997. *The Golgi Apparatus*. Birkhauser-Verlag, Basel, Switzerland.

Boevink, P., Oparka, K., Santa-Cruz, S., Martin, B., Betteridge, A. and Hawes, C. 1998. Stacks on tracks: the plant Golgi apparatus traffics on an actin/ER network. Plant Cell 15: 441-447.

Bogorad, L. and Vasil, I.K. 1991. *The Molecular Biology of Plastids*. Academic Press, San Diego, CA.

Brandizzi, F., Snapp, E.L., Roberts, A.G., Lippincott-Schwartz, J. and Hawes, C. 2002. Membrane protein transport between the endoplasmic reticulum and the Golgi in tobacco leaves is energy dependent but cytoskeleton independent: Evidence from selective photobleaching. Plant Cell 14: 1293-1309.

Bremmer, C. and Krohner, G.. 2000. Mitochondria. The death signal integrators. Science 289: 1150-1151.

Brittle, E. and Waters, M.G. 2000. ER-to-Golgi traffic—This bud's for you. Science 289: 403-404.

Carpita, N.C. and Gibeaut, D.M. 1993. Structural models of primary cell walls in flowering plants: consistency of molecular structure with the physical properties of walls during growth. Plant Journal 3: 1-30.

Carpita, N.C., McCann, M. and Giffing, L.R. 1996. The plant cell extracellular matrix. News from the cell's frontier. Plant Cell 8: 1451-1463.

Cassab, G.J. 1998. Plant cell wall proteins. Ann. Rev. Pl. Physiol. Molec. Biol. 49: 281-309.

Cech, T.R. 2000. The ribosome is a ribozyme. Science 289: 878-879.

Cheung, A.L., Chen, C.Y-h., Glaven, R.H., de Graaf B.H.J., Vidali,, L, Hepler., P.K. and Wu, H. 2002. Rab2 GTPase regulates trafficking between the endoplasmic reticulum and the Golgi bodies and is important for pollen tube growth. Plant Cell. 14: 945-962.

Chrispeels, M.J. and Crawford, N.M. and Schroeder, J.D. 1999. Proteins for transport of water and mineral nutrients across the membranes of plant cells. Plant Cell 11: 661-675.

Cooper, G.M. 1996. Protein sorting and transport. The endoplasmic reticulum, Golgi apparatus and lysosomes. In: *The Cell. A Molecular Approach*, pp.347-387. ASM Press, Washington, DC.

Cosgrove, D.J. 2001. Wall structure and wall loosening. Plant Physiol. 125: 131-134.

Dashek, W.V. 2000. Plant cells and tissues. In: *Methods in Plant Electron Microscopy and Cytochemistry*, pp. 1-25. Dashek, W.V. (ed.). Humana Press Inc., Totowa, NJ, USA.

Deepesh, N. De. 2000. *Plant Cell Vacuoles. An Introduction*. SCIRO, Collingwood, Victoria, Australia.

Edelman, M., Hallick, R.P. and Chua, N.H. 1982. *Methods in Chloroplast Molecular Biology*. Elsevier Biomedical Press, New York, NY.

Engelman, D.M. 1996. Crossing the hydrophobic barrier. Insertion of membrane proteins. Science 274: 1850-1851.

Etkin, L.D. 1997. A new face for the endoplasmic reticulum: RNA localization. Science 276: 1092-1093.

Fahn, A. 1990. *Plant Anatomy*. Pergamon Press, Elmsford, New York, NY.

Fleming, A.J., McQueen-Mason, S., Mandel, T. and Kuhleimer, C. 1997. Induction of leaf primordia by the cell wall protein expansin. Science 276: 1415-1417.

Fosket, D.E. and Morejohn, L.C. 1992. Structure and function of tubulin. Ann. Rev. Pl. Physiol. Pl. Molec. Biol. 43: 201-240.

Fosket, D.E. 1994. *Plant Growth and Development. A Molecular Approach*. Academic Press, New York, NY.

Gavin, R.H. 2000. *Cytoskeleton. Methods and Protocols*. Humana Press, NJ, USA.

Geitmann, A., Cresti, M. and Heath, I.B. 2000. *Cell Biology of Plant and Fungal Tip Growth*. IOS Press, Amsterdam, The Netherlands.

Hadjiolov, A.A. 1985. *The Nucleus and Ribosome Biogenesis*. Springer-Verlag, New York, NY.

Halliwell, B. 1984. *Chloroplast Metabolism. The Structure and Function of Chloroplasts in Green Leaf Cells*. Clarendon Press, New York, NY.

Hayaski, M. 2000. Plant Peroxisomes. Molecular basis of the regulation of their function. J. Plant Res. 113: 103-109.

Heath, I.B. 1990. *Tip Growth in Plant and Fungal Cells*. Academic Press, San Diego, CA.

Herman, E.M. and Larkins, B.A. 1999. Protein storage bodies and vacuoles. Plant Cell 11: 601-613.

Highley, T.L. and Dashek, W.V. 1998. Biotechnology in the study of brown- and white-rot decay. In: *Forest Products Biotechnology*. Bruce, A. and Palfreyman, J. (eds.). Taylor and Francis, London, England.

Hinz, G., Hillmer, S., Bäumer, M. and Hohl, I. 1999. Vacuolar storage proteins and the putative vacuolar sorting receptor BP-80 exit the Golgi apparatus of developing pea cotyledons in different transport vesicles. Plant Cell. 11: 1509-1524.

Hofte, H. and Chrispeels, M.J. 1992. Protein sorting to the vacuolar membrane. Plant Cell 4: 995-1004.

Hoober, J.K. 1984. *Chloroplasts*. Plenum Press, New York, NY.

Howell, S.H. 1998. *Molecular Genetics of Plant Development*. Cambridge Univ. Press, Cambridge, England.

Hyans, J.S. and Lyod, C.W. 1994. *Microtubules*. John Wiley, New York, NY.

Jacobson, K., Sheets, E.D. and Simson, R. 1995. Revisiting the fluid mosaic model of membranes. Science 268: 1441-1442.

Jordan, E.G. and Collins, C.A. 1982. *The Nucleolus*. Cambridge Univ. Press, New York, NY.

Jurgens, G. and Geldner, N. 2002. Protein secretion in plants: from the trans-Golgi network to the outer space. Traffic 3: 605-613.

Kaneko, T.S., Sato, M. and Osumi, M. 2000. Visualization of Golgi apparatus by zinc iodide-osmium tetroxide (ZIO) staining. In: *Methods in Plant Electron Microscopy and Cytochemistry*, pp. 223-235. Dashek, W.V. (ed.). Humana Press, Totowa, NJ, USA.

Kaneko, T.S., Watanabe, R., Sato M., Osumi, M. and Takatuki, A. 1994. Morphological changes in Golgi apparatus induced by Brefeldin A on cell wall regeneration in tobacco protoplasts. Plant Morph. 6: 1-11.

Kindl, H. and Lazarow, P.B. 1982. *Peroxisomes and Glyoxysomes*. NY Acad. Sci., New York, NY.

Kohorn, B.D. 2000. Plasma membrane cell wall contacts. Plant Physiol. 124: 31-38.

Kragler, F., Lucas, W.J. and Marzer, J. 1998. Plasmodesmata: dynamics, domains and patterning. Ann. Bot. 81: 1-10.

Laisson, C. and Moller, I.M. 1990. *The Plant Plasma Membrane: Structure, Function and Molecular Biology*. Springer-Verlag, Berlin, Germany.

Lee, C. and Chen, L.B. 1988. Dynamic behavior of endoplasmic reticulum in living cells. Cell 57: 37-46.

Leigh, R.A., Sanders, D. and Callow, J.A. 1997. *The Plant Vacuole. Advances in Botanical Research*. Acad. Press, San Diego, CA.

Leshem, Y.Y., Shewfelt, R.L., Willner, C.M. and Pantojoa, O. 1991. *Plant Membranes: A Biological Approach to Structural Development and Science*. Elsevier, New York, NY.

Levings, CS. III and Vasil, I.K. 1995. *The Molecular Biology of Plant Mitochondria. Advances in Cellular and Molecular Biology of Plants*, vol. 3. Kluwer Acad., Norwell, MA, USA.

Lucas, W.J. and Wolf, S. 1993. Plasmodesmata the intracellular organelles of green plants. Trends Cell Biol. 3: 308-315.

Lyndon, R.F. 1998. *Tle Shoot Apical Meustem: Its Growth and Development*. Cambridge University Press, Cambridge, UK.

Mackenzie, S. and L. McIntosh. 1999. Higher plant mitochondria. Plant Cell 11: 571-585.

Marmé, D., Marre, E. and Hertel, R. 1982. *Plasmalemma and Tonoplasts. Their Functions in the Plant Cell*. Elsevier Biomedical, Amsterdam, The Netherlands.

Marty, F. 1999. Plant vacuoles. Plant Cell 11: 587-599.

Maseuth, J.D. 1991. *Botany. An Introduction to Plant Biology*. Saunders College Publ., Philadelphia, PA, USA.

Maurel, C. and Chrispeels, M.J. 2001. Aquaporins: A molecular entree into plant water relations. Plant Physiol. 125: 135-138.

Mayfield, J. and Dashek, W.V. 1997. Methods for analysis of plant cell and tissue ultrastructure: *Methods in Plant Biochemistry and Molecular Biology*. pp. 3-11. Dashek, W.V. (ed.). Humana Press, Boca Raton, FL, USA.

Mazzarello, P. and Bentivoglio, M. 1998. The centenarian Golgi apparatus. Nature 392: 543-544.

McLean, B.G., Hempel, F.D. and Zambryski, P.C. 1997. Plant intracellular communication via plasmodesmata. Plant Cell 9: 1043-1054.

Moore, A.L. and Beechey, R.B. 1987. *Plant Mitochondria: Structural, Functional and Physiological Aspects*. Plenum Press, New York, NY.

Moore, R. and Clark, W.D. 1995. *Botany. Plant Form and Function*. Wm. C. Brown Publ., Dubuque, IA, USA.

Morré, D.J. 1990. Endomembrane system of plants and fungi. In: *Tip Growth in Plant and Fungal Systems*. Heath, B. (ed.). Academic Press, San Diego, CA.

Morré, D.J. and Keenan, T.W. 1997. Membrane flow revisited. What pathways are followed by membrane molecules moving through the Golgi apparatus ? Bio Sci. 47: 489-498.

Neumann, U., Brandizzi, F. and Hawes, C. 2003. Protein transport in plant cells: in and out of the Golgi. Ann. Bot. 92: 167-180.

Phillipson, B.A., Pimple P., Lamberti Pinto daSilva L., Crofts, A. J., Taylor, J.P., Movafeghi, A., Robinson, D.G. and Denecks, J. 2001. Secretory bulk flow of soluble proteins is efficient and COPII dependent. Plant Cell 13: 2005-2020.

Pimple, P. and Denecke, J. 2002. Protein-protein interactions in the secretory pathway, a growing demand for experimental approaches in vivo. Plant Molec. Biol. 50: 887-902.

Plant Cells and Organelles Special Issue (1999) Plant Cell 11: 507-761.

Pyke, K.A. 1999. Plastid division and development. Plant Cell 11: 549-556.

Quader, H. 1998. Cytoskeleton: microtubules. In: *Progress Botany*. Esser, K., Kadereit, J.W., Luttge, U., Rurge, M. (eds.). Springer Verlag, Berlin, Germany.

Raven, P.l.N., Evert, R.F., and Eichhorn, S.E. 1992. *Biology of Plants*. Worth Publ., New York, NY.

Robards, A.W. and Lucas, W.J. 1990. Plasmodesmata. Ann Rev. Pl. Physiol. Molec. Biol. 41: 369-419

Saifer, D. 1986. *Dynamic Aspects of Microtubules Biology*. NY Acad. Sci., New York, NY.

Sanderfoot, A. A., Ahmed, S. U., Marty-Mazars, D., Rapoport, I., Kirchhausen, T., Marty, F. and Raikhel, N.V. 1998. A putative vacuolar cargo receptor partially colocalizes with AtPEP12p on a prevacuolar compartment in *Arabidopsis* roots. Proc. Natl. Acad. Sci. USA 95: 9920-10025.

Science Special Issue (1999). Mitochondria. 283: 1475-1497.

Singer, S.J. and Nicolson, G.L. 1972. The fluid mosaic model of the structure of cell membranes. Science 175: 720-731.

Sitia, R. and Meldolesi, J. 1992. Endoplasmic reticulum: A dynamic patchwork of subregions. Molec. Biol. Cell 3: 1067-1072.

Smallwood, M., Knox, J.P. and Bowles, D.J. 1996. *Membrane Specialized Functions in Plants*. Bios, Scientific Publ., Oxford, UK.

Spirin, A.S. 1986. *Ribosomes Structure and Protein Biosynthesis*. Benjamin Cummings Publ. Co., Menlo Park, CA.

Staehelin, L.A. 1997. The plant ER: A dynamic organelle composed of a large number of discrete functional domains. Plant J. 11: 1151-1165.

Staehelin, L.A., and Moore, I. 1995. The plant Golgi apparatus: structure, functional organization and trafficking mechanisms. Ann. Rev. Plant Physiol. Plant Molec. Biol. 46: 261-288.

Steeves, T.A. and Sussex, I.M. 1989. *Patterns in Plant Development. A Molecular Approach*. Academic Press, New York, NY.

Sussman, M.R. and Harper, J.F. 1989. Molecular biology of the plasma membrane of higher plants. Plant Cell 1: 953-960.

Tu, B.P., Ho-Schleyer, S.C., Travers, K.J. and Weissman, J.S. 2000. Biochemical basis of oxidative protein folding in the endoplasmic reticulum. Science 290: 1571-1574.

Vallee, R.B. 1991. *Molecular Motors and the Cytoskelton*. Acad. Press, San Diego, CA.

Vitale, A. and Derecke, J. 1999. The endoplasmic reticulum—gateway of the secretory pathway. Plant Cell 11: 615-628.

Vitale, A. and Galili, G. 2001. The endomembrane system and the problem of protein sorting. Plant Physiol. 125: 115-118.

Volkman, D. and Baluska, F. 1999. Actin cytoskeleton in plants: From transport network to signaling networks. Microsc. Res. Tech. 47: 135-154.

Wink, M. 1993. The plant vacuole: A multifunctional compartment. J. Exper. Bot. 44: 231-246.

Zhang, G.F. and Staehelin, L.A. 1992. Functional compartmentation of the Golgi apparatus in plant cells. Plant Physiol. 99: 1070-1083.

Zimmerman, R.A. 1995. Ins and outs of the ribosome. Nature 376: 391-392.

CHAPTER 7

Movement of Molecules Across Membranes

Susanna Malmström

INTRODUCTION

In the plant cell, molecules and ions are constantly moved in and out and between compartments across membranes in order to sustain regular physiological functions and to respond to environmental stress. These movements are called **transport** and are usually accomplished by specific transport proteins embedded in the membrane.

In this chapter, we examine the basic concepts of plant membrane transport and learn about the different transport proteins, how they work and where they are located in the cell. We also look at some of the common research methods for the study of transport proteins and processes.

BASIC CONCEPTS OF MEMBRANES, PERMEABILITY, AND TRANSPORT

Ion Concentrations and Homeostasis in the Plant Cell

For normal growth and development, plants are dependent on the light energy of the sun, and on water and mineral nutrients taken up from the soil by their roots. With these three **essential elements** present, plants are able to synthesize all other molecules they need in order to grow and to reproduce themselves. An essential element is defined as one that has a clear physiological role and whose absence prevents a plant from completing a life cycle. Hydrogen, carbon, and oxygen are obtained from water and carbon dioxide (Fig. 7.1).

The mineral nutrients obtained from the soil which plants require are traditionally listed as macro- or micronutrients, depending on their relative concentration in plant tissue (Table 7.1). As the name infers, macronutrients are needed in higher amounts than micronutrients. The reason is that macro- and micronutrients have different functions in the plant. In general, macronutrients are constituents of organic compounds (proteins, nucleic acids) or are important in the regulation of cellular osmosis (e.g. K^+). Micronutrients, on the other hand, are mostly essential components of enzymes and so the amount required by the plant is much smaller (Marschner, 1995). Macronutrients such as nitrogen or sulfur exist as inorganic ions and as such they are transported into the cell (SO_4^{2-}, NO_3^-, NH_4^+, ...). Others, e.g. potassium or calcium, often remain as simple ions (K^+, Ca^{2+}...).

Inside the individual plant cell, the predominant cation is K^+, with a cytoplasmic concentration of between 60 and 150 mM (Leigh and Wyn-Jones, 1984; Walker *et al.*, 1996; Fig. 7.1). The cell is always trying to maintain the internal "normal" ionic concentrations (i.e. "at rest") so that usual physiological functions can take place. Ionic concentrations are in dynamic equilibrium, that is, they can fluctuate but the cell then regulates the level of the ion back to resting values. This balance of salts and water is called **homeostasis**.

All the essential nutrients have to be taken up by the roots and then distributed throughout the plant and inside the cells. Carbohydrates produced in the photosynthetic reactions have to be transported from the leaves to the other parts of the plant. In addition, toxic compounds may be taken up if present in the soil, or produced by the plant, and they must be exported out from the cell or sequestered in the vacuole. *All these processes involve transport of ions and solutes at the cellular level across membranes by specific membrane transport proteins.*

Limited Permeability and Semipermeability, Diffusion

Biological membranes form barriers between cells and between intracellular compartments. Every living cell in the plant is surrounded by the plasma membrane and inside the cell, the various organelles are

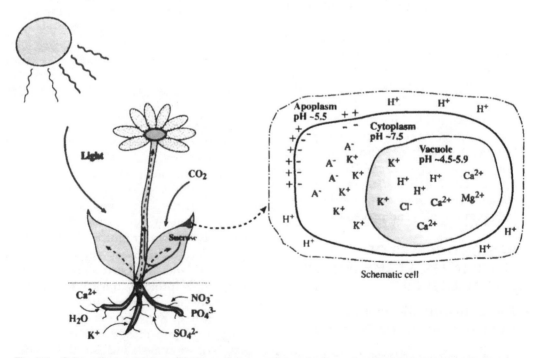

Fig. 7.1 Schematic presentation of the essential elements required by a plant (left) and of a plant cell, showing the distribution of charges and some ions (right)

TABLE 7.1 Average contents of mineral nutrients in (a higher) plant shoot dry matter that are sufficient for adequate growth and their principal functions (adapted from Marschner, 1995)

Element	Abbreviation	%	Function
Micronutrients			
Molybdenum	Mo	0.00001	Cofactor for enzymes involved in nitrogen metabolism (e.g. nitrogenase).
Nickel	Ni	0.00001	Metal component of the enzyme urease, essential for its structure and catalytic function.
Copper	Cu	0.0006	Participates in redox reactions as a component of enzymes (e.g. plastocyanin in chloroplasts).
Zinc	Zn	0.002	Catalytic and structural component of many enzymes involved in metabolism, DNA replication and gene expression. Also activator of enzymes involved in transcription ("zinc fingers"), cell division, protein synthesis and carbohydrate and auxin metabolism. Required for maintenance of integrity of biomembranes.
Manganese	Mn	0.005	Component of two enzymes: one in photosystem II, and one superoxide dismutase. Acts as cofactor for ~35 different enzymes, mostly involved in redox-, decarboxylation-, and hydrolytic reactions.
Boron	B	0.002	Cell wall formation (structural role). (Plasma) membrane integrity and functioning. Role in phenol metabolism (detoxifies by forming complexes).
Iron	Fe	0.01	Component of heme proteins (like cytochromes, catalases and peroxidases) involved in redox reactions, and of iron-sulphur proteins (like ferredoxin: acts as an electron transmitter in many basic metabolic processes). Required for the biosynthesis of chlorophyll.
Chlorine	Cl	0.01	(Indirectly) involved in the splitting of water (O_2 evolution) by photosystem II. Stimulates the V-type H^+-ATPase. Role in opening and closing of stomata in guard cells. Important in osmoregulation: accumulates in vacuoles.
Macronutrients			
Sulfur	S	0.1	Constituents of the amino acids cysteine and methionine, thus of proteins, in which cysteine has a structural role by forming disulfide bonds. Other sulfur-containing compounds, such as glutathione, phytochelatins, biotin, vitamin B_1, coenzyme A, and glucosinolates (secondary metabolite) are derived from these amino acids. Component of sulfolipids in biological membranes (abundant in thylakoid membranes).
Phosphorus	P	0.2	Structural element of DNA and RNA, and of phospholipids. Active component of phosphate ester compounds (energy-rich pyrophosphate bonds), such as ATP and glucose-6-phosphate, essential in energy transfer. Inorganic phosphate, Pi, has important regulatory role in carbon metabolism (carbon partitioning between chloroplasts and cytosol).
Magnesium	Mg	0.2	Central atom of the chlorophyll molecule. Essential for the proper aggregation of ribosome subunits and required for RNA polymerases (hence for both RNA and protein synthesis).

(Contd.)

(Contd. Table 7.1)

			Mg·ATP complex activates or is substrate for many enzymes in phosphorylation reactions (like ATPases). Mg^{2+} also required by photosynthetic enzymes (carbohydrate metabolism). Vacuolar Mg^{2+} important for the cation-anion balance and turgor regulation of cells.
Calcium	Ca	0.5	Essential in cell wall stabilization, by binding to pectins in the middle lamella. Important in cell extension (e.g. pollen tube growth) and for formation of secretory vesicles. Fundamental role in membrane stability: binds to and bridges phosphate and caboxylate groups of phospholipids and proteins at membrane surfaces. Acts as a counterion for organic and inorganic anions in the vacuole, where it often forms complex with oxalate which is important for osmoregulation of cells. Has a function as a second messenger in response to environmental stimuli (e.g. light, pathogens, and injury).
Potassium	K	1.0	Activates a large number of enzymes (e.g. in carbohydrate metabolism) by inducing conformational changes. Also activates the plasma membrane H^+-ATPase. Required for protein synthesis. Important in photosynthesis and phloem transport (sucrose export). Essential in osmoregulation (cell extension, stomatal movements, leaf movements) and plant-water relations, since K^+ is the major inorganic solute in cells. Important counterion in the cation-anion balance in the cytoplasm, chloroplasts, vacuoles, xylem, and phloem.
Nitrogen	N	1.5	Essential component of proteins (amino acids), nucleic acids and other nitrogen compounds such as coenzymes, secondary products and membrane constituents.

surrounded by their specific membrane. Each of these membranes differs, with respect to lipid composition and what kind of proteins are embedded in them or attached to them. The membranes serve to separate and protect the content of the cell and the organelles. Since the membrane is composed of lipids, water-soluble organic molecules and charged ions cannot freely pass across it and the membrane is considered **"semipermeable"**. Proteins embedded in the membrane let water, ions and other solutes pass through the membrane at different speeds. Some solutes and water molecules can also slowly diffuse across a membrane. As dealt with in the next section, the semipermeable property of the membrane is the reason for the build-up of the membrane potential.

Diffusion is the spontaneous transport of material or spreading of a substance which is caused by the random movements of the particles (atoms, ions, or molecules) in the medium. The rate of spreading is low. Diffusion tends to level out concentration gradients in, e.g. a liquid solution, i.e. particles in the medium tend to diffuse until their concentration is homogeneous. The direction of diffusion is from an area with high concentration to areas with lower concentrations.

Diffusion of a substrate s is described by the first law of Fick:

$$J_s = -D_s(\partial c_s/\partial x) \rightarrow \text{ often translates into}$$
$$J_s = -D_s(\Delta c_s/\Delta x)$$

where J_s is the flux density (flow of diffusion or rate of transport), in particles $m^{-2} s^{-1}$; D_s

the diffusion coefficient, indicating the mobility of the particles, in $m^{-2}\ s^{-1}$; c_s concentration of the particle, in particles m^{-3}; and x = the distance measured along the concentration gradient, in meters.

The minus sign indicates that the flow goes in the direction of decreasing concentration.

The meaning of Fick's law is that *the rate of diffusion is directly proportional to the concentration gradient, $(\partial c_s/\partial x)$.* Because it is a slow process, diffusion is efficient for transport only over short distances, e.g. inside a small cell, but for longer distances other mechanisms such as transport proteins have to be used.

Membrane Potential, Electrochemical Potential, Nernst Equation

To understand the transport processes in a plant cell, one must first consider some basic thermodynamics. Accumulation of a solute, from the outside of the plasma membrane into the cell, requires energy. The cell uses different forms of energy to perform the work it takes to drive transport. There are two basic forms of energy:

(1) Potential energy: stored energy, that can be used to perform work;

(2) Kinetic energy: the energy contained in a substance (or a body) in movement.

The plant cell uses potential energy to transport molecules and ions across the different biological membranes. There are four different forms for potential energy:

(1) **Chemical** potential energy
(2) **Osmotic** potential energy
(3) **Electrical** potential energy
(4) **Pressure** potential energy

The total potential energy of a specific compound is the sum of the four different forms of potential energy mentioned above, that are contained within this compound. *The cell uses all four forms of potential energy and hence, four different forms of work, to transport solutes.*

(1) Chemical work—a chemical compound contains energy within its chemical bonds, and this energy can be liberated when one or more bonds are broken. This energy can be used to transport solutes, e.g. ion pumps use ATP (see section below: Proteins regulate specific transport...).

(2) Osmotic work—when a solute moves from an area with a high concentration to an area with a low concentration, it performs osmotic work. Cotransporters (see section: Proteins regulate...) use osmotic energy to transport ions. Contrary movement, from a low to a high concentration area, requires extra work.

(3) Electrical work—when a chemical compound moves within an electrical field, from one electrical potential to another, it can either perform work or need additional work, depending on the charge and direction of the electrical field.

(4) Pressure work—if a system's volume changes, it can perform work. This kind of work is of less importance for membrane transport.

Different forms of energy can hinder or facilitate membrane transport

Two forms of energy may hinder the transport of a solute across a membrane:

(1) If the solute moves against its own concentration gradient, it has to overcome a barrier of **osmotic energy** and

(2) If a charged solute moves against an electrical gradient, it has to overcome the barrier of **electrical energy**. For example, as the plasma membrane is negatively charged on the inside, an anion has to overcome this barrier to come into the cell, while a cation has to overcome it to get out of the cell (see Fig. 7.1).

On the other hand, three of the forms of energy discussed above can be used to facilitate solute transport:

(1) **Chemical energy** liberated in a chemical reaction, may be used to transport the solute (e.g. ion pumps).

(2) A transport protein (e.g. some channels) can take advantage of the **osmotic energy** contained within the concentration (chemical) gradient of the transported solute to transport this solute from a compartment with high concentration to a compartment with low concentration. Other transporters (e.g. cotransporters) will use the osmotic energy contained within another solute's concentration gradient to transport the ion/solute of interest. This is called **coupled transport**: the transport of one solute is coupled to the simultaneous transport or diffusion of another solute, for which it is energetically favorable to cross the membrane. *In plant cells, this other solute is almost always H$^+$*. This is because there is generally, a large H$^+$ gradient across the plant plasma membrane, with a lower pH (higher H$^+$ concentration) on the outside of the cytosol in the apoplastic (cell wall) region. Hence, strong forces are acting on the H$^+$ ion to come into the cell.

(3) A transport protein (e.g. some channels) can also take advantage of the **electrical energy** that is contained within the electrical gradient that exists across the membrane: a cation can move with the electrical gradient from the outside of the cell to the inside, and an anion in the opposite direction. This is because the inside of the plasma membrane is more negative than the outside (see Fig. 7.1).

Membrane potential is able to drive uptake of cations

Electrical energy stored in the electrical gradient across the membrane is called **the membrane potential, ΔE**. The membrane potential means potential electrical energy (the difference between the electrical potentials outside and inside the membrane) which may be used to do transport. Since the electrical potential is relative, it cannot be measured, unless it is measured against something else. The membrane potential can be measured because it indicates the difference in electrical potentials between each side of the membrane. In a plant cell, the membrane potential is negative on the side of the membrane facing the cytosol, i.e. on the inside of the plasma membrane and on the outside of the vacuolar membrane (tonoplast). The negative charge on the inside is partly due to all the proteins in the cytosol, which are negatively charged at the neutral pH maintained there. The membrane potential is measured in **millivolts (mV)** and can vary between −20 and −200 mV across plant cell membranes (Allen and Sanders, 1997; Chrispeels *et al.*, 1999). It is higher at the plasma membrane (-100 to -200 mV) than at organellar membranes such as the vacuole.

The membrane potential is able to *drive* transport, usually into the cell against a concentration gradient, and it is also able to *stop* transport, usually out of the cell (with a concentration gradient). Thus, the membrane potential is highly important to the cell. Around the plasma membrane, the membrane potential attracts cations and repels anions from the cell.

The existence of a membrane potential indicates that there is an asymmetric distribution of charges over the membrane. This asymmetric distribution is the result of the membrane being semipermeable; cations and anions cross the membrane with different rates and so the charge distribution will be uneven, giving rise to a membrane potential. The membrane potential builds up as a result of—

- Unspecific diffusion: accumulated anions or cations can leak out of the cell (or into organelles). If they do so with different speed, i.e. if they have different membrane permeability, a membrane potential is created since the charge distribution will be asymmetric. The membrane potential formed by diffusion is called the **diffusion potential.**

- When cations (or anions) are transported against or with their own concentration gradients faster than anions (or cations) are able to follow.

Transported solutes move with or against the electrochemical gradient

The rate and direction of ion transport is governed by the electrical and concentration gradients that exist in and around the membrane of interest. These gradients are collectively called the **electrochemical gradient**.

The electrochemical gradient of a solute is thus composed of two components:

(1) The electrical component → the electrical potential, i.e. the electrical gradient across the membrane. This is a function of the membrane potential (ΔE).

(2) The chemical component → the osmotic potential for the solute. This is a function of the solute's concentration gradient across the membrane. (Be aware of the terminology: a chemical gradient = concentration gradient and contains potential *osmotic* energy, not chemical energy!)

This electrochemical gradient for the solute is also called its **electrochemical potential** (difference), and is written $\Delta\mu$.

The electrochemical potential for a particular solute indicates the potential energy it has to be transported. It can be calculated (with certain restrictions) to show whether the compound will be able to be transported or not. Here, we shall deduce the formula for that calculation.

The electrical potential for a solute is dependent on two variables: charge and electrical potential. Thus the electrical potential for a certain solute increases linearly with increase in number of charges and increase in electrical potential; this is written:

$$zFE$$

where z is number of charges of the ion (e.g. +1 for K^+, –2 for SO_4^{2-}); F the Faraday constant = 96485 joule/volt/mol ($J\ V^{-1}\ mol^{-1}$); E the electrical potential inside or outside the membrane.

As mentioned above, the electrical potential cannot be measured unless it is measured against something else. The membrane potential, ΔE, can be measured because it indicates the difference in electrical potential on each side of the membrane.

The osmotic potential for a chemical compound is dependent on the temperature and the concentration. The osmotic potential increases linearly with increasing temperature, but logarithmically with increasing concentration:

$$RTlnC$$

where R is the gas constant = 8.314 joule/Kelvin/mol ($J K^{-1} mol^{-1}$); T the absolute temperature (in Kelvin); C the ionic concentration inside or outside the membrane.

The electrochemical potential, μ, for a chemical compound then writes as:

$$\mu = zFE + RTlnC$$

The electrochemical potential (difference), $\Delta\mu$, is the difference between the electrochemical potential of the solute on each side of the membrane:

$$\Delta\mu = \mu_o - \mu_i$$

where o = outside and i = inside.

If an ion is equally distributed across the membrane, so that there is no net transport of the ion, there is equilibrium and

$$\Delta\mu = \mu_o - \mu_i = 0$$

$$\rightarrow \mu_o = \mu_i$$

$$\rightarrow RTlnC_o + zFE_o = RTlnC_i + zFE_i$$

where C_o is the outside (apoplastic or inside an organelle) concentration of a solute; C_i the inside (cytosolic) concentration of a solute; E_o the electric potential on the outside of the membrane; E_i the electric potential on the inside of the membrane

$$\rightarrow zFE_i - zFE_o = RTlnC_o - RtlnCi$$

$$\rightarrow zF(E_i - E_o) = RT(lnC_o - lnC_i)$$

$$\rightarrow zF(E_i - E_o) = RTln(C_o/C_i)$$

$$\rightarrow E_i - E_o = RT/zF \times ln(C_o/C_i)$$

$$\rightarrow \Delta E = RT/zF \times ln(C_o/C_i)$$

$$\rightarrow \Delta E = 2.3RT/zF \times log(C_o/C_i)$$

$$\rightarrow \Delta E = (2.3 \times 8.314 \times 298/z \times 96490) \times log(C_o/C_i) \text{ if the temperature is } 25°C = 298°K$$

$$\rightarrow \rightarrow \Delta E = 59/z \times log(C_o/C_i) \text{ mV}$$

This formula is called **the Nernst equation** and hence this specific ΔE is called **the Nernst potential**. The Nernst potential is the membrane potential that will arise when an ion is in equilibrium over the membrane, if it is supposed that this ion is the dominating one (and the temperature is +25°C). When this ion diffuses in or out of the cell according to its electrochemical gradient, at a given time the net transport of the ion will be stopped by the membrane potential that has built up; this membrane potential is the Nernst potential.

The Nernst equation can be used to calculate whether the membrane potential will be able to drive the transport of a given ion.

If the membrane potential is not large enough, or if it is acting as a hindrance for the accumulation of a solute, then other sources of energy have to be used to drive transport, or at least complement the energy source that the membrane potential is.

BOX 7.1 Will the membrane potential facilitate or hinder transport? (modified freely from Marschner, 1995)

The membrane potential across, e.g. the plasma membrane can be measured, as well as the concentration of ions on the inside (symplast) and on the outside of the cell (in the soil or in solution) (Fig. 7.2). We may now calculate whether the membrane potential is large enough to accumulate a given solute, or if it will act as a barrier instead. In the following simplified examples it is assumed that the ion of interest determines the membrane potential and that it is in equilibrium, and also that the temperature is +25°C. In reality, ΔE is determined by several

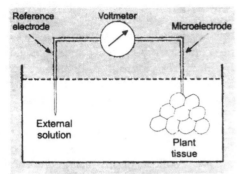

Fig. 7.2 Diagram of the system for measuring membrane potentials across plant cell membranes. The electrode in the plant tissue is usually inserted in the vacuole or the cytoplasm. The external solution serves as reference.

ions and the concentration of each has to be taken into consideration (see also below).

Example 1: K^+ in wheat roots

$$\Delta E = -110 \text{ mV}$$

$$C_o = 1 \text{ mM}$$

$$z = +1$$

We use the Nernst equation, assuming equilibrium: $\Delta E = 59/z \times \log(C_o/C_i)$

$$\rightarrow 59/1 \times \log(1/C_i) = -110$$

$$\rightarrow \log(1/C_i) = -110/59 = -1.86$$

$$\rightarrow 1/C_i = 10^{-1.86}$$

$$\rightarrow 1/C_i = 0.0138$$

$$\rightarrow C_i = 1/0.0138$$

$$\rightarrow C_i = 72 \text{ mM,}$$

This is the theoretical concentration of K^+ in the symplast, with the assumptions made above. The experimental concentration of K^+ in the symplast was 73 mM, therefore a membrane potential of -110 mV suffices to drive transport of K^+ into the cell. This usually takes place through a channel.

Example 2: K^+ in maize roots

$$\Delta E = -84 \text{ mV}$$

$$C_o = 1 \text{ mM}$$

$$\rightarrow 59/1 \times \log(1/C_i) = -84$$

$$\rightarrow \log(1/C_i) = -84/59 = -1.42$$

$$\rightarrow 1/C_i = 10^{-1.42} = 0.0038$$

$$\rightarrow C_i = 1/0.0038$$

$$\rightarrow C_i = 26 \text{ mM}$$

The experimental concentration in the root was 66 mM. Thus, a membrane potential of -84 mV is not large enough to drive all the uptake of K^+, but only up to 26 mM (through a channel). The rest of the uptake has to be through another transport protein that uses another source of energy, e.g. a symporter that uses the proton gradient.

In fact, the Nernst equation can quickly determine whether cations such as K^+ can be taken up or not: *for each −59 mV of membrane potential, K^+ can be accumulated against an additional 10-fold concentration gradient* (Ward, 1997). So, at a membrane potential of -59 mV, the outside concentration of K^+ can be 10 times lower but not more for uptake. For accumulation to take place with an outside concentration that is 100 times lower, the membrane potential must be $2 \times (-59) = -118$ mV.

If there is a membrane potential of -150 mV, the outside concentration of K^+ can be ~350 times lower in the external solution and still be taken up:

$$\Delta E = -150 \text{ mV} = 59/1 \times \log (C_o/C_i)$$

$$\log (C_o/C_i) = -150/59$$

$$C_o/C_i = 10^{-150/59} = {\sim}2.87 \times 10^{-3}$$

$$C_i = C_o \times 1/2.87 \times 10^{-3} = C_o \times 349$$

Example 3: Ca^{2+} in maize roots

$\Delta E = -84$ mV

$C_o = 1$ mM

$z = +2$

$\rightarrow 59/2 \times \log(1/C_i) = -84$

$\rightarrow \log(1/C_i) = -84 \rightarrow 2/59 = -2.85$

$\rightarrow 1/C_i = 10^{-2.85} = 0.0014$

$\rightarrow C_i = 1/0.0014$

$\rightarrow C_i = 704$ mM

When measured, the Ca^{2+} concentration in the cell shows to be 2 mM, much less than expected (704 mM) at equilibrium with a membrane potential of -84 mV. Ca^{2+} has to be transported out of the cell, by a transport protein that uses an energy source other than the membrane potential. This might be an antiporter that uses the proton gradient, or an ATP-driven Ca^{2+} pump.

Example 4: SO_4^{2-} in pea roots

$\Delta E = -110$ mV

$C_o = 0.25$ mM

$z = -2$

$\rightarrow 59/-2 \times \log(0.25/C_i) = -110$

$\rightarrow \log(0.25/C_i) = -110 \times -2/59 = 3.73$

$0.25/C_i = 10^{3.73} = 5370$

$C_i = 0.25/5370$

$C_i = 0.000047$ mM $= 47$ nM

The measured ion concentration in the plant tissue was 19 mM, which is by far a higher concentration than expected when this ion would be in equilibrium with a membrane potential of −110 mV. The membrane potential acts like a great barrier for the uptake of SO_4^{2-}, and has to be transported into the cells by a transport protein able to use another energy source, e.g. the proton gradient or ATP.

All the examples above take only one ion into account. In the cell, however, many ions are present. This fact has to be taken into consideration if one wants to calculate the membrane potential (diffusion potential) at equilibrium correctly. The **Goldman equation** (later further developed by Hodgkin and Katz) resolves this problem. Here, the Nernst equation is integrated over the thickness of the membrane, which separates two compartments each containing a different ionic solution in which the electrochemical potential for each ion is constant. The only existing electrochemical gradient is across the membrane and supposedly linear.

Recall the Nernst equation: $\Delta E = 59/z \times \log(C_o/C_i)$ (can also be written $\Delta E = RT/zF \times \ln(C_o/C_i)$).

Here we skip the quite complicated calculations and directly write the Goldman equation for the most abundant ion in the plant cell, K^+ and an anion, A^- (for example NO_3^- or SO_4^{2-}). The Goldman equation can include more than two ions if needed. Thus the membrane potential for KA at equilibrium is:

$\rightarrow \Delta E = RT/F \times \ln \left((P_K C_K^o + P_A C_A^i) / (P_K C_K^i + P_A C_A^o) \right)$

P_K is the permeability coefficient for K^+

P_A the permeability coefficient for A^-

C_K^o, C_K^i the concentration of K^+ on the outside and inside respectively

C_A^o, C_A^i the concentration of A^- on the outside and inside respectively

The permeability coefficient indicates how permeable the biological membrane is for that particular ion.

If $P_K = P_A = P$, one will get:

$\Delta E = RT/F \times \ln \left((PC_K^o + PC_A^i)/(PC_K^i + PC_A^o) \right)$

$$= RT/F \times ln \ (P(C_K^o + C_A^i)/P \ (C_K^i + C_A^o))$$

$$= RT/F \times ln \ ((C_K^o + C_A^i)/(C_K^i + C_A^o))$$

If the only ions of importance are K^+ and A^-, then $C_K^o = C_A^o$ and $C_K^i = C_A^i$, since it is an ion pair (KA). Then one gets:

$$\Delta E = RT/F \times ln \ ((C_K^o + C_K^i)/(C_K^i + C_K^o))$$

$$= RT/F \times ln \ (1)$$

$$= 0$$

That is, no membrane potential will be created and the ions will equilibrate on each side of the membrane, as was expected. The conclusion that can be drawn from this is that *a membrane (diffusion) potential will only be created when the particular ions have different permeability coefficients.*

Proton motive force used by cotransporters to drive uptake of anions and neutral molecules

The **proton motive force (pmf or Δp)** is the force that protons outside the cell – or inside the vacuole – are exerting on the biological membrane to come back into the cell. It is a function of the electrochemical potential for H^+.

The proton motive force is defined as the electrochemical potential for H^+ divided by Faraday's constant: $\Delta p = \Delta \mu \ / \ F$.

By dividing with Faraday's constant, Δp can be expressed in volts (V) and compared to the membrane potential.

Thus: $\Delta p = \Delta \mu / F$

$$\rightarrow \Delta p = zF\Delta E/F + RT/F \times ln \ ([H^+]^i/[H^+]^o), \quad \text{with } z = +1$$

$$\rightarrow \Delta p = \Delta E + (RT/F) \times ln \ ([H^+]^i/[H^+]^o)$$

$$\rightarrow \Delta p = \Delta E + (2.3RT/F) \times log \ ([H^+]^i/[H^+]^o)$$

$$\rightarrow \Delta p = \Delta E + (2.3RT/F) \times (log \ [H^+]^i - log \ [H^+]^o)$$

As $pH = -log \ [H^+]$:

$$\rightarrow \Delta p = \Delta E + (2.3RT/F) \times (- pH^i + pH^o)$$

$$\rightarrow \Delta p = \Delta E + (2.3RT/F) \times - (pH^i - pH^o)$$

$$\rightarrow \Delta p = \Delta E - (2.3RT/F) \times \Delta pH$$

At 25°C, this will write as

$$\rightarrow \Delta p = \Delta E - 59 \times \Delta pH$$

where Δp is expressed in millivolts (mV).

Example: a membrane potential of −120 mV and a pH difference of 2 at the plasma membrane (typical for many plant cells) will give a $\Delta p = -120 - 59 \times 2 = -238$ mV. This is *the effective electrochemical potential difference for H^+* over the plasma membrane. This potential can be used by cotransporters to drive different transport processes, because these proteins are able to couple the transport of one solute to the simultaneous transport of H^+. The proton motive force is thus the force that drives the uptake of anions and neutral solutes. The cotransport of H^+ may also drive the uptake of cations when the membrane potential is not large enough, as well as the transport of cations out of the cell or into the vacuole.

The proton motive force constitutes a force capable of driving the transport of most solutes in and out of the cell. As shall be seen later, primarily two types of H^+-transport proteins are responsible for building up the proton motive force.

Chemiosmosis occurs when chemical energy is converted to osmotic and electric energy and vice versa

Processes that convert one form of energy to another are called **chemiosmosis**. The Nobel laureate Peter Mitchell "invented" this expression to describe his hypothesis on energy coupling to solute transport (Briskin, 1990; Mitchell, 1976, 1985). The chemiosmotic hypothesis suggests that the electrochemical gradients of protons established across selectively permeable biological membranes can be used as an energy source to perform work, for example for translocation of solutes or for synthesis of ATP.

For example, the electrogenic H^+ pumps convert chemical energy (ATP) to osmotic and electric energy, and the ATP synthases in the mitochondria and the chloroplast do the reverse, i.e. they convert osmotic and electric energy to chemical energy.

Different sources of energy used for primary and secondary transport respectively

Primary and secondary transport are terms often used when solute transport is discussed. The two forms of transport differ functionally, with respect to the energy used:

(1) **Primary transport** (sometimes called active or uphill transport) uses the energy that results from chemical reactions, e.g. the hydrolysis of ATP for ion pumps, or the electrical energy from oxidation-reduction reactions, to drive transport of a solute. During primary transport the solute is transported *against* its electrochemical gradient, resulting in the *creation* of an electrochemical gradient of that particular solute.

(2) **Secondary transport** (sometimes called downhill or passive transport, which is not correct, since all transport requires energy) *uses* the energy stored in electrochemical gradients across the membrane. The secondary transport can be driven by the osmotic energy in the concentration gradient of the solute

to be transported, or be coupled to the use of the osmotic energy in the H^+ gradient, or be driven by the electrical energy stored in the electrical gradient across the membrane, i.e. the membrane potential.

So, note the difference between the two forms of transport:

- Primary transport *creates* electrochemical gradients → thus *primary*.
- Secondary transport *uses* electrochemical gradients → thus *secondary*.

There are two kinds of secondary transport:

(1) Facilitated diffusion: this means specific diffusion through a membrane protein, such as a channel or a permease. It is specific because each kind of transport protein permits passage of just one kind (or family) of solute, e.g. K^+ channels only let K^+ ions pass through. Therefore, there are many different kinds of transport proteins in the plant cell; one (or several) for each kind of molecule (or group of related molecules) that needs to cross a membrane. Facilitated diffusion is strictly regulated.

(2) Simple diffusion: unspecific diffusion through a biological membrane. This type of diffusion cannot be regulated.

PLASMODESMATA CONNECT CELLS

Plant cells are not, like animal cells, isolated individually. The cytoplasm of neighboring cells is connected via channel-like structures in the cell wall, called **plasmodesmata** (PD), allowing for an exchange of solutes between cells. Nevertheless, the transport processes through plasmodesmata are highly regulated. The domain of common cytoplasm contained within the plasma membrane is called the **symplast**. The cell walls surrounding the cells form another continuous space, called the **apoplast**. Ions and small molecules can move in the plant tissue either in the apoplastic space or enter a cell and move between cells in the symplast. In a higher plant embryo, all cells are interconnected by PDs. As the plant grows and develops, individual cells or groups of cells become isolated, forming so-called symplastic domains (McLean *et al.*, 1997).

The general structure of a PD is a complex plasma membrane-lined pore containing appressed membranes of endoplasmic reticulum (ER) termed the desmotubule, in its center (Lucas and Wolf, 1993; Overall and Blackman 1996; Fig. 7.3). The desmotubule is thus a continuation of the endoplasmic reticulum. Researchers have studied the function and regulation of PDs by looking at the movement of plant

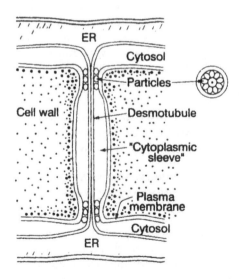

Fig. 7.3 Schematic presentation of a plasmodesma and some of its structural components (ER = endoplasmic reticulum). (From Marschner, p. 67, 1995).

viral movement proteins, which are large viral molecules able to pass through the PDs. Other techniques used include injection of fluorescently labeled dextrans and other fluorescent tracer molecules, and confocal laser scanning microscopy. Experiments involving expression of proteins marked with green fluorescent protein (GFP) have also been used to investigate the mechanisms of PD targeting and trafficking (Heinlein, 2002).

Passive, diffusion-driven symplastic transport of small molecules is proposed to occur in the space between the plasma membrane and the desmotubule. This transport is limited only by the hydrodynamic size of the molecule. The size of the pore accessible for diffusion-driven movement of molecules is termed the plasmodesmal size exclusion limit (SEL). Larger molecules such as proteins and RNAs may also be transported through the PD, but they require special mechanisms, which will not be described here.

The PD connection therefore represents one mode of transport of water, small molecules and ions. This passive, diffusion-driven symplastic transport is always governed by the electrochemical potential gradient for the solute, i.e. it moves down along its electrochemical gradient.

PROTEINS REGULATE SPECIFIC TRANSPORT PROCESSES ACROSS MEMBRANES

In the model plant *Arabidopsis thaliana* (Thale Cress) whose genome is fully sequenced, a total number of 833 transporters have been identified. For comparison, the whole genome of *Arabidopsis* is 125 megabases big and contains approximately 25,000 genes (*Arabidopsis* Genome Initiative, 2000). Recently, the draft genome sequences of two rice (*Oryza sativa*) varieties were completed, which will further improve our knowledge about plant transporters (Goff et al., 2002; Yu et al., 2002). There are three different main kinds of transport proteins in the plant cell: **channels, carriers**, and **pumps** (Fig. 7.4). All of them are intrinsic membrane proteins. Here, we shall go through each class of transporter with respect to their mode of function and give some specific examples. A general overview of transport systems in a plant cell is given in Fig. 7.5.

Ion and Water Channels Mediate Selective Diffusion

Channels can be thought of as protein pores spanning across the membrane. These pores, or "protein tubes", are either open or closed: they are said to be **gated** and work in an all-or-none fashion. A specific area in one end of the pore, controlling which ions will come through, is called the **selectivity filter**. The **gate** can be situated in the pore or in other parts of the protein, which then influence the opening and closing of the pore via conformational changes. The pore controls the selectivity independent of the gate. Different channels have different gating mechanisms; for example the opening and closing of voltage-gated channels depend on the membrane voltage sensed by the fourth transmembrane helix, hence acting as a voltage gate (see Box 7.2). Others are activated by the binding of a ligand-like Ca^{2+}, by a change in temperature or by mechanical stress (stretching of the membrane). When a channel is open, a large number of molecules or ions (millions per second) are able to freely pass through its central pore, provided it is the ion for which the channel is specific. When the concentration of the ion increases on one side of the membrane, the transport rate will also increase. Therefore, channels cannot have a defined V_{max}, and thus no K_m, like carriers

have (see below). Furthermore, compared with carriers the ion transport through channels does not involve a big conformational change of the protein. Because of the mentioned differences with carriers, channels do not act like enzymes. On the other hand, they do not act like simple pores either, since they show specificity for the transported ion and their activity is strictly regulated. In plant cells, a large variety of channels exists, because most channels are highly specific with regard to the ion transported (see Fig. 7.5).

Recently, the Nobel prize for chemistry in 2003 was awarded to Roderick MacKinnon and Peter Agre for their excellent work on the structure and function of K^+ channels and water channels respectively (see Box 7.2 for details). These pioneering works are also important for understanding how similar channels function in plants. In plant cells a large number of channels are known either by their sequence or their electrophysiological characteristics, or both. All subcellular membranes investigated so far (plasma membrane, tonoplast, plastidial, and mitochondrial membranes) are equipped with a variety of channels exhibiting different ion selectivities and specific regulation mechanisms (Barbier-Brygoo et al., 2000). Here we shall look closer at water channels and potassium channels, the latter the best studied ion channel family so far.

Water channels, aquaporins, allow rapid transmembrane water flow

Until recently, movement of water across biological membranes was a matter of debate among researchers. It was suggested to happen by simple diffusion, but the existence of channels mediating water flow was also proposed as early as the mid-19th century (Brücke, 1843; Pfeffer, 1877). The nature of these pores was not known but

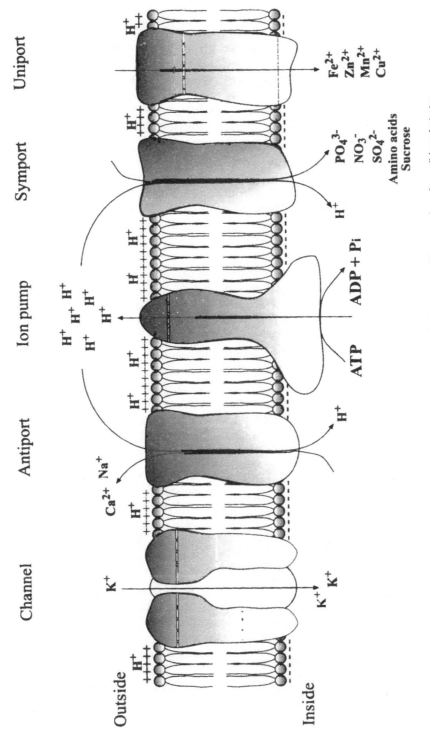

Fig. 7.4 Schematic presentation of the different types of transport proteins that exist in plants, with examples of possible substrates

physical, (bio)chemical, and pharmacological investigations led researchers to suggest that proteins could be involved in water transport. It was not until 1991 that the first protein shown to have water transport activity was cloned from the human erythrocyte plasma membrane. This protein was called CHIP28, for CHannel-forming Integral Protein of 28 kDa, now renamed AQP1 for aquaporin 1 (Preston and Agre, 1991). The proof for CHIP28 being an aquaporin came from its heterologous expression in frog oocytes. When placed in a hypotonic solution (i.e. less concentrated on the outside), the oocytes expressing CHIP28 in their plasma membrane swelled and eventually ruptured much faster than control oocytes (Preston et al., 1992). This implied that water was able to permeate the plasma membrane much faster when CHIP28 was present, thus proving that CHIP28 was a water channel. The same happened when CHIP28 was reconstituted into liposomes (Zeidel et al., 1992).

The oocyte expression system has since then been used to show the water transport capacity of a number of related proteins from different organisms, especially from plants. Some of these proteins transport small neutral solutes such as glycerol or urea instead of water, and some have mixed specificity (e.g. both glycerol and water). In fact, aquaporins belong to a superfamily called MIPs, for Major Intrinsic Proteins, which facilitate the transport of small polar molecules across membranes. The MIP family is an ancient family of proteins present in all living organisms: bacteria, fungi, plants and animals (Johanson et al., 2001).

The first plant aquaporin was identified in the vacuolar membrane of seeds, and was named γ-TIP, for Tonoplast Intrinsic Protein (Maurel et al., 1993). Since then, several other plant MIP homologs have been cloned and functionally characterized in the oocyte

system. In the fully sequenced *Arabidopsis* genome, 35 different MIP-encoding genes have been identified (Johanson et al., 2001). They can be divided into four different subfamilies based on sequence similarity and are present in the plasma membrane (PIPs, for Plasma membrane Intrinsic Proteins), in the vacuolar membrane (TIPs), and in the peribacteroid membrane of nitrogen-fixing nodules (NIPs, for NOD26-like Intrinsic Proteins). The last subfamily was very recently identified (SIPs, for Small basic Intrinsic Proteins) and their subcellular location as well as their substrate specificity are not yet known.

Water channels in plants are involved in water absorption in the root and in maintaining the water balance in the plant. At the cellular level, aquaporins are important for buffering osmotic fluctuations of the cytosol (Kjellbom et al., 1999; Johansson et al., 2000). Expression of some aquaporins is regulated by environmental factors such as drought stress, and the activity of certain isoforms is influenced by phosphorylation and by the cytosolic pH (Johansson et al., 1996; Tournaire-Roux et al., 2003).

BOX 7.2 Transport mechanisms of aquaporins and ion channels revealed and rewarded.

Peter Agre was awarded with the Nobel prize in chemistry in 2003 for his discovery and subsequent characterization of water channels. He was the first person to purify and then clone an aquaporin (Preston and Agre, 1991). Actually, he was working on other proteins in the red blood cell membrane, when he isolated a membrane protein of unknown function. He managed to obtain sequence information and realized that this protein might represent the water channel people had so long

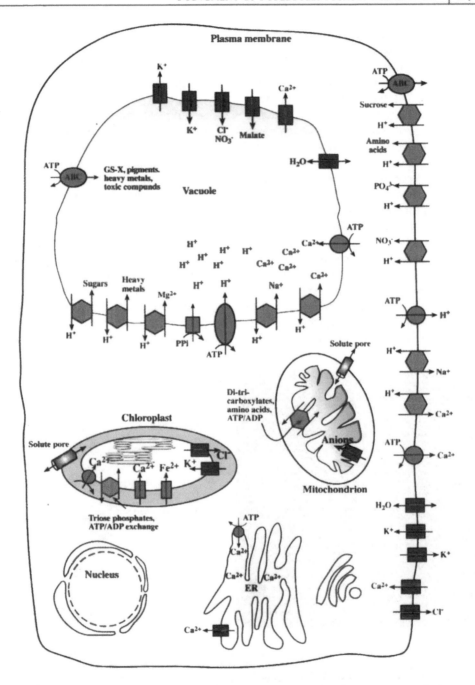

Fig. 7.5 Schematic overview of the major known membrane transport functions in a plant cell. Channels in blue, carriers in green, and pumps in red. The various organelles are not drawn to scale. ER = endoplasmic reticulum, GS-X = glutathione conjugate. The products of ATP hydrolysis, ADP + Pi, are omitted for the sake of clarity.

sought for. Later, Peter Agre became involved in the high-resolution determination of the 3D structure of the human AQP1 (Murata *et al.*, 2000) (Fig. 7.6). At the same time, another group published the structure of the *E. coli* glycerol facilitator, which also belongs to the MIP family (Fu *et al.*, 2000). These discoveries greatly enhanced comprehension of MIP function, i.e. how water and other small polar molecules are actually transported through these proteins. Surprisingly, the two proteins are highly similar in overall structure, despite the huge evolutionary distance and different substrate specificities. Based on sequence similarities, the plant aquaporins are suggested to have about the same structure as the AQP1.

Aquaporins are composed of four subunits, each of which consists of six transmembrane-spanning helices. Due to ancient gene duplication, the first half of the protein is homologous to the second half but in the opposite direction. There are two conserved motifs, NPA (= asparagine, proline, and alanine) which are like a "signature" for aquaporins. The actual channel or pore wherein the water molecules pass is surrounded by the six helices and formed by the two NPA-containing loops. Thus, each monomer contains a pore and water passes through all four subunits. The pore is so constructed that water molecules can only pass in a single file while larger molecules and protons are repelled. About halfway through the pore, the water molecules

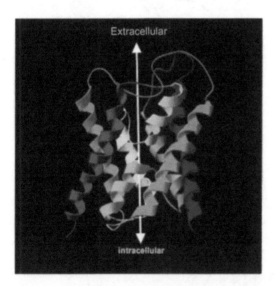
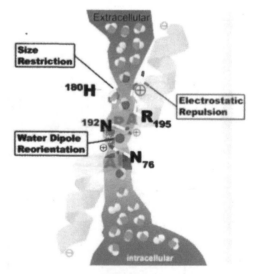

Fig. 7.6 Left panel—Ribbon model of the structure of AQP1 subunit, showing six tilted bilayer-spanning domains and two pore-forming loops with short transmembrane α-helices entering the membrane from the extracellular and the intracellular surface.

Right panel—Schematic diagram showing the aquaporin channel pore in the same orientation as to the left. Size restriction and electrostatic repulsion are proposed to prevent the passage of protons (in the form of hydronium ions, H_3O^+), as is the water dipole reorientation farther down the pore. (With permission of Federation of the European Biochemical Societies) (from Agre and Kozono, 2003).

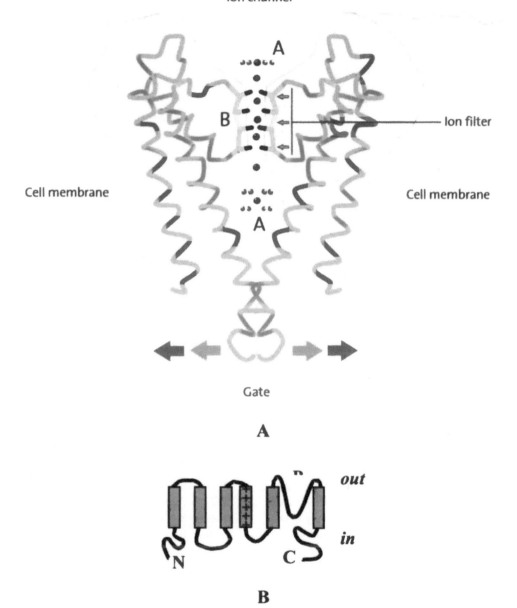

Fig. 7.7 **A** – Simplified model of a bacterial K⁺ channel structure. Two subunits are shown in the membrane. The ion selectivity filter is indicated, with the oxygen atoms as black sticks. See text for details (Illustration: Typoform. Courtesy Royal Swedish Academy of Sciences).

B – Topology model for the plant Shaker-like K⁺ channels, with six transmembrane spanning domains and pore (P)-loop dipping into membrane (*courtesy of Dr. S. Zimmermann, INRA Montpellier, France.*)

change orientation, which is thought to further block the passage of protons (Agre and Kozono, 2003).

The K⁺ channels, on the other hand, are also composed of four subunits but the pore is formed by all these four subunits. No one doubted the existence of ion channels as experimental evidence was abundant. Already in the 1950s it had been demonstrated that K⁺ ions move through the membrane in single file (Hodgkin and Keynes, 1955). Over the years, numerous studies using biochemical and biophysical methods have shown that many ion channels usually have a selectivity filter near one end of the pore and a separate gate near the other end (or sometimes elsewhere, see above). Despite the enormous difficulties in determining high-resolution 3D structures for membrane proteins, Roderick MacKinnon was able to report the first high-resolution structure of an ion channel, the KcsA K⁺ channel, in 1998 (Doyle *et al.*, 1998). He was successful because he used a bacterial K⁺ channel which was a little easier to handle than a eukaryotic one, but KcsA still has high homology to eukaryotic K⁺ channels. Roderick MacKinnon has continued to explore the mechanisms of K⁺ channels by looking at the structures of KcsA and other channels in different conformations (Jiang *et al.*, 2002a, b; Jiang *et al.*, 2003; Nishida and MacKinnon, 2002; Zhou *et al.*, 2001).

The resolved structures confirmed many experimental data and explained how the channel can be so highly selective for K⁺ over Na⁺, and how the ions can be transported so rapidly (high conduction rate) (Fig.7.7A). These two features are typical for K⁺ channels, both voltage-gated and ligand-gated ones. Selectivity comes from the configuration of the selectivity filter; it contains several binding sites that mimic the hydration shell of the K⁺ ion in solution. In the cytosol, the K⁺ ions are hydrated, i.e. surrounded by water molecules. In the channel, the water molecules are stripped off but the selectivity filter of the channel has oxygen atoms arranged in much the same way as the water molecules around the K⁺ ion in the cytosol. The amino acids (TVGYG = Threonine, Valine, Glycine, Tyrosine, Glycine) in the selectivity filter are conserved in all K⁺ channels in all organisms and are the "signature" of K⁺ channels. It was found that the glycine residues are extremely important for the structure of the selectivity filter.

In the pore, there are always at least two K⁺ ions present. When one enters the channel from the outside, another one exits the channel on the opposite side of the membrane. This is one reason for the high conduction rates. Another reason is that the filter's structure changes with binding of a K⁺ ion and as a result the ion binding is relatively weak, which facilitates movement of K⁺ ions down their electrochemical gradient (MacKinnon, 2003). Once the ions have left the selectivity filter, they are hydrated again.

The gating mechanism, how the channel opens and closes, involves conformational changes of the helices in the membrane (Fig. 7.7A). Other parts of the protein, such as the fourth transmembrane helix in voltage-gated channels, are said to be sensor or gating domains which communicate change in voltage or binding of a ligand to effect opening of the channel.

Shaker-like K⁺ channels best characterized plant ion channels

Potassium, K^+, is the most abundant cation in the cell, where it is important in osmoregulation (regulation of cell turgor, e.g. in stomatal movements) and in control of cell membrane polarization (Marschner, 1995; Véry and Sentenac, 2003). Furthermore, proteins in the cytosol are mostly negatively charged and K^+ is needed as an "electrical neutralizer" to balance these negative charges, in order for the proteins to maintain their hydration and conformation. Without K^+, many enzymes would not function properly (also see Table 7.1).

K^+ transport in plants has become a model transport system for researchers investigating the general mechanisms of solute uptake and transport in plants. Numerous physiological and electrophysiological studies on plant K^+ transport and its uptake in roots during the last century preceded the first molecular identification of two K^+ channels in the early 1990s (first plant ion channels whatsoever, see below). It was thanks to electrophysiological analyses of root K^+ transport that researchers were able to demonstrate that solute transport across the plant plasma membrane could be explained by the chemiosmotic model worked out by Peter Mitchell (Cheeseman and Hanson, 1979).

Cloning of the two K^+ channels AKT1 (Sentenac et al., 1992) and KAT1 (Anderson et al., 1992) was followed by molecular identification of several other K^+ transport systems, both channels and transporters (probably working as cotransporters, although the exact mechanism is not known). There are four major gene families for K^+ transport in plants: the Shaker (see below), KCO, HKT, and KUP/HAK/KT families, the first two being channel proteins. In addition, other less characterized gene families are likely to transport K^+ too.

For both AKT1 and KAT1, the cloning strategy was based on functional complementation of a yeast mutant, defective in its own K^+ uptake system. The yeast functional complementation technique has since been used to clone and characterize a wide array of plant transporters (see section below: Baker's yeast…).

AKT1 and KAT1 belong to the so-called Shaker channel family, the first members of which were cloned from a mutant fruit fly (*Drosophila melanogaster*) that was shaking (Kamb et al., 1987; Papazian et al., 1987; Tempel et al., 1987). These channels are highly selective for K^+ and strongly regulated (gated) by membrane voltage. They are composed of four subunits, each of which consists of a single polypeptide with six membrane-spanning domains and a so-called P-loop (P = pore) (see Fig. 7.7B). The P-loop contains the amino acids TVGYG that are the hallmark for K^+ selective channels (see Box 7.1). There are nine members of this family in *Arabidopsis*, which can be further divided into three functional subfamilies according to the voltage range within which they are active: this is called their **rectification** property. The plant Shaker channels are either inward, weakly inward or outward rectifiers. Thus, inward rectifiers are involved in K^+ uptake (K^+ rushing into the cell from the outside) because they are activated (gated open) by membrane hyperpolarization, i.e. membrane potentials going more negative than the K^+ equilibrium potential, E_K (or Nernst potential for K^+, see earlier). Physiologically, these channels open at a threshold even more negative than the E_K. Above this threshold potential (i.e. at less negative membrane potentials), this kind of Shaker channel is closed, thus avoiding loss of K^+ out of the cell.

The weak inward Shaker channels are also activated by membrane hyperpolarization, but they are never really closed within the physiological membrane potential range. Therefore, these channels are thought to function in both uptake and release of K^+ (Marten *et al.*, 1999; Lacombe *et al.*, 2000). The third group, outward rectifiers, activate by membrane depolarization (i.e. at membrane potentials more positive than the E_K). The outwardly rectifying channels are thus involved in K^+ release (Gaymard *et al.*, 1998). All Shaker channels studied so far are active at the plasma membrane.

The other channel family, KCO, resembles Shaker channels in that they have the conserved TVGYG motif in the pore and hence are highly K^+ selective. On the other hand, their overall structure is not the same and they have a Ca^{2+}-binding motif (so-called EF hand) in their C terminus. The only functionally characterized member of this family (KCO1) is localized in the vacuolar membrane and is an outwardly rectifying K^+ channel activated by Ca^{2+} (Czempinski *et al.*, 1997; Czempinski *et al.*, 2002; Schönknecht *et al.*, 2002).

The rectification properties and other electrophysiological characteristics of the Shaker and other ion channels have been (and still are) extensively studied by their expression in heterologous systems: *Xenopus laevis* (a frog) oocytes, insect and mammalian cell lines and yeast, and recently also in plant cells.

With K^+ ions intervening in many different important physiological processes, its uptake and distribution throughout the plant via K^+ transporters have to be tightly regulated. The molecular identification of K^+ transporter gene families has greatly aided in the study of how these proteins are regulated.

The *Arabidopsis* Shaker channels are regulated both at transcriptional and translational levels. The mRNA expression of some Shaker channels is influenced by hormones such as abscisic acid, cytokinins and auxin (Philippar *et al.*, 1999; Pilot *et al.*, 2003). Other channels show regulation of expression in response to light and sugars and also to the circadian clock (Deeken *et al.*, 2000; Moshelion *et al.*, 2002).

The functional Shaker channel in animal cells is made up of four subunits, often from different isoforms, and it is very likely that the plant counterparts function in the same way. Different experimental approaches, such as yeast two-hybrid experiments, coexpressions in oocytes, and biochemical analyses support this hypothesis. This **heterotetramerization** represents another level of regulation of these channels. It might provide the plant a way of even higher diversity of K^+ channels than represented by the nine members of the Shaker family.

At the protein level, all characterized plant Shaker channels are regulated by voltage, like their animal counterparts. Results from numerous studies on animal Shaker channels show that the positively charged fourth transmembrane segment, called S4, is involved in voltage sensing. However, while the animal Shaker channels are mostly outward rectifiers the corresponding plant channels are also inward rectifiers. Interestingly, in both plant inward and animal outward channels the S4 segment moves in the same direction in response to voltage change. The S4 movement in a given direction thus results in opposite effects on channel open probability in inward and outward rectifiers (Zei and Aldrich, 1998; Chen *et al.*, 2002). The molecular basis for this difference, or how the direction is decided, is a matter of interest.

Other factors that can modulate K^+ channel activity are pH, Ca^{2+}, cyclic nucleotides, and phosphorylation-dephosphorylation (Véry and Sentenac, 2003). All

Shaker channels characterized so far are strongly regulated by pH. As mentioned above, the KCO1 channel is activated by Ca^{2+}. Phosphorylation-dephosphorylation is a common mode of regulation of many proteins in the cell. Using the yeast two-hybrid screening approach, a phosphatase (AtPP2CA) interacting with the Shaker-like K^+ channel AKT2 was identified (Cherel et al., 2002). It was shown to modulate the activity of AKT2 in heterologous expression analyses.

Other proteins, e.g. kinases, 14-3-3 proteins, G proteins, farnesyl transferase, syntaxins, and the so-called β-subunits have been shown to be involved in the regulation of K^+ channel activity, by electrophysiological analyses and reverse-genetics approaches. Proofs of physical interaction are still missing, however. Thus it seems like an intricate signaling network is set up by plants to control the flux of K^+ in and between cells and there is still much to discover (Zimmermann et al., 1999)!

Other ion channels in plants

CNGCs—In animal cells, ion channels similar in sequence and structure but differing in function from Shaker channels play an important role in olfactive and visual signaling processes. These channels are gated by cyclic nucleotides and consequently are called Cyclic Nucleotide Gated Channels, or CNGCs. The CNGCs also have 6 transmembrane segments, a pore domain between the 5th and 6th transmembrane helices, and function as (hetero)tetramers in the (plasma) membrane. However, the amino acids of the selectivity filter of the CNGCs are not exactly the same as in K^+ channels and as a consequence, CNGCs are not selective only for K^+, but also for Na^+ and Ca^{2+}. Since they are activated by the binding of cyclic nucleotides to the C-terminal region, CNGCs are classified as ligand-gated chan-

nels. They are also regulated by the calcium-binding protein calmodulin (usually written CaM), thereby providing a link for two important cellular signaling pathways: Ca^{2+} and cyclic nucleotides.

A cDNA with sequence similarity to animal CNGCs was identified in a screen for CaM-fixing proteins in barley (Schuurink et al., 1998). Later, similar sequences were identified in Arabidopsis and tobacco (e.g. Köhler et al., 1999). With completion of the Arabidopsis genome sequence, the Arabidopsis CNGC gene family was shown to comprise 20 members. These proteins all have a putative cyclic nucleotide binding site and a CaM binding site in their C termini (Talke et al., 2003). Despite the great interest the research community has taken in these channels, plant CNGCs have been resistant to a convincing functional characterization by heterologous expression. A few studies suggest some isoforms are permeable to both K^+ and Na^+ and maybe Ca^{2+} (Leng et al., 1999, 2002; Balagué et al., 2003). Based on reverse genetic experiments, plant CNGCs are proposed to be involved in different signaling processes, e.g. plant defense reactions to pathogen attacks (Clough et al., 2000; Balagué et al., 2003). So far however, very little is known about the function and membrane localization of plant CNGCs.

Glutamate receptors—A family of proteins related to animal ionotropic glutamate receptors has recently been identified in plants (Lam et al., 1998). Twenty members were found in Arabidopsis (Lacombe et al., 2001). Based on sequence similarity, they are suspected to form nonselective cation channels permeable to K^+, Na^+, and/or Ca^{2+}, but their role in plants are not known and functional characterization has not succeeded to date.

Anion channels—As in animal cells, plant cell membranes are also equipped with channels for the permeation of anions. The major mineral anions in plant tissues are nitrate, chloride, sulfate, and phosphate (Barbier-Brygoo *et al.*, 2000). Another important anion is carbonate, although present in much lower concentrations. When inorganic anions are insufficient in the external medium, the concentration of organic anions derived from cell metabolism may increase. The most abundant organic anion in plant cells is malate. The highest anion concentrations are found in the vacuole, which acts like a storage compartment. In the cytosol, concentrations are maintained in the millimolar range, while the abundance in the apoplastic compartment and in the soil solution is even lower. Because of this concentration gradient and because of the negative plasma membrane potential, anion plasma membrane channels function as efflux channels when they are open. Similarly, passive anion fluxes into the tonoplast may be driven by the negatively charged tonoplast. On the other hand, anions must be taken up actively into the cytosol by anion/proton symport systems in the plasma membrane.

Functionally, most anion channels are regulated by voltage but their activity is also influenced by e.g. Ca^{2+}, ATP, phosphorylation or membrane stretching. Experimental evidence supports their role in cell signaling, osmoregulation, plant nutrition, and metabolism. For example, since the plasma membrane is negative on the inside, opening of anion channels permitting an efflux of anions will depolarize the membrane. This in turn may activate other voltage-dependent channels (e.g. Ca^{2+} channels) and contribute to electrical and/or Ca^{2+} signaling, as well as regulation of membrane potential and pH gradient across the plasma membrane (Barbier-Brygoo *et al.*, 2000).

Not much is known about plant anion channels at the molecular level. Approaches

such as biochemical isolation of proteins or expression cloning in heterologous systems have yielded no results so far. The most successful approach has been to search for homologs of animal anion channels in plant genomes. Using this method, members of the family of voltage-dependent chloride channels (ClC) have been cloned from tobacco and *Arabidopsis*. In the complete sequence of the *Arabidopsis* genome, seven putative anion channels belonging to the ClC family have been identified. Functional expression of these channels has been problematic however, and it has not been proven that these proteins can function as anion channels in plant cells (Barbier-Brygoo *et al.*, 2000). Also, their cellular (membrane) localization is not known, something which would help in elucidating their function. A few experiments though, point to an intracellular localization for some of the plant ClCs (Lurin *et al.*, 2000). An intracellular localization might explain why functional expression is so difficult, since the channel has to be expressed in the plasma membrane to be accessible for electrophysiological recordings.

Mystery of plant Ca^{2+} channels

The calcium ion, Ca^{2+}, is an essential macronutrient for plants, required for normal growth and development. Calcium is required extracellularly as a structural component of cell walls, but is also important inside the cell as a cytoplasmic secondary messenger, and in the vacuole as a countercation for inorganic and organic anions (Marschner, 1995; White, 1998). The cytosolic Ca^{2+} concentration is maintained 1,000–10,000 times lower compared to the outside of the cell or to intracellular Ca^{2+} stores, such as the endoplasmic reticulum and the vacuole. There is thus a steep electrochemical gradient for Ca^{2+} over these membranes (especially the plasma membrane), due to the negative inside membrane potential difference and the concentration difference. Therefore, it is supposed that Ca^{2+} enters the cell via channels in the plasma membrane, which would facilitate rapid Ca^{2+} influx down its electrochemical gradient.

A number of electrophysiological and biochemical studies have proven the existence of Ca^{2+} channels in different plant species and membranes, and also characterized these channel activities (White, 2000; White et al., 2002). The channels are classified as influx channels if they exist in the plasma membrane, while those found in intracellular membranes are called release channels. There is evidence for both voltage-gated and ligand-gated Ca^{2+} channels. Different techniques have been used for their characterization, e.g. ^{45}Ca flux measurements in isolated vesicles, or electrically by incorporating vesicles into planar lipid bilayers or by patch-clamping different protoplasts, especially from root cells (see section Methods and techniques... below).

The molecular identities of these channels have been surprisingly difficult to determine, even with completion of the *Arabidopsis* genome sequence (and now also the rice genome) and the large number of EST (expressed sequence tag) sequences available from many species. Thus, no genes with obvious sequence similarity to the animal Ca^{2+} channels exist in plants. Recently, some candidate genes were proposed, such as AtTPC1 (for Two-Pore Channel 1) from *Arabidopsis*, a putative channel with some structural similarities to animal Ca^{2+} channels, which might function as a depolarization-activated Ca^{2+} channel (Furuichi et al., 2001). Other proposed candidates for voltage-regulated Ca^{2+} channels include members of the annexin gene family (White et al., 2002). Concerning the ligand-gated Ca^{2+} channels, the CNGC family and the GLR family have both been proposed as possible candidates (White et al., 2002; Talke et al, 2003).

Carriers Mediate Uptake and Efflux of Solutes and Act Like Membrane-bound Enzymes

Strictly speaking, carriers include pumps (ATPases/PPi-ases) and permeases. Permeases can then be divided into cotransporters (coupled permeases) and to uniporters (uncoupled permeases). Lastly, cotransporters are divided into symporters and antiporters (see Fig. 7.4). A brief description of these is given below.

Ion pumps (ATPases/PPi-ases): involved in primary transport → they create electrochemical gradients when transporting ions. They are energized by metabolites, usually ATP. These proteins are said to be **electrogenic**, that is, they are able to create a membrane potential.

Permeases: are involved in secondary transport → they use electrochemical gradients to do transport.

—**Cotransporters** (also called secondary transporters): the transport is coupled to a simultaneous transport of H^+ (Na^+ in animal cells). The membrane potential can help in transport.

—**Symporters**: the direction of H^+ transport is the same as for the other ion transported.

—**Antiporters**: the direction of H^+ transport is opposite to the direction of the transport of the other ion.

—**Uniporters**: the transport is not coupled to H^+ transport, but driven by the membrane potential and/or the concentration gradient of the transported ion.

Traditionally, however, researchers usually talk about pumps as one group and carriers as another, meaning cotransporters and uniporters.

All carriers, both pumps and cotransporters, are characterized by a change in conformation during transport. This is to prevent the ions from leaking out (or in) again. The ion/solute to be transported is fixed to a specific ion binding site on the carrier. Once the ion is bound, the energy required by this specific carrier is used to change the conformation, thereby allowing the ion to be transported to the other side of the membrane. Finally, the ion is released and the carrier returns to its initial conformation.

In most respects, carriers behave like enzymes:

- They have a specific ion ("substrate") binding site.
- They have a characteristic V_{max}: when all the ion binding sites are occupied the transport velocity is maximal.
- They have a characteristic K_m: this indicates the affinity for the ion.
- They can be inhibited by specific inhibitors, either competitive or noncompetitive inhibitors.

There is, however, a difference with normal enzymes: the transported substrate (the ions/solutes) is not modified during the catalytic cycle. In the next sections, we shall discuss the different ion pumps and carriers that exist in plant cells, and also give some specific examples.

Ion Pumps Transport Solutes Against their Electrochemical Gradient

As discussed above, membrane transport of an ion or solute against its electrochemical gradient (often called uphill transport) must be linked directly or indirectly to an energy-consuming mechanism, such as a pump in the membrane (Marschner, 1995). An ion pump can be compared to the pump used to inflate a tyre on a bike: you have to use energy to get the air into the tyre. Ion pumps also need energy and most of them are energized by ATP and some by inorganic pyrophosphate (PPi). As ion pumps change conformation and go through a catalytic cycle with both substrate and ATP (or PPi) binding for every one or two ion transported, the rate of transport is relatively slow, about a few hundred molecules per second. Remember that channels usually transport millions of ions per second.

Four major groups or families of ion pumps can be distinguished (Fig. 7.8). These are:

(a) P-type ATPases
(b) V-type ATPases
(c) H^+-pumping pyrophosphatases
(d) ABC transporters.

All but the H^+-pumping pyrophosphatase are energized by ATP. The ABC transporters and the P-type ATPases are the largest families with diverse substrate specificities. The V-type ATPases and the H^+-pumping pyrophosphatase are only involved in pumping protons.

P-type ATPase family pump cations

The P-type ATPase superfamily is found in all three branches of life. The reason for the name is that these enzymes bind ATP and form a phosphorylated intermediate as part of their reaction cycle. The P-type ATPases can be divided into five major evolutionary related subfamilies which group according to the ions they transport (Axelsen and Palmgren, 1998, 2001). In *Arabidopsis*, group 1 has been shown to be involved in heavy metal transport, group 2 in calcium transport, group 3 consists of the plasma membrane H^+-ATPases, group 4 is proposed to transport phospholipids while the substrate specificity of group 5 is not yet known (Fig. 7.8A). Some substrate specifici-

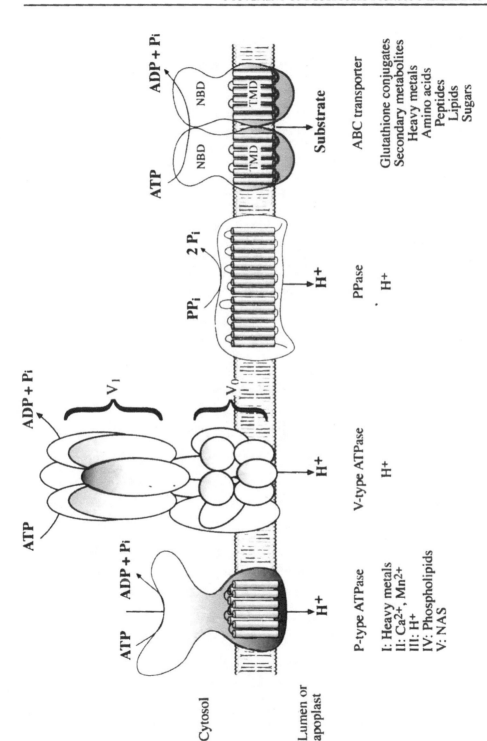

Fig. 7.8 The different types of pumps in plant cells. For the P-type ATPase, the pyrophosphatase and the ABC transporter, the proposed number of transmembrane domains is indicated. The substrates transported are shown below each class of transporter. For the P-type ATPases, the phylogenetic groups corresponding to the transported substrate are also indicated. NBD = Nucleotide Binding Domain, TMD = Transmembrane Domain.

ties are not present in plants, such as the animal Na⁺/K⁺-ATPase or the yeast Na⁺-pump which belong to group 2. On the other hand, group 3 (H⁺-ATPases) exists only in plants, yeast, and archae.

P-type ATPases have been thoroughly studied in animals, such as the Ca^{2+}-pump and the Na⁺/K⁺ -pump (Nobel Prize, 1997). The latter pump corresponds to the H⁺ pump in plants in terms of energization of membrane transport. Much of the knowledge from the animal field has also advanced the understanding of plant P-type ATPases. In fact, the largest gene family of P-type ATPases found in any organism so far is from *Arabidopsis* where a total of 46 genes have been identified in the complete genome sequence (Axelsen and Palmgren, 2001; Baxter *et al.*, 2003). Recently, completion of two draft sequences of rice allowed

identification of 43 similar isoforms of P-type ATPases in rice (Baxter *et al.*, 2003).

P-type ATPases usually have a single subunit consisting of 10 transmembrane helices and several cytosolic domains (see Fig. 7.9). Some of them can be regulated by inhibitory subunits. Most of the P-type ATPases studied in plants possess an autoinhibitory domain in their N- or C-termini and no regulatory subunits. The plant P-type ATPases have been localized to the plasma membrane, to the vacuolar membrane, and to the ER, perhaps also to the chloroplast inner envelope. The direction of the ion transport is usually away from the cytosol (efflux), either to the apoplast or into the lumen of organelles, but some of them may function as high affinity uptake transporters (Palmgren and Harper, 1999). They have a conserved motif (DKTGT = aspartic

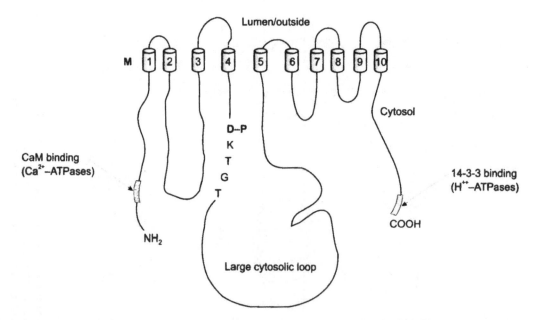

Fig. 7.9 Schematic representation of a typical plant P-type ATPase with 10 transmembrane domains, and the cytosolic domains. N and C termini may be shorter or longer. The conserved sequence diagnostic for P-type ATPases is indicated, as is the location for the binding of regulatory proteins. P = phosphate group.

acid, lysine, threonine, glycine, threonine) which can be used to identify a protein as belonging to the P-type ATPase family and which is found in the large cytosolic loop (Fig. 7.9). During the catalytic cycle, the phosphate group from ATP is transferred to the conserved aspartic acid. In the next two sections, we take a closer look at two P-type ATPases, the Ca^{2+}-ATPase and the plasma membrane H^+-ATPase.

Ca^{2+}-ATPases important for maintaining cytoplasmic Ca^{2+} concentration at low level

As mentioned in connection to Ca^{2+} channels, the intracellular, cytosolic Ca^{2+} concentration is 1,000-10,000 times lower than in the apoplast or in the vacuole in plant cells. This is because high concentrations of Ca^{2+} are toxic to the cell, since it will then interfere with the phospho-compounds in the cell (e.g. ATP), forming insoluble Ca^{2+}-phosphate precipitates. Plant cells (as well as animal, bacterial, and fungal cells) therefore have evolved systems for the control of cytoplasmic Ca^{2+} concentration. One of the components for the control of the cytoplasmic Ca^{2+} concentration is the Ca^{2+}-ATPases, belonging to the P-type ATPase family.

In animal cells, Ca^{2+}-ATPases are found mainly in the plasma membrane (called PMCA) and in the sarco/endoplasmic reticulum (called SERCA). A few years ago, a high-resolution crystal structure was resolved (at 2.6 Å) for one SERCA with two Ca^{2+} ions bound to it (Toyoshima et al., 2000). The same group has recently published the crystal structure of the same protein, but with an inhibitor bound to it (Toyoshima and Nomura, 2002). The structures have not only confirmed the results of numerous biochemical studies on how most P-type ATPases work and function, but also

brought new insight into the structure-function relationships for these enzymes (Fig. 7.10). The substrate for Ca^{2+}-ATPases is Mg^{2+}-ATP, as a complex. The reaction cycle includes two conformational (or energy) states called E1 and E2. In the first step, Ca^{2+} binds to a site on the cytoplasmic side of the membrane. The binding of Ca^{2+} is necessary for the next step which is the binding of Mg^{2+}-ATP and subsequent phosphorylation of the ATPase. All this results in a large change in conformation so that the Ca^{2+} ion(s) now are buried (occluded) in the membrane. Finally, another conformational change brings about the release of Ca^{2+} to the other side of the membrane (Toyoshima et al., 2003).

There are two types of plant Ca^{2+}-ATPases, those that are stimulated by the small, soluble Ca^{2+}-binding protein calmodulin (CaM), called ACA for autoinhibited Ca^{2+}-ATPase (PMCA in animals), and those that are not, called ECA for ER-type Ca^{2+}-ATPase (SERCA in animals) (Geisler et al., 2000a). The sequencing of the Arabidopsis genome identified ten genes encoding ACA-type Ca^{2+}-ATPases and four ECA genes. In plants, ACA-type Ca^{2+}-ATPases have been found in the plasma membrane, in the vacuolar membrane and in the ER membranes, maybe also in the chloroplast inner envelope (see Fig. 7.5). The plant CaM- Ca^{2+}-ATPases have an extended N terminus containing the site for CaM fixation, and functioning as an autoinhibitory domain (Malmström et al., 1997, 2000; Harper et al., 1998; Bonza et al., 2000; Geisler et al., 2000b). The N-terminal location of the autoinhibitory domain is in contrast to the plant plasma membrane H^+-ATPase and to the animal PMCA, which have a long, autoinhibitory C terminus that is the target for 14-3-3 and CaM binding respectively (see below; Carafoli, 1997; Penniston and Enyedi, 1998).

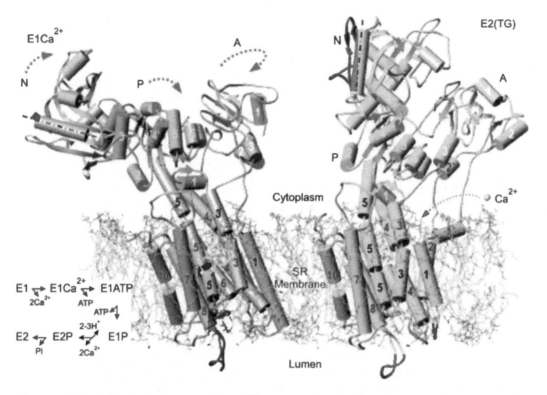

Fig. 7.10 Model of the sarcoplasmic reticulum Ca^{2+}-ATPase (SERCA) in two different conformations. $E1Ca^{2+}$ = Ca^{2+}-bound form in the E1 state in the simplified reaction scheme on the left. E2(TG) = SERCA with the inhibitor thapsigargin (TG) bound to it, corresponding to the E2 state in the reaction scheme. Cylinders represent α-helices and arrows β-strands. N = nucleotide binding domain, P = phosphorylation domain, A = actuator or anchor domain. Arrows in $E1Ca^{2+}$ indicate direction of movement of the cytoplasmic domains during the change from $E1Ca^{2+}$ to E2(TG). The proposed pathway for Ca^{2+} to enter into the binding cavity is indicated with an arrow in E2(TG) (from Toyoshima, Nomura, Sugita, 2003).

As in animal cells, Ca^{2+} ions are thought to function as secondary messengers in plant cells for the transduction of signals from the outside. Many different physiological stimuli have been shown to increase the intracellular concentration of Ca^{2+} (Sanders *et al.*, 1999; Anil and Rao, 2001; Knight and Knight, 2001). The Ca^{2+} ions enter the cell via different channels in the plasma membrane or the vacuolar membrane. Then, Ca^{2+} binding proteins such as CaM bind Ca^{2+} and are activated. These in turn can bind to e.g., the N terminus of a Ca^{2+}-ATPase and activate it to extrude the Ca^{2+} from the cytosol so as to restore resting Ca^{2+} levels. However, the exact role of Ca^{2+}-ATPases in plant signal transduction has not yet been clarified.

The ECA-type of Ca^{2+}-ATPases has mainly been found in the ER membranes; it is not known how they are regulated at the protein level (Sanders *et al.*, 2002). One of the isoforms has been shown to also transport manganese (Wu *et al.*, 2002).

Plasma membrane H⁺-ATPase: essential for circulation of ions and solutes in plant cells

The plasma membrane H^+-ATPase is intensively studied because of its role in nutrient uptake and efflux. There are 11 plasma membrane H^+-ATPase isoforms in *Arabidopsis* and 10 isoforms were recently identified in the rice genome (Baxter *et al.*, 2003). Several isoforms are expressed in a tissue-specific manner and the fact that there are so many isoforms is usually explained by the need to express ranging numbers of pumps in different tissues.

The major physiological role of the H^+-ATPase is to energize secondary transport by the creation of the electrochemical gradient for H^+ across the plasma membrane. One example is the H^+-coupled transport of sucrose into phloem vessels. Indirect evidence for this role is the expression pattern of the H^+-ATPase isoforms: they are highly abundant in the phloem, in root hairs and epidermis, and in guard cells. All these cell types are specialized in active transport and accumulation of solutes. The role in solute transport is supported by genetic studies (e.g. analysis of mutant plants) which also point to a role for this enzyme in salt tolerance, opening and closing of the stomatal pores (osmoregulation), and in cellular expansion (Morsomme and Boutry, 2000; Palmgren, 2001).

Like for the Ca^{2+}-ATPase, the substrate of the H^+-ATPase is Mg^{2+}-ATP. While Mg^{2+} is essential for functioning of the enzyme, K^+ and Ca^{2+} are able to stimulate its activity (Marschner, 1995). Since the plasma membrane H^+-ATPase plays a central role in the establishment of the proton motive force across the plasma membrane, it is thought to be tightly regulated to be able to respond quickly to cell needs. A large number of biochemical investigations have been performed to study regulation of the plasma membrane H^+-ATPase. These have shown that the C-terminal region is highly important for regulation of this enzyme: it can function as an autoinhibitory domain, and is the target for phosphorylation and binding of the regulatory protein 14-3-3 (Morsomme and Boutry, 2000; Palmgren, 2001) (see Fig. 7.9). First, it was demonstrated that when the C terminus is removed or displaced, the H^+-ATPase is activated with an increased V_{max}, a lower K_m for ATP, a more alkaline pH optimum, and an improved coupling between ATP hydrolysis and H^+ pumping (i.e. the enzyme is less "leaky") (Palmgren *et al.*, 1990, 1991; Regenberg *et al.*, 1995). More recently, several research groups have shown that H^+-ATPase is activated when one of the last amino acids (a threonine) is phosphorylated, thus creating a binding site for the regulatory protein 14-3-3 and displacement of the C terminus (Olsson *et al.*, 1998; Fuglsang *et al*, 1999; Svennelid *et al.*, 1999). These results were obtained thanks to earlier studies involving the fungal toxin fusicoccin. It has long been known that fusicoccin is able to cause wilting of plant leaves by an irreversible opening of their stomata and other physiological effects (Marré, 1979). All of these effects could be ascribed to activation of the H^+-ATPase involving the C-terminal autoinhibitory domain (Olsson, 2000). Initially, it was thought that fusicoccin bound directly to the H^+ pump but was later shown that its binding must involve a protein belonging to the 14-3-3 family of proteins (Korthout and de Boer, 1994; Marra *et al.*, 1994; Oecking *et al.*, 1994). These were exciting results because 14-3-3 proteins had only been shown thus far to play a role in the animal brain. Now it is known that 14-3-3 proteins regulate the activity of several plant proteins involved in primary metabolism. Another compound able to activate the H^+-ATPase is the phospholipid lysophosphatidylcholine (lyso-PC), probably by displacing the autoinhibitory C terminus (Palmgren and Sommarin, 1989). Lyso-PC is a product of the enzyme phospholipase A_2 (PLA_2).

Finally, the expression of H^+-ATPase is regulated at both the transcriptional and the translational level (Morsomme and Boutry, 2000). The transcriptional regulation involves the tissue-specific expression of certain isoforms mentioned above, and also regulation by environmental factors.

H^+ pumps from different gene families

The different H^+ pumps are of extreme importance for the cellular energetics and ion currents in plant cells. There are three distinct electrogenic H^+ pumps from three different (gene or protein) families that generate proton electrochemical gradients: the plasma membrane H^+-ATPase (P-type ATPase family), the vacuolar-type H^+-ATPase (V-type ATPase), and the vacuolar H^+ pumping pyrophosphatase (H^+-PPase). In addition, a fourth type of H^+ pump exists in the cell: it is the F-type H^+-ATPase in the inner membranes of mitochondria and the thylakoid membrane of chloroplasts. Normally, the F-type H^+-ATPase (or F_0-F_1 ATPase) operates in the opposite direction, i.e. it catalyzes the synthesis of ATP using the proton motive force created by the respiratory and the photosynthetic electron transport. The F-type H^+-ATPase is similar in structure to the V-type H^+-ATPase (see below), and they are evolutionarily related. On the other hand, the plasma membrane H^+-ATPase and the vacuolar H^+-ATPase are probably examples of independent evolution: their structure and mode of action are not at all the same (see below) but their function is (to pump protons).

The first two pumps in particular are extremely important for the cell to be able to perform secondary transport:

- The plasma membrane H^+-ATPase uses ATP as energy source to pump protons from the cytoplasm out to the apoplast across the plasma membrane.

- The vacuolar H^+-ATPase uses ATP as energy source to pump protons from the cytoplasm into the vacuole across the vacuolar membrane (tonoplast).

The electrogenic H^+ pumps create two different gradients over the membrane:

(1) An electric gradient = the membrane potential

(2) A concentration (chemical) gradient of H^+.

As shown above, a great electrochemical potential resides within the electrochemical gradient for H^+; this is the proton motive force.

In contrast to the single subunit of the plasma membrane H^+-ATPase, the composition of the V-type ATPase is highly complex (see Fig. 7.8B). It is composed of a peripheral V_1 cytoplasmic sector consisting of five to eight subunits, and of a membrane integral V_0 subcomplex consisting of three to five subunits. ATP hydrolysis takes place in the V_1 part while the V_0 subcomplex is responsible for the H^+ translocation (Ratajczak, 2000). V-type ATPases are found in all eukaryotes and are thought to be essential not only for the energization of ion and metabolite fluxes, but also for receptor cycling and protein targeting, trafficking of membrane proteins (protein sorting) and in membrane fusion as a result its acidification of different intracellular compartments (Sze and Palmgren, 1999). In fact, V-type ATPases have been shown to be localized with a variety of biological membranes in plant cells, e.g. the ER, Golgi, coated vesicles, provacuoles, vacuoles, and even the plasma membrane. Therefore the term V-ATPase refers to common biochemical and molecular features rather than membrane localization (Sze and Palmgren, 1999).

Similarly, pyrophosphatase activities have been described not only in plant vacuolar membranes, but also in plant submitochondrial particles (SMPs) and in thylakoid membranes (Drozdowicz et al., 2000). H^+-PPases exist only in plants, some bacteria, and in some archae and protozoa (Maeshima, 2000). The most studied plant H^+-PPase is the one located in the vacuolar membrane. It consists of a single 80 kD polypeptide, and pumps H^+ across the membrane by hydrolyzing PPi (Maeshima, 2000; Fig. 7.8C). It is considered to contribute to establishment of the transmembrane H^+-electrochemical potential difference which is required for the secondary transport of solutes in and out of the vacuole (Rea and Poole, 1993). Recently, a second H^+-PPase isoform was identified and characterized in *Arabidopsis*; it appears to represent a novel category of H^+-PPases and the authors hypothesized that this enzyme might function in PPi synthesis (thus operate in the reverse direction) in the inner membrane of mitochondria (Drozdowicz et al., 2000).

ABC transporters: transporting multiple substrates

The last group of ion pumps, the ATP Binding Cassette or ABC transporters, are mostly known for being involved in drug resistance processes such as the multidrug resistance transporter (MDR subfamily) in mammalian cells. Similarly, plant ABC transporters are able to not only transport toxic compounds (usually as glutathione conjugates) out from the cell or into the vacuole, but also a wide variety of other substrates: lipids, heavy metal ions, sugars, amino acids, peptides and secondary metabolites (Theodoulou, 2000). ABC transporters are members of the ABC protein family which is the largest protein family known and larger in plants than in animals or yeast: in *Arabidopsis* 131 genes have been identified as belonging to this family (Sanchez-Fernandez et al., 2001; Jasinski et al., 2003)! Members of this family are found in all organisms and can be divided into several subfamilies, 12 in *Arabidopsis*. The best characterized ones are the pleiotropic drug resistance (PDR), multidrug resistance (MDR), and multidrug resistance-associated protein (MRP) subfamilies. Not all of them are membrane proteins, but are instead found in the cytosol. In *Arabidopsis*, of the 131 ABC genes identified, about 100 are predicated to be membrane proteins (Sanchez-Fernandez et al., 2001).

A typical feature of ABC transporters is their modular structure (Fig. 7.8D). Each functional transporter is composed of four core domains, two cytosolic nucleotide binding domains (NBDs), and two transmembrane domains (TMDs) made of four to six membrane-spanning helices (Theodoulou, 2000; Jasinski et al., 2003). The core domains may be on four separate polypeptides or fused in different combinations on one or two polypeptides. In eukaryotes the most common organization is the "half-molecule" with one TMD and one NBD, and the "full molecule" containing all four domains in a single polypeptide (Sanchez-Fernandez et al., 2001). Different subfamilies have differently organized domains: either $(TMD-NBD)_2$ or $(NBD-TMD)_2$, with some variations. In *Arabidopsis*, of the 100 predicted membrane ABC proteins approximately 50% are "full-size" and 50% "half-size" transporters (Sanchez-Fernandez *et al.*, 2001). A few plant ABC transporters have been characterized from some subfamilies. These have mostly been localized to plasma or vacuolar membranes, but also to peroxisomal membranes and in mitochondria (Footitt *et al.*, 2002; Jasinski *et al.*, 2003). Even if the function of most plant ABC transporters is not yet known, it is agreed that these proteins are of major importance for plants not only for cellular detoxification, but also e.g. for regulation of ion fluxes and plant growth and development (Martinoia *et al.*, 2002). The proposed roles of ABC transporters come both from studies of membrane vesicles and from analysis of mutant plants (see section Methods and techniques... below). A major future challenge for "animal" researchers is also to elucidate the three-dimensional structure of ABC transporters, for better understanding how these versatile transporters are able to transport such a large variety of substrates.

Cotransporters and Uniporters Facilitate Secondary Transport

In plants an enormous amount of various carrier-type proteins exists, representing the most common mode of transport (it would be too costly and too slow to use only ATP-driven transport). In the model plant, *Arabidopsis*, the total number of putative carriers is 546, which is 65.5% of the total number of transport proteins identified in its genome. The high number of transporters is required because of the lifestyle of plants, with uptake of mineral nutrients from the soil and photosynthetic production of sugars. All these compounds need to be distributed throughout the plant, mineral nutrients from the roots to the shoot, and sugars from the leaves to the rest of the plant. Furthermore, many solutes need to be transported within the cell, for example from the chloroplast to the cytosol for photosynthetic products, or toxic compounds and metals have to be compartmentalized in the vacuole. The function of several carriers has been established but the majority have not been characterized. Table 7.2 describes known and putative secondary transporters mainly from *Arabidopsis*.

In general, uptake systems function as symporters, coupling the transport of a solute against its electrochemical gradient to the influx of protons down its electrochemical gradient. These proteins are mostly localized in the plasma membrane. Efflux systems can be localized both at the plasma membrane and in organellar membranes such as the vacuolar membrane. They often function as antiporters (also called exchangers), i.e. the direction of the solute transport is opposite to the direction of the flow of protons.

Uniporters are energized by the membrane potential and/or the concentration gradient of the transported ion. In fact, they function like a channel but much more slowly.

The rate of transport for cotransporters is in between that of channels and pumps, about 100–10,000 molecules per second. Sometimes cotransporters may behave like channels and transport ions much faster. This seems to depend on the conditions under which they are assayed. Hence animal researchers no longer strictly distinguish between ion channels and cotransporters (in which this phenomenon has been studied in greated detail). In plants some studies suggest that the wheat HKT1 Na^+/K^+ transporter may also function as a channel (Rubio *et al.*, 1995; Gassmann *et al.*, 1996). Indeed, the membrane topology of the yeast homolog of HKT1, the TRK K^+ transporter, was recently reanalyzed (Durell and Guy, 1999; Kato *et al.*, 2001;

TABLE 7.2 Known and putative carriers in *Arabidopsis thaliana* (not exhaustive). Data mainly from TransportDB - Transporter Protein Analysis Database: http://66.93.129.133/transporter/wb/index2.html

Family	Subfamilies/ Examples	No. of genes	Possible substrate	Proposed transport scheme	Comments	References
AAA	AtAATP1-2	2	ATP/ADP exchange (chloroplast)	Counter-exchange		Neuhaus et al., 1997
AAAP	AtAAP1-5, AtAUX1	43	Amino acids, Auxin	Symport		Williams and Miller, 2001
AE	AtBOR1	7	Boron	?	BOR 1 involved in boron efflux to xylem and shoots.	Takano et al., 2002
AEC	AtPIN1	8	Auxin	Antiport	Auxin efflux	Gälweiler et al., 1998
Amt	AtAMT1, AtAMT2	6	NH_4^+ or methylNH4	Uniport	NH_4^+ uptake	Ninnemann et al., 1994
APC	AtAAP1	12	Cationic amino acids	Symport		Ortiz-Lopez et al., 2000
	N.C.	5	Organic acid/Na^+?			
CaCA	CAX1,2,3,4	12	Ca^{2+}/H^+ (Mn^{2+}/H^+)	Antiport		Hirschi et al., 1996
CCC	Axi4 (*Nicotiana tabacum*)	1	NaCL and/or KCl	Symport		Harling et al., 1997
CDF	ZAT (AtMTP1)	8	Zn^{2+}, other heavy metals	Antiport	Metal efflux	Van der Zaal et al., 1999

(Contd.)

(Contd. Table 7.2)

CPA1	AtNHX1, AtSOS1/AtNHX7	8	Na$^+$/(K$^+$, Li$^+$, Cs$^+$)/H$^+$	Antiport	Na$^+$ efflux at VM, PM	Apse et al., 1999
CPA2	AtKEA1	32	Na$^+$ and/or K$^+$/H$^+$	Antiport		Véry and Sentenac, 2003
DASS	SodiT1 (Spinacia oleracea), AtSDAT	5	Malate:Na$^+$ antiporter (Ath), oxoglutarate: malate symporter (So)	Antiporter or symporter depending on species	Chloroplast inner envelope (spinach)	Weber et al., 1995
DMT	TPT; POP	39	Organic phosphate esters/Pi (TPTs), Organocations (POPs)	Exchange with Pi	Chloroplasts, non-green plastids.	Flügge, 1999
ENT	N.C.	8	Nucleosides	?	N.C. in plants	
FBT	N.C.	7	Folates/pteridine	?	N.C. in plants	
GPH	AtSUC1, AtSUC2	9	Sucrose	Symport	Uptake of sugars, plasma membrane	Saier and Stolz, 1994; Lemoine, 1994
HAAAP	N.C.	1	Amino acids	?	N.C. in plants	
KUP	AtKUP/HAK/KT	13	K$^+$	Symport	PM, maybe also VM? Both high and low affinity K$^+$ uptake	Fu and Luan, 1998
MC	AtDTC, AtAAC, AtBAC1, AtUCP1	52	Di-and tricarboxylates, ATP/ADP, basic amino acids, H$^+$ uncoupling	Counter-exchange: Uncoupling proteins	Many plant genes/proteins not characterized.	Picault et al., 2004

(Contd.)

Family	Genes	Number	Substrate	Mechanism	Function	Reference
MFS	HUP1; PT1	93	Hexose; Pi; other sugars, organic cations	Hexose: H^+ symporter; Pi uptake	Uniporter, symporter or antiparters	Leggewie et al., 1997
MOP	MATE subfamily	56	Drugs, alkaloids, antibiotics, Cd^{2+}	Antiport	Mechanism not known in detail	Li et al., 2002
NCS1	N.C.	1	Uracil?	?		
NCS2	N.C.	12	Nucleobases, Na^+/Ascorbic Acid?	?		
NhaD	N.C.	2	Na^+/H^+	?		
Nramp	AtNramp1-4; EIN2	7	Metal ions (Mn^{2+}, Fe^{2+}, Cd^{2+})	Antiport	VM, also PM?	Curie et al., 2000
OPT	YS1	15	Fe^{3+}, Zn^{2+}, Cu^{2+}, Ni^{2+}, Mn^{2+} and Cd^{2+}	Symport	Uptake of metal ions as phytosiderophore complex	Curie et al., 2001
PiT Transporter	AtPht2;1	1	Phosphate	Symport	Low affinity uptake	Daram et al., 1999
POT	AtCHL1, AtPTR	49	Peptides; nitrate; histidine	Symport	High + Low affinity for nitrate	Tsay et al., 1993
RND	N.C.	2	Drug efflux	?		
SSS	N.C.	1	Urea	?		
SulP	SUT1, SUT3 Nod70	11	Sulfate	Symport	Sulfate uptake	Smith et al., 1995
TDT	N.C.	4	Tellurite? Dicarboxylates?	?		
Trk	AtHKT1, Hkt1	1	K^+	K^+/Na^+ Symport	Role in salt tolerance	Schachtman as Schroeder, 1994
ZIP	AtIRT1, AtIRT2	13	Fe^{2+}, Zn^{2+}, Mn^{2+}, maybe also Cd^{2+}, CO^{2+}	?	Role in iron uptake from the soil (IRT1)	Eide et al., 1996

N.C. = Not characterized
VM = Vacuolar membrane
PM = Plasma membrane

Zeng *et al.*, 2004). Previously, the TRK protein was suggested to have 12 transmembrane segments, as do many cotransporters. After comparison with the bacterial KcsA K⁺ channel (see Box 7.1) and immunocytochemical experiments, the authors proposed a more channel-like structure of this family of transporters (Kato *et al.*, 2001; Zeng *et al.*, 2004).

Na⁺/H⁺ antiporters: role in salt tolerance

Sodium is toxic to many higher plants, including most crop species, having both osmotic effects and Na⁺-specific toxic effects. In nonsaline growth conditions, the cytosolic K⁺ concentration is 100–200 mM and that of Na⁺ 1–10 mM. At higher Na⁺ concentrations in the soil, Na⁺ can therefore enter cells "downhill" with its concentration gradient. When the K⁺/Na⁺ ratio is disturbed or when the concentration of total salts is too high, many enzymes are inactivated and protein synthesis is inhibited. In chloroplasts, photosynthesis is inhibited by high concentrations of Na⁺ and/or Cl⁻. These are the Na⁺-specific toxic effects (Marschner, 1995).

One reason for the osmotic effects of Na⁺ (and other salts) is its accumulation in the apoplast, which leads to dehydration, turgor loss, and death of leaf cells (Marschner, 1995). Also, a high Na⁺ concentration (and salt) in the soil leads to osmotic stress because the soil water potential is lowered; hence the plant needs to develop an even lower water potential to allow for water influx along a "downhill" gradient. This osmotic stress resembles drought stress, i.e. insufficient water in the soil.

Plants have evolved two solutions for survival in a saline environment: either exclude Na⁺ ions at the plasma membrane or sequester them in the large intracellular vacuole. In general, the more salt-sensitive plants are "excluders" while the more salt-tolerant ones are "includers".

One interesting example of all the different cotransporters is the family of Na⁺/H⁺ antiporters. In *Arabidopsis*, eight genes have been identified as belonging to the family called CPA1, for Monovalent Cation:Proton Antiporter 1 (Mäser *et al.*, 2001; see Table 7.2). The first plant CPA1 gene was identified in the genome sequence of *Arabidopsis*, based on sequence homology with a yeast Na⁺/H⁺ antiporter and also bacterial and mammalian homolog (AtNHX1; Apse *et al.*, 1999; Gaxiola *et al.*, 1999); an antiporter activity had already been shown to exist in vacuolar membranes by biochemical methods (Blumwald and Poole, 1985).

When AtNHX1 was cloned, it was shown by different biochemical assays to be localized in the vacuolar membrane and to function as an electroneutral Na⁺/H⁺ exchanger (Apse *et al.*, 1999). Most exciting was that overexpression of this gene in *Arabidopsis* led to a high degree of salt tolerance in this normally salt-sensitive plant. Transgenic plants were unaffected when treated with up to 200 mM NaCl, while wildtype plants showed progressive growth inhibition (Apse *et al.*, 1999). Surprisingly, it appeared that one single gene could be responsible for such a high degree of salt tolerance. These findings might open up new possibilities to engineer crop plants with improved salt tolerance and some attempts have been made in tomato and rape (Apse *et al.*, 1999; Zhang *et al.*, 2001a, b).

At least one more of the CPA1-family proteins is localized to the tonoplast (AtNHX2; Yokoi *et al.*, 2002) while another resides in the plasma membrane (AtSOS1; Qiu *et al.*, 2003). Also these transporters have been shown to be important for salt tolerance (Yokoi *et al.*, 2002; Shi *et al.*, 2003). In fact, the SOS1 gene was identified in a screen for plants with a deregulated response to salt stress (SOS = $\underline{S}$alt $\underline{O}$verly $\underline{S}$ensitive pathway) and its protein product shown to function as an electroneutral Na^+/H^+ exchanger (Shi *et al.*, 2000; Qiu *et al.*, 2002, 2003). The mRNA levels of several *NHX* genes are increased by salt stress and phosphorylation is one of the regulatory mechanisms for their transport activity (Shi *et al.*, 2000; Shi and Zhu, 2002; Yokoi *et al.*, 2002; Qiu *et al.*, 2004). It has also been proposed that the vacuolar Na^+/H^+ exchangers could have a role in regulation of the vacuolar pH in nonsaline conditions (Yamaguchi *et al.*, 2001; Shi and Zhu, 2002).

Transporter Genes Mostly Present in Genome as Large Gene Families

When the *Arabidopsis* genome sequence was finished and analyzed for putative transporter genes, it became apparent that most of the transporters are encoded by multigene families. Expression patterns of the respective family members often overlap and it seems that functional redundancy is common. For example, many isolated null mutations do not have an obvious phenotype. In addition, when it comes to carriers, many inorganic nutrients are transported by more than one transporter family. Why is this?

Because plants are immobile, they cannot escape from adverse environmental conditions; instead they have to adapt. This fact may explain why so many different transporter genes are present. Ions need to be taken up and redistributed throughout the plant under many different conditions, and different tissues may express different isoforms of a family. In addition to a role in plant nutrition, many ion transporters are also involved in the transductions of the signals telling the changes in conditions or in detoxification processes. The unusually high number of ABC transporter genes may be attributable to the fact that plants do not have excretory organs and thus need to export toxic compounds at the cellular level. Plants also have to cope with a wide variety of secondary metabolites, most of which have to be transported out from the cell that synthesizes them or sequestered in organelles because they are toxic (Sánchez-Fernández *et al.*, 2001). Furthermore, the soil represents a complex chemical environment for plants, and many kinds of transporters have to be present in the root cells. Also, the same ion often has to cross both the plasma membrane and intracellular, organellar membranes, something which may necessitate a diversity of isoforms (localization to different membranes). It has been shown in some cases that different isoforms have different substrate affinities or kinetics (e.g. the plasma membrane H^+-ATPases; Palmgren and Christensen, 1994; Luo *et al.*, 1999). For ion channels such as the K^+ channels, different isoforms can form heterotetramers, thus contributing to a fine tuning of regulation and function.

Genetic analyses have shown that a polyploidization event in an ancestral plant around 150 million years ago resulted in genetic redundancy, and also that many genes exist as tandem gene arrays because of local gene duplications (Axelsen and Palmgren, 2001). Thus, plants have evolved into organisms having a large set of genes rapidly ready for transcription and translation, thereby being able to quickly adapt to a changing environment.

Recently, a transcriptome approach was taken in order to assess how the activity of specific transporters is integrated into a response to fluctuating conditions (Maathuis *et al.*, 2003). An oligonucleotide array (oligonucleotides fixed on a glass slide) representing 1096 Arabidopsis transporter genes was constructed and the root transporter transcriptome analyzed with respect to different conditions of cation stress. A transcriptome can be defined as the complete collection of transcribed elements (mRNAs) of the genome. In this case, the mRNAs were isolated from roots coming from plants that had been subject to cation stress. For each treatment, the mRNAs were then allowed to hybridize to the oligonucleotides on the microarray. Genes that were better expressed with a certain treatment for example would then give an increased signal compared to untreated plants. The most important results of this analysis were that the plant response was even more complex than previously thought: nutrient stress did not affect specific gene families but rather a collection of genes across all transporter families (Maathuis et al., 2003).

PLANT MEMBRANES WITH THEIR SPECIFIC TRANSPORT SYSTEMS

Methods for isolation of high purity plant membrane vesicles (so-called "transport-competent" vesicles) were developed in the late 1970s and early 80s (see section Ion Transport… below). This greatly helped in the study of the transport capacities of the different plant membranes. Likewise, development of electrophysiological methods has allowed for detailed characterization of many transport processes in especially the plasma membrane (see section Electrophysiological techniques… below). Later, molecular cloning, forward and reverse genetics and whole genome analysis have additionally increased our understanding of plasma membrane transport (see sections Baker's yeast… and Transgenic plants… below).

Plasma Membrane, Interface for Nutrient Uptake and Signal Transduction between Cell Interior/Exterior

The plasma membrane represents the outermost barrier of the cell, and forms a living barrier between the plant cell and its surroundings. It is the site for regulation of the composition of the cell's interior and also for reception of a variety of signals that must be communicated to the cell from the exterior. One of the major, and most important components of the plasma membrane, is the electrogenic H^+-ATPase. In fact, the first plant transporter ever to be cloned was the plasma membrane H^+-ATPase, by three independent groups in the same year (Boutry et al., 1989; Harper et al., 1989; Pardo and Serrano, 1989). As discussed above, its activity allows for establishment of an electrochemical H^+ gradient across the membrane, thus providing the energy for solute uptake or efflux by secondary transporters. This enzyme is exclusively localized in the plasma membrane, in contrast to many other transport proteins that exist both at the plasma membrane and in intracellular membranes. All the transport systems (carrier type) needed for the uptake of solutes from the extracellular environment are of course localized in the plasma membrane, e.g. for uptake of nitrate, potassium, phosphate, sulfate, divalent cations such that calcium, iron and zinc, and other essential macro- and micronutrients (see Fig. 7.5). These uptake systems are most often expressed in root cells (but not exclusively). The solutes taken up in the root must be relocated to the shoot; hence transport systems exist in the plasma membrane of e.g. xylem parenchyma cells for the loading of solutes into the xylem. Likewise, in other cells transport systems for both the efflux and uptake of certain solutes exist in the plasma membrane, like sugar transporters for the distribution of photosynthetic products from the leaves to the rest of the plant (sugar translocation, phloem transport or loading and unloading).

Ion channels are also involved in the uptake or release of ions, e.g. all characterized members of the K^+ channels of the Shaker family are localized in the plasma membrane. Several different types of Ca^{2+} channels have been characterized in the plasma membrane by electrophysiological methods, as have at least two types of voltage-dependent anion channels, allowing for the efflux of anions (White, 1998; Barbier-Brygoo, 2000). In the plasma membrane of stomatal guard cells, a third type of anion channel which is mechanosensitive has been characterized (Cosgrove and Hedrich, 1991).

Aquaporins, members of the MIP family (Major Intrinsic Proteins) are as the name indicates, highly abundant in the membranes of plant cells and may represent as much as 20% of the total protein in a membrane (Johansson et al., 1996). The PIP subfamily of aquaporins has a function in the osmoregulation of the cell and probably for regulation of transcellular water flux (Chrispeels et al., 1999; Johansson et al., 2000).

Finally, systems for the efflux of toxic compounds, e.g. ABC transporters and proteins of the MATE subfamily (= Multidrug And Toxic Extrusion; see Table 7.2) also exist in the plasma membrane (e.g. Sidler et al., 1998; Li et al., 2002).

Vacuolar Membrane Contains Transport Systems for Accumulation of Ions and Toxic Compounds

The vacuole has many functions essential for plant life, such as generating turgor (by the uptake of water) which is necessary for cell elongation and for hydraulic stiffness of the stem, storage of reserve proteins in seeds and fruits, and pigments in flowers, as reservoirs of ions and metabolites, and in sequestration of toxic compounds (Wink, 1993; Marty, 1999). In fact, several types of vacuoles exist within the plant cell; these can be distinguished by their content of soluble proteins and by the class of aquaporins localized in their respective membranes (Jauh et al., 1999). Most of the mature vacuoles are acidic (pH 4.5–5.9), containing hydrolytic enzymes, and function as digestive organelles analogous to the lysozymes in animal cells. The vacuole thus plays an important role in the general cell homeostasis and detoxification processes. Vacuoles are also involved in cellular responses to environmental and biotic factors that provoke stress (Marty, 1999); one example is the release of vacuolar Ca^{2+} to the cytosol upon exposure of the cell to external stimuli.

Several transport systems have been identified in the vacuolar membrane (Fig. 7.5; Martinoia et al., 2000). A typical feature of plant vacuolar membranes is the existence of two systems for the transport of protons from the cytosol into the vacuolar lumen. These are the vacuolar H^+-ATPase and the vacuolar H^+-PPase (described earlier) important for acidification of the lumen and for the establishment of an electrochemical gradient of H^+ that can be used for secondary transport of other solutes into and out of the vacuole.

Other ATP-driven, primary transport systems identified in the vacuolar membrane are the Ca^{2+}-ATPases and ABC transporters. As described above, ABC transporters function in the sequestration of a wide array of different compounds, many toxic ones, but also for example secondary metabolites such as pigments. Since the vacuole is the organelle that accumulates these compounds, several ABC transporters are expressed in its membrane (Martinoia *et al.*, 2000).

Ca^{2+}-ATPases are thought to be important for maintaining a low cytoplasmic Ca^{2+} concentration, and to provide the vacuole with relatively high amounts of Ca^{2+} (around 1 mM) that can then be released by channels and used as a second messenger upon an external stimuli (Geisler *et al.*, 2000a). Similarly, several Ca^{2+}/H^+ antiporters (e.g. CAX1) have been localized in the vacuolar membrane, also functioning in the sequestration of Ca^{2+} into the vacuole, but with a lower affinity for Ca^{2+} (Hirschi *et al.*, 1996; Ueoka-Nakanishi *et al.*, 1999). Some members of the CAX family seem to transport Mn^{2+} or metal ions rather than Ca^{2+} (Hirschi *et al.*, 2000). Other cotransporters that have been identified in the vacuolar membrane are the Na^+/H^+ exchanger (e.g. AtNHX1) and metal ion transporters of the Nramp family (= Metal Ion Transporters, Natural resistance-associated macrophage proteins) (see Table 7.2; Apse *et al.*, 1999; Curie and Briat, 2003). Transport systems for the uptake and release of, for instance, inorganic anions, amino acids, organic acids and carbohydrates have been described but not identified. Interestingly, leaf cells seem to use a uniport mechanism for the transport of sucrose into vacuoles, while in root cells its uptake may be via an antiporter (Martinoia *et al.*, 2000).

As in the plasma membrane, a variety of ion channels exist in the vacuolar membrane. So far only one ion channel has been identified at the molecular level: the outwardly rectifying K^+ channel (Arabidopsis KCO1; Czempinski *et al.*, 1997; Schönknecht *et al.*, 2002). KCO1 was cloned and later localized to the vacuolar membrane by fusing the protein to the fluorescent protein GFP and expressing it in a tobacco cell line (Czempinski *et al.*, 2002).

Both cation and anion channels have been characterized however by electrophysiological methods applied to the vacuolar membrane. It is possible to biochemically isolate pure vacuoles, which are then accessible for electrophysiological recordings. The membrane potential difference between the cytosol and the vacuolar lumen is estimated to be 20–50 mV, negative at the cytosolic side (Allen and Sanders, 1997). It is known that there are several different channels permeable for Ca^{2+}, K^+, Cl^-, and malate. Cation channels are usually involved in release of the respective ion, while the flow of anions is directed into the vacuole. The channels are either regulated by voltage or by ligands such as Ca^{2+}, IP_3 or cADPR (cyclic ADP-ribose), while one channel is pressure- or stretch-activated (Allen and Sanders, 1997).

In addition to the ion channels, several isoforms of water channels, aquaporins, are localized in the vacuolar membrane; these are the TIP-type of aquaporins. Usually, TIPs are highly abundant proteins (Karlsson *et al.*, 2000). Like the PIPs, the TIPs are important for osmoregulation of the plant cell, but they may also have specific functions in the formation of new vacuoles and their increase in volume (vacuole biogenesis; Marty, 1999). As mentioned above, different TIP isoforms may serve as markers for different types of vacuoles (Jauh *et al.*, 1999).

Chloroplasts have Three-Membrane Systems and Need to Export Photosynthetic Products

Chloroplasts are the sites for photosynthetic reactions in plant leaf cells. There are several membrane systems in chloroplasts: the outer and inner envelope membranes, and the thylakoid membrane, each with its specific transport capacity (Neuhaus and Wagner, 2000). The light-dependent electron transport chain in the thylakoid membrane results in accumulation of protons in the thylakoid lumen. Dissipation of this H^+ gradient is coupled with production of ATP by the ATP synthase (F_0-F_1 ATPase). In the stroma, reduction of CO_2 to carbohydrates takes place and the products of these reactions need then to be transported to the cytosol for further conversion into sucrose. This is accomplished by the triose phosphate transporter, also called the phosphate translocator, an antiporter localized to the inner envelope membrane of chloroplasts (Flügge, 1999). In the same membrane, at least two different ATP/ADP translocators have been cloned and characterized from *Arabidopsis* (Table 7.2; Möhlmann *et al.*, 1998; Neuhaus *et al.*, 1997).

When chloroplasts are illuminated they take up calcium. This light-dependent uptake is believed to be mediated by a uniport-type carrier in the inner envelope membrane and to be energized by the membrane potential difference across this membrane (Kreimer *et al.*, 1985; Roh *et al.*, 1998). An ACA-type of Ca^{2+}-ATPase has also been associated with the inner envelope membrane (Huang *et al.*, 1993). More recently, an active Ca^{2+}/H^+ antiporter was identified in the thylakoid membrane, where it would supply necessary Ca^{2+} to the lumen for the assembly of different photosynthetic proteins (Ettinger *et al.*, 1999).

The membrane potential may be built up by the activity of an H^+-ATPase. In pea the activity of an H^+-ATPase has been measured in isolated chloroplast inner envelopes (Berkowitz and Peters, 1993; Shingles and McCarty, 1994).

The membrane potential gradient has also been shown to drive the uptake of Fe^{2+} across the inner envelope by a uniport mechanism (Shingles *et al.*, 2002). Other metal ions, e.g. Zn^{2+}, Cu^{2+}, and Mn^{2+} are also necessary for biochemical processes within the chloroplast and must be transported across the inner envelope, but it is not known whether the Fe^{2+} uniport also transports these ions or if there are other transporters in action.

Ion channels permeable for K^+ or anions have been characterized in the inner envelope membrane (Heiber *et al.*, 1995; Wang *et al.*, 1993; Neuhaus and Wagner, 2000).

In the outer envelope, three different solute channels (so-called porins) have been characterized, each with different permeabilities to amines, amino acids, 3-carbon compounds, sugars, ATP, and phosphate (Pohlmeyer *et al.*, 1997, 1998; Bölter *et al.*, 1999). There are also protein import channels in the outer chloroplastic membrane, as chloroplasts are not capable of synthesizing all the proteins they need by themselves (e.g. Hinnah *et al.*, 1997).

Mitochondria are Surrounded by Two Membranes that Contain Transport Systems for Metabolic Products

Plant mitochondria contribute to the cellular energy supply and have, like the animal ones, an F_0-F_1 ATPsynthase in the inner membrane that produces ATP with the energy built up by the electron transport chain during oxidative phosphorylation (or respiration). In addition, plant mitochondria are important in plant-specific metabolic pathways like photorespiration and photosynthesis in certain types of plants (C4 and CAM = crassulacean acid metabolism) (Picault *et al.*, 2004). Therefore, metabolites, nucleotides, and cofactors need to be actively transported into and out of mitochondria (across the inner membrane), since the inner membrane is impermeable to charged or polar molecules. Indeed, a high number (almost 60) of putative mitochondrial carrier proteins have been identified in the *Arabidopsis* genome (Picault *et al.*, 2004). A few of these have been cloned and functionally characterized as reconstituted proteins in lipid vesicles: examples are DTC (transporting di- and tricarboxylates) from *Arabidopsis* and tobacco, AtBAC (Basic Amino acid Carrier), an uncoupling protein (AtPUMP1 or AtUCP1) and AtAACs, exchanging ADP for ATP (Fig. 7.5, Table 7.2; Borecky *et al.*, 2001; Haferkamp *et al.*, 2002; Picault *et al.*, 2002; Hoyos *et al.*, 2003).

As in the outer membrane of chloroplasts, there are pore-forming channels also known as VDACs (Voltage-Dependent Anion Channels) in the outer membrane of plant mitochondria. They are permeable to small molecules up to 8 kD. The voltage dependence has only been observed *in vitro* (in similar bacterial channels) and is intriguing since no membrane potential difference exists across the mitochondrial outer membrane. Recent studies indicate that *in vivo*, these channels may instead be regulated by metabolites, substrates and nucleotides (Bölter and Soll, 2001). In general, the traditional view of the outer membrane pores in chloroplasts and mitochondria as generally open nonselective diffusion pores is becoming more complex with the pores being more regulated and specific than previously thought.

A few other transport proteins have been identified in mitochondria. An anion channel of the ClC family has been localized to the inner mitochondrial membrane of tobacco by immunological colocalization experiments on purified membranes (Lurin *et al.*, 2000). Reverse genetics identified a half-sized ABC transporter (STA1) which is involved in maturation of cytosolic Fe/S proteins and was localized to mitochondria by fusion with GFP (Kushnir *et al.*, 2001)

Few Transporters are Characterized in Endoplasmic Reticulum and Other Membranes

The smooth endoplasmic reticulum is the site of lipid synthesis and membrane assembly, whereas in the rough ER synthesis of membrane proteins and proteins to be secreted outside the cell or to the vacuole takes place. Not much is known about the "intrinsic" composition of the ER membranes regarding ion channels, carriers and pumps. A few transport proteins have however, been characterized in or localized to these membranes.

Two different types of Ca^{2+}-ATPases from *Arabidopsis*, one stimulated by CaM (ACA2), the other not (ECA1), have been localized to the ER membranes by expression of a GFP fusion protein (ACA2) or by immunolocalization (ECA1) (Liang *et al.*, 1997; Hong *et al.*, 1999). Another pump, the V-type H^+-ATPase has also been immunologically associated with ER (Herman *et al.*, 1994).

There are some reports on Ca^{2+} channel activities associated with the ER: a voltage-sensitive channel from *Bryonia dioica* tendrils (Klüsener *et al.*, 1995), and an IP_3-sensitive Ca^{2+} release activity (Muir and Sanders, 1997).

Even less is known about transporters in other types of membranes, like the Golgi, the nuclear envelope, the peroxisomes and glyoxysomes. Recently, an ABC transporter was localized to peroxisomal membranes (Footitt *et al.*, 2002). The nucleus is proposed to play a role in calcium-based signaling. An ECA-type Ca^{2+}-ATPase was immunolocalized around the nuclear envelope in tomato (Downie *et al.*, 1998). Another study showed that ATP-dependent Ca^{2+} transport systems are associated with the plant nucleus of carrot cells (Bunney *et al.*, 2000).

Finally, patch-clamp experiments have detected ion channel activities in the nuclear envelope (Matzke *et al.*, 1992; Grygorczyk and Grygorczyk, 1998).

Surely, future research will identify many more transporters as residents of the different organellar membranes, thus contributing to the proper functioning of the respective organelle. In this respect, the complete genome sequences of *Arabidopsis* and rice, in combination with different imaging techniques, will greatly aid in gaining new knowledge and comprehension.

METHODS AND TECHNIQUES FOR STUDYING PLANT ION TRANSPORT

Ion Transport Often Studied in Membrane Vesicles

Ion and solute transporters are often studied *in vitro* in their native membranes, purified from the plant cells as small spherical (about 1 μm in diameter) vesicles. Isolated vesicles represent a very convenient experimental system for the study of transport processes, as a high level of control over the experimental conditions is achieved, e.g. with regard to buffer composition and energy supply (Briskin, 1990).

Membrane vesicles are formed when the plant tissue is broken into smaller pieces by grinding, slicing or homogenization and the membranes thereby also broken into smaller pieces (fractionated). Mechanical treatment is necessary to remove the cell walls. Sometimes enzymatic digestion of the cell wall is used. During the 1980s, methods for the further separation of plant membrane vesicles of different cellular origin were greatly improved (Sze, 1980, 1985). Membrane vesicles are usually separated by centrifugation through a **sucrose density gradient** (i.e. different

sucrose concentrations in the bottom and top of the centrifuge tube), which separates membranes according to their density (Fig. 7.11). For example, vacuolar membranes have a relatively low protein/lipid ratio and hence a low density, while plasma membranes are denser and can be collected in the high-density fraction. Plasma membrane vesicles of high purity are also commonly isolated with a technique called **aqueous two-phase partitioning**, which instead is based on the surface properties of the membrane (Larsson *et al.*, 1994).

Initially, the membrane vesicles obtained were a mixture of "right-side out" and "inside-out" (cytoplasmic-side out) vesicles which complicated the assays. Now techniques for obtaining highly pure fractions of inside-out vesicles exist (e.g. Johansson *et al.*, 1995). Furthermore, the vesicles must not be leaky to enable for example a pH gradient to build up: one speaks of sealed or transport-competent vesicles. Sealed inside-out plasma membrane vesicles, for example, have been extremely useful in the study of the plasma membrane H^+-ATPase because the ATP binding site is then available on the external side. The ATP driven proton pumping by this enzyme can be measured by loading the vesicles with a pH sensitive fluorescent indicator. The ATPase activity can be assayed with other probes on the outside of the vesicles, probes that usually change spectral properties in the reaction and thus the activity can be measured in a spectrophotometer. Ca^{2+}-ATPase transport activity can be assayed by measuring the uptake of radioactive Ca^{45} into membrane vesicles or with fluorescent calcium indicators (Askerlund, 1996). Similarly, cotransporters are usually characterized in membrane vesicles. One example is the plasma membrane Na^+/H^+ exchanger whose activity was studied by first letting a pH gradient

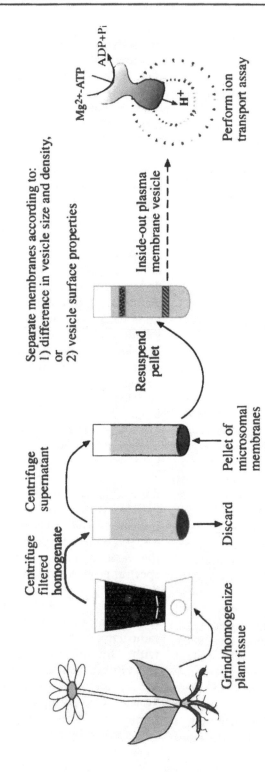

Fig. 7.11 Simplified scheme for preparation of membrane vesicles from plants. Only the major steps are shown. The first centrifugation removes large particles such as plastids and mitochondria and starch grains. The second centrifugation removes soluble proteins. Alternatively, gel filtration can be used to remove soluble proteins. The last step is either 1) a high-speed centrifugation for separation based on density (sucrose density gradient centrifugation), or 2) separation of the membrane vesicles in two different aqueous solutions, according to their surface properties (aqueous two-phase partition).

build up across the vesicle membranes; subsequent addition of Na^+ dissipated the pH gradient (Qiu *et al.*, 2003).

An alternative to vacuolar membrane vesicles is to isolate intact vacuoles. This again is achieved through centrifugation through a sucrose gradient. Intact vacuoles can be used for transporter studies such as membrane vesicles, or they can be used in patch-clamp experiments (see below). When the transport activities of a single, specific protein is to be studied, the purified protein (if purification is possible) can be reconstituted into artificial vesicles called **liposomes**.

In addition to being useful for transport studies, separation of membrane vesicles is also used as a method for determining in which membrane a specific transporter is localized: this is called **immunolocalization**. After sucrose density centrifugation, the different membrane fractions are separated by SDS-PAGE and transferred to a nylon or nitrocellulose membrane. Antibodies to the protein of interest and to different membrane markers (usually membrane proteins thought to be specific for a certain membrane) are allowed to react with the different fractions. If the protein of interest is in the plasma membrane for example, fractions corresponding to the plasma membrane marker will be highlighted.

From a more physiological point of view, transport processes can be studied using **tracer flux measurements** (see MacRobbie, 2000). In this technique cells (in a small piece of tissue, e.g. guard cells in leaves) are loaded with a radioactive compound of interest, e.g. $^{86}Rb^+$ for the study of K^+ channels. These channels are capable of letting rubidium ions pass through their pores instead of K^+. The cells are then treated with different stimulating agents, and the efflux of $^{86}Rb^+$ to the external medium quantified, giving an indication of the activity of ion channels.

Ion fluxes can also be measured with ion selective electrodes; an electrode equipped with a membrane only permeable to a specific ion is planted in a tissue. In this way, different ion activities in a plant tissue at different physiological conditions can be estimated, allowing for an assessment of the ion channel activities and the sequence of events.

Water channel activity is commonly tested in frog oocytes. The water channel of interest is heterologously expressed in *Xenopus laevis* oocytes and the oocytes placed in a hypotonic solution (less concentrated on the outside of the cell than inside). The active water channels will transport water from the outside to the inside, to equilibrate the ionic solutions, and when enough water has come into the cell it will swell and eventually burst. Like other transporters, water channels can be reconstituted and studied in artificial liposomes.

Electrophysiological Techniques Allow for Functional Characterization of Ion Channels

Classical electrophysiological methods allow for assessment of the electrical properties of a membrane and have been employed in plant experiments for more than 50 years. In the last 20 years, a much better electrophysiological method with higher resolution, the patch-clamp technique, has been developed and used to characterize in detail many membrane transport activities. The patch-clamp technique was developed in the 1970s by the German scientists Neher and Sakmann, later awarded the Nobel Prize (Neher and Sakmann, 1976; Hamill *et al.*, 1981). Some years later, the first recordings of plant ion channels were achieved on protoplasts from leaf and guard cells respectively (Moran et al., 1984; Schroeder et al., 1984). Plant cells are more difficult to work with than animal cells because a cell wall has to be digested before the experiment. Additionally, vacuolar and plasma membranes are often close to each other and the cytoplasmic compartment is relatively small (Hedrich et al., 1990). Therefore, direct, in planta measurements of electrical phenomena are quite complicated. So, the patch-clamp technique has been applied instead and plant protoplasts and vacuoles successfully employed.

The basic principle of the patch-clamp technique is to measure the current carried by the ions passing through one or many ion channel(s). A small glass pipette (around 1 μm in diameter at the tip), containing an electrode, is applied to the membrane surface of a cell. By applying a negative pressure (usually by mouth suction), the membrane is tightly sealed to the pipette so that the only current that can pass is through the membrane attached to the pipette. In this way, the background noise is very low and the small currents resulting from ion channel activities, or even from a single ion channel, can be resolved. The electrode in the pipette is connected to electrical devises to amplify and digitalize the signal (Fig. 7.12A). Different modes of recordings are possible: cell attached, whole cell, inside-out patch or outside-out patch (Fig. 7.12B). In all modes but the cell attached, the composition of the solution can be controlled both at the cytoplasmic side and on the outside of the membrane. The membrane potential (voltage) of the cell or the patch is usually clamped (set or controlled) to different specific values; this is termed applying a voltage protocol. In addition, substances supposed to activate the channels, such as ligands or ions, can be applied to the patch/cell. The activity (amplitude of the current) of the channels in the patch can thus be measured in relation to the voltage (or ligand) applied. A typical output from a whole cell mode recording on a guard cell protoplast is shown in Fig. 7.12C. The voltages applied are shown above the corresponding current output. In the figure, the membrane potential was held at –40 mV (called holding potential) and then stepped progressively backward to more negative potentials, adding –20 mV for each step. Between each step the holding potential was applied. It can be seen that the biggest currents were induced when the voltage was –180 mV. When the membrane potential was set to zero, the ion channels deactivated. By convention, the currents elicited that are directed downward indicate inward currents, i.e. cations moving into the cell or anions moving out of the cell, or both (a net flow of positive ions into the cell). Outward currents are therefore directed upward, indicating cations moving out of the cell, anions moving in, or both. In Fig. 7.12C, the activity of the channels elicits downward currents which change with the membrane potential applied. Thus, the currents are primarily caused by inward-rectifying, voltage-dependent K^+ channels in the plasma membrane of this guard cell.

As shown in Figure 7.12C, the data from such current measurements are often presented in a current-voltage curve (called an I-V curve), such as that in Figure 7.12D. Here the total current measured for each voltage is plotted against the voltage applied (the membrane potential). From this type of curve it is possible to see that the activity of the channels is voltage dependent because the relation is nonlinear (at least when the ionic concentrations on either side of the membrane are relatively similar, which is the case here). It can also be

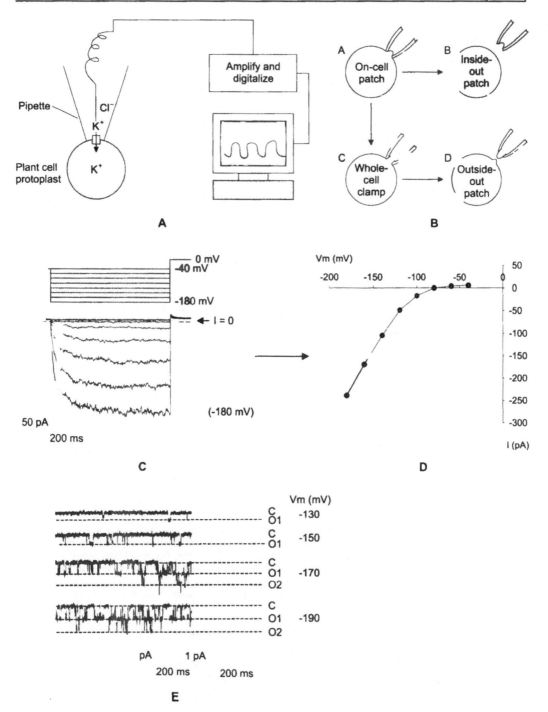

A

B

C

D

E

(Contd. Figure 7.12)

seen that the channels in this guard cell protoplast become active at –80 to –100 mV.

By using other kinds of voltage protocols in combination with various ionic solutions (different compositions and/or concentrations), the **selectivity** of an ion channel can be determined. Selectivity is the ability of different ions to enter the channel pore. As already seen (see Box 7.2), K^+ channels are highly selective for K^+ over Na^+: this ion doesn't fit well in the pore structure and therefore does not enter. Another term often used is **permeability** of a channel: this refers to which ions are able to pass through the channel pore, but does not specify the affinity (selectivity). Some ions may be permeant in experimental conditions, but not *in vivo* (*in planta*). By comparing the permeability of different ions, the selectivity of a channel can be estimated.

When recording in the inside-out or outside-out mode (when the membrane patch is excised), currents from a single ion channel protein in the patch can be resolved. The transport rate must be at least 1×10^6 monovalent ions per second for the current to be resolved and not measured as noise (Ward, 1997). Hence, pumps and cotransporters cannot be studied in excised patches since their transport rates are much lower than for channels. A typical example of single-channel currents from an inside-out patch is presented in Fig. 7.12E. Like for whole-cell recording, an I-V curve can be constructed and used to analyze the properties of the channel, in combination with the use of solutions with different ionic compositions. The Nernst equation is often used to make calculations e.g., for selectivity, both for single-channel analysis and for whole-cell mode recordings.

(Contd. Figure 7.12)

Fig. 7.12 Electrophysiological analysis of plant cells

A—Setup for the electrophysiological recordings of plant cells. A small glass pipette (1 μm in diameter at the tip) containing an electrode is applied to the membrane surface of a cell. When the membrane is tightly sealed to the pipette ("giga-ohm seal"), small currents can be resolved and analyzed in a computer.

B—From the initial cell-attached, or on-cell patch configuration, the investigator can excise a small patch to be in the inside-out mode (cytosolic side facing the external solution), measuring currents from channels only in the patch. Alternatively, to analyze the channels in the whole cell, the membrane inside the pipette can be ruptured to pass from cell-attached to whole-cell configuration. Finally, an outside-out patch (cytosolic side toward the pipette solution) can be obtained from the whole-cell mode.

C—Example of whole-cell currents recorded on a guard cell protoplast from maize, showing the activity of inward-rectifying, voltage-dependent K^+ channels. Voltages are applied for a little less than a second and are shown above the corresponding current output. The membrane potential was held at –40 mV (holding potential) and was then stepped to progressively more negative potentials, adding –20 mV for each step. Between each step, the holding potential was applied. Currents were measured in picoampere (pA, see scalebar) and the largest currents were recorded when the most negative voltage was applied (–180 mV).

D—Current-voltage relation for the data in C (I-V curve). The total current measured for each voltage was plotted against the voltage applied (the membrane potential).

E—Single-channel recording from an inside-out patch excised from a protoplast of a cortical root cell of *Arabidopsis*. The membrane potentials applied are indicated to the right. Downward currents indicate opening of a channel. C, O1 and O2; closed and open states for the first and second ion channel in the patch respectively. Channels are activated by hyperpolarization. Amplitude of the current and the time the channels are open can be estimated by using the scalebar.

(Figure. 7.12B, courtesy of Dr. S. Zimmermann (INRA, Montpellier, France). Data in Figures 7.12C, D and E kindly provided by Dr. A-A Véry (INRA, Montpellier, France).

Patch-clamp techniques have been extensively applied to study transport processes in isolated plant cells or vacuoles from "wild-type" plants. One especially well-studied cell type is the guard cell, two of which make up the stomatal pore. In fact, the guard cell has become a model for transport studies. With development of techniques for making transgenic plants and the ability to isolate plants mutant in a specific gene (e.g. a "knockout" plant, see below), the transport "behavior" of such plants can be studied with electrophysiological methods and compared to wild-type plants. This might give indications on the physiological role(s) of a certain transporter.

Another very commonly used approach for the functional characterization of individual plant transporters is to express them in other types of cells, e.g. *X. laevis* oocytes, cultured insect or mammalian cells, or in plant suspension cells. Overexpression of a given ion channel is a convenient approach since *in vivo* hundreds of channels are present (and often active) simultaneously. The electrophysiological signature of a single channel would therefore be extremely difficult to detect. By using heterologous cells for electrophysiological studies it is usually known which protein is in the membrane, i.e. the molecular identity is known. The advantage with oocytes is that they are big (about 1 mm in diameter) and relatively easily manipulated, while the other cells are much smaller. The transporter in question needs to be cloned and then complementary RNA transcripts or DNA produced and introduced into the host cells, which are able to produce large amounts of the foreign protein and often to insert it into the plasma membrane. The currents produced by the expressed channels (or pumps or cotransporters) can be analyzed in the whole cell configuration. For oocytes, two electrodes are needed to make whole cell recordings since the oocytes are so large. This special arrangement is called the **two-electrode voltage clamp** technique. As for plant protoplasts, single ion channels (but not pumps or cotransporters) can be studied in patches excised from the membrane of the different cell types in the inside-out or outside-out mode. The proteins can be tagged in different ways (e.g. with GFP), so as to know which cells are expressing the transporter.

Baker's Yeast Used to Clone Low Abundance Transporters

Most ion channels and many cotransporters (and some pumps) are present in low amounts in the membranes. So, it has proven almost impossible to clone them by classical biochemical methods, i.e. to isolate the protein from the membrane and to extract peptide sequence information so as to find the corresponding DNA sequence in a collection of complementary DNAs (a cDNA library). Only some H^+-and Ca^{2+}-AT-Pase genes have been cloned using this approach. Instead, most ion channels and carriers have been cloned by so-called **expression cloning** in the yeast *Saccharomyces cerevisiae* (Baker's yeast). Yeast is eukaryotic but a single cell that contains many of the same transport systems as found in plants. For example, the primary transport system that generates the driving force for secondary transporters is a plasma membrane H^+-ATPase, as in plants. Yeast is also easy to manipulate and to make mutations in. Various yeast strains mutant in a specific transport system, e.g. K^+ uptake, can be constructed. These yeast strains cannot survive on normal growth media unless a gene, in this case a plant gene from a cDNA library which restores the transport function, is reinserted: this is called **functional complementation**.

The first K^+ channels were cloned using this powerful technique (Anderson *et al.*, 1992; Sentenac *et al.*, 1992) and also cotransporters for sucrose, amino acids, NH_4^+, SO_4^{2-} and others (e.g. Riesmeier *et al.*, 1992; Frommer *et al.*, 1993; Hsu *et al.*, 1993; Ninnemann *et al.*, 1994; Smith *et al.*, 1995).

The yeast expression system can also be used to characterize already cloned transporters. For example, several *Arabidopsis* Ca^{2+}-ATPases have been functionally characterized in a yeast strain that lacked its own Ca^{2+}-ATPases (Liang *et al.*, 1997; Harper *et al.*, 1998; Sze *et al.*, 2000). The plant Ca^{2+}-ATPase was able to restore yeast growth in defined conditions, thus indicating that it actually functions as a Ca^{2+} pump. Once expressed in yeast, the Ca^{2+} pump can be isolated in membrane vesicles from the yeast and its transport function assayed, as in vesicles from plant membranes (see above). Similarly, the plasma membrane H^+-ATPase has been extensively studied in the yeast expression system (e.g. Palmgren and Christensen, 1994; Regenberg *et al.*, 1995; Morsomme *et al.*, 1996; Luo *et al.*, 1999). Mutated forms of plant transporters (with altered functions) can be expressed in yeast, and their transport characteristics studied. Results are then compared with results from the "wild-type" transporter expressed in the same way and a more detailed picture of the mechanism of transport (and kinetics, regulation, structure-function relationship) can be obtained.

Transgenic Plants and Mutants as Tools to Study Transport Processes

The ultimate goal of research on plant membrane ion transporters is to understand their physiological role(s). Many different experimental approaches are used toward obtaining this goal, such as functional characterization of individual transporters, expression and localization studies of both single transporters and gene families, and different biochemical analyses. Within the last 10 years, **transgenic plants** have become another important tool for the study of plant transporters. Ever since *Arabidopsis* was chosen as a model species by the research community, much work and many investigations have been done on this plant. The common effort to sequence the entire genome of *Arabidopsis* was a big step forward for science in general, and the results of the *Arabidopsis* (and now also rice) genome sequencing are of enormous importance to plant scientists.

Arabidopsis is a good model plant because of its small genome and short regeneration time. It is also easy to transform (to introduce foreign DNA into), and a relatively simple technique for this was developed in the early 1990s (Bechtold *et al.*, 1993). A special kind of bacterial DNA, T-DNA from the soil bacterium *Agrobacterium tumefasciens*, was introduced randomly into the genome by infiltrating flowering *Arabidopsis* plants with a liquid culture of the bacteria under vacuum. Later on, an even simpler "floral dip" method without vacuum was developed (Clough and Bent, 1998). Simplified transformation methods greatly aided creation of collections of so-called **T-DNA tagged** mutant lines (homologous recombination difficult to achieve in plants). The T-DNA inserted in a specific gene could then be found by PCR screening, and the phenotype of the mutant line of plants analyzed and compared to wild-type plants. Plants created in this way are called "**knockout**" plants or loss-of function mutant: the gene of interest is not expressed and thus knocked out (Krysan *et al.*, 1996). This kind of approach is called **reverse genetics**: the gene of interest is known but not the phenotype. If a phenotype is found, conclusions may be drawn concerning the normal physiological role of the gene product, e.g. a transporter. An example of this is shown in Fig 7.13A for IRT1, a high-affinity iron carrier (Vert *et al.*, 2002).

In **forward genetics**, the investigator has a mutant plant with a known phenotype and searches for the mutated gene responsible for the phenotype. Both approaches have helped researchers obtain indications on the physiological role of different transporters. Other plants can also be transformed, e.g. maize, wheat, rice, tobacco, but sometimes the procedure is more complicated and takes more time.

Some researchers use transposable elements (transposon tagging) instead of T-DNA as a way to knock out a gene (e.g. Bhatt *et al.*, 1996; Cowperthwaite *et al.*, 2002).

If the knockout of a gene proves lethal, other approaches have to be employed. One commonly used approach is to reduce the expression of a certain gene by transforming plants with an **antisense** construct: this is RNA produced in the opposite direction to the "normal" RNA of the gene, and so expression of the gene will decrease to different degrees. The opposite result may be obtained when a certain gene is **overexpressed**: it is expressed in the whole plant to a higher degree than normally. Effects of the overexpression (and antisense expression) are then studied and analyzed.

A quite common problem with T-DNA knockout plants is that they do not have an apparent phenotype (Bouché and Bouchez, 2001). This might be due to the often large number of members of a gene family in plants. Other transporters of the same family might thus have a function similar to the missing one; this is called functional redundancy. In other cases the phenotype is extremely subtle and only discovered under special conditions. More recently, a technique called RNAi (for RNA interference, a natural phenomenon for regulation of gene expression; Novina and Sharp, 2004) was developed as a tool to "knock out" (suppress) one or several genes simultaneously. Thus, more than one gene of a gene family can be silenced and a stronger phenotype may be obtained.

The transformation approach is also commonly used to study in which part of the plant a protein is expressed, and in which membrane in the cell. This can provide information about cellular and developmental functions. The promoter of the gene of interest is fused to a **reporter gene**, for expression studies usually the so-called GUS gene (beta-glucuronidase) whose gene product produces a bluing in the tissues in which it is expressed (if treated with a chemical substrate). For membrane localization studies, transporter proteins can be "tagged" in either their N- or C-terminal end by the fluorescent protein GFP (green fluorescent protein) or its derivatives, or with specific peptides recognized by an antibody (epitope tagging). The plant cells can then be examined with a confocal microscope. If everything works well, the membrane in which the protein is usually localized will be colored (see Fig. 7.13B and C).

It is important to remember however, that genetic modifications are never neutral, but may result in up- or downregulation of other genes. Genetic approaches have to be used with caution and in combination with many other experimental approaches.

Current Trends in Study of Plant Ion Transport

In the last decade, the study of plant ion transport has evolved from the biochemical study of a single transporter to integration of a wide variety of strategies (Assman, 2001; Barbier-Brygoo *et al.*, 2001).

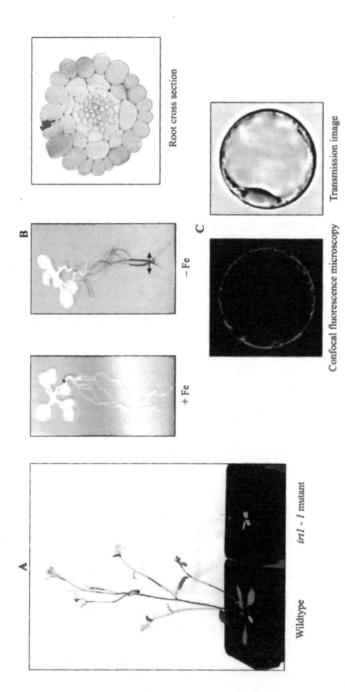

Wildtype   *irt1 - 1* mutant

Root cross section

B   + Fe   − Fe

C

Confocal fluorescence microscopy   Transmission image

Fig. 7.13 Transgenic plants as tools to learn about the physiological function of a transporter

A—Reverse genetics approach. The gene coding for the IRT1 iron transporter was knocked out with a T-DNA insertion. The phenotype of the mutant *irt1-1* plant is lethal: the plants are sterile and die after 3-4 weeks. The leaves are chlorotic because of iron deficiency. From this and other experiments, it could be concluded that IRT1 is essential for the iron homeostasis *in planta*.

B—Reporter gene approach for tissue localization. The promoter of the *IRT1* gene was fused to the beta-glucuronidase (*GUS*) gene, and transgenic plants expressing the fusion protein were analyzed by *GUS* histochemical staining, giving a blue color where *GUS* is expressed. The *IRT1* promoter is only active during iron deficiency, and it is expressed primarily in the epidermal cells of roots, indicating a role of the *IRT1* protein in iron absorption from the soil solution.

C—Reporter gene approach for subcellular localization. The green fluorescent protein (GFP) was fused to the C terminus of IRT1 and expressed transiently in *Arabidopsis* protoplasts from a cell suspension culture. The fusion protein expression was under the control of the strong 35S promoter of *Cauliflower mosaic virus* (commonly used in such experiments). The cells were analyzed with a confocal microscope, showing that *IRT1* is localized in the plasma membrane, which is in agreement with its role in iron uptake

(Figure modified from Vert *et al.*, 2002. Pictures kindly provided by Dr. C. Curie, INRA, Montpellier, France)

Complete sequences of Arabidopsis and rice have opened up for whole genome comparisons and for transcriptome analyses of transporters with **microarray** approaches (see above; e.g. Baxter *et al.*, 2003; Maathuis *et al.*, 2003). Transcriptome analyses allow for assessment of how a given stimulus affects transcript levels of multiple ion transport genes and their regulators. **Bioinformatic** analyses of the huge amount of data present in different databases have therefore become an important tool.

Information in databases is also important for the **proteomics** approach, which can provide a basis for further functional studies. In this approach, membrane proteins are purified and parts of their peptide sequences determined with different techniques, such as mass spectrometry. The sequences obtained can then be compared with sequences in the databases. Proteomics is not without problems for membrane-bound transporters, but the field is rapidly developing (e.g. Ferro *et al.*, 2003; Marmagne *et al.*, 2004).

In conclusion, a combination of molecular, genetic, and physiological approaches, frequently with high-throughput methods, is now used by plant researchers to elucidate the physiological function and regulation of plant ion transporters.

ACKNOWLEDGMENTS

Drs. Sabine Zimmermann and A-A Véry are gratefully acknowledged for providing helpful suggestions for the manuscript and for providing some of the Figures.

Some Useful Weblinks

- A P-type ATPase database: http://biobase.dk/~axe/Patbase.html
- Arabidopsis Membrane Protein Library: http://www.cbs.umn.edu/arabidopsis/
- ARAMEMNON—Arabidopsis membrane protein database (release 1.6): http://aramemnon.botanik.uni-koeln.de/
- Chemistry 2003: http://www.nobel.se/chemistry/laureates/2003/index.html
- Nature web focus: Ion channels: structure and function: http://www.nature.com/nature/focus/ionchannel/
- PlantsT: Functional Genomics of Plant Transporters: http://plantst.sdsc.edu/
- TAIR Homepage: http://www.arabidopsis.org/
- The TIGR Rice Genome Project: http://www.tigr.org/tdb/e2k1/osa1/
- TransportDB—Transporter Protein Analysis Database: http://66.93.129.133/transporter/wb/index2.html

References

Agre, P. and Kozono, D. 2003. Aquaporin water channels: molecular mechanisms for human diseases. FEBS Lett. 555: 72-78.

Allen, G.J. and Sanders, D. 1997. Vacuolar ion channels of higher plants. Adv. Bot. Res. 25: 218-252.

Anderson, J.A., Huprikar, S.S., Kochian, L.V., Lucas, W.J., and Gaber, R.F. 1992. Functional expression of a probable *Arabidopsis thaliana* potassium channel in *Saccharomyces cerevisiae*. Proc. Natl. Acad. Sci. USA 89: 3736-3740.

Anil, V.S. and Rao, K.S. 2001. Calcium-mediated signal transduction in plants: A perspective on the role of Ca^{2+} and CDPKs during early plant development. J. Plant Physiol. 158: 1237-1256.

Apse, M.P., Aharon, G.S., Snedden W.A., and Blumwald, E. 1999. Salt tolerance conferred by overexpression of a vacuolar Na^+/H^+ antiport in *Arabidopsis*. Science 285: 1256-1258.

Arabidopsis Genome Initiative. 2000. Analysis of the genome of the flowering plant *Arabidopsis thaliana*. Nature 408: 796-815.

Arazi, T., Sunkar, R., Kaplan, B., and Fromm, H. 1999. A tobacco plasma membrane calmodulin-binding transporter confers Ni^{2+} tolerance and Pb^{2+} hypersensitivity in transgenic plants. Plant J. 20: 171-182.

Askerlund, P. 1996. Modulation of an intracellular calmodulin-stimulated Ca^{2+}-pumping ATPase in cauliflower by trypsin. The use of Calcium Green-5N to measure Ca^{2+} transport in membrane vesicles. Plant Physiol. 110: 913-922.

Assmann, S.M. 2001. From proton pump to proteome. Twenty-five years of research on ion transport in higher plants. Plant Physiol. 125: 139-141.

Axelsen, K.B. and Palmgren, M.G. 1998. Evolution of substrate specificities in the P-type ATPase superfamily. J. Mol. Evol. 46: 84-101.

Axelsen, K.B. and Palmgren, M.G. 2001. Inventory of the superfamily of P-type ion pumps in *Arabidopsis*. Plant Physiol. 126: 696-706.

Balagué, C., Linn, B., Alcon, C., Flottes, G., Malmström, S., Köhler, C., Neuhaus, G., Pelletier, G., Gaymard, F., and Roby, D. 2003. HLM1, an essential signaling component in the hypersensitive response, is a member of the cyclic nucleotide-gated channel ion channel family. Plant Cell 15: 365-379.

Barbier-Brygoo, H., Gaymard, F., Rolland, N., and Joyard, J. 2001. Strategies to identify transport systems in plants. Trends Plant Sci. 6: 577-585.

Barbier-Brygoo, H., Vinauger, M., Colcombet, J., Ephritikhine, G., Frachisse, J.-M., and Maurel, C. 2000. Anion channels in higher plants: functional characterization, molecular structure and physiological role. Biochim. Biophys. Acta 1465: 199-218.

Baxter, I., Tchieu, J., Sussman, M.R., Boutry, M., Palmgren, M.G., Gribskov, M., Harper, J.F., and Axelsen, K.B. 2003. Genomic comparison of P-type ATPase ion pumps in *Arabidopsis* and rice. Plant Physiol. 132: 618-628.

Bechtold, N., Ellis, J., and Pelletier, G. 1993. *In planta Agrobacterium* mediated gene transfer by infiltration of adult *Arabidopsis thaliana* plants. Mol. Biol. Genet. 316: 1194-1199.

Berkowitz, G.A. and Peters, J.S. 1993. Chloroplast inner-envelope ATPase acts as a primary H^+ pump. Plant Physiol. 102: 261-267.

Bhatt, A.M., Page, T., Lawson, E.J., Lister, C., and Dean, C. 1996. Use of Ac as an insertional mutagen in Arabidopsis. Plant J. 9: 935-945.

Blumwald, E. and Poole, R.J. 1985. Na^+/H^+ antiport in isolated tonoplast vesicles from storage tissue of *Beta vulgaris*. Plant Physiol. 78: 163-167.

Bölter, B. and Soll, J. 2001. Ion channels in the outer membranes of chloroplasts and mitochondria: open doors or regulated gates? EMBO J. 20: 935-940.

Bölter, B., Soll, J., Hill, K., Hemmler, R., and Wagner, R. 1999. A rectifying ATP-regulated solute channel in the chloroplastic outer envelope from pea. EMBO J. 18: 5505-5516.

Bonza, M.C., Morandini, P., Luoni, L., Geisler, M., Palmgren, M.G., and De Michelis, M.I. 2000. At-ACA8 encodes a plasma membrane-localized calcium-ATPase of *Arabidopsis* with a calmodulin-binding domain at the N terminus. Plant Physiol. 123: 1495-1506.

Borecky, J. *et al.* 2001. Functional reconstitution of *Arabidopsis thaliana* plant uncoupling mitochondrial protein (AtPUMP1) expressed in *Escherichia coli*. FEBS Lett. 505: 240-244.

Bouché, N. and Bouchez D. 2001. *Arabidopsis* gene knockout: phenotypes wanted. Curr. Opin. Plant Biol. 4: 111-117.

Boutry, M., Michelet, B., and Goffeau, A. 1989. Molecular cloning of a family of plant genes encoding a protein homologous to plasma membrane H^+-translocating ATPases. Biochem. Biophys. Res. Commun. 162: 567-574.

Briskin, D.P. 1990. Transport in plasma membrane vesicles—Approaches and perspectives. In: *The Plant Plasma Membrane. Structure, Function and Molecular Biology*, pp. 154-181. Larsson, C. and Møller, I.M. (eds.). Springer-Verlag, Berlin, Germany.

Brücke, E. 1843. Beiträge zur Lehre von der Diffusion tropfbarer Flüssigkeiten durch poröse Scheidewände. Amer. Phys. Chem. 58: 77-94.

Bunney, T.D., Shaw, P.J., Watkins, P.A.C., Taylor, J.P., Beven, A.F., Wells, B., Calder, G.M., and Drøbak, B.K. 2000. ATP-dependent regulation of nuclear Ca^{2+} levels in plant cells. FEBS Lett. 476: 145-149.

Carafoli, E. 1997. Plasma membrane calcium pump: structure, function and relationships. Basic Res. Cardiol. (Suppl.) 1: 59-61.

Cheeseman, J.M. and Hanson, J.B. 1979. Energy-linked potassium influx as related to cell potential in corn roots. Plant Physiol. 64: 842-845.

Chen, J., Piper, D.R., and Sanguinetti, M.C. 2002. Voltage sensing and activation gating of HCN pacemaker channels. Trends Cardiovasc. Med. 12: 42-45.

Chérel, I., Michard, E., Platet, N., Mouline, K., Alcon, C., Sentenac, H., and Thibaud, J.-B. 2002. Physical and functional interaction of the Arabidopsis K^+ channel AKT2 and phosphatase AtPP2CA. Plant Cell 14: 1133-1146.

Chrispeels, M.J., Crawford, N.M., and Schroeder, J.I. 1999. Proteins for transport of water and mineral nutrients across the membranes of plant cells. Plant Cell 11: 661-676.

Clough, S.J. and Bent, A. 1998. Floral dip: a simplified method for Agrobacterium-mediated transformation of Arabidopsis thaliana.. Plant J. 16: 735-743.

Clough, S.J., Fengler, K.A., Yu, I., Lippok, B., Smith, R.K., Jr., and Bent, A.F. 2000. The Arabidopsis dnd1 "defense, no death" gene encodes a mutated cyclic nucleotide-gated ion channel. Proc. Natl. Acad. Sci. USA 97: 9323-9328.

Cosgrove, D.J. and Hedrich, R. 1991. Stretch-activated chloride, potassium, and calcium channels coexisting in plasma membranes of guard cells of Vicia faba L. Planta 186: 143-153.

Cowperthwaite, M., Park, W., Xu, Z., Yan, X., Maurais, S.C., and Dooner, H.K. 2002. Use of the transposon Ac as a gene-searching engine in the maize genome. Plant Cell 14: 713-726.

Curie, C. and Briat, J.F. 2003. Iron transport and signaling in plants. Annu. Rev. Plant Biol. 54: 183-206.

Curie, C., Alonso, J.M., Le Jean, M., Ecker, J.R., and Briat, J.F. 2000. Involvement of NRAMP1 from Arabidopsis thaliana in iron transport. Biochem. J. 347: 749-755.

Curie, C., Panaviene, Z., Loulergue, C., Dellaporta, S.L., Briat, J.-F., and Walker, E.L. 2001. Maize yellow stripe1 encodes a membrane protein directly involved in Fe(III) uptake. Nature 409: 346-349.

Czempinski, K., Frachisse, J.M., Maurel, C., Barbier-Brygoo, H. and Mueller-Roeber, B. 2002. Vacuolar membrane localization of the Arabidopsis 'two-pore' K^+ channel KCO1. Plant J. 29: 809-820.

Czempinski, K., Zimmermann, S., Ehrhardt, T., and Mueller-Roeber, B. 1997. New structure and function in plant K^+ channels: KCO1, an outward rectifier with a steep Ca^{2+}-dependency. EMBO J. 16: 2565-2575.

Daram, P., Brunner, S., Rausch, C., Steiner, C., Amrhein, N., and Bucher, M. 1999. Pht2;1 encodes a low-affinity phosphate transporter from Arabidopsis. Plant Cell 11: 2153-2166.

Deeken, R., Sanders, C., Ache, P., and Hedrich, R. 2000. Developmental and light-dependent regulation of a phloem-localised K^+ channel of Arabidopsis thaliana. Plant J. 23: 285-290.

Downie, L., Priddle, J., Hawes, C., and Evans, D.E. 1998. A calcium pump at the higher plant nuclear envelope? FEBS Lett 429: 44-48.

Doyle, D., Cabral, J., Pfuetzner, R., Kuo, A., Gulbis, J., Cohen, S., Chait, B., and MacKinnon, R. 1998. The structure of the potassium channel: Molecular basis of K^+ conduction and sensitivity. Science 280: 69-77.

Drozdowicz, Y.M., Kissinger, J.C., and Rea, P.A. 2000. AVP2, a sequence-divergent, K(+)-insensitive H(+)-translocating inorganic pyrophosphatase from Arabidopsis. Plant Physiol. 123: 353-362.

Durell, S.R., and Guy, H.R. 1999. Structural models of the KtrB, TrkH, and Trk1, 2 symporters based on the structure of the KcsA K^+ channel. Biophys. J. 77: 789-807.

Eide, D., Broderius, M., Fett, J., and Guerinot, M.L. 1996. A novel iron-regulated metal transporter from plants identified by functional expression in yeast. Proc. Natl. Acad. Sci. USA 93: 5624-5628.

Ettinger, W.F., Clear, A.M., Fanning, K.J., and Peck, M.L. 1999. Identification of a Ca^{2+}/H^+ antiport in the plant chloroplast thylakoid membrane. Plant Physiol. 119: 1379-1385.

Johansson, U., Karlsson, M., Johansson, I., Gustavsson, S., Sjövall, S., Fraysse, L., Weig, A.R., and Kjellbom, P. 2001. The complete set of genes encoding major intrinsic proteins in *Arabidopsis* provides a framework for a new nomenclature for major intrinsic proteins in plants. Plant Physiol. 126: 1358-1369.

Kamb, A., Iverson, L.E., and Tanouye, M.A. 1987. Molecular characterization of Shaker, a *Drosophila* gene that encodes a potassium channel. Cell 50: 405-413.

Karlsson, M., Johansson, I., Bush, M., McCann, M.C., Maurel, C., Larsson, C., Kjellbom, P. 2000. An abundant TIP expressed in mature highly vacuolated cells. Plant J. 21: 1-8.

Kato, Y., Sakaguchi, M., Mori, Y., Saito, K., Nakamura, T., Bakker, E.P., Sato, Y., Goshima, S., and Uozumi, N. 2001. Evidence in support of a four transmembrane-pore-transmembrane topology model for the *Arabidopsis thaliana* Na^+/K^+ translocating AtHKT1 protein, a member of the superfamily of K^+ transporters. Proc. Natl. Acad. Sci. USA 98: 6488-6493.

Kjellbom, P., Larsson, C., Johansson, I.I., Karlsson, M., and Johanson, U. 1999. Aquaporins and water homeostasis in plants. Trends Plant Sci. 4: 308-314.

Klüsener, B., Boheim, G., Liss, H., Engelberth, J., and Weiler, E.W. 1995. Gadolinium-sensitive, voltage-dependent calcium release channels in the endoplasmic reticulum of a higher plant mechanoreceptor organ. EMBO J 14: 2708-2714.

Knight, H. and Knight, M.R. 2001. Abiotic stress signalling pathways: Specificity and crosstalk. Trends Plant Sci. 6: 262-267.

Köhler, C., Merkle, T., and Neuhaus, G. 1999. Characterisation of a novel gene family of putative cyclic nucleotide- and calmodulin-regulated ion channels in *Arabidopsis thaliana*. Plant J. 18: 97-104.

Korthout, H.A. and de Boer, A.H. 1994. A fusicoccin binding protein belongs to the family of 14-3-3 brain protein homologs. Plant Cell 6: 1681-1692.

Kreimer, G., Melkonian, M., and Latzko, E. 1985. An electrogenic uniport mediates light-dependent Ca^{2+} uptake by intact chloroplasts. FEBS Lett. 180: 253-258.

Krysan, P.J., Young, J.C., Tax, F., and Sussman, M.R. 1996. Identification of transferred DNA insertions within *Arabidopsis* genes involved in signal transduction and ion transport. Proc. Natl. Acad. Sci. USA 93: 8145-8150.

Kurdjian, A. and Guern, J. 1989. Intracellular pH: measurement and importance in cell activity. Annu. Rev. Plant Plant Physiol. Plant Molec. Biol. 40: 271-303.

Kushnir, S., Babiychuk, E., Storozhenko, S., Davey, M.W., Papenbrock, J., De Ricke, R. *et al.* 2001. A mutation of the mitochondrial ABC transporter Sta1 leads to dwarfism and chlorosis in the *Arabidopsis* mutant *starik*. Plant Cell 13: 89-100.

Lacombe, B., Becker, D., Hedrich, R., DeSalle, R., Hollmann, M., Kwak, J.M., *et al.* 2001. The identity of plant glutamate receptors. Science 292: 1486-1487.

Lacombe, B., Pilot, G., Michard, E., Gaymard, F., Sentenac, H., and Thibaud, J.B. 2000. A Shaker-like K^+ channel with weak rectification is expressed in both source and sink phloem tissues of *Arabidopsis*. Plant Cell 12: 837-851.

Lam, H.M., Chiu, J., Hsieh, M.H., Meisel, L., Oliveira, I.C., Shin, M., and Coruzzi, G. 1998. Glutamate-receptor genes in plants. Nature 396: 125-126.

Larsson, C., Sommarin, M., and Widell, S. 1994. Isolation of highly purified plasma membranes and the separation of inside-out and right-side-out vesicles. Methods Enzymol. 228: 451-469.

Leggewie, G., Willmitzer, L., and Riesmeier, J.W. 1997. Two cDNAs from potato are able to complement a phosphate uptake deficient yeast mutant: Identification of phosphate transporters from higher plants. Plant Cell 9: 381-392.

Leigh, R.A., and Wyn-Jones, R.G. 1984. A hypothesis relating critical potassium concentrations for growth to the distribution and function of this ion in the plant cell. New Phytol. 97: 1-13.

Lemoine, R. 2000. Sucrose transporters in plants: update on function and structure. Biochim. Biophys. Acta. 1465: 246-262.

Leng, Q., Mercier, R.W., Yao, W., and Berkowitz, G.A. 1999. Cloning and first functional characterization of a plant cyclic nucleotide-gated cation channel. Plant Physiol. 121: 753-761.

Leng, Q., Mercier, R.W., Hua, B.G., Fromm, H., and Berkowitz, G.A. 2002. Electrophysiological analysis of cloned cyclic nucleotide-gated ion channels. Plant Physiol. 128: 400-410.

Li, L., He, Z., Pandey, G.K., Tsuchiya, T., and Luan, S. 2002. Functional cloning and characterization of a plant efflux carrier for multidrug and heavy metal detoxification. J. Biol. Chem. 277: 5360-5368.

Liang, F., Cunningham, K.W., Harper, J.F., and Sze, H. 1997. ECA1 complements yeast mutants defective in Ca^{2+} pumps and encodes an endoplasmic reticulum-type Ca^{2+}-ATPase in Arabidopsis thaliana. Proc. Natl. Acad. Sci. USA 94: 8579-8584.

Lucas, W.J. and Wolf, S. 1993. Plasmodesmata: the intercellular organelles of green plants. Trends Cell Biol. 3: 308-315.

Luo, H., Morsomme, P., and Boutry, M. 1999. The two major types of plant plasma membrane H^+-ATPases show different enzymatic properties and confer differential pH sensitivity of yeast growth. Plant Physiol. 119: 627-634.

Lurin, C., Güclü, J., Cheniclet, C., Carde, J-P., Barbier-Brygoo, H., and Maurel, C. 2000. CLC-Nt1, a putative chloride channel protein of tobacco, co-localizes with mitochondrial membrane markers. Biochem. J. 348: 291-295.

Maathuis, F.J.M., Filatov, V., Herzyk, P., Krijger, G.C., Axelsen, K.B., Chen, S. et al. 2003. Transcriptome analysis of root transporters reveals participation of multiple gene families in the response to cation stress. Plant J. 35: 675-692.

MacKinnon, R. 2003. Potassium channels. FEBS Lett. 555: 62-65.

MacRobbie, E.A. 2000. ABA activates multiple Ca^{2+} fluxes in stomatal guard cells, triggering vacuolar $K^+(Rb^+)$ release. Proc. Natl. Acad. Sci. USA 97: 12361-12368.

Maeshima, M. 2000. Vacuolar H^+-pyrophosphatase. Biochim. Biophys. Acta. 1465: 37-51.

Malmström, S., Askerlund, P., and Palmgren, M.G. 1997. A calmodulin-stimulated Ca^{2+}-ATPase from plant vacuolar membranes with a putative regulatory domain at its N-terminus. FEBS Lett. 400: 324-328.

Malmström, S., Åkerlund, H.-E., and Askerlund, P. 2000. Regulatory role of the N terminus of the vacuolar Ca^{2+}-ATPase in cauliflower. Plant Physiol 122: 517-526.

Marmagne, A., Rouet, M.A., Ferro, M., Rolland, N., Alcon, C., Joyard, J., Garin, J., Barbier-Brygoo, H., and Ephritikhine, G. 2004. Identification of new intrinsic proteins in Arabidopsis plasma membrane proteome. Mol Cell Proteomics. 3: 675-691.

Marra, M., Fullone, M.R., Fogliano, V., Pen, J., Mattei, M., Masi, S., and Aducci, P. 1994. The 30-Kilodalton protein present in purified fusicoccin receptor preparations is a 14-3-3-like protein. Plant Physiol. 106: 1497-1501.

Marré, E. 1979. Fusicoccin: A tool in plant physiology. Annu. Rev. Plant Physiol. 30: 273-288.

Marschner, H. 1995. Mineral Nutrition of Higher Plants. Academic Press Inc., San Diego, CA.

Marten, I., Hoth, S., Deeken, R., Ache, P., Ketchum, K.A., Hoshi, T., and Hedrich, R. 1999. AKT3, a phloem-localized K^+ channel, is blocked by protons. Proc. Natl. Acad. Sci. USA 96: 7581-7586.

Martinoia, E., Massonneau, A., and Frangne, N. 2000. Transport processes of solutes across the vacuolar membrane of higher plants. Plant Cell Physiol. 41: 1175-1186.

Martinoia, E., Klein, M., Geisler, M., Bovet, L., Forestier, C., Kolukisaoglu, Ü., Müller-Röber, B., Schulz, B. 2002. Multifunctionality of plant ABC transporters - more than just detoxifiers. Planta 214: 345-355.

Marty, F. 1999. Plant vacuoles. Plant Cell 11: 587-600.

Mäser, P., Thomine, S., Schroeder, J.I., Ward, J.M., Hirschi, K., Sze, H. et al. 2001. Phylogenetic relationships within cation transporter families of Arabidopsis. Plant Physiol. 126: 1646-1667.

Matzke, A.J., Behensky, C., Weiger, T., and Matzke, M.A. 1992. A large conductance ion channel in the nuclear envelope of a higher plant cell. FEBS Lett. 302: 81-85.

Maurel, C., Reizer, J., Schroeder, J.I., and Chrispeels, M.J. 1993. The vacuolar membrane protein gamma-TIP creates water-specific channels in Xenopus oocytes. EMBO J. 12: 2241-2247.

McLean, B.G., Hempel, F.D., and Zambryski, P.C. 1997. Plant intercellular communication via plasmodesmata. Plant Cell 9: 1043-1054.

Mitchell, P.M. 1976. Vectorial chemistry and the molecular mechanism of chemiosmotic coupling: power transmission by proticity. Biochem. Soc. Trans. 4: 399-430.

Mitchell, P.M. 1985. The correlation of chemical and osmotic forces in biochemistry. J. Biochem. 97: 1-18.

Möhlmann, T., Tjaden, J., Schwöppe, C., Winkler, H.H., Kampfenkel, K., and Neuhaus, H.E. 1998. Occurrence of two plastidic ATP/ADP transporters in *Arabidopsis thaliana:* Molecular characterisation and comparative structural analysis of homologous ATP/ADP translocators from plastids and *Rickettsia prowazekii*. Eur. J. Biochem. 252: 353-359.

Moran, N., Ehrenstein, G., Iwasa, K., Bare, C., and Mischke, C. 1984. Ion channels in the plasmalemma of wheat protoplasts. Science 226: 835-838.

Morsomme, P. and Boutry, M. 2000. The plant plasma membrane H⁺-ATPase: structure, function and regulation. Biochim. Biophys. Acta. 1465: 1-16.

Morsomme, P., de Kerchove d'Exaerde, A., De Meester, S., Thines, D., Goffeau, A., and Boutry, M. 1996. Single point mutations in various domains of a plant plasma membrane H(+)-ATPase expressed in *Saccharomyces cerevisiae* increase H(+)-pumping and permit yeast growth at low pH. EMBO J. 15: 5513-5526.

Moshelion, M., Becker, D., Czempinski, K., Mueller-Roeber, B., Attali, B., Hedrich, R., and Moran, N. 2002. Diurnal and circadian regulation of putative potassium channels in a leaf moving organ. Plant Physiol. 128: 634-642.

Muir, S.R. and Sanders, D. 1997. Inositol 1,4,5-trisphosphate-sensitive Ca²⁺ release across nonvacuolar membranes in cauliflower. Plant Physiol. 114: 1511-1521.

Murata, K., Mitsuoka, K., Hirai, T., Walz, T., Agre, P., Heymann, J. B., Engel, A., and Fujiyoshi, Y. 2000. Structural determinants of water permeation through aquaporin-1. *Nature* 407: 599-605.

Neher, E. and Sakmann, B. 1976. Single-channel currents recorded from membrane of denervated frog muscle fibre. Nature 260: 799-802.

Neuhaus, H.E. and Wagner, R. 2000. Solute pores, ion channels, and metabolite transporters in the outer and inner envelope membranes of higher plant plastids. Biochim. Biophys. Acta. 1465: 307-323.

Neuhaus, H.E., Thom, E., Möhlmann, T., Steup, M., Kampfenkel, K. 1997. Characterization of a novel ATP/ADP transporter from *Arabidopsis thaliana* L. Plant J. 11: 73-82.

Ninnemann, O., Jauniaux, J.C., and Frommer, W.B. 1994. Identification of a high affinity NH₄⁺ transporter from plants. EMBO J. 13: 3464-3471.

Nishida, M. and MacKinnon, R. 2002. Structural basis of inward rectification: cytoplasmic pore of the G protein-gated inward rectifier GIRK1 at 1.8 Å resolution. Cell 111: 957-965.

Novina, C.D. and Sharp, P.A. 2004. The RNAi revolution. *Nature* 430: 161-164.

Oecking, C, Eckerskorn, C, and Weiler, EW. 1994. The fusicoccin receptor of plants is a member of the 14-3-3 superfamily of eukaryotic regulatory proteins. FEBS Lett. 352: 163-166.

Olsson, A. 2000. The plant plasma membrane H⁺-ATPase. Regulation by phosphorylation and 14-3-3 proteins. PhD diss., Lund Univ., Sweden.

Olsson, A., Svennelid, F., Ek, B., Sommarin, M., and Larsson, C. 1998. A phosphothreonine residue at the C-terminal end of the plasma membrane H⁺-ATPase is protected by fusicoccin-induced 14-3-3 binding. Plant Physiol. 118: 551-555.

Ortiz-Lopez, A., Chang, H.-C., and Bush, D.R. 2000. Amino acid transporters in plants. Biochim. Biophys. Acta. 1465: 275-280.

Overall, R.L. and Blackman, L.M. 1996. A model of the macromolecular structure of plasmodesmata. Trends Plant Sci. 1: 307-311.

Palmgren, M.G. 2001. Plant plasma membrane H⁺-ATPases: Powerhouses for nutrient uptake. Annu. Rev. Plant Physiol. Plant Molec. Biol. 52: 817-845.

Palmgren, M.G. and Sommarin, M. 1989. Lysophosphatidylcholine stimulates ATP dependent proton accumulation in isolated oat root plasma membrane vesicles. *Plant Physiol.* 90: 1009-1014.

Palmgren, M.G. and Christensen, G. 1994. Functional comparisons between plant plasma membrane H(+)-ATPase isoforms expressed in yeast. J. Biol. Chem. 269: 3027-3033.

Palmgren, M.G. and Harper, J.F. 1999. Pumping with P-type ATPases. J. Exper. Bot. 50: 883-893.

Palmgren, M.G. Larsson, C., and Sommarin, M. 1990. Proteolytic activation of the plant plasma membrane H(+)-ATPase by removal of a terminal segment. J. Biol. Chem. 265: 13423-13426.

Palmgren, M.G., Sommarin, M., Serrano, R., and Larsson, C. 1991. Identification of an autoinhibitory domain in the C-terminal region of the plant plasma membrane H(+)-ATPase. J. Biol. Chem. 266: 20470-20475.

Papazian, D.M., Schwarz, T.L., Tempel, B.L., Jan, Y.N., and Jan, L.Y. 1987. Cloning of genomic and complementary DNA from Shaker, a putative potassium channel gene from Drosophila. Science 237: 749-53.

Pardo, J.M. and Serrano, R. 1989. Structure of a plasma membrane H$^+$-ATPase gene from the plant *Arabidopsis thaliana*. J. Biol. Chem. 264: 8557-8562.

Penniston, J.T. and Enyedi, A. 1998. Modulation of the plasma membrane Ca^{2+} pump. J. Membr. Biol. 165: 101-109.

Pfeffer, W. 1877. Verlag von Wilhelm Engelmann, Osmotische Untersuchungen. Studien zur Zellmechanik. Leipzig.

Philippar, K., Fuchs, I., Lüthen, H., Hoth, S., Bauer, C.S., Haga, K. *et al.* 1999. Auxin-induced K$^+$ channel expression represents an essential step in coleoptile growth and gravitropism. Proc. Natl. Acad. Sci. USA 96: 12186-12191.

Picault, N., Hodges, M., Palmieri, L., and Palmieri, F. 2004. The growing family of mitochondrial carriers in *Arabidopsis*. Trends Plant Sci 9: 138-146.

Picault, N., Palmieri, L., Pisano, I., Hodges, M., and Palmieri, F. 2002. Identification of a novel transporter for dicarboxylates and tricarboxylates in plant mitochondria. Bacterial expression, reconstitution, functional characterization, and tissue distribution. J. Biol. Chem. 277: 24204-24211.

Pilot, G., Gaymard, F., Mouline, K., Cherel, I., and Sentenac, H. 2003. Regulated expression of *Arabidopsis* shaker K$^+$ channel genes involved in K$^+$ uptake and distribution in the plant. Plant Molec. Biol. 51: 773-787.

Pohlmeyer, K., Soll, J., Steinkamp, T., Hinnah, S., and Wagner, R. 1997. Isolation and characterization of an amino acid-selective channel protein present in the chloroplastic outer envelope membrane. Proc. Natl. Acad. Sci. USA 94: 9504-9509.

Pohlmeyer, K., Soll, J., Grimm, R., Hill, K., and Wagner, R. 1998. A high-conductance solute channel in the chloroplastic outer envelope from Pea. Plant Cell. 10: 1207-1216.

Preston, G.M. and Agre, P. 1991. Isolation of the cDNA for erythrocyte integral membrane protein of 28 kilodaltons: member of an ancient channel family, Proc. Natl. Acad. Sci. USA 88: 11110-11114.

Preston, G.M., Carroll, T.P., Guggino, W.B., and Agre, P. 1992. Appearance of water channels in *Xenopus oocytes* expressing red cell CHIP28 protein, Science 256: 385-387.

Qiu, Q.S., Guo, Y., Dietrich, M.A., Schumaker, K.S., and Zhu, J.K. 2002. Regulation of SOS1, a plasma membrane Na$^+$/H$^+$ exchanger in *Arabidopsis thaliana*, by SOS2 and SOS3. Proc. Natl. Acad. Sci. USA 99: 8436-8441.

Qiu, Q.S., Barkla, B.J., Vera-Estrella, R., Zhu, J.K., and Schumaker, K.S. 2003. Na$^+$/H$^+$ exchange activity in the plasma membrane of Arabidopsis. Plant Physiol. 132: 1041-1052.

Qiu, Q.S., Guo, Y., Quintero, F.J., Pardo, J.M., Schumaker, K.S., and Zhu, J.K. 2004. Regulation of vacuolar Na$^+$/H$^+$ exchange in *Arabidopsis thaliana* by the salt-overly-sensitive (SOS) pathway. J. Biol. Chem. 279: 207-215.

Ratajczak, R. 2000. Structure, function and regulation of the plant vacuolar H$^+$-translocating ATPase. Biochim. Biophys. Acta. 1465: 17-36.

Rea, P.A. and Poole, R.J. 1993. Vacuolar H$^+$-translocating pyrophosphatase. Annu. Rev. Plant Physiol. Plant Mol. Biol. 44: 157-180.

Regenberg, B., Villalba, J.M., Lanfermeijer, F.C., and Palmgren, M.G. 1995. C-terminal deletion analysis of plant plasma membrane H$^+$-ATPase: Yeast as a model system for solute transport across the plant plasma membrane. Plant Cell 7: 1655-1666.

Riesmeier, J.W., Willmitzer, L., and Frommer, W.B. 1992. Isolation and characterization of a sucrose carrier cDNA from spinach by functional expression in yeast. EMBO J. 11: 4705-4713.

Roh, M.H., Shingles, R., Cleveland, M.J., and McCarty, R.E. 1998. Direct measurement of calcium transport across chloroplast inner-envelope vesicles. Plant Physiol. 118: 1447-1454.

Rubio, F., Gassmann, W., and Schroeder, J.I. 1995. Sodium-driven potassium uptake by the plant potassium transporter HKT1 and mutations conferring salt tolerance. Science 270: 1660-1663.

Saier, N. and Stolz, J. 1994. SUC1 and SUC2: two sucrose transporters from *Arabidopsis thaliana*; expression and characterization in baker's yeast and identification of the histidine-tagged protein. Plant J. 6: 67-77.

Sanchez-Fernandez, R., Davies, T.G., Coleman, J.O., and Rea, P.A. 2001. The *Arabidopsis thaliana* ABC protein superfamily, a complete inventory. J. Biol. Chem. 276: 30231-30244.

Sanders, D., Brownlee, C., and Harper, J.F. 1999. Communicating with calcium. *Plant Cell* 11: 691-706.

Sanders, D., Pelloux, J., Brownlee, C., and Harper, J.F. 2002. Calcium at the crossroads of signaling. Plant Cell 14: S401-417.

Schachtman, D.P. and Schroeder, J.I. 1994. Structure and transport mechanism of a high-affinity potassium uptake transporter from higher plants. Nature 370: 655-658.

Schönknecht, G., Spoormaker, P., Steinmeyer, R., Brüggeman, L., Ache, P., Dutta, R. et al. 2002. KCO1 is a component of the slow-vacuolar (SV) ion channel. FEBS Lett. 511: 28-32.

Schroeder, J.I., Hedrich, R., and Fernandez, J.M. 1984. Potassium-selective single channels in guard cell protoplasts of *Vicia faba*. Nature 312: 361-362.

Schuurink, R.C., Shartzer, S.F., Fath, A., and Jones, R.L. 1998. Characterization of a calmodulin-binding transporter from the plasma membrane of barley aleurone. Proc. Natl. Acad. Sci. USA 95: 1944-1949.

Sentenac, H., Bonneaud, N., Minet, M., Lacroute, F., Salmon, J.-M., Gaymard, F., and Grignon, C. 1992. Cloning and expression in yeast of a plant potassium ion transport system. Science 256: 663-665.

Shi, H. and Zhu, J.K. 2002. Regulation of expression of the vacuolar Na$^+$/H$^+$ antiporter gene AtNHX1 by salt stress and abscisic acid. Plant Molec. Biol. 50: 543-550.

Shi, H., Ishitani, M., Kim, C., and Zhu, J.K. 2000. The *Arabidopsis thaliana* salt tolerance gene SOS1 encodes a putative Na$^+$/H$^+$ antiporter. Proc. Natl. Acad. Sci. USA 97: 6896-6901.

Shi, H., Lee, B.H., Wu, S.J., and Zhu, J.K. 2003. Overexpression of a plasma membrane Na$^+$/H$^+$ antiporter gene improves salt tolerance in *Arabidopsis thaliana*. Nat. Biotec. 21: 81-85.

Shingles, R. and McCarty, R.E. 1994. Direct measurements of ATP-dependent proton concentration changes and characterisation of a K$^+$-stimulated ATPase in pea chloroplasts inner envelope vesicles. Plant Physiol. 106: 731-737.

Shingles, R., North, M., and McCarty, R.E. 2002. Ferrous ion transport across chloroplast inner envelope membranes. Plant Physiol. 128: 1022-1030.

Sidler, M., Hassa, P., Hasan, S., Ringli, C., and Dudler, R. 1998. Involvement of an ABC transporter in a developmental pathway regulating hypocotyl cell elongation in the light. Plant Cell 10: 1623-1636.

Smith, F.W., Ealing, P.M., Hawkesford, M.J., and Clarkson, D.T. 1995. Plant members of a family of sulfate transporters reveal functional subtypes. Proc. Natl. Acad. Sci. USA 92: 9373-9377.

Svennelid, F., Olsson, A., Piotrowski, M., Rosenquist, M., Ottman, C., Larsson, C., Oecking, C., and Sommarin, M. 1999. Phosphorylation of Thr-948 at the C terminus of the plasma membrane H$^+$-ATPase creates a binding site for the regulatory 14-3-3 protein. Plant Cell 11: 2379-2392.

Sze, H. 1980. Nigericin-stimulated ATPase activity in microsomal vesicles of tobacco callus. Proc. Natl. Acad. Sci. USA 77: 5904-5908.

Sze, H. 1985. H$^+$-translocating ATPases: Advances using membrane vesicles. Annu. Rev. Plant Physiol. 36: 175-208.

Sze, H., Li, X., and Palmgren, MG. 1999. Energization of Plant Cell Membranes by H$^+$-Pumping ATPases: Regulation and Biosynthesis. Plant Cell 11: 677-690.

Sze, H., Liang, F., Hwang, I., Curran, A.C., and Harper, J.F. 2000. Diversity and regulation of plant Ca^{2+} pumps: insights from expression in yeast. Annu. Rev. Plant Physiol. Plant Molec. Biol. 51: 433-462.

Takano, J., Noguchi, K., Yasumori, M., Kobayashi, M., Gajdos, Z., Miwa, K., Hayashi, H., Yoneyama, T., and Fujiwara, T. 2002.

Arabidopsis boron transporter for xylem loading. Nature 420: 337-340.

Talke, I.N., Blaudez, D., Maathuis, F.J., and Sanders, D. 2003. CNGCs: prime targets of plant cyclic nucleotide signalling? Trends Plant Sci. 8: 286-293.

Tempel, B.L., Papazian, D.M., Schwarz, T.L., Jan, Y.N., and Jan, L.Y. 1987. Sequence of a probable potassium channel component encoded at Shaker locus of *Drosophila*. Science 237: 770-775.

Theodoulou, F. 2000. Plant ABC transporters. Biochim. Biophys. Acta. 1465: 79-103.

Tournaire-Roux, C., Sutka, M., Javot, H., Gout, E., Gerbeau, P., Luu, D.T., Bligny, R., and Maurel, C. 2003. Cytosolic pH regulates root water transport during anoxic stress through gating of aquaporins. Nature 425: 393-397.

Toyoshima, C. and Nomura, H. 2002. Structural changes in the calcium pump accompanying the dissociation of calcium. *Nature* 418: 605-611.

Toyoshima, C., Nomura, H., and Sugita, Y. 2003. Structural basis of ion pumping by Ca^{2+}-ATPase of sarcoplasmic reticulum. FEBS Lett. 555: 106-110.

Toyoshima, C., Nakasako, M., Nomura, H., and Ogawa, H. 2000. Crystal structure of the calcium pump of sarcoplasmic reticulum at 2.6 Å resolution. Nature 405: 647-655.

Tsay, Y.-F., Schroeder, J.I., Feldmann, K.A., and Crawford, N.M. 1993. The herbicide sensitivity gene *CHL1* of *Arabidopsis* encodes a nitrate-inducible nitrate transporter. Cell 72: 705-713.

Ueoka-Nakanishi, H., Nakanishi, Y., Tanaka, Y., and Maeshima, M. 1999. Properties and molecular cloning of Ca^{2+}/H^+ antiporter in the vacuolar membrane of mung bean. Eur. J. Biochem. 262: 417-425.

Van der Zaal, B.J., Neuteboom, L.W., Pinas, J.E., Chardonnes, A.N., Schat, H., Verkleij, J.A.C., and Hooykaas, P.J.J. 1999. Overexpression of a novel *Arabidopsis* gene related to putative zinc-transporter genes from animals can lead to enhanced zinc resistance and accumulation. Plant Physiol. 119: 1047-1055.

Vert, G., Grotz, N., Dedaldechamp, F., Gaymard, F., Guerinot, M.L., Briat, J.F., and Curie, C. 2002. IRT1, an *Arabidopsis* transporter essential for iron uptake from the soil and for plant growth. Plant Cell 14: 1223-1233.

Véry, A.-A. and Sentenac, H. 2003. Molecular mechanisms and regulation of K^+ transport in higher plants. Ann. Rev. Plant Biol. 54: 575-603.

Walker, D.J., Leigh, R.A., and Miller, A.J. 1996. Potassium homeostasis in vacuolate plant cells. Proc. Natl. Acad. Sci. USA 93: 10510-10514.

Wang, X., Berkowitz, G.A., and Peters, J.S. 1993. K^+-conducting ion channel of the chloroplast inner envelope: functional reconstitution into liposomes. Proc. Natl. Acad. Sci. USA 90: 4981-4985.

Ward, J.M. 1997. Patch-clamping and other molecular approaches for the study of plasma membrane transporters demystified. Plant Physiol. 114: 1151-1159.

Weber, A., Menzlaff, E., Arbinger, B., Gutensohn, M., Eckerskorn, C., and Fluegge, U.-I. 1995. The 2-oxoglutarate/malate translocator of chloroplast envelope membranes: molecular cloning of a transporter containing a 12-helix motif and expression of the functional protein in yeast cells. Biochemistry 34: 2621-2627.

White, P.J. 1998. Calcium channels in the plasma membrane of root cells. Ann. Bot. 81: 173-183.

White, P.J. 2000. Calcium channels in higher plants. Biochim. Biophys. Acta. 1465: 171-189.

White, P.J., Bowen, H.C., Demidchik, V., Nichols, C., and Davies, J.M. 2002. Genes for calcium-permeable channels in the plasma membrane of plant root cells. Biochim. Biophys. Acta. 1564: 299-309

Williams, L.E. and Miller, A.J. 2001. Transporters responsible for the uptake and partitioning of nitrogenous solutes. Annu. Rev. Plant Physiol. Plant Molec. Biol. 52: 659-688.

Wink, M. 1993. The plant vacuole: A multifunctional compartment. J. Exper. Bot. 44: 231-246.

Wu, Z., Liang, F., Hong, B., Young, J.C., Sussman, M. R., Harper, J.F., Sze, H. 2002. An endoplasmic reticulum-bound Ca^{2+}/Mn^{2+} pump, ECA1, supports plant growth and confers tolerance to Mn^{2+} stress. Plant Physiol. 130: 128-137.

Yamaguchi, T., Fukada-Tanaka, S., Inagaki, Y., Saito, N., Yonekura-Sakakibara, K., Tanaka, Y., Kusumi, T., and Iida, S. 2001. Genes encoding the vacuolar Na^+/H^+ exchanger and flower coloration. Plant Cell Physiol. 42: 451-461.

Yokoi, S., Quintero, F.J., Cubero, B., Ruiz, M.T., Bressan, R.A., Hasegawa, P.M., and Pardo, J.M. 2002. Differential expression and function of *Arabidopsis thaliana* NHX Na$^+$/H$^+$ antiporters in the salt stress response. Plant J. 30: 529-539.

Yu, J., Hu, S., Wang, J., Wong, G.K.-S., Li, S., Liu, B. *et al.* 2002. A draft sequence of the rice genome (*Oryza sativa* L. ssp. *indica*), Science 296: 79-92.

Zei, P.C. and Aldrich, R.W. 1998. Voltage-dependent gating of single wild-type and S4 mutant KAT1 inward rectifier potassium channels. J. Gen. Physiol. 112: 679-713.

Zeidel, M.L., Ambudkar, S.V., Smith, B.L., and Agre, P. 1992. Reconstitution of functional water channels in liposomes containing purified red cell CHIP28 protein. Biochemistry 31: 7436-7440.

Zeng, G.F., Pypaert, M., and Slayman, C.L. 2004. Epitope tagging of the yeast K$^+$ carrier TRK2p demonstrates folding that is consistent with a channel-like structure. J. Biol. Chem. 279: 3003-3013.

Zhang, H.X. and Blumwald, E. 2001a. Transgenic salt-tolerant tomato plants accumulate salt in foliage but not in fruit. Nat. Biotechnol. 19: 765-768.

Zhang, H.X., Hodson, J.N., Williams, J.P., and Blumwald, E. 2001b. Engineering salt-tolerant *Brassica* plants: characterization of yield and seed oil quality in transgenic plants with increased vacuolar sodium accumulation. Proc. Natl. Acad. Sci. USA 98: 12832-12836.

Zhou, Y., Morais-Cabral, J.H., Kaufman, A., and MacKinnon, R. 2001. Chemistry of ion coordination and hydration revealed by a K$^+$ channel-Fab complex at 2.0 Å resolution. *Nature* 414: 43-48.

Zimmermann, S., Ehrhardt, T., Plesch, G., and Mueller-Roeber, B. 1999. Ion channels in plant signalling. Cell. Molec. Life Sci. 55: 183-220.

CHAPTER 8

Mitosis in Plant Cells

Virginia Shepherd

INTRODUCTION

All cells come from pre-existing cells

It is difficult, perhaps impossible for us to imagine what it was like when scientists realized for the first time that there is a world inside the one we can see. Everything we know today about the processes of cell division began with observations by a small number of scientists working around 350 years ago. Science is a co-operative activity and every discovery is the result of accumulated observations, theorizing and inspired guessing on the part of a group of scientists focused on particular problems.

The word "cell" came from Robert Hooke, who published the classic work "Micrographia" in 1665. Hooke was one of the outstanding inventors and scientists of his day. He described cavities in cork, charcoal and plant tissues as "cells", rather like the tiny rooms in which monks and nuns lived. Although he did not see their contents, he guessed that cells contained a "nourishing juice" that might travel through pores from one cell to another.

The nucleus, mitosis and chromosomes were first observed in the nineteenth cen-tury. The Cell Theory (1838) of Matthias Schleiden and Theodor Schwann stated that all organisms are composed of cells, a great conceptual leap although it may seem obvious to us today. A few decades later Virchow coined the famous Latin phrase *"omnis cellula e cellula"* or "all cells arise from pre-existing cells". You cannot make a cell, even if you have the entire genetic sequence of the organism and all the raw material. The cell is a fundamental unit of existence.

Chromosomes were observed within the nucleus by a number of scientists, including Nageli (1842), who described the behavior of chromosomes during cell division. In 1882 Walther Fleming gave the name "mitosis" to this behavior. Figure 8.1 shows early illustrations of some stages of mitosis in the bioluminescent dinoflagellate *Noctiluca miliaris*. Eduard Strasburger (1875) recognised that " all nuclei come from pre-existing nuclei" or, in Latin, *"omnis nucleus e nucleo"*. You cannot make a nucleus, except from a pre-existing nucleus. Similarly, you cannot make mitochondria and chloro-plasts, and the nucleus of a cell cannot make them. These structures are fundamental.

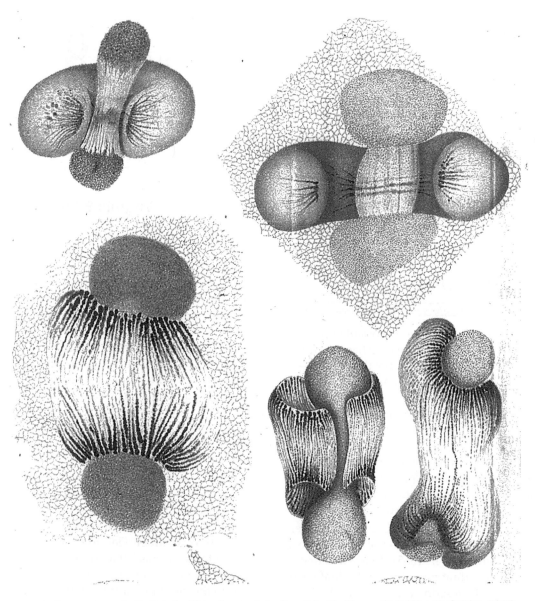

Fig. 8.1 Some stages of mitosis in the bioluminescent dinoflagellate *Noctiluca miliaris* (Gary M. Calkins, 1898). Anaphase is seen in the bottom left-hand corner.

Shortly after the beginning of the twentieth century the chromosome was shown to be the cellular structure responsible for heredity. The stage was then set for the extraordinarily rapid growth in understanding cell division and heredity in the twentieth and twenty-first centuries. In 1944 Oswald T. Avery and co-workers showed that DNA was the material that carried specific inherited characteristics, but it was not until 1953 that James Watson, Francis Crick and Rosalind Franklin showed that DNA encoded information by the precise sequence of nucleotide bases in the DNA molecule.

The idea then became established that genes encoding information were fixed, like beads on a string, and the gene sequence static and stable. From this perspective it was difficult to see how evolution could occur, except by random point mutations.

Working with corn, Barbara McClintock showed in the 1940s that parts of a chromosome could be released from one position and reinserted in a different one. She called this transposition and it had dramatic effects on genetic expression. Transposition led to the formation of unusually colored corn kernels that could not have developed in any other way. Her ideas met with scepticism for many years but transposition was rediscovered in the 1970s. Barbara McClintock was awarded the Nobel Prize in 1983. It is now known that transposition accelerates the evolution of genomes. There is a balance between stability of the genome, necessary for replication of DNA, and capacity of the genome for change, necessary for evolution and environmental adaptation. Large-scale chromosomal rearrangements provide much of the plasticity of the genome needed for evolution.

It remains true that all cells must come from a pre-existing cell, all nuclei must come from a pre-existing nucleus, and all chloroplasts and mitochondria must come from pre-existing chloroplasts and mitochondria. As Virchow recognized in 1858, there has been an uninterrupted sequence of cell divisions from the first appearance of cells on earth to all the organisms living on earth today.

STRUCTURE OF CHROMOSOMES

All living things are composed of cells. Another attribute of living things is that they grow and reproduce themselves, and their progeny reproduce themselves in turn. The seeds in a pea-pod contain an embryo and a nutritive tissue called endosperm. The embryo forms from a fertilized ovule (essentially an egg) and the seed germinates to grow into a pea-plant. Fertilized eggs can grow into pea-plants, goannas, human beings or a litter of puppies largely through the processes of cell division and differentiation. For these macroscopic processes to occur it is necessary that cells themselves must reproduce. Cells reproduce themselves by dividing. The genetic material of a cell is arranged in the form of chromosomes within the cell nucleus.

Chromosomes Made from Chromatin

The nucleus contains genetic material in the form of chromosomes. When a cell is preparing to divide the chromosomes are *condensed*, and when the cell is growing and performing its normal functions in the organism the chromosomes are *decondensed*.

When a cell is treated with a stain such as toluidine blue, the decondensed chromosomes are faintly visible inside the nucleus as very thin, very long and tangled strands of a material called *chromatin*. The stain binds to the negatively charged groups of the DNA in the chromatin so that we can see the latter. Chromatin is a complex of about 40% DNA and 60% proteins of a special type called histones.

Chromosomes are made from chromatin. The DNA of one chromosome exists as a single enormously long double-stranded molecule. Thus, a chromosome is a single very long DNA molecule associated with histone proteins.

Chromosomes are only readily visible under a light microscope when they condense as the cell is preparing to divide. In a *condensed* chromosome the chromatin has been compacted around 50 thousand times and it becomes shorter and a great deal thicker. In this form the chromosomes can be more easily moved around and sepa-

rated inside the cell during cell division, just as it is easier to carry a number of items bundled into a bag than it is to carry the same items as separate objects. However the DNA of the chromosome is no longer translated into proteins when the chromosome has been condensed. A diagram of a condensed, replicated chromosome is shown in Fig. 8.2.

When the cell is not dividing and is occupied with performing its role in the plant body, the chromosomes are so long,

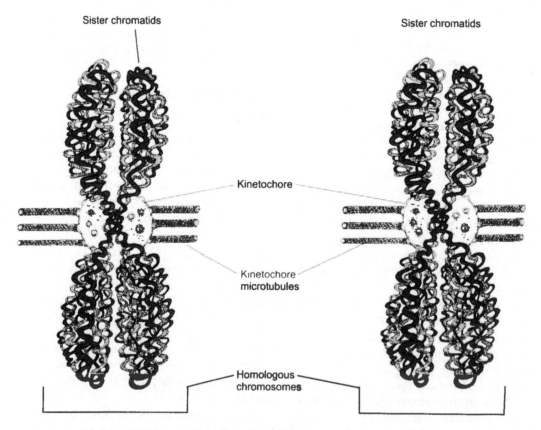

Fig. 8.2 Structure of homologous chromosomes in metaphase of mitosis. DNA was duplicated in the S phase of the cell cycle. Each chromosome consists of two virtually identical chromatids held together at a region of very condensed DNA called the centromere. Each chromatid is shown with its kinetochore. The kinetochores face in opposite directions. They each capture a spindle microtubule. The chromosomes are aligned along the metaphase plate at the equator of the cell. At anaphase the twin chromatids will be pulled by the spindle microtubules in opposite directions. This separates the chromatids, and they are then called daughter chromosomes.

thin and tangled that they are not readily visible under a microscope. At this time they are *decondensed*. The decondensed chromosomes are the form in which the DNA is read and translated into proteins.

The reasons for condensation and decondensation of chromosomes can easily be imagined if we compare the chromosome to a ball of wool. While it is easier to move a ball of wool around, it would not be possible to knit with the wool tightly coiled into a ball. On the other hand, knitting would be far easier with a long unravelled strand of wool, but such a strand would be far more difficult to move around.

Condensed chromosomes are easier to move around but their DNA not translated. The DNA of decondensed chromosomes can be translated but they are difficult to move around.

Heterochromatin and Euchromatin

The chromosome does not stain evenly with basic dyes. There are heavily stained and lightly stained regions and the density of staining shows how condensed or compact the DNA is in a particular region. The words *heterochromatin* and *euchromatin* are used to describe how condensed the DNA is in a region of a chromosome.

Heterochromatin is a condensed region of DNA that is not translated. The genes may be inactive or missing. *Euchromatin* is far less condensed. Most of the active genes are located in euchromatin.

Structure of Chromosomes

The DNA of each chromosome is replicated (duplicated) in the S phase of the cell cycle. It forms *sister chromatids* held together by a *centromere*. The centromere is a pinched or constricted region of the chromosome. It is made from heterochromatin (very condensed DNA). The centromere separates the chromosome into two "arms". It is often shown as situated in the center of the chromosome but may be towards the top or bottom, dividing the chromosome into short and long "arms". The position of the centromere is used when classifying and identifying different chromosomes.

The ends of both chromosome arms are called *telomeres*. Telomeres are probably the region where the chromosome is anchored in the nuclear membrane. Like the centromere, telomeres also contain repetitive sequences of DNA base pairs and are made from heterochromatin.

Chromatids are virtually identical. Before the cell divides, it temporarily contains two copies of the normal amount of DNA. For example, a cell from a rye plant with a diploid number of chromosomes (2n = 14), would have 14 replicated chromosomes, each made up of two chromatids, after completing the S phase. This is a total of 28 chromatids. The number of chromosomes can be worked out by counting the centromeres. If there are 14 centromeres, there are 14 replicated chromosomes.

Chromosomes are usually found as pairs. Each chromosome in a pair originally came from one of the parents during fertilization. These chromosome pairs are called *homologous chromosomes* (Fig. 8.2).

How is so much DNA Packed into so Small a Nucleus?

Compared with prokaryotic cells, the nucleus of every eukaryotic cell contains a very large amount of DNA. Almost 2 meters of DNA can be contained within a plant cell nucleus whose diameter is only about 10 μm. That is, the nucleus has a diameter about 200,000 times smaller than the length of the DNA that has to fit inside! In the case of human beings, the DNA of the shortest chromosome in a human nucleus spans a length of about 14 mm when decondensed. Yet the nucleus itself is only about 10 µm in diameter, and there are a total of 46 human chromosomes! It has been said that if all the DNA of all the cells of a human being were taken out and stretched to their full length, the DNA would reach to the moon and back, not once but several times over. How is such an enormous length of DNA compressed into so small a space?

Three levels of packing enables the DNA to fit into the relatively small space of the nucleus (Fig. 8.3) and each is considered in turn.

First, the DNA is coiled around protein spools to form structures resembling a string of beads called *nucleosomes*.

Second, the nucleosomes are twisted into a helix called the *30 nm fiber*. This can be imagined as a twisted string of beads.

Third, the 30 nm fibers are twisted to form a thicker fibre 700-1,000 nm in diameter. This can be imagined as a looped, twisted string of beads.

DNA Coils Around Histone Proteins to make Nucleosomes

When a Persian rug or carpet is woven, the weaver must keep many meters of differently coloured threads on spools so that they can be packed into a small space, easily moved around, and used in the right place at the right time. The spools act as the structures on which a large length of thread is wound, and the spools themselves are controlled so that the thread is released as needed.

The cell has similar mechanisms for organizing and controlling the enormously long DNA molecules that make up the chromosomes. The proteins of the chromatin are of two types, called *histones* and *non-histone proteins*. Histones can be

imagined as the spools on which DNA is wound, while the *non-histone proteins* control the functioning of the spools and the behavior of the thread.

There are five kinds of histone protein, called H1, H2A, H2B, H3, and H4. The cores around which DNA is wound are made from histones H2A, H2B, H3, and H4. The core is made from two molecules each of H2A, H2B, H3, and H4, making a group of eight called an *octamer*. Each core functions in a manner similar to a spool but unlike an ordinary spool, the histone core has the advantage of attracting negatively charged phosphate groups of DNA thread. This is because histones are positively charged and hence attract the negatively charged phosphate groups of the DNA molecules. Many proteins have an overall negative charge, but histones contain a large amount of the basic amino acids arginine and lysine, which gives them their positive charge.

When DNA is coiled around a histone core it forms a structure called a *nucleosome* that slightly resembles a bead.

Coiling of the DNA into nucleosomes is very regular and highly organized. The DNA is coiled around a histone spool or core. It is wound twice around the histone core and this part of the DNA is called *core DNA*. It is around 146 base pairs long. The DNA that remains interconnects the histone cores and is called *linker DNA*. It is about 200 base pairs long. Viewed with an electron microscope the nucleosome "beads" are about 10 nm in diameter. The string of linked nucleosomes compacts the full length of the DNA about six times.

Nucleosomes Supercoiled into Helix to make 30 nm Fiber

Nucleosomes are twisted into a helix that is three times thicker then they are, about 30 nm in diameter, called *the 30 nm fiber*. The

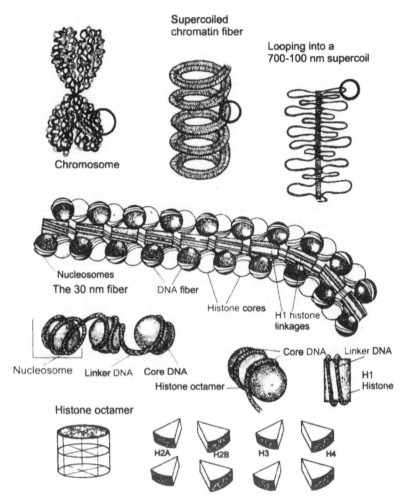

Fig. 8.3 Schematic diagram showing multiple levels of DNA packaging in a chromosome. Packaging compresses the DNA molecule about 50,000 times.

A metaphase chromosome is shown, consisting of condensed replicated DNA.

The magnified region of the chromosome shows the supercoiled chromatin fibre, known as the 700 nm fibre. There is further coiling within this supercoil.

The furry-looking fibre of 700 to 1,000 nm diameter is made up of 30 nm fibres looped and coiled around an axis.

The 30 nm fiber consists of packed nucleosomes. A nucleosome is a length of DNA wrapped around a histone core or spool. This DNA is called core DNA.

The nucleosomes are strung together like beads on a string by connecting strands of DNA, called linker DNA. The linker DNA is bound to a histone called H1.

The core DNA is wound twice around each histone spool to make a nucleosome. The DNA in between, wound around histone H1, is the linker DNA.

The DNA wound around the nucleosome core is about 146 nucleotide base-pairs long. The entire core is about 11 nm wide.

Each core consists of two each of the histone proteins H2A, H2B, H3, and H4. These are associated in an octameric form.

fifth histone, called H1, joins the nucleosomes together in a repeating array by binding to the linker DNA. In 1976, John T. Finch and Aaron Klug called this the *solenoid model* for the superstructure of chromatin. This is a good description, because a solenoid is a long length of wire that has been wound around a small core. There are six nucleosomes per turn in the 30 nm fiber. This compacts the DNA about 40 times.

Twisting and Looping of 30 nm Fibres Makes a 700-1,000 nm Fibre

In the 1970s Ulrich Laemmli and colleagues of the University of Geneva showed electron micrographs of chromosomes in dividing cells from a famous cancer cell line called HeLa cells. They found that the 30 nm fibers of DNA were looped and appeared to emerge from and return to the same place on a protein scaffolding. Looping of the 30 nm fibers makes the chromosome arm into a furry-looking fiber of 700 to 1,000 nm diameter.

In the condensed chromosomes the thick fuzzy loops and scaffolds are twisted still further. A condensed chromosome in a dividing cell has two sister chromatids, each of which is around 700 nm wide. The double helix of DNA is only 2 nm wide. This explains why chromosomes are barely visible when they are decondensed, whereas the condensed chromosomes of a dividing cell are 700 times thicker and easily visible.

Non-histone Proteins Regulate Genes

The majority of cells in the plant body contain the same genes, yet clearly the genes translate differently in cells of different types at different times. If this were not the case every cell would function identically and the body would be a formless blob. While the histones seem to be quite passive in their role as spools, nonhistone proteins are involved in regulating and activating genes. There are thousands of nonhistone proteins. Nonhistone proteins can control the structure of the nucleosomes and thus which regions of the DNA are exposed to be transcribed. Some important nonhistone protein families include the High Mobility Group proteins (HMGPs), which bind to and bend the DNA, and the chromatin remodelling proteins (CHa, CHb, and so on), which use energy to move the nucleosomes around on the DNA strands. The Histone Acetyltransferases (HACs) can change histone spool structure by adding acetyl groups to the spools, while Histone Deacetylases (HDAs) can remove the acetyl groups. Nonhistone proteins can also enable DNA to uncoil from the nucleosome. The DNA Methyltransferases (DMTs) can add a methyl group to the DNA; this is associated with silencing the genes in that region. There are many diverse and complex roles for the nonhistones in controlling chromosome structure and the expression or silencing of particular genes.

Sequencing Plant Genomes

The order of bases in the genes of an organism can be determined by a process called sequencing. The Polymerase Chain Reaction (PCR) is used to make millions of copies of DNA in a test-tube. The PCR uses the same enzyme (DNA Polymerase) that the cell uses to duplicate DNA. The first step in PCR is "unzipping" the DNA by heating it to 90-95°C, which breaks the hydrogen bonds between the nucleotides. The unzipped DNA is then cooled to 55°C so that a DNA primer can bind to the strands. Then the temperature is raised to 72°C and the DNA polymerase uses the four nucleotides (A, C, T, G) supplied in the test-tube to make complementary DNA strands. The high temperature is suitable because the DNA polymerase comes from a bacterium that lives in hot springs. The nucleotides are given a fluorescent tag so that a genetic analyser can detect them later. The PCR process can make a billion copies of the DNA in a few hours. The copied DNA is passed through a genetic analyser. This machine uses a laser beam to excite the different fluorescent tags on the labelled nucleotides. They emit different wavelengths of light and the analyser detects and measures the amount and order of the bases from the fluorescent signals. Computer software then calculates the sequence of nucleotide bases in the DNA.

Recently all five chromosomes of the weed *Arabidopsis thaliana* were sequenced by an international collaboration of scientists from Japan, The United States and Europe (The Arabidopsis Genome Initiative). *Arabidopsis* is popular in plant cell research because it has a small genome. The scientists found 25,498 genes. Many of the genes were the same as those found in other organisms, including humans. Most of the genome (about 70%) consisted of multiple gene copies. This is one of the most interesting findings. While some genes did occur as single copies, most existed as hundreds or even thousands of copies of the same genes. The International Rice Genome Sequencing Project, led by Japan, is close to sequencing the rice genome. Rice has one of the smallest cereal genomes; there are plans however, to sequence maize, sorghum, wheat, barley, tomato, soybean, potato, lotus, and cotton in the near future.

RHYTHMS OF LIFE—CELL CYCLE

Cell Division in Plants Occurs in Meristems

Only in the science-fiction classic "The Day of the Triffids" was there a type of plant, the triffid, that was able to move around. Although real plants are capable of movements, such as the sleep movements of leaves, the closing of traps in insectivorous plants, or the looping of a vine around a fence, the movements are, to our perception, very slow and limited. This is because plant cells are enclosed in cell walls. Plant movements occur by controlling the direction or rate of growth.

In contrast to animal cells, cell division in plants mostly takes place in special regions called *meristems*. The cells produced in the meristem then expand, grow, and differentiate into all the cells of the plant body. A flowering plant begins its life as a single diploid cell. The main event during formation of an embryo is the development of a root-shoot axis (Fig. 8.4).

This is called a *polar axis* because the embryo is divided into two opposite ends that grow in opposite directions to become the shoot and root. The cells at the opposite ends of the embryo remain juvenile or embryonic as the plant develops and matures. Their role is to continue dividing and producing more cells for further differentiation. These special regions of dividing cells are the *meristems* and the embryonic cells located there are called *meristematic cells*. The meristematic regions of the root and shoot in the maturing plant are called *apices* (singular; apex). These meristems are called *apical meristems*. A shoot apical meristem can change from a vegetative phase

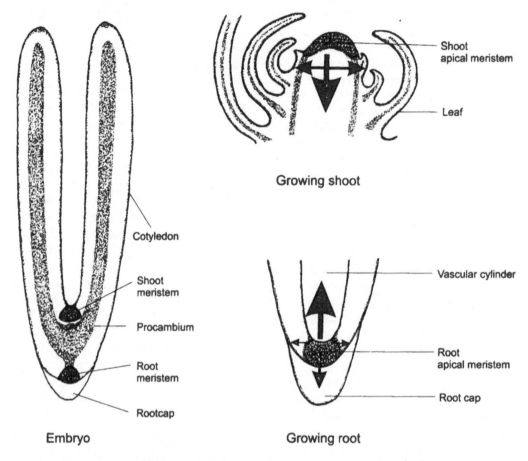

Fig. 8.4 Mitosis in plants takes place mainly in the root and shoot meristems. These regions of perpetually juvenile cells continue dividing throughout the life of the plant. The embryo of a flowering plant has a polar axis, with a meristem at each pole. One becomes the root meristem, the other the shoot meristem. The cells produced by the meristems then enlarge and differentiate. The procambium is a different type of meristem that can produce vascular cells, xylem and phloem. Arrows show the main directions of cell division in the shoot and root meristem. In the shoot cell divisions are directed downwards and laterally. The shoot with its meristem grows upwards. In the root the main directions of cell division are upwards, lateral, and downwards. Mitosis in the meristem produces the root-cap and cells that thicken the root. New cells are continually produced and enlarge above the root meristem. The roots grow downwards. (Figure based on Figs. 1.1 and 5.11, Esau, 1965.)

(producing leaves) into a reproductive phase, as occurs with flowering.

Buds are a kind of meristem in which all the cells are in a state of arrested development, or *dormancy*. The arrested growth of buds is controlled by the environment and by plant growth regulators. Buds can be dormant over winter (seasonal dormancy). The shoot apical meristem controls growth of the buds beneath it, called *lateral buds*. This is called *apical dominance*. Pruning the top of a plant leads to bushy growth from the buds below because removing the shoot apical meristem releases the dormancy of the buds beneath. The root and shoot meristems continue to produce new cells, which enlarge and differentiate. This results in growth of the plant. Thus the plant contains a series of cells that are increasing older as one moves upwards from the root meristem and downwards from the shoot meristem.

The growth resulting from divisions in the apical meristems produces what is called the *primary plant body*. Many plants, for example monocotyledons (e.g. grasses) retain a primary body throughout their lives. However most gymnosperms and flowering plants develop a secondary body. The secondary body is produced by a different type of meristem, called the *vascular cambium*. The vascular cambium produces the vascular tissue, xylem and phloem. The secondary plant body includes familiar plant structures such as bark and wood. The vascular cambium develops from meristematic cells present in the embryo, called the *procambium*.

Overview of Five Phases of Cell Cycle

Another way in which plants differ from animals is in the importance of the planes of cell division. As seen above, most active cell division takes place in the meristems. Because plant cells are encased in rigid cell walls, the plane or direction of cell division is very important in determining the shape of the plant. Once a cell has divided in a particular direction there is no turning back, no possibility of the cell moving into a different position, as can occur with animals. In both plants and animals, cell division takes place as part of a cycle, called the *cell cycle* (Fig. 8.5).

Duration of the cell cycle varies greatly according to the type of plant and the type of tissue within it. In the early embryo cell division occurs throughout the body, but as the embryo matures cell division is restricted on the whole to the meristems, and the other cells take on different and specialized roles. The cell cycle is controlled, so that some cells become arrested in a non-dividing state while others, the meristematic cells, pass through the cell cycle throughout their lives. The meristematic cells of plants divide once every 18 to 24 hours and the process of division requires 4 to 6 hours. Most of a meristematic cell's time is spent in the processes of growth and performing functions, such as transporting ions and nutrients. This portion of the cell cycle is called the **interphase** because it occurs between cell divisions.

Cell division involves both division of the nucleus and division of the cytoplasm. *Mitosis* is the process by which the nucleus divides, resulting in two new nuclei, each with the same chromosome number as the parent cell. *Cytokinesis* is the process by which the cytoplasm is divided into the two new daughter cells. The stages of mitosis in the meristematic region of an onion root are shown in Figure 8.5.

The cell cycle is a continuous process but it is convenient to break it up into phases. The five phases of the cell cycle are called $G1 \rightarrow S \rightarrow G2 \rightarrow M \rightarrow C$. *Interphase* includes the stages from $G1 \rightarrow S \rightarrow G2$. Cell division includes the *M phase (mitosis)* and *C phase (cytokinesis)*. The cell is metabolically active only during interphase. The decondensed form of DNA acts as a template for RNA synthesis and thus for protein synthesis. During cell division the condensed DNA cannot be "read" and RNA and proteins are not synthesised.

After mitosis and cytokinesis the two daughter cells start their own cell cycles at *G1*.

G1 Most Important Growth Phase

DNA is not replicated immediately after a cell divides. There is a gap between cell division and DNA replication. This is called the *G1* phase, a growth phase. A cell doubles its size during *G1* and also actively synthesises proteins, including those of the cytoskeleton (actin filaments and microtubules), ribosomes, and enzymes. The mitochondria and chloroplasts also divide during this phase. They have their own bacteria-like circular DNA and reproduce like bacteria, by binary fission. The *G1* phase is the longest part of the cycle for most plant cells. If the meristematic cells stop dividing over winter (in winter dormancy) the cells become arrested in the *G1* phase.

During *G1* the cell carries out its functions in the plant body. It synthesises proteins and enzymes appropriate to its role and carries out its particular functions. For example the role of a cell in the leaf palisade mesophyll is predominantly to photosynthesise; it produces sugars which will be used as an energy source by the whole plant. On the other hand a cell in the root may be involved in the uptake of ions that contribute to the mineral nutrition of the plant. Divisions of these cells are co-ordinated so that the size of the root system is in proportion to the size of the leaf canopy.

S Phase. DNA Replicated

The cell begins its preparations for division by synthesising and replicating its DNA. This is called the *S* or synthesis phase. If this DNA duplication did not take place, every cell division would halve the amount of DNA in the daughter cells until there was none left, and clearly this does not happen. Histones are also synthesised in the *S* phase. At the end of the *S* phase there are two copies of the genome in the form of chromosomes consisting of paired sister chromatids.

G2 Phase. Cell Prepares for Mitosis

The *G2* phase comes immediately before mitosis. The nucleus moves to the centre of the cell. The chromosomes begin to condense. *Microtubules* are structures made from the protein tubulin, and they are a part of the plant cytoskeleton. During G2 extra tubulin is synthesized and extra microtubules are made by the cell. Some of the newly synthesized microtubules are organised into a structure called a *pre-prophase band*. This is a dense ring-like band of microtubules encircling the nucleus. The pre-prophase band encircles the middle region of the cell. *The pre-prophase band shows the plane where cell division is going to occur.*

M Phase. Mitosis Divides Nuclei

During mitosis the DNA that was doubled in the *S* phase is divided into two poles of the parent cell. The cell must equally portion its replicated DNA into two daughter cells, so that each contains the same genome. Mitosis involves condensing the chromosomes so that they can be moved around, separating the sister chromatids, and moving one set into each pole of the parent cell. One of each kind of chromosome is then present at each of the poles of the parent cell.

Mitosis is a continuous process, but it is so important that biologists have divided it up into four phases (Fig. 8.6).

(a) *Prophase: chromosome condensation*

In *prophase*, the chromatin condenses into coiled chromosomes, each consisting of identical sister chromatids coiled around

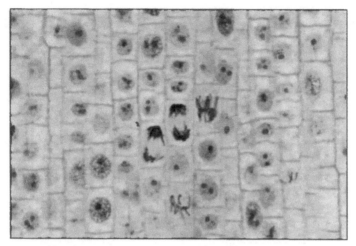

Mitosis in onion root cells

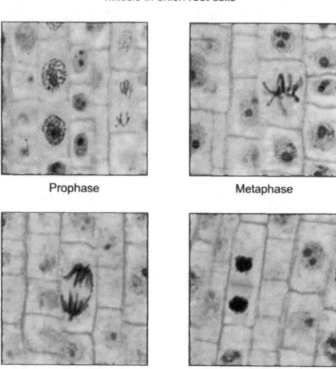

| Prophase | Metaphase |
| Anaphase | Telophase |

Fig. 8.5 Mitosis in an onion root. The cells were stained with hematoxylin so that they and the chromosomes appear colored. Normally plant cells are colorless, unless they contain pigments in their vacuoles or chloroplasts. These root cells do not contain chloroplasts. The top picture shows a section of the root above the root meristem. Note that all the cells are in different stages of the cell cycle. Use the four smaller photographs to identify stages of mitosis in different cells in the larger photograph.

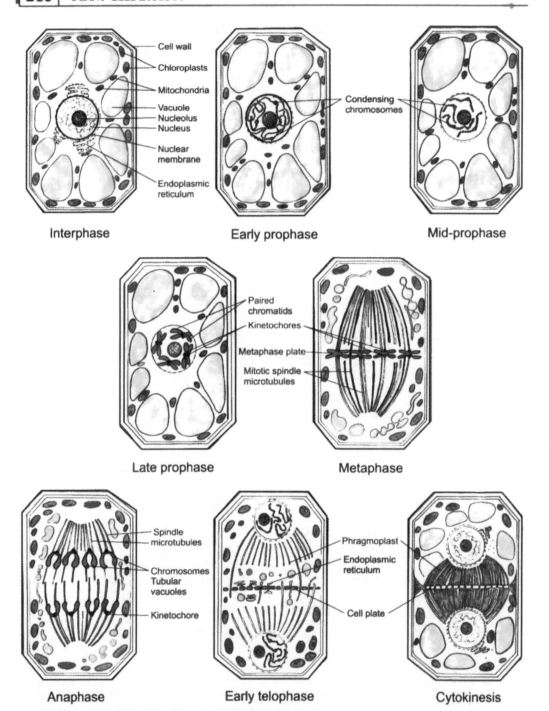

Fig. 8.6 Schematic representation of different stages of the cell cycle in an idealized plant cell

(Contd. Figure 8.6)

each other and held together at the centromere. The pre-prophase band disappears. In late prophase, special protein complexes called *kinetochores* are assembled on the *centromere* of each chromatid. Each chromatid has its own kinetochore and the kinetochores face opposite poles of the parent cell.

(b) *Metaphase: chromosome alignment at the spindle equator*

In *metaphase,* the nuclear membrane suddenly disintegrates and the nucleolus disappears. This frees the chromosomes. The *mitotic spindle* forms in the centre of the cell.

The mitotic spindle is made from microtubules. It lies at a right angle to the position occupied by the pre-prophase band in $G2$. The spindle is wider at the equator of the cell and constricted at the two poles of the cell.

The kinetochores of the chromosomes attach to the mitotic spindle. The chromosomes line up in the centre or equator of the cell. Each chromosome reverses its movements a number of times before settling at the cell equator. Not all the spindle microtubules attach to chromosomes. Some extend from pole to pole of the parent cell.

(Contd. Figure 8.6)

In interphase the chromatin is barely visible and found in the form of euchromatin or decondensed DNA. The genes of the DNA are actively read. The outer nuclear envelope is a specialized part of the endoplasmic reticulum and is continuous with it. The rough endoplasmic reticulum is studded with ribosomes, the cellular machinery that synthesizes proteins from the template of messenger RNA. This travels out of the nucleus through pores in the nuclear membrane. The nucleolus is rich in ribosomal RNA and is a sort of ribosome factory where the subunits to make the ribosomes are synthesized. The nucleolus is part of the chromosome and composed of DNA and protein. About 90% of the volume of the interphase cell is occupied by a vacuole. The vacuole appears to be separate compartments because it is divided by strands of cytoplasm that pass through it. Mitochondria and chloroplasts divide by binary fission during interphase.

In early prophase the DNA begins to condense and become visible. The replicated homologous chromosomes are shown condensing (red and blue). This cell has only two pairs of homologous chromosomes.

In mid-prophase the chromosomes continue condensing.

In late prophase the nuclear membrane begins to disintegrate and the kinetochores are assembled on each chromatid of the fully condensed chromosome.

In metaphase the nuclear membrane disintegrates and the nucleolus disappears. The mitotic spindle, made from microtubules, forms in the centre of the cell. The kinetochores of the chromatids attach to the spindle microtubules and the chromosomes are aligned in the centre of the cell. The kinetochores of each chromatid face in opposite directions. The vacuole of the cell breaks up into smaller vesicles and tubular elements that can move around.

In anaphase the microtubules attached to the kinetochores shorten. This pulls the chromatids apart and they become separate chromosomes. They move to opposite poles of the cell along the spindle microtubules. The cell vacuole system is transformed into a motile system of tubular and vesicular elements.

In telophase the chromosomes reach the poles of the parent cell. A nuclear membrane forms around them. The nucleolus reappears in each daughter cell. The chromosomes start to decondense. The motile elements of the vacuolar system actively move across the equator of the cell. The phragmoplast, a barrel-shaped cluster of microtubules and actin filaments begins to form between the two daughter nuclei.

Cytokinesis follows mitosis. Cytokinesis is when the cytoplasm is divided and a new cell wall is synthesized between the daughter cells. The phragmoplast becomes a solid band of microtubules encircling the equator of the cell and guides vesicles derived from the Golgi and the endoplasmic reticulum to the centre of the cell. The vesicles fuse together to make a cell plate. The cell plate expands from the centre of the phragmoplast towards the walls of the parent cell. At first the cell plate is composed of a material called callose, carried by the vesicles. As the new cell wall is completed cellulose replaces the callose.

The kinetochores have a base with "bristles" arising from it; the bristles are connected to special proteins called *microtubule-associated proteins* (MAPS). These proteins help each chromatid to attach to and move along the microtubules of the spindle.

The daughter cells need to inherit a vacuole from the parent cell but the vacuole does not simply split in two. Instead it breaks up into small spheres and tubular elements that can move around. Later these will fuse together to make a new vacuole for each daughter cell.

(c) *Anaphase: separation of chromatids into daughter chromosomes*

Anaphase begins suddenly when the sister chromatids separate at their centromeres. The *kinetochores* of each sister chromatid are attached to a spindle microtubule. The chromatids are pulled apart by the spindle microtubules to which they are attached. Since the kinetochores face in opposite directions, each chromatid of a pair moves in the opposite direction. Each chromatid then becomes an independent *daughter chromosome*.

The chromosomes move with their "arms" bent behind them and are dragged along the spindle to opposite poles of the parent cell. At the end of anaphase two replicated sets of chromosomes are distributed with one set at each pole.

(d) *Telophase: nuclear membrane reformation*

In *telophase*, a nuclear membrane forms around the bunches of daughter chromosomes, enclosing them in separate nuclei. The nucleolus reforms in each nucleus. The mitotic spindle disappears. The chromosomes begin to decondense and become difficult to see. A barrel-shaped cluster of microtubules and actin filaments forms between the two daughter nuclei. This is called the *phragmoplast*. Like the spindle,

the phragmoplast consists of two oppositely directed clusters of microtubules. The phragmoplast bisects the cell. The role of the phragmoplast is to position the vesicles that carry materials for making the cell plate in cytokinesis.

Small vesicles and tubules of the fragmented parental vacuole move actively across the equator where the cell plate will develop. This enables each daughter cell to rebuild a large central vacuole of its own when cell division is completed.

C Phase. Cytokinesis Divides Cell Cytoplasm

The C phase of the cell cycle is called *cytokinesis* (see Fig. 8.6). Cytokinesis is actual cell division as opposed to nuclear division. Cytokinesis is more complex in plant cells than in animal cells because the dividing cell must build a new cell wall within itself.

The plane of cytokinesis is predetermined by the pre-prophase band. The microtubules of the pre-prophase band probably alter the properties of the cell wall at the equator, where they form a band that encircles the nucleus.

The microtubules of the phragmoplast guide membrane vesicles derived from the Golgi into position at the equator of the cell. These membranes cluster in the centre of the cell and then expand radially out to the edges of the cell. The membranes contain materials that form a partition called the *cell plate*.

The cell plate is synthesized in the centre of the phragmoplast by the fusing of these vesicles. The vesicles carry the polysaccharide materials from which the cell plate is made. If they were not controlled by a solid structure such as the phragmoplast the vesicles might balloon or adopt the wrong shapes. Instead the vesicles are squeezed into dumbbell-shaped structures by a protein called phragmo-plastin that is associated with the phragmoplast.

The cell plate begins its existence in the centre of the phragmoplast but grows like a disc towards the walls of the parent cell. At first the cell plate is composed of a material called callose. Like cellulose, callose is a polysaccharide. Chemically it is a 1,3-β glucan whereas cellulose is a 1,4-β glucan. As the new cell wall is completed, callose is replaced by cellulose.

Later, as the cell plate begins to consolidate, tubular elements of the endoplasmic reticulum become trapped in the gaps of the cell plate. Cell plate material is deposited around them, forming holes containing trapped endoplasmic reticulum.

Membranous vesicles and tubules fuse together to make a new large central vacuole for each daughter cell.

(e) *Plasmodesmata—channels for cell-to-cell communication*

No plant cell stands alone. Plant cells are in communication with one another through small channels that pierce the cell walls. The cytoplasm and endoplasmic reticulum of neighbouring cells passes through these small pores. These minute cytoplasmic and membranous bridges that interconnect the neighbouring cells are called **plasmodesmata**. Nutrients, ions, small molecules, electrical and chemical signals can all travel from one cell to another through plasmodesmata.

Plasmodesmata are formed during cytokinesis. Strands of tubular smooth endoplasmic reticulum are trapped within the developing cell plate. In a way that is not completely understood, the Golgi-derived vesicles and the tubular endoplasmic reticulum are positioned together as the cell plate consolidates. Membranes of the vesicles provide the continuous plasma-membrane

that lines the plasmodesmata, and the endoplasmic reticulum is trapped inside the plasmodesmata. The plasmamembranes, the cytoplasm and the endoplasmic reticulum are continuous between cells as parts of the plasmodesmata. For this reason plants are sometimes called supracellular organisms.

Individual plasmodesmata are cylindrical structures and the compressed tube of trapped endoplasmic reticulum inside is called the desmotubule. The desmotubule is surrounded by a very thin coating of cytoplasm. This narrow cytoplasmic channel is continuous with the cytoplasm in the neighbouring cells. Recent research shows that actin molecules may also pass through plasmodesmata, so the actin cytoskeleton is also continuous between plant cells.

It has long been known that viruses such as the tobacco mosaic virus can pass through plasmodesmata and travel from cell to cell. Such viruses are larger than the dimensions of the plasmodesmata and use special proteins, called movement proteins, to change the internal structure of the plasmodesmata and to traffic their RNA from cell to cell.

Plant cells in the plant body are a continuum of linked cytoplasm, endoplasmic reticulum, plasma membrane and cytoskeleton. This is called the **symplast**. The continuum of plant cell walls, which is external to the living cytoplasm, is called the **apoplast**.

(f) *Microtubule arrays throughout mitosis*

Microtubules are involved at all stages of mitosis, as summarized below.

    (i) The first sign that mitosis is approaching is formation of the pre-prophase band, made up of bundles of microtubules encircling the cell centre. The pre-prophase band sets the position wherein the cell plate will develop. It probably does this by changing the cell wall properties in the region where it is located.

(ii) The pre-prophase band disappears and the nuclear envelope disintegrates at the same time that the mitotic spindle appears during metaphase. The spindle is made from microtubules.

(iii) The phragmoplast forms as anaphase moves into telophase. The microtubules of the phragmoplast control the direction of movement of the Golgi-derived vesicles that contain the polysaccharide materials for making the cell plate. The phragmoplast guides the developing cell plate towards the edges of the cell, to a position already established by the pre-prophase band.

(iv) In interphase the microtubules are mainly situated close to the plasma lemma where they control the direction of cellulose deposition and hence the direction in which the cell expands before preparing to enter mitosis. These interphase microtubules are called *cortical microtubules*.

Chromosome Numbers and Polyploidy

The number of chromosomes bears little relation to the complexity, size, or type of organism whose cells contain them. Nonetheless, the most widespread process of evolutionary change in plants is due to changes in the number of chromosomes. *Polyploidy* is the process whereby the number of chromosomes is multiplied. Polyploidy is a kind of *mutation*. It takes place when meiosis fails to halve the chromosome number and produces diploid, instead of haploid gametes. If a diploid (2n) gamete is fertilized by a haploid gamete (n), the progeny are triploid (3n) and contain three genomes. If a diploid gamete fertilizes another diploid gamete, the result is tetraploid (4n) and contains four genomes. Wheats (*Triticum* sp) are a good example. The haploid chromosome number is 7, but there are species with 14, 21, 28, 42, and 56 chromosomes.

Polyploidy is very common in plants. About 1/3 of flowering plants and even more of ferns are polyploid. The biologist C.D. Darlington suggested that a basic chromosome number of n = 7 could give rise by polyploidy to all the higher chromosome numbers found within woody flowering plants. There are two main types of polyploids. *Autopolyploidy* occurs when the chromosome number of a diploid individual is doubled. *Allopolyploidy* occurs when a hybrid is formed between two different species.

Polyploidy increases the size of cells but it also slows down cell division. Polyploidy can reduce fertility because the chromosomes may not be able to segregate evenly during meiosis. Triploids are sterile because an odd number of genomes cannot be divided into two equal parts during meiosis. The same is true of all odd ploidy numbers. One of the most important reasons for the high degree of polyploidy in plants is selective breeding by human beings.

Table 8.1 shows the chromosome numbers of a variety of different plants. It is clear that the chromosome number is not related to the size or shape of the plant. The sporophyte of the giant redwood *Sequoia sempervirans* is hexaploid. Its basic chromosome number is 10, plus a chromosome fragment making n = 11. The far smaller sporophyte of *Lemna* (each plant is less than

TABLE 8.1 Chromosome numbers of some well-known plants

Common name	Botanical name	Botanical family	Chromosome number of sporophyte: type of ploidy	Chromosome number, gametophyte
Corn	**Zea mays**	Poaceae (Cereal)	Diploid, 2n = 20	10
Rye	Secale cereale	Poaceae (Cereal)	Diploid, 2n = 14	7
Oats	Avena sativa	Poaceae (Cereal)	Polyploid, 6n = 42, with polyploid origin	7
Rice	Oryza sativa	Poaceae (Cereal)	Diploid, 2n = 24 (other ploidys exist for other rices)	12
Barley	Hordeum vulgare	Poaceae (Cereal)	Diploid, 2n = 14	7
Wheat	Triticum sp.	Poaceae (Cereal)	Polyploid, with polyploid or diploid origin, 2n =14, 3n = 21, 4n = 28, 6n = 42, 8n = 56	7
Sugar-cane	Saccharum sp.	Poaceae (Source of sugar; possibly the most complex plant genome)	Autopolyploid, interspecific hybrids, 2n = 80 (most common); also from 70 to 120	40 (most common)
Faba bean	Vicia faba	Fabaceae (Pulse)	Diploid, 2n = 12	6
Cabbage, cauliflower, Brussels sprouts, broccoli	Brassica oleracea	Brassicaceae (Cruciferous vegetables)	Diploid, 2n = 18	9
Potato	Solanum tuberosum	Solanaceae (Tubers)	Polyploid, (tetraploid) 4n = 48	
Pineapple	Ananas sp.	Bromeliaceae (Fruit)	Triploid, 3n = 48	16
Orange	Citrus sp.	Rutaceae (fruit)	Diploid, 2n =18; triploid, 3n = 27; tetraploid, 4n = 36	9
Banana	Musa sp.	Musaceae (Fruit)	Complex hybrids, usually triploid. 2n = 22 (AA), 3n (AAA) = 33, 4n (AAAA)= 44; also hybrids, 2n = AB, 3n = AAB or ABB, 4n = ABBB	
Blackberry/raspberry	Rubus sp.	Rosaceae	2n = 14 to 98	7
Strawberry	Fragaria	Rosaceae (Fruit)	Polyploids, from ancient diploid origin: octaploids most common. 2n = 14, 4n = 28, 6n = 42, 8n = 56	7
Magnolia	Magnolia grandiflora	Magnoliaceaea (Flowering plant)	114	57

(Contd.)

(Contd. Table 8.1)

Cactus	*Opuntia violacea*	Cactaceae	44	
Sundew	*Drosera banksii*	Droseraceae (Insectivorous plant)	12	
Giant redwood	*Sequoia sempervirans*	Taxodiaceae (World's largest tree)	Hexaploid, 6n = 66	
Duckweed (tiny flowering plant)	*Lemna minor*	Lemnaceae (Floating aquatic flowering plant)	Polyploid, 2n= 20; 3n =30; 4n = 40, 5n = 50, 6n = 60, 7n = 70	
Deadly nightshade	*Solanum nigrum*	Solanaceae (Poisonous berries)	Polyploid, 4n = 48; other ploidys reported: 72, 60, 36, 18	12, 24, 36
Gymnosperm	*Araucaria cunninghammi*	Araucariaceae (Conifer)	26	
Gymnosperm	*Agathis australis*	Araucariaceae (Conifer)	26	
Maidenhair tree	*Gingko biloba*	Gingkoaceae	24	12
Cycad	*Macrozamia moorei*	Cycadaceae	18	
Moss	*Bryum* sp.			10
Adder's tongue fern	*Ophioglossum vulgatum*	Ophioglossaceae (Fern)	Polyploid. 2n = 240; 10n = 1200	108, 116, 120, 212

4 mm long) is also polyploid, with different observers counting 20 (diploid), 30 (triploid), or 40 (tetraploid), etc. chromosomes. However, like *Sequoia*, *Lemna* has a basic chromosome number of ten.

Plants such as the cereals and pulses are cultivated by humans for their seeds. They tend to be either diploid or functionally diploid (see Table 8.1) so that seeds can be easily produced. Maize and rice are diploid, whereas bread wheat is hexaploid. Hexaploid bread wheat is *functionally diploid* and has been bred from a wild tetraploid crossed with another diploid.

When the seed is less important than other parts of the plant, humans have tended to propagate the plants asexually. *Apomixis* is the name given to all types of asexual reproduction that replace sexual reproduction. It has been advantageous for human beings to cultivate sterile triploid and other polyploid plants. The polyploid potato and strawberry propagate asexually and form clones. Apomixis is the major method of propagating sugarcane, *Citrus*, blackberries/raspberries, pineapples and bananas. Most bananas come from sterile triploid hybrid plants. The common banana *Musa paradisiaca* is a hybrid between two species, *Musa acuminata* and *Musa balbisiana*. If the genome for diploid *M. acuminata* is described AA and *M. balbisiana* is BB, the sterile triploid hybrid may be AAB or ABB or other combinations. The fruits are seedless and develop without fertilisation. This is called *parthenocarpic* development.

Plants Show Marked Alternation of Generations

Just as there is a plant cell cycle, so there is a plant life-cycle. Plants show to different extents an alternation between diploid and haploid bodies, and this is called *alternation of generations*. The haploid body is called the *gametophyte* and the diploid body called the *sporophyte*. The two can be remarkably different in appearance.

During evolution from "lower" (e.g. mosses, ferns) to "higher" plants (e.g. flowering plants) the gametophyte tended to reduce in size and in the proportion of the life-cycle it occupies. The gametophyte or haploid body predominates in lower plants, while the sporophyte or diploid body predominates in the higher.

In the bryophytes (mosses and liverworts) the gametophyte is the green "leafy" plant that dominates the life-cycle. The sporophyte of the moss is the stalked structure topped by a capsule or sporangium, often seen growing out of the leafy gametophyte. In the so-called "higher plants" (including angiosperms, the flowering plants) the gametophyte is a small dependent structure consisting of only a small number of cells. The sporophyte is the tree or shrub or herb that we think of as the plant.

In the lower plants the gametophyte begins life as a haploid spore, which divides by mitosis to form the multicellular gametophyte body. The reproductive cells (gametes) are formed by mitosis of the haploid cells of the gametophyte in structures called *antheridia* (producing sperm or male gametes) and *archegonia* (producing eggs or female gametes). The haploid gametes fuse, often after the sperm has swum to the egg in the archegonium through a film of moisture, to form a *zygote*. This grows into the diploid sporophyte, which may last only long enough to produce haploid spores, following meiotic cell divisions. The spores germinate and then divide by mitosis to form the multicellular gametophyte.

In mosses the gametophyte produces haploid gametes by mitosis in the antheridia and archegonia. The sperm swims to the egg and the gametes fuse to make the diploid zygote. The zygote divides by mitosis to form the stalked sporophyte. The sporophyte produces spores in a structure called a capsule. Meiotic divisions produce the haploid spores. These can be very resistant to drying out. When they germinate, they produce the leafy gametophyte again.

The fern gametophyte is a tiny heart-shaped green structure called the prothallus that is not often noticed. It produces haploid gametes that fuse to make a diploid zygote. The zygote grows into the multicellular "leafy" sporophyte most often thought of as a fern. This produces spores by meiotic divisions (brown spots on the underside of fern "leaves") which germinate to produce the gametophyte.

The diploid body of the angiosperms (flowering plants) dominates the life-cycle. The haploid gametes (pollen and ovules) are formed by meiosis and occupy only a small portion of the plant's life.

The male gametophyte is the *pollen grain* and *pollen tube*. The female gametophyte is called the *embryo sac* and is contained within the ovule. Unlike the situation with bryophytes and ferns, the gametophytes do not live independently and do not photosynthesise for themselves.

When the haploid gametes (sperm) are released from the pollen grain one sperm fertilizes the egg cell and produces a diploid zygote. The zygote develops into the *embryo*. Two of the haploid cells within the female gametophyte fuse to form a diploid nucleus, and this is fertilized by a haploid sperm to form a triploid nutritive tissue called the *endosperm*. Fertilization of two cells, one haploid, the other diploid, by two different sperms is called *double fertilization*. The entire ovule develops into the *seed*, containing embryo and endosperm.

Alternation of generations and the different types of nuclear division, produce the wonderful and diverse forms of the world of plants that surrounds and sustains us, from the green slimy *Spirogyra* in a pond to ferns, cycads, the tiny *Wolffia* duckweed to the giant *Sequoia* redwood that may be thousands of years old.

CONTROL OF CELL CYCLE: THE DEVELOPMENT OF FORM

Cell Division Finely Controlled to Generate Plant Form

The cell cycle clearly needs to be controlled in order for plant bodies to achieve their complex and beautiful forms. Uncontrolled cell division, with every cell dividing at the same rate, or at uncontrolled rates, would lead to an undifferentiated mass of tissue like a cancer or tumour. The cell cycle must run through its steps in the correct order, $G1 \rightarrow S \rightarrow G2 \rightarrow M \rightarrow C$. DNA must be replicated before cell division takes place. Cells must double in size before dividing, or inevitably the cells would become smaller with every division, and clearly this does not happen. Some cells must divide rapidly, especially those in the meristems, while others must divide very slowly or cease to divide entirely. Cell divisions in the meristem must be capable of responding to changes in the environment or flowers might be continuously produced, or never appear.

Control of Cell Cycle by Checkpoints

The cell cycle is controlled by a *checkpoint system*

(a) *G1-S checkpoint*

A checkpoint at *G1* ensures that the cell has approximately doubled its volume and its DNA is undamaged. Plants and animals appear to be similar in the way they progress through this checkpoint. Cells can only pass on from *G1* if other conditions are satisfied. Correct messages must be received from plant growth regulators and sugars must be available. If the cell passes this checkpoint, it moves on to the *S* phase and the DNA is replicated. If not, the cell is arrested in *G1*.

(b) *G2- M checkpoint*

A second checkpoint at the end of *G2* ensures that all the DNA has been accurately replicated before the cell is allowed to pass into mitosis (*M* phase).

(c) *Protein phosphorylation and dephosphorylation as control mechanisms*

The cell cycle is intricately controlled by a feedback system involving phosphorylation and dephosphorylation of key proteins. Proteins are often regulated by *phosphorylation* and *dephosphorylation*. When a protein is phosphorylated, a phosphate group is taken from an ATP molecule and added to the protein. This is done by another protein, an enzyme called a *protein kinase*. Adding a phosphate group can drastically change the shape of the protein molecule and thus change its function. Phosphorylation can activate a protein to perform a particular function or it can prevent a protein from carrying out a function.

A protein activated by phosphorylation can be deactivated by the reverse process of dephosphorylation wherein the phosphate group is removed. The phosphate group is removed and transferred to an ADP (adenosine diphosphate) molecule by an enzyme called a *protein phosphatase*.

Cell Cycle Centrally Controlled by Cyclic Activation of Special Proteins

There are two important classes of proteins that interact to allow the cell to either pass a checkpoint and move on to the next part of the cycle, or arrest the cell at a particular stage. These are called *cyclins* and *cyclin-dependent kinases*. Cyclins were so named because they were first discovered in frog cells, where they accumulated to a high concentration during interphase and then disappeared altogether during mitosis. Cyclins bind to and activate the cyclin-dependent kinases.

The different cyclin/cyclin-dependent kinase complexes then trigger each step in the cell cycle by phosphorylating specific proteins and activating them. These controlling elements have been conserved throughout evolution and strong similarities exist between the way animals and plants control their cell cycles.

(a) *Cyclin-dependent proteins*

Cyclin-dependent protein kinases (CDKs) are enzymes that phosphorylate proteins crucial to each stage of cell division. CDKs are activated by binding to cyclins and also by phosphorylation or dephosphorylation of some of their amino acids.

(b) *Cyclins regulate cyclin-dependent proteins and control checkpoints*

Cyclins (Cycs) are proteins that associate with cyclin-dependent protein kinases. Alone cyclins have no ability to act as enzymes. However when they bind to cyclin-dependent proteins the resultant complexes function as enzymes.

The complex of CDK-Cyc functions as an enzyme and activates a stage of the cell cycle by phosphorylating a target protein.

The rising-falling concentration of cyclins at specific times in the cell cycle is what actually controls the cycle. Cyclins are rapidly synthesized during one phase (e.g. interphase) and then destroyed (e.g. during mitosis) by another class of proteins called **ubiquitins**. They are targeted for destruction by having a special chemical sequence called a "destruction box" that enables ubiquitin to recognize them. Ubiquitins take the cyclins to organelles called proteosomes, in which the cyclins are broken down.

There are numerous different cyclins and cyclin-dependent kinases involved in controlling the cell cycle. As the cell passes through each of the G1 and G2 checkpoints, the CDK protein associates with a cyclin protein. Then the activated CDK-cyclin complex is able to phosphorylate the proteins essential for the cell to pass through each stage. When the phase is completed the cyclins are broken down, inactivating the CDK until more cyclins are made.

(c) *Control of G1-S checkpoint in plants*

A cyclin called **cyclin-D** becomes active in the *G1 to S transition*. Cyclin-D can be controlled by plant growth regulators (auxin, cytokinins, abscisic acid, gibberellins) and sugars. Cyclin-D binds to **cyclin-dependent kinase A** (CDK-A). The activated CycD-CDK-A complex phosphorylates a protein called **retinoblastoma protein** (RB).

RB is very important. Before RB is phosphorylated it puts the brakes on DNA synthesis. It does this by binding to two proteins called E2F and DP. Proteins E2F and DP can trigger the genes that set off DNA replication and do this by complexing together. However, unphosphorylated RB binds to them, and thereby precludes their binding together, stopping stimulation of DNA synthesis. When RB is phosphorylated (by Cyc-D/CDK-A) it can no longer bind to E2F and DP. E2F and DP can then complex together and trigger DNA synthesis.

The "stress hormone" abscisic acid can prevent the cell from moving into the S phase and replicating its DNA. Abscisic acid (ABA) stimulates production of another protein called CDK-inhibitory protein (CIK). If CIK binds to the CycD/CDK complex, the complex cannot function as an enzyme. The cell cannot then proceed to the S phase.

If the cell is undergoing stressful conditions, such as lack of water, ABA is produced, followed by CIK, and the cell does not proceed to the S phase.

Auxins and cytokinins are necessary for progressing through this part of the cell cycle. They increase the activity of CDK-A.

(d) *Control of G2-M checkpoint in plants*

The G2-M checkpoint provides a check that the cell is ready to proceed to mitosis. Two kinds of CDKs (CDK-A and CDK-B) and two kinds of associated cyclins (Cyc-A and Cyc-B) are involved. Growth regulators auxin and cytokinin can increase expression of the genes for these proteins.

FROM WHERE DO CHLOROPLASTS AND MITOCHONDRIA COME?

Theory of Endosymbiosis

In order for cell division to produce functioning daughter cells not only the nucleus, but the organelles within the cell, the chloroplasts and mitochondria, must also divide.

Prokaryotic cells, such as bacteria, lack a membrane-bound nucleus, although they do have a true chromosome, a large molecule of DNA existing in a circular form. Bacterial cells are very small (from 1-10 μm in size) and are often thought of as single-celled organisms. However, bacteria can nonetheless live in the form of multicellular filaments (cyanobacteria, myxobacteria, actinobacteria) and some (gliding bacteria) can interact with one another to form complex physical structures, such as bacterial cities or bacterial trees.

Eukaryotic cells are much larger (1-100 μm; with some plant cells of the family Characeae being centimeters in length) and have a double-membrane-bound nucleus that contains the chromosomes. Eukaryotic cells have thousands of times more genetic material than bacteria. Furthermore, eukaryotic cells can themselves be considered aggregates, composites, or fusions of a number of different

bacteria that were once free living. This is the so-called *theory of endosymbiosis (endo,* within and *symbiosis,* living together).

In this theory the chloroplasts and mitochondria within a plant cell are the descendants of bacteria that long ago either invaded or were engulfed by another bacterial cell. Both mitochondria and chloroplasts contain their own chromosomes which, like those of bacteria, are circular and not bound by a membrane. Both mitochondria and chloroplasts like bacteria, reproduce by fission, and do so independently of the cell in which they live. Both have their own ribosomes, which differ from those of the home cell, and construct their own outer membranes.

However, chloroplasts and mitochondria can no longer exist outside a cell. In the long course of development of their relationship with cells, they lost parts of their own chromosome and these functions were taken over by the cell nucleus. Chloroplasts and mitochondria have to be inherited through generations. When a cell divides, a fair share of the chloroplasts and mitochondria must be passed into the daughter cells.

Margulis and Sagan (1986) painted a vivid picture of how the eukaryotic cell might long ago have been produced by endosymbiosis. The nucleus of eucaryotic cells results from the joining together of genetic material from two different ancestral procaryotes. A bacterium began living as a guest inside a larger host archaebacterium. Their genomes then fused together. The Archaean host provided the cytoplasm and the bacterial guest enabled mitosis. The earliest eucaryotic or nucleated cell could not breathe oxygen. The collaboration later gained the ability to use oxygen by engulfing the ancestor to mitochondria, an oxygen-using eubacterium, possibly an α-proteobacterium.

The host was possibly similar to the archae bacterium *Thermoplasma*, which lives in hot springs today. It was probably able to live in hostile environments and like *Thermoplasma*, its genetic material was wrapped in proteins, as is the case for most eukaryotic cells, but not for most bacteria. In the case of eukaryotic cells and mitochondria the symbiosis has continued over 1000 million years.

About a hundred million years later some of the ancestors of eukaryotic cells acquired a new partner, in this case not through invasion, but by engulfing. The mitochondria-containing ancestor of plants probably ate bacterial cells that were capable of photosynthesis. Some of these cells were not digested; then survived in the cytoplasm with their capacities for photosynthesis remaining intact. Indeed, a very large cynobacterium, *Prochloron*, is grass green because it contains the same chlorophyll found in plants. The ancestor of the chloroplast may well have resembled the large green *Prochloron* bacterium that lives on the outer surfaces of sea squirts. The chloroplast-cell relationship sustains all animal life on earth. All food chains start with a green plant. Energy enters the biosphere by the interaction of light with chloroplasts. Through acquiring the ancestor of chloroplasts the plants achieved the ability to make food from water, carbon dioxide and light, to make complex carbohydrates from simple inorganic materials, and thus to generate order from chaos in a way that non-living systems cannot do.

MAKING AND BREAKING IDENTITY: COMPARISON OF MEIOSIS AND MITOSIS

Mitosis produces cells with essentially the same genetic make-up and these constitute the whole plant body. Each mitotic division produces two daughter cells, each with a nucleus containing an identical set of chromosomes and genes. *Mitosis maintains the identity of the organism.*

In contrast to mitosis there are two nuclear divisions during meiosis, resulting in four daughter cells, each with a haploid nucleus. That is, each daughter cell has half the chromosome number of the parent cell. While the individual cells produced after a mitotic division are genetically identical, cells produced by meiosis possess combinations of chromosomes that differ from those of the parent cell. *Meiosis is an identity breaker.*

During meiosis the DNA is replicated once, as in mitosis. Chromosome combinations inside the cells produced by meiosis differ from those of the parent cell for two reasons.

First, the homologous chromosomes pair up and twine around each other in prophase of meiosis. This is called synapsis. The entwined homologous chromosomes are called bivalents. A bivalent consists of four chromatids.

In metaphase 1 of meiosis the bivalents line up at random on the metaphase plate. In anaphase 1 the bivalents are pulled apart and move to opposite poles of the cell. Whereas in mitosis the identical chromatids separate, in meiosis it is the homologous chromosomes. Since each bivalent consists of four chromatids, there are two chromatids of each chromosome at the cell poles following anaphase 1 of meiosis. After anaphase of mitosis however, there is a single chromatid of each chromosome at the cell poles.

There are normally two homologous chromosomes, one of which came from each of the parents. The number of ways the chromosomes can separate in anaphase 1 of meiosis is 2^n, where n is the chromosome number. If the diploid chromosome number of the parent cell was 4, there would be $2^4 = 16$ possible combinations of chromosomes that can move to each cell pole in anaphase 1.

The second reason why the cells produced by meiosis have chromosome combinations that differ from the parent cell is that the chromatids that make up the homologous chromosomes are no longer identical. In prophase of meiosis an exchange of parts of the chromosomes takes place, called *crossing over*. The cross-shaped region where the chromatids cross over is called a *chiasma*. Chromosomes exchange pieces with each other at the chiasma.

Crossing over means the chromatids are no longer identical. Each of the four cells produced by meiosis has a new assortment of chromosomes, and these chromosomes have exchanged pieces with each other. The genetic structure of each of the resultant cells is unique.

Because of meiosis and fertilization, a population of diploid plants contains individuals which differ genetically from one another and from their parents.

In 1858 Virchow pointed out that all individuals today exist because of an unbroken line of cell divisions stretching far back into the past. Mitosis maintains the identity of the individual, while meiosis assures the uniqueness of each individual produced by sexual reproduction.

OVERVIEW OF CONCEPTS

This section provides a summary of the concepts discussed in greater depth above.

Chromosome Structure

- The genetic material of a cell is arranged in the form of *chromosomes* within the cell nucleus.

- A chromosome is a single DNA molecule, combined with special proteins called *histones* and *nonhistones*. This material is called *chromatin*. Chromosomes are made from chromatin.

- Histone proteins are organized as cores around which the long DNA molecules are wound so that they can be packed into the cells.

- DNA is the fundamental genetic material and contains *genes*.

- Genes are sequences of a DNA molecule that encodes for proteins that act as enzymes, structural elements and regulators of cell function.

- The total DNA of a cell, consisting of at least one chromosome, is called the *genome*.

Cell Division

Cell division involves division of both the nucleus and the cell cytoplasm.

- *Mitosis* and *meiosis* are the processes by which the nucleus divides.

- *Cytokinesis* is the process whereby the cytoplasm is divided into two new daughter cells.

- *Mitosis* results in two daughter nuclei with the same number of chromosomes and an amount of genetic material identical to the parent cell. Mitosis generates numerous cells and these build the body of the plant.

- *Meiosis* produces daughter cells that each have half the amount of genetic material of the parent cell. For this reason meiosis is called a "reduction division". Meiosis produces reproductive cells, called gametes, eggs, and sperm.

• *Cytokinesis* divides the cell cytoplasm, and each daughter cell then contains a nucleus as well as its share of mitochondria, chloroplasts, and vacuoles. A new cell wall is synthesised between the two new daughter cells.

Cell Cycle in Plants

Cell division is only part of the life-cycle of cells. The life-cycle of cells is called the *cell cycle*.

• The cell cycle consists of *cell division*, when the cell is occupied with dividing, and *interphase*, when the cell carries out its functions in the plant body, grows and increases in volume, and prepares for the next cell division.

• The cell cycle is called a cycle because the cells pass through various phases and then the cycle starts again.

• Duration of the cell cycle varies in different parts of the plant and in different cells. Meristematic cells of plants divide once every 18 to 24 hours, and the actual process of division takes 4 to 6 hours.

• The cell cycle has five phases that flow smoothly into one another.

• The phases can be diagrammed as $G1 \rightarrow S \rightarrow G2 \rightarrow M \rightarrow C$.

• Phases $G1 \rightarrow S \rightarrow G2$ are together called *interphase* because they lie between cell divisions.

• Phases $M \rightarrow C$ are mitosis (nuclear division) and cytokinesis (cytoplasmic division).

Interphase cells carry out all the roles of cell life such as photosynthesis, respiration, protein synthesis, growth, storage of nutrients and so on. These functions are suspended while a cell is dividing.

(i) *Interphase (G1, S and G2 phases)*

• The *G1 phase* follows cell division.

• DNA is not synthesised during *G1*.

• *G1* is a growth phase, when the cell doubles size and synthesizes extra proteins, ribosomes, and enzymes.

• The *S phase* is a "synthesis phase" in which DNA is replicated in preparation for cell division.

• Histones are also synthesised during the *S* phase.

• DNA is replicated so that there are two versions of each chromosome, held together by a structure called a centromere.

• Each version of the chromosome is called a sister chromatid.

• The *G2 phase* is that in which the chromosomes begin to condense.

• The cell prepares itself for cell division in *G2*.

(ii) *Nuclear division and cell division (M and C phases)*

The *M phase* is that in which actual mitosis takes place. Mitosis is so important it is itself divided into four phases.

• In *prophase*, the nuclear membrane disintegrates and chromosomes condense.

• In *metaphase* the duplicated chromosomes (sister chromatids) are attached to the mitotic spindle by structures called kinetochores. Chromosomes are lined up in the centre of the cell.

• In *anaphase* the sister chromatids are separated and become daughter chromosomes. They are pulled along the spindle fibers to opposite ends of the cell.

- In *telophase* the nuclear membrane forms around the bunches of daughter chromosomes, making them into separate nuclei.

The *C phase* of the cell cycle constitutes cytokinesis (division of the cytoplasm).

- Following cytokinesis each daughter cell contains about half the cytoplasm of the parent cell, as well as a fair share of chloroplasts, mitochondria and small vacuoles derived from the vacuole of the parent.
- Each cell contains a nucleus with a full number of chromosomes.
- Daughter cells are enclosed within cell walls formed from the cell plate synthesized in the center of the dividing parent cell and growing towards the edges.

Cell Cycle Controlled and Regulated at Checkpoints

Some cells must divide rapidly (for example, in the meristems) while other cells divide very slowly or not at all. In order to achieve this the cell cycle must be regulated and controlled.

- The cell cycle is controlled by a *checkpoint system*.
- There are two control points, one in *G1* and one in *G2*.
- Cells can only pass *G1* and synthesise DNA if they have doubled in volume.
- Cells can only pass *G2* if their DNA is correctly synthesized and other conditions are met, such as external temperature being warm enough and light and nutrients sufficient.
- Special proteins called *cyclins* and *cyclin-dependent kinases* interact to allow the cell to either pass a checkpoint and move on to the next part of the cycle, or to arrest the cell at a particular stage.

Plant Cells Divide with Cell Plates

Cytokinesis in plants is more complicated than in animal cells due to the presence of cell walls.

- During cytokinesis plant cells assemble special membranes in the centre of the dividing cell.
- These membranes expand from the centre of the dividing cell to its edges, where they fuse with the plasma membrane. The membranes form a partition called the *cell plate*. Cellulose is deposited on the cell plate and a new cell wall is constructed.

Mitosis in Plants Takes Place in Meristems

- In plants mitosis takes place mainly in specialized growing regions called *meristems*.
- The embryo consists of two opposite ends, one of which grows upward to become the shoot, and the other grows downward to become the root. The special regions at each of the opposite ends remain throughout life and are called *meristems*.
- *Meristematic cells* are perpetually juvenile cells whose role is to continue dividing and to produce more cells for further differentiation.

Chromosome Numbers

The chromosome number is referred to as the *ploidy*.

- Cells that have two of each type of chromosome have a *diploid* chromosome number, called 2*n*. Mitosis of diploid cells results in diploid daughter cells.

- Cells containing only one of each type of chromosome have a *haploid* chromosome number, called *n*. When diploid cells undergo meiosis they produce haploid daughter cells. When haploid cells fuse during fertilization they produce a diploid chromosome number. Haploid cells can also divide by mitosis and produce daughter cells with a haploid chromosome number.

Many plants are *polyploid*. This is due to a kind of mutation that occurs when meiosis fails to halve the chromosome number and produces diploid, instead of haploid gametes.

Alternation of Generations

Plants show to different extents an alternation between diploid and haploid bodies; this is called *alternation of generations*.

- Diploid and haploid bodies may differ markedly in appearance.
- The haploid body is called the *gametophyte* and the diploid body the *sporophyte*.
- The so-called "lower plants", such as mosses, spend most of their life-cycle as a gametophyte. Ferns divide their life-cycles evenly between the gametophyte and sporophyte bodies.
- The diploid body dominates the life-cycle of flowering plants. The male gametophyte consists only of pollen and the pollen tube. The female gametophyte is called the embryo sac. These gametophytes do not have an independent life and do not photosynthesise for themselves.

- The pollen releases sperm that fertilize both the haploid egg cell and a diploid cell in the embryo sac, producing the triploid nutritive endosperm. The diploid zygote becomes the embryo and is nourished by the triploid endosperm within the seed.

From Where do Chloroplasts and Mitochondria Come?

The cell and nucleus are fundamental structures derived from previously existing cells and nuclei. The same is true for the choroplasts and mitochondria.

The theory of endosymbiosis states that chloroplasts and mitochondria are "endosymbionts" living within cells, supported by the cellular environment, and supporting the cell in turn by photosynthesising and respiring.

- These descendants of bacteria were incorporated long ago by another ancestral cell and this gave rise to the photosynthetic, oxygen-respiring plant cell.
- Mitochondria and chloroplasts have their own chromosomes.
- These are like bacterial chromosomes. They are circular and are not enclosed by a nuclear membrane.
- Mitochondria and chloroplasts reproduce by fission, like bacteria.
- Mitochondria and chloroplasts construct their own outer membranes and their own ribosomes.

Further Reading

Alberts, B., Bray, D., Johnson, A., Lewis, J., Raff, M., Roberts, K., and Walter, P. 1998. *Essential Cell Biology. An Introduction to the Molecular Biology of the Cell.* Garland Publ. Inc., NewYork, NY.

Esau, K. 1965. *Plant Anatomy.* John Wiley and Sons, Inc. New York, NY. (2nd ed.).

Gabriel, M.L. and Fogel, S. (eds.). 1955. *Great Experiments in Biology.* Prentice- Hall, New York, NY.

Keller, E.F. 1983. *A Feeling For the Organism. The Life and Work of Barbara McClintock.* W.H. Freeman and Co., New York, NY.

Margulis, L. and Sagan, D. 1986. *Microcosmos.* Summit Books, New York, NY.

Raven, P.H. and Johnson, G.B. 2001. Biology. McGraw-Hill, New York, NY. (6th ed.).

Raven, P.H., Evert., R.F., and Eichhorn, S.E. 1999. Biology of Plants. W.H. Freeman and Co., San Francisco, CA (7th ed.).

CHAPTER 9

Meiosis in Plants

Renata Śnieżko

INTRODUCTION: MEIOSIS IN LIFE CYCLE OF PLANTS WITH DIFFERENT PHYLOGENETIC POSITIONS

Alternation of Generations

During billions of years of evolution plants have developed a specific mode of living as autotrophic organisms distributed in different environments in both water and on land. They have changed their morphologies from unicellular algae to multicellular flowering organisms and modified their reproductive processes. Relatively early, they developed sexual reproduction, which is associated with meiosis, a process of two division cycles with accompanying recombination of genetic material. After meiosis four haploid daughter cells arise from one diploid cell. Each daughter cell exhibits little change in gene arrangement comparised to the previous generation. There is a difference between the sexual reproduction of plants and animals. In animals, meiosis occurs within a sexual gland producing gametes, so meiosis is directly connected with sexual reproduc-

tion in every generation. In plants, meiosis also occurs in diploid cells, but results in spores, which develop into a haploid generation. As a result of evolution, plants undergo alternation of generations in their life cycle with meiosis occurring between the diploid and haploid phases (Fig. 9.1A).

The haploid generation is called the gametophyte because it produces gametes, cells capable of fusing during fertilization. The latter leads to the diploid generation which produces spores after meiosis. These postmeiotic spores develop into the haploid generation. This scheme of haplo- and diplophase alternation is differently realized depending on the systematic level. In the life cycle, the haplophase or diplophase can be the state in which the plant lives longer and develops its vegetative organs. The proportion between haplo- and diplophase is dependent upon systematic position and the environment. In the early plants, living in water or in wet places the haplophase prevailed over the diplophase in their life cycles so they are haplobionts (Fig. 9.1B and C). Plants living on land belong to diplobionts—the diplophase prevails in their life cycles (Fig. 9.1D and E).

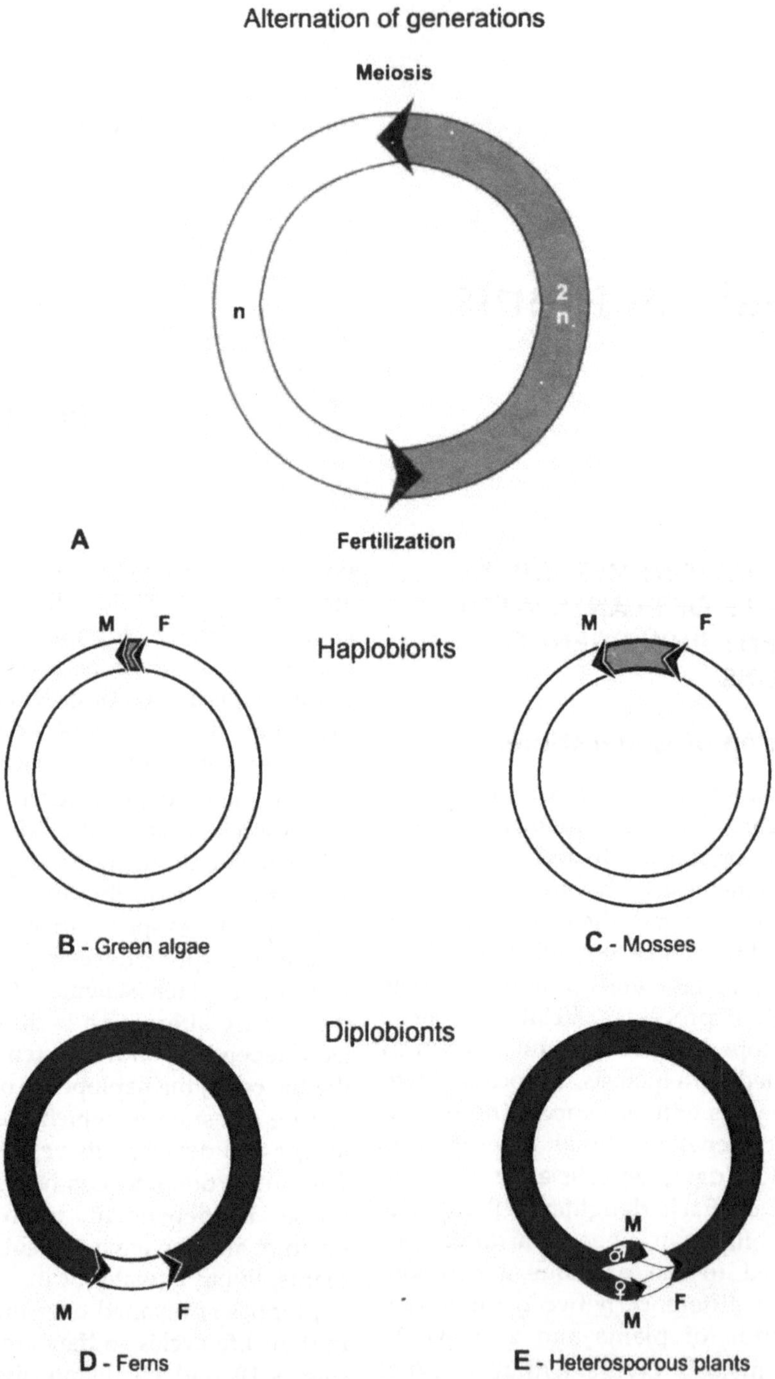

Fig. 9.1 Scheme of alternation of generations in plants showing position of meiosis in different life cycles and shortening of haplophase during evolution

Meiosis in Life Cycles

Sexual reproduction of plants occurs at three levels of evolutionary advancement: isogamy, anisogamy, and oogamy. The type of reproduction is defined on the basis of morphological and functional differences between gametes, but is also associated with the systematic position of the plant. In unicellular plants, considered the simplest and evolutionary old, evolution from isogamy to oogamy has already occurred.

Isogamy

Isogamy is observed in unicellular green algae resembling primitive plants. They live in water as haploid organisms (haplobionts), propagate by mitosis many times during the vegetation season, then undertake sexual reproduction before the period of unfavorable conditions in the environment, e.g. winter or drought. Unicellular organisms differentiate into morphologically similar gametes. Gametes differ in physiology, but both sexes possess flagella and are very similar in size, appearance, and behavior. Gametes "+" and "–" fuse into one diploid cell during fertilization. The diploid zygote is called the zygospore because it is covered with a thick cell wall and remains dormant in unfavorable conditions. At the beginning of the following vegetation season, it undergoes meiosis and four haploid cells possessing flagella are released from its cell wall, float away, and divide by mitosis. In these green algae, the haplophase takes several generations of cells but the diplophase is limited to just one cell – the zygospore.

Anisogamy

The more evident dimorphism between male and female gametes is defined as anisogamy. Gametes differentiate from unicellular plants but in a manner charac-

teristic for the sex. When they become mature enough for fusion, they differ in size and amount of storage materials but both gametes can move. The male gamete is small, devoid of starch grains and floats relatively quickly. The slowly floating female gamete enlarges and contains a lot of storage materials in the cytoplasm. Its cell wall is thick, except for one place at the tip. There the male gamete can adhere and fuse with the female cell during fertilization. In anisogamy, the gametes play two different roles important for the success of reproduction. The small male cell is well adapted to spread its genotype, penetrate the new area of environment, and compete with other male gametes. The most vivid, effectively floating male gamete has the best chance to find a female one and fertilize it. Competition in the haplophase is a mechanism of selection: organisms representing the best genetic and phenotypic features, i.e. the most vital and properly adapted male cells become the donors of genetic information for the next generations.

The role of the female gamete on the other hand, is to store nutrients for the offspring, and then to feed and protect them during the early stages of their development. Another role of the female gamete is to send signals attracting the male partner, as chemotropism is the most probable mechanism of communication between gametes. In anisogamy, the thick cell wall of the female gamete is completed after fertilization and protects the zygote during the period of dormancy. The storage materials are used before the zygote undergoes meiosis and produces haploid cells of a new generation. At this level of sexual reproduction, the difference between male and female organisms is already discernible and during the following processes of evolution leads to a great specialization in the male and female roles.

Oogamy

Among unicellular green algae there are also species which have developed the highest level of sexual reproduction – oogamy. They differentiate into small male cells with flagella (spermatozoids) or much larger female cells (settled eggs). Eggs are devoid of flagella, store starch grains and lipid globules, and are enveloped in a thick cell wall. Spermatozoids are more numerous than eggs and compete strongly during the search for a female gamete. They are probably attracted by some chemical signals sent from the eggs but this likelihood has not, yet been particularly researched. When the spermatozoid fuses with the egg, a zygospore is formed. It remains for some time in dormancy and then undergoes meiosis at the beginning of the new vegetation season. As a result of meiosis, four haploid daughter cells are released. In this life cycle, the diploid zygospore representing the diplophase undergoes meiosis immediately when environmental conditions become favorable. In all green algae, the haplophase dominates the diplophase in the life cycle. The haplophase takes numerous generations of cells proliferating by mitosis. The diplophase is limited to the one-cell zygospore.

Haplobionts

Among haplobionts, there are multicellular algae, the Characeae, and mosses or Bryophyta. Gametophytes of these plants develop a morphologically differentiated body with vegetative and generative organs—gametangia. Gametangia differ in structure and physiology because of the separation of male and female sex organs. The male gametangium is called the antheridium. Inside it numerous small cells differentiate into spermatozoids. Spermatozoids float from the antheridium and search for the female gamete to fertilize it. The female gametangium of Characeae is the oogonium with a single egg cell. The spermatozoid penetrates and fertilizes the egg. The zygote of Characeae is dropped from the maternal plant and settles on the bottom of the pond. It stays in dormancy until the next vegetation season. In favorable conditions, the zygote undergoes meiosis, which changes the life cycle into the haplophase represented by the haploid spores with flagella. They float for some time in the pond, then settle and develop a thallus. In the multicellular organisms of Characeae the diplophase is also limited to the zygospore, which undergoes meiosis.

In mosses, fertilization occurs inside the archegonium on the gametophyte. The zygote develops into sporophyte, a diploid organism dependent on the gametophyte at least at the beginning of its development. It is protected by the somatic cells of the gametophyte and takes nutrients from it. The mature sporophyte has a simple body structure, a small stem with a reproductive organ (sporangium) at the top. The sporangium is a place of meiosis and postmeiotic spores formation. Spores are released from the sporangium during low humidity conditions and spread by wind. They germinate in the haploid body of the gametophyte. In Bryophyta, the haplophase predominates in the life cycle but the diplophase is more advanced than in algae. The sporophyte is differentiated on the vegetative and generative organ and adapted to land environment.

Diplobionts

Vascular plants (ferns, gymnosperms, and angiosperms) are terrestrial and display an opposite pattern of generation dominance in the life cycle, i.e. they are diplobionts. In these plants the gametophyte is reduced compared to the sporophyte, which develops many different vegetative and also generative organs. The structure of the generative organs is dependent on the systematic position of the plant and the sex line. In any type of sporangium there are always diploid cells determined to undergo meiosis—archesporial cells or archesporial tissue.

Evolution of Sporogenesis

Isospory

In Pteridophyta, there are two patterns of sexual reproduction: isospory and heterospory. Isosporial plants (some ferns and horsetails) develop one kind of sporangia. The sporangium has a somatic wall enveloping sporogenic cells located inside. The number of archesporial cells depends upon the plant species. Four similar spores arise after meiosis from one archesporial cell. The spores mature in the sporangium and become covered by a thick cell wall of sporopollenin, a very resistant polymer protecting the spore during the period of spreading in a dry environment. Spores require humidity for germination. The gametophyte (prothallium) is a relatively small and simple haploid plant. It lacks differentiated vegetative organs. The spores are morphologically similar but can differ in physiology. In ferns, the spores develop into a monoecious gametophyte with archegonia and antheridia. In most ferns, the antheridia reach maturity at a different time than the archegonia and fertilization by selfing is relatively rare. In horsetails, the dioecious gametophytes develop from similar spores. Their sex depends on the conditions of development and the amount of nutrients in the soil.

Heterospory

In Selaginellales, the life cycle is of a more advanced type. The gametophytes are more reduced than the sporophytes. A diploid plant develops vegetative and generative branches with sporophylls. Sporophylls with male sporangia are situated in the upper part of the branch and female sporophylls in the lower part. Male and female lines are separated to the different sporangia. Sexual difference is apparent in the structure of the sporangium already before meiosis. Male sporangia are smaller and spores arising in them are also small compared to the female line. Hence they are named microsporangia and microspores. The structure of the microsporangium is very similar to the eusporangium. The archespore is located in the center. It consists of numerous cells, which become meiocytes, i.e. microspore mother cells. They differentiate very synchronecously and nearly all of them enter meiosis concurrently. In the microsporangium hundreds of microspores are produced because four haploid cells arise from each meiocyte. The least number is 64 microspores from 16 meiocytes. Microspores remain inside the microsporangium and develop into male gametophytes. Mitotic divisions lead to the formation of a haploid prothalium possessing a simple antheridium. In the antheridium, spermatogeneous cells divide by mitosis and afterwards, each daughter cell differentiates into sperm equipped with two flagella.

In the female line, the megasporangia start to develop similar to the microsporangium, but with far fewer archesporial cells. The latter begin to differentiate into meiocytes but this process is inhibited in all cells except one. This cell becomes a megaspore mother cell and undergoes meiosis. Regular division of genetic material is followed by unequal cytokinesis. One spore takes most of the cytoplasm and three sister megaspores take much less. They degenerate quickly when the large megaspore develops into the female prothalium. Entire development from megaspore to mature female gametophyte occurs inside the megasporangium on the maternal plant. Microsporangia and megasporangia are collected on the same stem and open simultaneously on a rainy day. Sperms realized from microsporangia swim down to the megasporangia and fertilize egg cells.

In the life cycle of *Selaginella*, both gametophytes remain on the maternal sporophyte during their development and fertilization. They are reduced compared to the independently living gametophytes of isosporic plants.

Heterospory is a type of sexual reproduction of seed plants (gymnosperms and angiosperms). Structure and development of the sporangia differ depending on the sex. The male sporangium is located in an anther. It contains numerous archesporial cells which differentiate into meiocytes. Initially, meiotic cells are connected by cytomictic canals, then the canals are closed and each meiocyte undergoes meiosis. In angiosperms, meiocytes are covered with a callose wall before they enter meiotic division. From each meiocyte, four haploid microspores develop inside the callose wall forming a tetrad. The callose is digested after meiosis and the microspores separate. They develop into male gametophytes, pollen grains. Here and there the somatic cells of the anther wall differentiate into a structure helping to open the organ in conditions suitable for spreading the pollen.

In seed plants, the female organ containing one or several archesporial cells is called an ovule and does not resemble the structure of any sporangium. In contemporary plants, no intermediate form of an organ between the typical sporangium and the ovule is found. This analogous organ in most plants contains a single meiocyte. In gymnosperms, the meiocyte (megaspore mother cell) is situated in the central part of the ovule, the nucellus. The female meiocyte undergoes meiosis and four unequal megaspores are formed in a linear tetrad. The megaspore in the chalazal position is larger than its sister cells. It is a functional megaspore that develops into haploid prothallium tissue. The three smaller megaspores degenerate. Inside the ovule the prothallium grows and matures for fertilization. On its micropylar pole, two or three archegonia develop, each with one egg cell. Fertilization and zygote development occur inside the ovule, which becomes the seed.

Modifications of meiosis during megasporogenesis in flowering plants

In angiosperms, meiosis in the female line (megasporogenesis) can undergo the following three types: monosporic, bisporic or tetrasporic. The differences among each type are associated with cytokinesis following meiotic divisions. There are some additional modifications in each type of megasporogenesis described as a pattern named after the plant in which this pattern occurs. The most popular is the monosporic type of megasporogenesis called *Polygonum*. The megasporogenesis starts with archesporial cell differentiation. In plants considered more primitive, e.g. Rosaceae or Ranunculaceae, several cells differentiate into archespore inside the nucellus, but most become inhibited in development. The deepest position inside the nucellus is favorable for effective development and one cell located there can develop quicker and dominate the others. This cell becomes the meiocyte and undergoes meiosis. In the *Polygonum* type, cytokinesis follows the first and second meiotic divisions and four megaspores may be arranged in a linear or T-shaped tetrad. Cytokinesis is unequal and the chalazal megaspore obtains more plastids and mitochondria than the others. It is larger and dominates over its sister cells. The chalazal megaspore becomes functional and develops into the female **gametophyte**—the embryo sac. During this development three cycles of mitosis produce eight nuclei, which are distributed to seven cells in the process of cellularization.

The other patterns of megasporogenesis involve an inhibition of cytokinesis during meiosis. In the bisporic type, the first meiotic division is followed by cytokinesis and a dyad is formed. In the dyad, both cells or only one of them can undergo a second division. More often meiosis is completed in the chalazal cell of the dyad, where two haploid nuclei are formed. The other cell of the dyad can be inhibited in further development or can complete the division II and start to degenerate immediately afterwards. As a result, one cell of the dyad containing two haploid nuclei becomes a functional megaspore and develops into an embryo sac. Two cycles of mitosis producing eight nuclei followed by cellularization lead to a properly formed embryo sac of the *Allium* type, similar to the *Polygonum*.

In the tetrasporic type of megasporogenesis, meiotic karyokinesis occurs until four haploid nuclei formation but is not followed by cytokinesis. The haploid nuclei are located in one cell, the coenomegaspore, which contains all the cytoplasm of the meiocyte and is surrounded by the same cell wall. The coenomegaspore develops into the embryo sac after one cycle of mitosis, which suffices to produce the eight nuclei needed for seven cells of embryo sac formation (Adoxa type). There are many other types of tetrasporic embryo sac development but they do not differ in meiotic pattern. In the tetrasporic type of megasporogenesis all the energy and

contents of the meiocyte are saved during coenomegaspore formation. The entire cytoplasm and each haploid nucleus obtained after meiosis are engaged in embryo sac formation, so this type is considered the most progressive. For comparison, the monosporic and bisporic type of megasporogenesis consumes more energy for cytokinesis and cell wall formation between megaspores. The functional megaspore obtains only part of the cytoplasm inherited from the meiocyte. The rest is located in the sister megaspores predestined to degenerate. Contemporarily, this process is described as apoptosis, i.e., genetically controlled and energy-consuming disintegration of cells arising in one stage, but not demanded in subsequent development of the organism.

This short and simplified review of life cycles in the plant kingdom shows that meiosis is a very important moment in plant reproduction and associated with sexual processes (Fig. 9.1). Alternation of generations is a process persisting throughout evolution. In angiosperms, the most advanced and progressive plants, it is also present, even if the gametophytes are very small, simple structures dependent for development and functions on the maternal plant.

MEIOSIS AS A PROCESS OF GENETIC RECOMBINATION: KARYOKINESIS

Archesporial cells differentiate from somatic tissue, which initially proliferates by mitosis. After that period, the archesporial cells enlarge and become isodiametric in shape. In eusporangia, microsporangia and anthers, the archesporial cells are connected by cytomictic canals. This has a strong influence on integration of the archesporial tissue and nearly all cells achieve the same physiological state at the same time. In most plants change from the archesporial phase into meiocytes occurs synchronously. In each cell, the nucleus takes the central position. It is relatively large (very often occupies 2/3 of the cell volume) with an eminent nucleolus. Metabolism in premeiotic cells is very vivid. Before meiosis, a full replication of DNA must be completed and also synthesis of the ribosomes and proteins to enrich the cytoplasm. Premeiotic cells increase their respiration to store the energy very essential for the two cycles of meiotic division. Meiocytes stop growing and close all cytomictic canals. In angiosperms they are enveloped in a callose wall, which isolates each cell as a separate unit (Fig. 9.2B, E-G). At that phase, the meiocytes with a callose cell wall or without it seem not to communicate with the surrounding tissue. They do not take up macromolecular substances, but this has yet to be proven for all plant species.

Prophase I

The onset of meiosis (prophase I) is demonstrated by chromatin condensation in the nucleus. This process is relatively long and can take a few days. It can be interrupted by periods of metabolic dormancy. Sometimes, prophase I begins in autumn, is arrested for all of winter, and then proceeds again in spring. Prophase I can last two-three months, whereas all the stages needed to complete the divisions may take just a few hours. During prophase I five substages are distinguished, depending on chromatin condensation and arrangement: leptotene, zygotene, pachytene, diplotene, and diakinesis.

Leptotene

Chromatin condensation begins in the leptotene (Fig. 9.2A and B). At that time the chromosomes look like thin threads attached to the nuclear envelope at the telomere point. Sister chromatids are longitudinally linked by a special protein so the chromosome represents a unit structure at that time. The site of the future centromere is already distinguishable.

Zygotene

Chromosomes lose contact with the nuclear envelope and change their position inside the nucleus. At this stage homologous chromosomes begin to pair. In the diploid nucleus one chromosome from the maternal set couples with a homolog from the paternal set. The process occurs step by step. The mechanism of initiation of pairing the homolog chromosomes is not completely clear. Some parts match well and join earlier than others. As a rule alignment spreads along the whole homolog chromosomes but sometimes there are parts that do not match and do not join, forming small loops. Homologous portions lie alongside each other to form synapses. In these places, a special protein strengthens the adhesion. The protein forms a ladderlike structure between the two chromosomes called a synaptonemal complex. Synaptonemal adhesion spreads along the whole bivalent and stabilizes it. Proteins of the synaptonemal complex are connected to short sequences of DNA called zygDNA because these sequences are synthesized during zygotene. The length of the synaptonemal complex is related to the decondensation and condensation of chromatin occurring in subsequent stages.

The premeiotic chromosome consists of two sister chromatids as before mitosis, but in meiosis the chromosome, instead of splitting, joins a homologous one in a pair (bivalent). The bivalent consists of four homologous chromatids and sometimes the term "tetrad" is used to describe the struc-

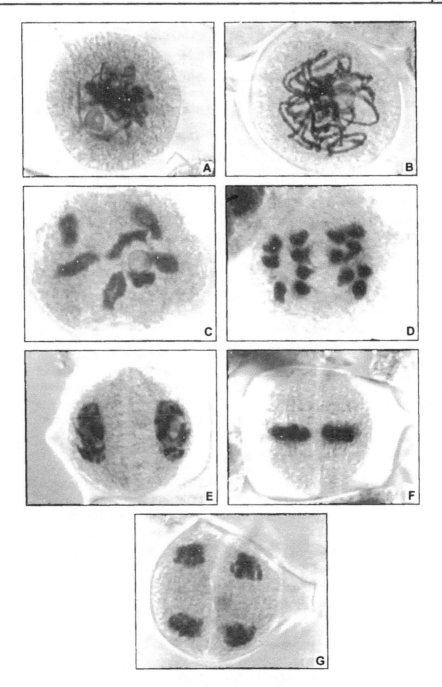

Fig. 9.2 Meiotic stages during microsporogenesis in rye (*Secale cereale*). Chromosomes stained with acetoorcein. A, B – prophase I (leptotene - pachytene), C – diakinesis, D – anaphase I; E – telophase I; F – metaphase II; G – telophase II (Photos J. Cebrat, Agricultural Academy, Wroclaw, Poland; 900 X).

ture. Each chromatid is partitioned to one of the four daughter cells. This results in transferral of full genetic information to the spores and the next generation of plants.

Pachytene

This can last for a few days when special events lead to genetic recombination. The alignment of sister chromatids is broken in some places and they cross with the chromatids of another chromosome. The places of crossing are called chiasmata. They are the points at which the chromatids can be broken and linked again with part of a homologous chromatid. In this way, fragments of homologous chromatids are exchanged. This mechanism of exchanging some genes between homologous chromosomes is called crossing over. After an exchange of chromatid fragments, full synapsis of chromosomes is restored. The places of crossing over remain invisible until the next stage, diplotene.

Diplotene

In most species crossing over proceeds similarly in subsequent generations. Some parts of chromosomes are not engaged in crossing over and other parts are exchanged in every generation. On the chromatids there are "hot" places, which break easily and link again, and some stable places that either do not or very rarely break.

After crossing over condensation of chromatin proceeds. Then the synaptinemal proteins are digested except where chiasmata have occurred. During the late diplotene, the chiasmata are terminalized, i.e. shifted toward the ends of chromosomes. Synaptinemal proteins are completely destroyed and the chromatids remain apart, connected only in the centromere. Otherwise, they have free arms. At that stage chromatids are not completely

condensed and some gene expression is possible. It takes place especially in the cells of the female line where storage material is still being gathered in the megaspore mother cell. In the male line, the diplotene is shorter and passes into diakinesis.

Diakinesis

Diakinesis is a time of strong condensation of each chromatid arm. At that stage the chromosomes are shortest. They often look like crosses because the chromatid tetrad is separated except at the centromere. Then the number of chromosomes and ploidy of the plant can be easily estimated (Fig. 9.2C).

In diakinesis, karyokinetic spindle formation begins in close vicinity of the nucleus. First the organelles move away from the central region of the cell. Next the nuclear envelope disappears and the bivalents occupy the middle of the spindle. In many handbooks, the stage of premetaphase I is distinguished wherein the bivalents find their position in the equatorial plate of the karyokinetic spindle. They segregate on the spindle: the shorter toward the middle and the longer outside with their arms in the plate. The microtubules of the karyokinetic spindle become attached to the centromeres by a kinetochore protein. The kinetochores of two sister chromatids are joined by the additional protein molecules, which are present only during meiosis until anaphase II. The centromeres of homologous chromosomes become separated and turn toward the opposite spindle poles. This moment is the end of prophase I and the beginning of metaphase I.

From a genetic point of view, prophase I is a very important period because it is a time of genetic recombination between the chromosomes originating from maternal and paternal sets.

Metaphase I

The following stages of meiosis are of short duration. In metaphase I there is a moment in which the bivalents look inactive in the equatorial plate but in fact the synaptonemal complex disappears in each of them and entire chromosomes become ready to be pulled toward the spindle pole.

Anaphase I

Anaphase I is a stage of chromosome movement (Fig. 9.2D). The mechanism of chromosome movement is complicated because the microtubules and also actin microfilaments of the spindle are actively involved. The centromere of each chromosome is the location where the microtubules and microfilaments are attached to the kinetochore and create pulling strength. The arms of chromosomes move passively, following the centromeres. The chromosomes are segregated to spindle poles at random, which is also a factor of genetic rearrangement.

Telophase I

When the centromeres reach the spindle pole, anaphase I ends and telophase I begins (Fig. 9.2 E). During this last phase of meiotic division I in the newly formed daughter nuclei, some chromosomes originating from maternal and some from paternal sets become decondensed and the nuclear envelope is restored. In each nucleus the number of chromosomes is reduced to the "n" level, but the amount of chromatin is similar to that in the daughter nuclei after mitosis, i.e. two sets of chromatids. In the postmitotic nucleus one set represents full maternal and the other full paternal genetic information. After meiosis I the genetic information is mixed and creates a new composition differing from that after mitosis.

Meiotic Division II

Reduction of the chromatin amount to the haploid level is realized in the second meiotic division. Meiosis II is not proceeded by replication of DNA and can occur immediately after telophase I or after cytokinesis and cell wall deposition between dyad cells. In the female line cytokinesis and dyad formation take place in more than 80% plants. In microsporogenesis and sporogenesis more plants display simultaneous cytokinesis after karyokinesis II.

Meiosis II starts with simultaneous condensation of chromosomes and spindle formation in both daughter nuclei. The condensed chromosomes gather at the equatorial plate and stay there during metaphase II (Fig. 9.2F), then break into centromeres. In anaphase II single chromatids are pulled to the spindle poles, after which in telophase II (Fig. 9.2G) they undergo decondensation and are enclosed in a new nuclear envelope. This stage is final for meiotic karyokinesis. The rearrangement of genes is realized not only in meiotic division I, but also during anaphase II when the chromatids are segregated and transported to the poles of the karyokinetic spindles at random. This means that in most plants the daughter genomes in the spores consist of some chromatids originating from the maternal and others the paternal set. The genome in the four postmeiotic nuclei is composed of the same genes in a new arrangement (Fig. 9.3).

CYTOKINESIS

In meiosis, cytokinesis is as important as karyokinesis because both processes affect haploid cell viability and transfer of genetic material to the next generation. In plants, genetic information, besides the nucleus, is also deposited in plastids and mitochondria. Organelles and cytoplasm are inherited from the meiocyte and have to be distributed to the isospores and microspores in such proportions that all four cells contain all the organelles needed for future development. In the female line (megasporogenesis), distribution of cytoplasm and organelles to the sister megaspores may be unequal. The functional megaspore may be better equipped because of unequal cytokinesis or a specific process of organelle segregation. The pattern of cytokinesis and organelle distribution is genetically determined, realized by action of the cytoskeleton.

Simultaneous Cytokinesis

Cytokinesis may be a single process ending meiosis, the simultaneous type, or two successive processes may take place after the first and second meiotic division. To some degree, the type of cytokinesis is associated with the sex line.

In sporogenesis and microsporogenesis, simultaneous cytokinesis (Fig. 9.4) occurs more often than the successive type. In megasporogenesis, successive cytokinesis occurs in more than 80% plants (Figs. 9.5, 9.6 and 9.7). The type of cytokinesis can also be considered a systematic feature of the plant species, signaling the origin and relationship of the taxon. The simultaneous type of cytokinesis is regarded as more primitive by most plant embryologists and taxonomists, but their point of view is questioned by others. There are plants in which both types of cytokinesis occurr in the same sporangium or microsporangium.

Successive Cytokinesis

In successive microsporogenesis, division I is completed by cell wall formation between the daughter nuclei. Division II occurs in both cells of the dyad simultaneously

(Fig. 9.2F and G). The direction of sister karyokinetic spindles may be perpendicular. Meiosis II also ends in cytokinesis with cell wall deposition, leading to formation of four microspores. This pattern of divisions results in the most common, the tetrahedral shape of the microspore tetrad. If the karyokinetic spindles are parallel, the tetrad becomes quadripolar. T-shaped or linear tetrads of microspores have been observed in some plant species, e.g., in Asclepiadacae, but are not common.

In the female line, division I is followed by successive cytokinesis in the monosporic and bisporic type of megasporogenesis. A dyad is formed and meiotic division II takes place in the two cells independently (Fig. 9.5B and C, Fig. 9.7C, Fig. 9.8B and C). During megasporogenesis, division II may be delayed or completely inhibited in one of the dyad cells, while it occurs in a regular manner in the sister cell. One cell of a dyad may divide into two equal parts, the other unequally into larger and smaller megaspores. The linear tetrad arises from a dyad in which both spindles are in the micropyle—chalaza axis (Fig. 9.5C, Fig. 9.6B, Fig. 9.7A and B, Fig. 9.8C). In the sister cells of a dyad, the axes of karyokinetic spindles can often be perpendicular to each other. Such a direction of two sister spindles leads to the formation of a T-shaped megaspore tetrad, which is as common as the linear type. Given the common types of tetrads mentioned above, triads or quadripolar tetrads are rarely formed.

POLARIZATION OF MEIOTIC CELLS

The type of cytokinesis is associated with the phenomena of polarization and segregation of organelles before and during meiosis.

Prophase I

Metaphase I

Anaphase I

Dyad
Prophase II

Tetrad
of spores

Division I

Division II

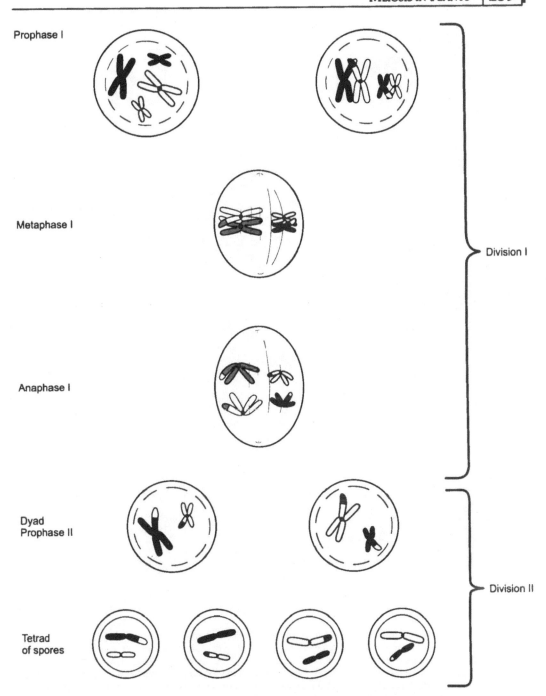

Fig. 9.3 Scheme of genetic recombination during meiosis showing crossing over in prophase I; at random distribution of maternal and paternal chromosomes during anaphase I and II; genes differently rearranged in each of 4 haploid genomes of the spores

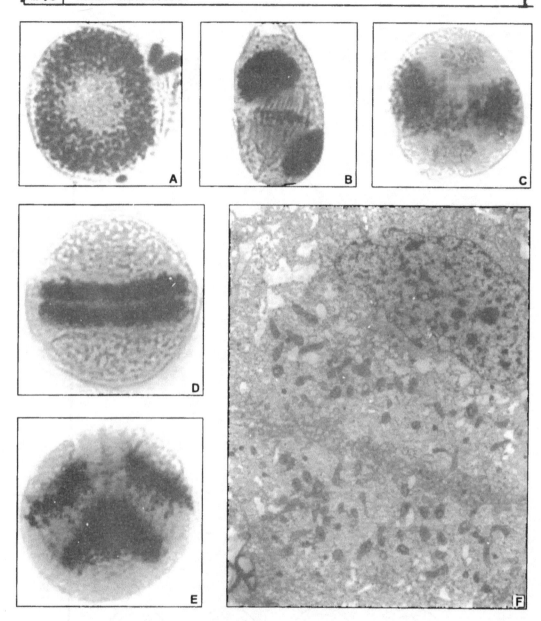

Fig. 9.4 Distribution of organelles during sporogenesis in *Equisetum palustre*: A to E–Plastids with starch stained by PAS, seen as black grains (1,400 X). A – ring of organelles around nucleus in the meiocyte; B – two groups of organelles on the side of karyokinetic spindle at metaphase I; C – movement of organelles from the groups to the middle plate between the daughter nuclei at early telophase I; D – sandwich-like layer of mitochondria and plastids seen from the late telophase I to the end of anaphase II; E – distribution of organelles to 4 spores at late telophase II; F – ultrastructure of organelles layer analogous to the stage shown in picture D, plastids (dark irregular grains) moved outside, mitochondria (light round vesicles) stay in middle layer (Photos I. Gielwanowska, Warmian-Mazurian University in Olsztyn, Poland; 5,000 X).

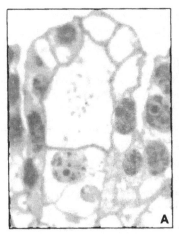
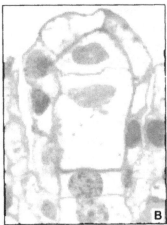
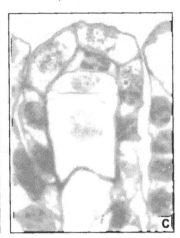

Fig. 9.5 Megasporogenesis of *Polygonum* type in *Epipactis palustre*: unequal division I and II; functional megaspore on chalazal position is much larger than sister megaspores. A – meiocyte in diakinesis stage; B – dyad with small micropylar cell; C – delay of meiosis II in micropylar cell of dyad, chalazal cell divided on larger chalazal megaspore and smaller middle one. Semithin sections stained in toluidine blue (Photos J. Bednara, Maria Curie-Sklodowska University, Lublin, Poland; 600 X).

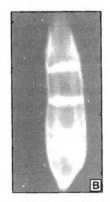

Fig. 9.6 Callose fluorescence in cell walls of meiotic cells during megasporogenesis in *Epilobium roseum*. A – meiocyte; B – heteropolar tetrad with a functional micropylar megaspore; pole free from callose. Stained by aniline blue, fluorescence microscope. (Photos J. Bednara, Maria Curie-Sklodowska University, Lublin, Poland; 1,200 X).

Fig. 9.7 Homopolar megaspore tetrads of *Oenothera biennis*. Callose present in cross walls:, both polar megaspores equipped with plastids. Traces of polarization visible as: A – thinner cross cell wall on chalazal side as effect of delay of meiosis II; B – callose spot on chalazal pole of the tetrad while micropylar pole completely free from callose. A and B – aniline blue staining and fluorescence microscope; C – homopolar tetrad: simultaneous division II on chalazal and micropylar side and similar amount of plastids in both polar megaspores; stained by PAS (Photos R. Śnieżko, Maria Curie- Sklodowska University, Lublin, Poland; 1,200 X).

Callose in Meiotic Cell Wall

In microsporangia and in ovules with monosporic megasporogenesis, the meiocyte is covered by a callose cell wall in early prophase I. Callose is a polysaccharide, which can be very quickly synthesized and dissolved. In other tissues, callose appears in the cell wall in the process of closing connections between cells, healing wounds after mechanical disruption, or in the reaction evoked by infection or incompatibility. In the meiocyte, callose can isolate the cell when it changes destination from somatic development to the generative. It is a time of diminished contact with the surrounding cells. Distribution of callose in the meiocyte cell wall is variable depending on: sex line, plant species, and pattern of polarization of the meiocyte and future tetrad. Several patterns of callose deposition have been found. All reflect the polarization phenomenon.

In microsporogenesis, meiotic cells are evidently not polarized before meiosis. A callose wall envelopes a round meiocyte. Cell walls formed after cytokinesis are also built of callose and every microspore inside a tetrad is separated as an individual cell (Fig. 9.2E-G). There are some traces of microspore polarization in the tetrad, i.e. the outer callose wall is thicker then the inner wall and the position of the nucleus and organelles also shows the outer versus inner arrangement. In microsporogenesis as in isosporial meiosis, polarization is not very strongly expressed. The spores are similar to each other and all develop into male gametophytes, the pollen grains.

The female meiocyte is strongly polarized before meiosis. It is elongated in the micropyle-chalaza axis (Fig. 9.6A and Fig. 9.8A). After meiosis, the megaspores are arranged along this axis and their position in the tetrad determines their further development. Most gymnosperm and flowering plants have linear or T-shaped megaspore tetrads with the functional megaspore on the chalazal pole (Fig. 9.5C). This pole in the meiocyte is predetermined

to become the locus of future development of the female gametophyte. It is marked by the polarization phenomenon associated with cell wall formation, organelle segregation visible before and during meiosis, and delay or inhibition of meiosis II on the nonfunctional side.

Callose wall formation around the meiocyte has been observed in many plants. It covers the enveloping wall and is deposited in cross walls formed after divisions I and II (Fig. 9.6). During megasporogenesis connections through plasmodesmata occur between somatic and meiotic cells at early prophase I, but later most of the connections become closed by the callose layer. Meiotic cells covered by callose are isolated from the surrounding tissue, which can be a factor decreasing nutrition and hormonal stimulation. However, in female meiocytes, complete isolation is relatively short as in Oenotheraceae (megasporogenesis of the *Oenothera* type), Caryophylaceae and Papaveraceae (megasporogenesis of the *Polygonum* type). In these plants, callose disappears from the functional pole of the meiocyte in late prophase I. In Orchidaceae, the megasporocyte appears not be completely isolated by callose, because the chalazal cell wall is sieve-like with many pores in the callose layer. Plasmodesmata connecting the female meiocyte (megaspore mother cell) with somatic cells of the tissue on the chalazal side have been found. They persist until the end of meiosis and supply the functional megaspore during its future development. On the micropylar side of the meiocyte, the plasmodesmata become closed before meiosis and the cells are tightly covered by a callose wall. After meiosis, three nonfunctional megaspores are isolated from contact with the surrounding tissue and begin to degenerate immediately after division II. The callose layer on the chalazal cell wall disappears

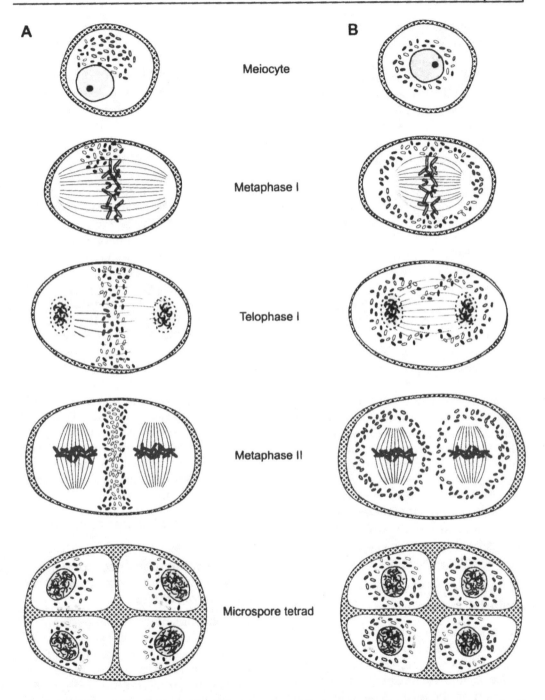

A   B

Meiocyte

Metaphase I

Telophase I

Metaphase II

Microspore tetrad

Fig. 9.8 Scheme of 2 types of organelle redistribution observed in flowering plants during microsporogenesis with simultaneous cytokinesis. Layers of organelles separate the karyokinetic spindles during meiotic division II.

when the functional megaspore starts to develop into the embryo sac.

In many plants with megasporogenesis of the *Polygonum* type, the same pattern of callose wall formation over meiocytes occurs. In early prophase I, a layer of callose covers the whole meiocyte but a little later the chalazal pole of the cell appears free from the isolating polysaccharide. Plasmodesmata can be preserved until the end of meiosis in this place and supply the functional megaspore formed at the chalazal pole of the tetrad. This position is favorable for the developing gametophyte because it is near the source of nutrients supplied by the vascular bundle end to the chalaza of the ovule.

In Oenotheraceae, polarization of the megasporocyte and future tetrad occurs at opposite ends. The micropylar pole is destined to locate the functional megaspore. In these plants the meiocyte is enveloped in a callose cell wall, but the micropylar pole becomes free from it in prophase I. This pole stays in contact with the nucellus and the functional megaspore is formed on this side (Fig. 9.6). The surrounding nucellar tissue is full of starch grains and undergoes a kind of lysis before the embryo sac begins to develop. It takes nutrients from nucellar tissue and grows in the micropylar direction. In homopolar tetrads of *Oenothera biennis* and related taxons the callose is deposited only in cross walls (Fig. 9.7A and B)

Delay of Cytokinesis as a Polarizing Factor

In Oenotheraceae, megasporocyte polarization demonstrated by callose wall formation around the chalazal part is associated with other factors establishing the prevailing position on the micropylar side. Domination of the micropylar megaspore over three sister megaspores is manifested by special plastid segregation, delay or inhibition of division II in chlazal cell of dyad, pattern of cytokinesis, and cross cell wall formation in the tetrad (Fig. 9.7A). On the less privileged side, delay or inhibition of karyokinesis and cytokinesis has been observed not only in monosporic megasporogenesis, but also in the bisporic one. In such a case, triads or dyads with only one viable cell are formed.

In the meiotic cells described above, polarization is evident and the result of meiosis is a heteropolar tetrad. In some species of *Oenothera* this heteropolar pattern is accompanied by a homopolar one. In a homopolar megasporocyte callose is not deposited in the surrounding cell wall. After divisions I and II, callose appears only in cross walls (Fig. 9.7A and B). As a result, both polar megaspores of the tetrad have contact with the surrounding nucellus. The homopolar pattern is strengthened by an equal distribution of plastids to both polar megaspores (Fig. 9.7C). In a homopolar tetrad, two polar megaspores are very often functional and start to develop. Polarization of the tetrad is revealed in a later stage by a more advanced development of the embryo sac on the micropylar side. At that stage, different postmeiotic factors determine polarization in the ovule.

Distribution of Organelles during Meiosis

Rearrangement and segregation of organelles during sporogenesis and microsporogenesis

The phenomenon of segregation and rearrangement of organelles begins before meiotic division. In the meiocyte, all plastids, mitochondria and most other cytoplasmic structures are removed from the vicinity of the nucleus, the future position of the meiotic spindle. There are several patterns of segregation of the organelles during meiosis in eusporangia and microsporangia (Fig. 9.9A). The organelles can be gathered in one group at the side of the nucleus, in two groups at the spindle poles or in a ring surrounding the nucleus. At the beginning of division I, all organelles are distributed outside the karyokinetic spindle until the end of telophase I, then become segregated and redistributed.

During sporogenesis in some mosses and ferns and also during microsporogenesis in many flowering plants, meiosis is followed by simultaneous cytokinesis. After meiosis I, organelles enter the equatorial plate of the phragmoplast. They spread in a layer separating daughter nuclei and the cytoplasm. In this type of cytokinesis a cell wall is not formed and division II begins immediately after telophase I (Fig. 9.4 D-F). The layer of organelles may be important for keeping the distance between karyokinetic spindles during division II. Spindle fusion is a known phenomenon, which occurs when the spindles remain close to each other, e.g. during the first mitosis in a heterokaryon or in a zygote. In the zygote of many plants, the nuclei of the egg cell and the sperm cell start division separately, but their spindles fuse into one during metaphase. Then, the chromosomes from both gametes can mix in one diploid set. During meiosis II two karyokinetic spindles have to remain separated because their fusion can greatly disturb meiosis, disabling a proper reduction of genetic material to the haploid level.

In the equatorial plate of the phragmoplast, the layer of organelles consists of mixed plastids and mitochondria. In this layer, plastids are segregated to the surface while the mitochondria occupy the middle (Fig. 9.4F). After division II, two phragmoplasts are formed in the telophase. The microtubules align not only between the daughter nuclei, but also between the poles of the sister spindles. The organelles spread from the previously arranged layer to the equatorial plates of the new phragmoplasts in a way leading to separation of each nucleus. Then the plastids move toward the nuclei; next the mitochondria divide into the four portions of the cytoplasm. In the gap between the organelles, a new cell wall is built and four haploid microspores are separated concomitantly. Each contains a similar or identical number of organelles. Recent observations under the confocal microscope support the very regular distribution of organelles to each microspore in the tetrad. Immediately after meiosis, each microspore is polarized. The nucleus is positioned at the distant side of the cell and the organelles distributed on the side of the center of the tetrad (Fig. 9.4E). During future development organelles disperse inside the microspore cytoplasm and surround the nucleus. Then the callose wall of the tetrad is dissolved and the four microspores set free in the loculus.

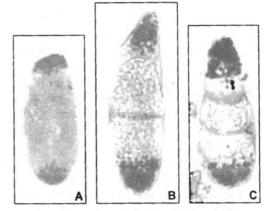

Fig. 9.9 Plastids with starch gathered at the poles of meiotic cells during megasporogenesis in *Epilobium palustre*. Plastids with starch stained by PAS, seen as dark grains. A – meiocyte; B – dyad; C – megaspore tetrad (Photos J. Bednara, Maria Curie-Sklodowska University in Lublin, Poland; 1,200 X).

The ring and sphere pattern of organelle distribution was found in meiotic cells of Malvaceae, plants with simultaneous cytokinesis. The organelles form a sphere around the nucleus before meiosis I and become redistributed after telophase I (Fig. 9.9B). Then, they make two spheres surrounding daughter nuclei and later two karyokinetic spindles. In telophase II, two spheres of plastids and mitochondria break and reorganize into four spheres around each haploid nucleus. Such an arrangement of the organelles leaves space for cell wall formation after meiosis is completed.

In plants with successive cytokinesis, distribution of organelles is similar at the beginning of meiosis. Organelles are removed from the position of the future karyokinetic spindle and gather in a group but later disperse in cytoplasm again until the end of meiosis. After divisions I and II, cytokinesis divides portions of cytoplasm with organelles and a cell wall is built. After successive cytokinesis the four microspores contain a similar number of plastids and mitochondria distributed over the cell.

During microsporogenesis in plants of an unbalanced or alloploid genotype, a very specific, unequal distribution of organelles has been observed. Karyokinesis in such plants can be irregular and the karyokinetic spindle may be multipolar or only tripolar. The chromosomes are sometimes not properly disjuncted and pull to the poles. In such genetically unbalanced taxons, different disturbances of karyokinesis are very often accompanied by irregular cytokinesis. As a result, a mixed population of microspores develops in the same loculus. During karyokinesis, besides the normal nuclei, a different number of various micronuclei arise, and after unequal cytokinesis small, handicapped microspores with a diminished amount of organelles are formed besides normal and viable ones.

Segregation of organelles during megasporogenesis

Meiosis in the female line involves successive cytokinesis of the monosporic and bisporic type. During megasporogenesis distribution of organelles depends on a pattern of megasporogenesis genetically determined in a manner characteristic of the plant species. Organelle distribution to the megaspores can be affected by polarity factors.

In megasporogenesis of the *Polygonum* type, the chalazal megaspore is predetermined to develop into an embryo sac. It is always much bigger than the sister cells as a result of unequal cytokinesis I and II. The chalazal megaspore takes a much bigger cytoplasmic portion and most of the plastids and mitochondria from the meiocyte. This prevailing chalazal position of the functional megaspore is also dependent on unequal cytokinesis in bisporic megasporogenesis, the *Allium* type. Development of the functional megaspore is shortened and takes only one unequal cytokinesis after meiosis I: already the micropylar cell of the dyad is small and the chalazal much larger. Development of the small cell may be inhibited so that it does not complete meiosis II. The bigger chalazal cell undergoes karyokinesis. As a result, the two daughter nuclei are surrounded by the greater part of the cytoplasm and organelles transferred from the meiocyte. In both types of megasporogenesis, unequal cytokinesis after division I ensures that the functional megaspore is better equipped than its sister cells, to the advantage of the future embryo sac. Unequal cytokinesis economizes the process of gametophyte development because a smaller part of the cytoplasm is destined for degeneration.

In some plants, a characteristic segregation and unequal distribution of organelles have been found. Segregation of plastids to the pole of the meiocyte has been observed in many plants from the Oenotheraceae family. As described earlier, in the ovule Oenotheraceae display polarization opposite to *Polygonum* and the *Allium* type. The best example is megasporogenesis in taxa of *Euoenothera*. The female meiocyte is an elongated cell with the nucleus shifted onto the micropylar side of the cell. Before prophase I, organelles are randomly distributed in the whole cytoplasm. When the callose wall becomes deposited around the cell, plastids containing starch grains gather at the meiocyte poles (Fig. 9.9A). They form tight groups, very often larger at the micropylar pole and smaller at the chalazal. During karyokinesis and cytokinesis I, the plastids remain pressed to the poles. In the dyad, each cell has its own group of plastids (Fig. 9.9B). Clusters of plastids also remain together during meiosis II. A linear tetrad of megaspores arises. In such a tetrad all megaspores are similar in size but only the micropylar and chalazal megaspores contain groups of plastids (Fig. 9.7C, Fig. 9.9B and C). Two middle megaspores are devoid of them. They degenerate relatively quickly because without plastids they are inviable and unable to become functional. Immediately after meiosis, both polar megaspores are equipped with these important organelles and both are potentially able to develop into an embryo sac. Later, the starch and plastids disintegrate in the chalazal megaspore. Concomitantly, the plastids in the micropylar one become more numerous and filled with starch grains. The presence of plastids is a necessary condition for viability and domination of the micropylar megaspore over its sister megaspores. This pattern of megasporogenesis and type of tetrad are called heteropolar—the micropylar pole is the only site of gametophyte development.

In Oenotheraceae family, heteropolar tetrads are formed in species from different genera, e.g. *Fuchsia, Clarkia, Godetia, Epilobium,* and many taxa of *Euoenothera*.

In other genera, unequal cytokinesis results in a larger size of micropylar megaspore, increasing the advantage of this megaspore in future development. Besides this heteropolar pattern of megasporogenesis, in some *Oenothera* taxa, the homopolar type of tetrad arises. The homopolar type of megasporogenesis has been found mainly in permanent heterozygotes. The meiocyte is similarly elongated as in other *Oenothera*, but is not covered by a callose wall in prophase I.

After divisions I and II, callose appears only in the cross walls formed between megaspores. Such a distribution of callose does not influence the connection between megaspores and the surrounding tissues. Isolation is not a polarizing factor in the tetrads. In homopolar tetrads, all megaspores are similar in size. Polar megaspores are equipped with plastids (Fig. 9.7C). In consequence, the embryo sac can develop from the megaspore in the micropylar and chalazal position or simultaneously from both polar megaspores. All these possibilities have been observed in ovules of heterozygotic *Oenothera*. The proportions between the developmental patterns are dependent on the plant taxon. Two embryo sacs developing simultaneously in the micropylar and chalazal position in the same ovule have a different genotype determined by a specific segregation of chromosomes during meiosis. This leads to repetition of the maternal and paternal sets but in the haploid number. The genetic factor modifies embryo sac development and in most ovules only one of the embryo sacs can reach maturity, the one with a genome suitable for the female gametophyte. In heterozygotic *Oenothera* taxa, special genetic selection in the haplophase is involved in mantaining the heterozygosity of these self-pollinated plants, e.g., *Oe. biennis*. In this and similar taxa, more often than in other species, double embryo sacs in one ovule have been found. The developmental success of one of them depends on their genomes and the polarization phenomena determined by the mother plant. In most taxa, these factors make the micropylar position more privileged. After a period of parallel development, the micropylar embryo sac grows quicker than the chalazal one, which can be inhibited before the first mitosis or later. In *Oe. suaveolens* 70% of ovules contain an embryo sac originating from the micropylar megaspore and nearly 20% from the chalazal one. In this species, very rarely there are two mature embryo sacs present in one ovule. The rest of the ovules (about 7-10%) abort because of lack of space or nutrients or some other reasons.

In Oenotheraceae, cytokinesis influences domination of the micropylar tetrad in many genera wherein the dvision is unequal. This domination is strengthened by isolation of the chalazal pole by a callose wall and segregation of plastids. All these factors make the micropylar position more priviliged. In heterozygotic *Euoenothera*, cytokinesis is not so influential. Polarization phenomena are also weaker. The homopolar type of megasporogenesis provides an equal development to both polar megaspores which, however, is corrected by additional specific phenomena affecting female gametophyte development. Development of two embryo sacs in one ovule is not advantageous for a plant because of the useless competition between gametophytes during their growth and fertilization. Heteropolar megasporogenesis provides better conditions for fertility in the female line.

Karyokinetic and cytokinetic disturbances can be equally harmful to the fertility of a plant. A greater amount of cytoplasm and a full set of organelles are distributed to functional spores (mega- or microspores) destined for further development. A diminished number of mitochondria decreases the vigor of the spores and a lack of plastids makes them inviable. These factors connected with cytokinesis decrease fertility to the same degree as genetic disorders (distortions). Both kinds of factors influence the fertility of many crop plants bred by numerous crossings between unrelated taxa.

In F_1 hybrids, an unbalanced genotype is created to obtain the demanded phenotype features but concomitantly sexual reproduction of the plant decreases. In wild plants, natural selection leads to elimination of sexually handicapped individuals and evolution leads to activation of new mechanisms maintaining the effectiveness of plant reproduction.

CYTOSKELETON DURING MEIOSIS

During meiosis all phenomena of movement and redistribution of organelles are influenced by the activity of the cytoskeleton. The meiocyte contains both types of cytoskeleton elements: microtubules with many kinds of associated proteins and actin microfilaments with myosin and other accompanying proteins. An array of enzymes cooperating with cytoskeletal fibrils is also active in meiotic cells. In meiocyte activity of calmodulin, a protein engaged in the actin-myosin interaction generating contraction in the cytoplasm has been found as well.

Before meiosis and during prophase I, the nucleus is surrounded by an array of microtubules and microfilaments (Fig. 9.10A and B). Two types of fibrils fix its position, then change the arrangement and gather at the future poles of the karyokinetic spindle—a special structure of the cytoskeleton in the dividing cells. Elements of the cytoskeleton and MTOC (microtubule organizing centre) proteins are anchored or in close contact with the nuclear envelope. In late prophase I, microtubules surrounding the nucleus disappear and a new generation of them is polymerized, joining kinetochores of the bivalents with poles of the karyokinetic spindle. In the kinetochores, MTOC and gamma-tubulin depositions localize. Future poles of the karyokinetic spindle are broad and dense. From the poles, synthesis of microtubules occurs toward the equatorial direction, especially during phragmoplast formation in telophase I (Fig. 9.2E). Most of the microtubules reach the equatorial plate of the spindle; some are aligned between the opposite poles.

In the karyokinetic spindle, the microfilaments protrude radially from the poles toward the middle of the spindle. During division, they are aligned between the centromeres of bivalents and spindle poles (Fig. 9.10C).

Both microtubules and microfilaments are elements of the karyokinetic spindle structure and are actively involved in chromosome movement during division. They determine the redistribution of bivalents at premetaphase I and act during disjunction in anaphase I and II (Fig. 9.2D-F). Disjunction is a mechanism of genetic material distribution and is regulated by cytoskeleton action. In metaphase I and II, the bundle of microtubules is connected with the centromere of each chromosome by the kinetochore protein and this place is actively pulled to the spindle pole. Recent investigations showed that the microtubules are aligned with actin microfilaments. Actually, the microfilaments are considered more important for generation of movement and microtubules responsible for direction of the chromosome movement to the spindle pole. Colchicine causes degradation of microtubules by inhibiting karyokinetic spindle formation and prevents karyokinesis and cytokinesis as well. Cytochalasin, a substance destroying actin functions, causes serious disturbances in the disjunction of chromosomes during anaphase.

Another cytoskeleton configuration typical of a dividing plant cell is the phragmoplast. As in the karyokinetic spindle, all types of cytoskeleton elements are engaged in phragmoplast formation but microtubules predominate. The phragmoplast is responsible for distribution of cytoplasm and organelles to daughter cells. Some microtubules of the phragmoplast originate from the spindle but most of them become polymerized as a radial arrangement on the surface of daughter nuclei in late telophase. Microfilaments are also present in the phragmoplast (Fig. 9.10D and E) but do not penetrate the equatorial plate. In the area free of microfilaments, cell wall synthesis begins by gathering and fusion of vesicles in the equatorial plate. The vesicles originate from the Golgi apparatus and contain pectin and hemicellulose precursors. The vesicles fuse into one flat cistern, the middle lamella, which spreads to the edge of the cell and becomes a plate supporting the cellulosic cell wall.

In cytokinesis of the successive type, meiosis II is followed by phragmoplast formation in each cell of the dyad. In simultaneous cytokinesis, meiosis II ends by formation of the phragmoplast on both sister spindles or by development of a four-polar phragmoplast structure. The microtubules align not only between the newly formed nuclei but also between the poles of two sister spindles. Sometimes the four-polar phragmoplast is followed by formation of a plate-like microspore tetrad.

The four-polar spindle and phragmoplast develop during meiosis in green algae and some mosses which have only one plastid in the cells. The meiocyte also contains only one plastid. The spindle is formed very early in prophase I as a four-polar structure. At the same time, the chloroplast is moved to the nuclear side and divides twice to produce four daughter plastids. These enter the spindle and are pushed to the poles of the spindle, one to each pole. Then the microtubules assemble around the plastids and fix them in the same position until the end of meiosis. After segregation of the plastids, the caryokineses of meiosis I and II proceed immediately one after the other and four haploid nuclei are formed. The four nuclei are distributed to the spindle poles near the plastids. Simultaneous cytokinesis of the cytoplasm occurs. In the process of meiosis all types of cytoskeleton elements are engaged. Microfilaments actively generate movement first of the plastids, later of the chromosomes, while the microtubules direct them to the four poles. The phragmoplast develops simultaneously between every pole of the spindle, and the new cell wall separates four haploid spores, each containing one plastid.

EFFECTS OF MEIOSIS

Effect of Meiotic Recombination on Offspring

The most important effect of meiosis is genetic recombination and chromosome number reduction. The latter is necessary to keep the balance in the genome in subsequent generations. Recombination is one of the most powerful factors of evolution. Recombination of genes occurs mainly during crossing over but, additionally, in anaphases I and II some genetic rearrangement is evoked by random distribution of chromosomes to the spindle poles. The mixed set of chromosomes in the spore repeats no set from the parental gametes (Fig. 9.3), but the contents of genetic information has to be complete for coding all processes during the plant life cycle. Meiosis as a process is regulated to fullfill this condition, but the effect of genetic recombination can be increased or diminished by changes in gene expression caused by the new arrangement. The meiotic rearrangement of genes: (1) alters their position in the chromosome and (2) sometimes doubles or removes some genes from the haploid set. Both changes have an influence on expression or cessation of genetic information and should influence the phenotypes of the next generation. However, it is not so evident as the population of offspring is very similar to the parents. Some individual differences in morphology are rather slight and generally, the species maintains its form and developmental patterns for a very long time.

This can be explained by the effect of consequences of the alternation of generations. The genetic variability of the postmeiotic spores appears in the gametophytes, which are selected before fertilization. Every lethal gene in the haploid genome leads to gametophyte death. Only vital and properly developed gametophytes take part in sexual mating. They pass their genotypes to the zygote by fertilization. Sometimes the combination of genes in the diploid genotype of the zygote can be lethal and then the process of postzygotic selection removes such unviable sporophytes at a very early stage of development. This selection can be easily observed in nature, e.g. apple or plum drop in May and June, when the fruits have just started to develop. The fruits are removed from the

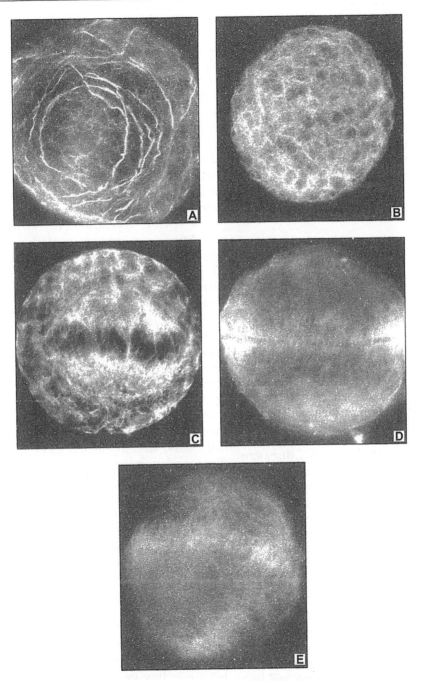

Fig. 9.10 Cytoskeleton of F-actin during sporogenesis in *Psilotum nudum*. A – archesporial cell before meio-sis; B – early prophase I; C – metaphase I; D – telophase I; E – telophase II. Staining of phalloidine/ rhodamine B, fluorescence microscope (Photos J. Bednara and D. Tchórzewska, Maria Curie-Sklodowska University, Lublin, Poland; 800 X).

plant due to lack of developing seeds inside—the zygotes or early embryos are aborted shortly after fertilization.

Most unfavorable arrangements in genetic information are blurred by fertilization, which brings balance to the diploid genotype of the sporophyte. If the zygote contains a balanced genotype, every gene has at least two copies originating from a different parent. If one of the genes appears to be lethal, its expression can be silenced and complemented by the expression of its allele.

Paradox of Sexuality

To reproduce sexually, a plant has to develop special organs such as sporangia or flowers, which is an effort demanding energy and materials. The organs are useless after the period of reproduction, so an individual organism's expenditure on its development seems very high. The special cells in sexual organs differentiate and undergo meiosis—a complicated and energy-consuming process. Full replication of chromosomes is associated with risk of mutations. A similar risk threatens anaphase I and II when disturbance of disjunction can cause chromosomal mutations. However, even if replication and disjunction occur in a proper way, with no disturbances, the short-term genetic profit of meiosis is doubtful. The rearrangement of genes during meiosis is also energy consuming and not always profitable. The new gene arrangement very often appears worse than the order present in the parental plant—a well-adapted organism with its genes organized in the best pattern. Changing this pattern brings some troubles to the next generation. The well-known problem of viability of gametes and embryos (pre- and postzygotic selection) is a case in point. Additionally, in plants the microspores and pollen grains are under

strong stress during the period of spreading, when the tiny organisms are exposed to drought, cold or excess humidity, to say nothing of the many occasions when they can be eaten by animals, attacked by fungi or bacteria. The toll of death in the haplophase is really very high. A phenomenon called competition in the haplophase describes the chances of the most vital partners participating in fertilization. Plants usually produce many more pollen grains than needed for fertilization of every developed ovule, because only with this condition can sexual reproduction be efficient. The only profit from sexual reproduction is that among the progeny created in a sexual manner a few individuals appear better adapted for unexpected stress than their parents could have been. This ensures that some individuals will be resistant to stress and able to reproduce and prolong the species' future existence. This chance for better adaptation of some plants in the next generation is worth all the efforts of parental organisms as well as the death toll in haplophase, postzygotic developmental disorders, developmental risks of mutations, and worse adaptation of many plants among the progeny. Sexual reproduction is very costly for the parents but assures survival of the offspring and development of the species in future.

Lack of genetic rearrangement can be considered a result of evolution which decreases the adaptative potential and in the long term is very damaging to plant species survival. Plants can overcome this danger of decreased adaptability by different additional mechanisms introduced in the life cycle. For plants displaying an *Oenothera*-like pattern of meiosis, such a mechanism consists of lack of incompatibility or very weak reactions after self- and cross-pollination. The two types of pollination lead to fertilization of numerous ovules and even postzygotic mortality is low. Most of *Oenothera* taxa are very fertile after self-pollination and after crossing with many different taxa. In many permanent heterozygote species, the seeds are larger and plants show heterosis in their phenotype. These plants are a good example of reproductive success even though meiosis does not rearrange genes.

In nature plants reproducing by apomixis sustain their adaptive potential because they can also set seeds by a regular sexual process. In many species, apomictic seeds represent only 70-80% of the crop. The rest of the seeds are developed in a sexual way after typical double fertilization proceeded by meiosis in the male and female lines. Thus the apomictic plant is not completely devoid of profits ensured by meiosis.

PLANTS WHICH AVOID GENE REARRANGEMENT IN THEIR LIFE CYCLE

Meiosis is a process of two consecutive divisions occurring one after the other, common to all sexually reproducing organisms so that the same genetic material rearrangement can be found in plants and in animals. The difference is that in animals meiosis occurs in a special type of cell, the germ cells destined to produce gametes. In plants, somatic cells which occupy a particular position in the organism change their physiology and become capable of meiotic division and producing haploid spores as a result. This difference between plants and animals does not affect the genetic consequences of meiosis, i.e., the gene rearrangement.

In angiosperm plants, development of generative organs occurs as the meristem changes its destination in order to become a flower bud. The process starts in the genera-

tive phase of the plant life cycle. In many plants, this physiological maturation is stimulated by changes in the environment and signals from the organism itself.

The reactions are associated with special gene expression. Genes determining the generative functions of the plant are presently under intense investigation. Research on the genes determining flower development and the differentiation of meiotic cells is based on mutants and transgenic plants, mainly *Arabidopsis*, *Petunia*, and *Zea mays*. A group of genes called MADs is responsible for the initiation of every organ and also for the sequence of developmental events, e.g. anther and ovule formation and meiosis. The particular genes have been separated, sequenced and recombined in genetic experiments. A model of effective MAD gene expression has been described and proven by evoking disorder in the initiation or inhibition of meiosis in the generative organs of mutant plants. Proper development of the flower, with functional anthers and ovules, is an important factor influencing plant fertility, thereby underscoring knowledge of MAD gene expression as necessary for crop improvement. However, there are some plants which have developed different mechanisms of sexual reproduction. Though meiosis is strongly modified or even inhibited in those plants, they remain fertile and produce viable seed. They are also an object of interest to geneticists and breeders as potentially helpful in crop improvement.

Meiosis without Crossing Over

Another strategy diminishing the effect of gene rearrangement depends on a special mechanism avoiding random crossing over and segregation of chromosomes to the poles of the karyokinetic spindle. After condensation, during premetaphase I chromosomes become arranged in a ring. In the ring, the chromosomes join each other only at the end tips instead of forming bivalents. In metaphase I, the ring is located in the equatorial plate on the karyokinetic spindle. In anaphase I, the chromosomes are segregated to the poles of the spindle in accordance with their maternal or paternal origin. As a consequence of such segregation, the maternal set is on one pole and the paternal on the other pole of the spindle in telophase I, and after cytokinesis every cell in the dyad contains a different set of chromosomes repeating the set obtained from one of the parental gametes. This type of meiosis is named *Oenothera*-like, because it was found in heterozygotic *Oenothera* taxa, e.g. *Oenothera biennis* L. Rings of chromosomes are also formed in some species of Comellinaceae. Oenotheraceae and Comellinaceae represent families which display evolutionary changes. In these families, new genotypes arise by mutations and sexual hybridization. The mutations consist in a translocation of chromosome segments inside the genome. Some chromosomes mutually exchange their segments, others lose fragments or accept fragments from other chromosomes. All these processes result in decreasing homology in the diploid genome, until the stage when every chromosome has an individual sequence and cannot form a pair with another. Bivalents cannot be formed and during meiosis chromosomes remain single, ready to form a ring. It is possible because at the chromosome ends similar sequences of DNA are repeated, so they are homologs and stick together. In some plants, mutations change chromosomes in such a way that they have parts homologous to one chromosome and parts homologous to another. During meiosis I such chromosomes conjunct in very variable assemblies resembling crossings and loops connected by bridges. In some species, the rings consist of two chromosomes joined at the ends but unable to conjugate along the arms.

Translocation of chromosomal segments relatively quickly causes loss of homology in chromosomal sets, and in the diploid state it is not possible to adjust the chromosomes from the maternal and paternal sets. This process may be more or less advanced in various species in the family Oenotheraceae; careful observations by eminent caryologists have led to the conclusion that all taxa on the continent of Eurasia originate from the taxon brought to Europe from North America in the fifteenth century (after Christopher Columbus), which is not long ago in terms of evolution. The rate of mutations produced a row of stable taxa with a different degree of lack of homology between the parental sets of chromosomes. In some species, considered more advanced, all chromosomes are arranged in a ring during meiosis, e.g. *Oenothera biennis* L. In less advanced species some chromosomes are still unchanged by translocation and are able to form some bivalents while the rest of the chromosomes form a ring, e.g. *Oe. lamarckiana* or *Oe. brevistylis*.

In the Oenotheraceae family, the most advanced species is *Oenothera biennis* L. and the most primitive (wild type) is *Oe. hookeri* L. In the latter, normal bivalents are formed and crossing over takes place during meiosis. The rearrangement of genes causes no noticeable changes in the phenotype of the offspring. In traditional genetics, *Oe. hookeri* used to be called a homozygote: its genome consists of two very similar sets of chromosomes and the plant produces male and female gametes of the same haploid genome called [h]hookeri. According to contemporary knowledge on the genetic constitution of *Oenothera* however, the genotype of the plant cannot be considered a complete homozygote nor are the forms named species. The term "taxon" for the form of a stable genotype and phenotype is suggested as a more cautious description of plants which have very specific and complicated sexual reproduction. In spite of the lack of chromosome homology in *Oenothera*, meiosis is an efficient process and both homo- and heterozygotic taxa produce a good amount of fertile pollen grains and embryo sacs. The plants are very fertile and can cross in different combinations with no decrease in fertility. To sustain genetic and phenotypic stability, *Oenothera* taxa develop specific mechanisms: (1) special turn of meiosis, competition and selection in the haplophase, and (2) postzygotic elimination of homozygotes in self-pollinating plants.

Similar modifications during meiosis with a ring of chromosomes and segregation of parental sets were observed in an animal, the cockroach *Periplaneta americana*. These insects live in mines where the populations are strictly separated. They are very well adapted to the conditions and have developed their own system of sexual reproduction, avoiding the rearrangement of genes. Evolution of isolated population was also very quick because of chromosome segments translocations. In isolated populations, all changes of gene expression caused by duplication of chromosomal segments or by the effect of the position in the genome overlap as a consequence of crossing between a limited number of individuals. The overlapping genetic effect has led to speciation of new taxa.

Apomixis—Significance in Crop Breeding Programs

Some plants are very well adapted to their environment. Their sets of genes are arranged in the best possible way for expression, ensuring successful development and good fertility. Such plants often live in small isolated niches with very stable environmental and climatic conditions (e.g. endemic plants in the mountains or in deserts) but can live also in big populations displaying quick evolutionary events such as speciation and divergence. Some plants have developed their own system of reproduction called apomixis. There are many different types of apomixis but in none of them is the embryo a product of fertilization; instead it originates from the diploid cell of the maternal plant. Apomixis yields the same genetic effect as vegetative propagation or cloning. In every generation the plants have a similar genome and phenotype. But apomixis is not simple vegetative reproduction because the generative organs are developed on the plant and the final results of the flowering process are fruits and seeds.

There are two main patterns of apomixis, which omit meiosis: apospory or apogamy. Apospory occurs when meiosis is inhibited and the female gametophyte develops directly from one of the somatic cells in the ovule. The cell is diploid and the embryo sac develops due to mitotic divisions, so every cell of it is also diploid. In this embryo sac, the egg cell and the central cell are formed as gametes which differentiate and look like those in a haploid gametophyte. In such an embryo sac, the egg cell develops into a diploid embryo without fertilization, but pollination of the flower and very often fertilization of the central cell are necessary to stimulate embryo development. The process in which the central cell is fertilized and the egg cell is not, is called semigamy. An aposporic plant without meiosis and fertilization can give normal diploid progeny.

In apogamy, the embryo sac can develop after meiosis or without it, but the egg cell is inhibited in development. The embryo develops as an adventitious one from a somatic cell located in the vicinity of the embryo sac. Very often, but not always, pollination and sometimes fertilization of the central cell are needed as a stimulus to start embryo development. In this pattern of reproduction, a gamete is not able to develop into an embryo, which is signified by the term apogamy. There are many different modes of aposporic and apogamic embryo development. There are particular differences in the moment of meiosis inhibition or embryo origin. In many theoretical dissertations the modes of apomictic development are described in detail but differ in terminology. It is beyond the purpose of this article to discuss all those details. What is important is the fact that meiosis can be completely inhibited or that the products of meiosis may not enter the process of embryo development. In both patterns of apomixis, the plant develops generative organs but the effect of genetic rearrangement during reproduction is avoided. In every type of apomixis, the new daughter plant has the same genotype and the same possibility of adaptation as the maternal plant.

In nature, apomixis is exceptional and profitable only in an unchanging environment. However, in agriculture where cultivars of crop plants are very often hybrids or mutants whose sexual reproduction can be handicapped giving very variable progeny, apomixis is a very desired feature. Meiosis in hybrid plants leads to a segregation of genes in gametes and fertilization creates new genetic combinations. In the phenotype of such recombinants (plants with new gene combinations) the desired features are very often lost. This effect of profligation is well known to every gardener and farmer. The solution to all these problems would be obtaining matromorphic progeny with a stabilized genotype and desired phenotype of the crop plant. Such a profitable mode of reproduction is characteristic of apomictic plants only, so there are international research projects concerning apomixis. The idea is to find the genetic factor switching the development in generative organs from the sexual mode to apomixis. Transgenic crop plants could be produced by introducing the gene of apomixis to the matromorphic seed set in order to change the mode of reproduction. Research has already brought interesting results, but there is still a long way to full success. Some genes coding apomixis have been discovered. They are expressed during the state of microsporocyte formation or a little later during meiosis. The effect of the expression of these genes consists in changing the ovule development and inhibition of the embryo sac formation. Transgenic plants of *Arabidopsis* show that megasporocyte formation is a process separate from meiosis and that the meiocyte remains diploid under the influence of a special protein, or needs some other protein to enter meiosis. This knowledge opens new vistas in the field of introducing apomixis, especially of the apogametic type, into many more crops.

Recommended Reading

Alberts, B. *et al.* 1994. *Molecular Biology of the Cell.* Garland publ. Inc., New York, NY, pp. 911-943 and 1011-1021 (3rd ed.).

Asker, S.E., Jerling, L. 1992. *Apomixis in Plants.* CRC Press, Boca Raton, FL, USA.

Bednara, J., Śnieżko, R., and Szczuka, E. 2000. Embryological researches in the Department of Plant Anatomy and Cytology, M.Curie-Sklodowska University in Lublin. In: *Botanical Guidebooks. Plant Embryology: Past, Present, Future.* Polish Acad. Sci., W. Szafer Institute Botany, Cracow, Poland, 24: 75-85.

Bednara, J., Gielwanowska, I., and Rodkiewicz, B. 1986. Regular arrangements of mitochondria and plastids during sporogenesis in *Equisetum.* Protoplasma 130: 145-152.

Bernard, J. 1990. *Meiosis.* Cambridge Univ. Press, Cambridge, England.

Bhatt, A.M., Canales, C., and Dickinson, H.G. 2001. Plant meiosis: the means to 1N. Trends in Plant Science vol. 6 (3): 114-121.

Brown, R.C. and Lemmon, B.E. 2001. Sporogenesis in eusporangiate ferns: I. Monoplastidic meiosis in Angiopteris (Marattiales). J. Plant Res. 114(1115): 223-235.

Brown, R.C. and Lemmor, B.E. 2001. Sporogenesis in eusporangiate ferns: II. Polyplastidic meiosis in Ophioglossum (Ophioglossales). J. Plant Res. 114(1115): 227-246.

Brown, R.C. and Lemmon, B.E. 2001. The cytoskeleton and spatial control of cytokinesis in the plant life cycle. Protoplasma 215 (1-4): 29-35.

Campbell, N.A., Reece, J.B., and Mitchell L.G., 1999. *Biology.* Addison Wesley Longman, Inc., England, pp. 28-30 (5th ed.).

Crane, C. 2001. Classification of the apomictic mechanism. In: The Flowering of Apomixis. From Mechanisms to Genetic Engineering, pp. 24-43. Savidan, Y., Carman, J.G., and Dresselhaus, T. (eds.). CIMMYT, IRD, European Commision DG VI, Mexico, D.F., Mexico.

Czapik, R. 1994. How to detect apomixis in angiospermae? Polish Bot. Studies 8: 13-21.

Dickinson, H.G. and Heslop-Harrison, J., 1977. Ribosomes and organelles during meiosis in Angiosperms. Phil. Trans. Roy. Soc. London 277: 327-342.

Dittmer, H.J. 1964. *Phylogeny and Form in the Plant Kingdom.* D.van Nostrand Co., Inc. Princeton, NJ.

Furness, C.A., Rudall, P.J., and Sampson, F.B. 2002. Evolution of microsporogenesis in angiosperms. Intl. J. Plant Sci. 163(2): 235-260.

Gonzalez, F., Rudall, P.J., and Furness, C.A. 2001. Microsporogenesis and systematics of Aristolochiaceae. Bot. J. Linn. Soc. 137(3): 221-242.

Grossinklaus, U., Nogler, G.A., and van Dijk, P. 2001. How to avoid sex: The genetic control of gametophytic apomixis. Plant Cell 13: 1491-1498.

Harte, C. 1994. *Oenothera. Contributions of a Plant to Biology.* Springer-Verlag, Berlin. pp. 177-200.

Johri, B.M. 1984. *Embryology of Angiosperms.* Springer-Verlag, Berlin, chaps. 4, 5, and 10.

Koltunow, A.M. and Tucker, M.R. 2003. Advances in apomictic research: can we fix heterosis? In: *Plant Biotechnology and Beyond.* Proc. 10th IAPTC&B Congress 2002, pp. 39-46. Vasil, I.K. (ed.). Orlando FL, USA. Kluwer Acad. Publ., Dordrecht, The Netherlands.

McLean, R.C. and Ivery-Cook, W.R. 1961. *Textbook of Theoretical Botany.* Longmans, New York, NY, vol. II (2nd ed.).

Mercier, R., Grelon, M., Vezon, D., Horlow, Ch., and Pelletier, G. 2001. How to characterize meiotic functions in plant? Biochemie 83: 1023-1028.

Moore, G. 2002. Meiosis in allopolyploids—the importance of "Teflon" chromosomes. Review. Trends in Genetics 18(9): 456-463.

Nogler, G.A. 1994. Genetics of gametophytic apomixis—a historical sketch. Polish Bot. Studies 8: 5-11.

Ottaviano, E., Mulcahy, D.L., Sari Gorla, M., and Bergamini-Mulcahy, G.B. 1992. *Angiosperm Pollen and Ovules.* Springer-Verlag, Berlin.

Raghavan, V. 2000. *Developmental Biology of Flowering Plants.* Springer-Verlag, New York, NY.

Rodkiewicz B., Sniezko, R., Fyk, B., Nieweglowska, B., and Tchorzewska, D. 1996. Embriologia Angiospermae rozwojowa i eksperymentalna [Developmental and Experimental Embryology of Angiosperms]. Wydawnictwo UMCS, Lublin: 34-85, 209-230 (in Polish).

Rodkiewicz, B. 1970. Callose in cell walls during megasporogenesis in angiosperms. Planta 93: 39-47.

Rodkiewicz, B., Duda. E., and Kudlicka, K. 1988. Organelle aggregations during microsporogenesis in Stangeria, Nymphaea and Malva. In: *Sexual Reproduction in Higher Plants.* Cresti, M., Gori, P., and Pacini, E. Springer-Verlag, Berlin.

Scott, R.J. and Stead, A.D. 1994. *Molecular and cellular aspects of plant reproduction.* Cambridge Univ. Press, Cambridge, England.

Spillane, C., Steimer, A., and Grossniklaus, U. 2001. Apomixis in agriculture: the quest for clonal seeds. Sexual Plant Reprod. 14: 179-187.

Summer, A.T. 2003. Chromosomes: organization and function. In: *Chromosome Botany.* Sharma,

A. and Sen, S. (eds.). Blackwell, Malden, M.A. Science Publ., Enfield, NH, USA.

Tung, S.H., Ye, X.L., Zee, S.Y., and Yeung, E.C. 2000. The microtubular cytoskeleton during megasporogenesis in the Nun orchid, *Phaius tankervilliae*. New Phytol. 146(3): 503-513.

Śnieżko, R. 2000. Structural and functional polarization in the ovules of flowering plants. In: *Botanical Guidebooks Plant Embryology: Past, Present, Future.* Polish Acad. Sci., W. Szafer Institute Botany, Cracow, Poland, 24: 87-107.

Śnieżko, R. and Harte C. 1984. Polarity and competition between megaspores in the ovules of Oenothera hybrids. Plant Systematics and Evolution 144: 83-97.

CHAPTER 10

Mendelian Genetics

G.S. Miglani

Genetics is the science of *heredity* and *variation*. Basteson coined the term genetics in 1905. Heredity is the cause of similarities between individuals. For this reason, brothers and sisters with the same parents resemble each other. Variation is the cause of the differences between individuals. This is the reason that brothers and sisters who do resemble each other are still unique individuals. Precisely, genetics is a branch of biology that deals with the mechanisms responsible for transmission of biological properties from one generation to the next. Thus, genetics deals with understanding the nature, molecular structure, organization, biological function, regulation, and manipulation of hereditary particles called genes, which play a central role in life processes. It also covers the development of ways and means to use the knowledge of genetics for the welfare of mankind.

There are certain well-known facts about life: (a) Life comes only from preexisting life; (b) Life exists in diverse forms: bacteria, fungi, plants, and animals; (c) Like begets like, e.g. dogs produce pups and cats produce kittens; (d) Traits, i.e. biological characteristics, show recombination. We generally say that the child has a nose like his father and eyes like his mother; (e) Variation exists in living forms. Members of a species in a natural population differ from one another; (f) Living things have individuality. Biological individuality has pattern, limitations, and characteristics that depend at least in part on the parents of an individual; (g) Similarities and differences are heritable.

The basis of genetics is variation and without variation, there would have been no science of genetics. The principles of genetics apply only to sexually reproducing organisms having biparental parentage, microorganisms, plants and animals, including man. The hereditary transmission of biological properties is obviously an aspect of *sexual reproduction*. Genes play a central role in different processes of life.

MECHANICS OF INHERITANCE

One important aspect in genetics is the study of the manner in which characters are transmitted from one generation to the next. This branch of genetics is termed *transmission, classical* or *Mendelian genetics*. Early ideas about transmission of traits included *"preformation"*. The main feature of this idea was a miniature man called *"homunculus"* by Swammerdam in 1694 (cited from Strickberger, 1986). The theory of encapsulation believed that a female included all hereditary information for all her immediate and remote progeny.

Darwin (1868) believed in the theory of *"pangenesis"*, i.e. very small exact but invisible copies of each body organ and component, called "gemmules", were transported by the blood stream to the sex organs and these were assembled into gametes. Upon fertilization, gemmules of the opposite sex mixed. During development, these gemmules then separated to different parts of the body. Thus, the individuals constituted a mixture of maternal and paternal organs. The pangenesis hypothesis was rejected on the basis of an experiment carried out by Galton (1909). He transfused the blood between white and black rabbits but obtained no rabbits with intermediate skin color in their progeny.

Lamarck (1809) tried to explain the extraordinary ability of hereditary factors to respond directly to the environment. He proposed the theory of *inheritance of acquired characteristics*. He assumed that each hereditary particle had a spiritual conscious-like property that could absorb and interpret messages from outside. He believed that variations in organisms are induced in response to an urgent need and regarded these variations as heritable. This theory could not be proven experimentally.

Heredity was supposed to be transmitted as though it were a miscible fluid like blood in animals and man. The "bloods" of the parents were assumed to mix, fuse or blend in their offspring. Adult structures of plants and animals develop from uniform embryonic tissues. Thus, development of an individual is epigenetic and genes act in *epigenetic systems*. The process of development

consists of more than growth. The sex cells are more or less drops of structureless liquid containing nothing whatsoever resembling the body that it is to develop from them. Development of an individual does not stop at some arbitrary stage in life. An individual that has stopped developing is considered dead.

Weismann (1885, 1987) established through his famous experiment on mice that *"pangenesis"* and *"inheritance of acquired traits"* could not be verified. According to his *germplasm theory*, multicellular organisms give rise to two types of tissues—*somatoplasm* and *germplasm*. Somatoplasm consists of tissues essential for functioning of the organism but lacking the property of entering into sexual reproduction. Changes that occur in somatic cells are not passed on from parents to offspring and are thus not heritable. Germplasm is set aside for reproductive purposes as the cells of these tissues divide meiotically to form gametes. Any change in the germplasm is transmitted from parents to offspring and could lead to altered inheritance.

Continuity of Life

There is a continuity of germplasm between all descendant generations. Continuity of life results from its inherent characteristic of reproduction. Reproduction is the basic necessity of a species to continue its existence on this planet. Continuity of life is perpetuation of a species to repeat its genetic possession in every subsequent generation. A species stores biological information in *DNA*. Each cell of a given species has a constant amount of DNA that is doubled by replication and cell division ensures equal distribution of parent cell DNA to daughter cells. Gametes have half the amount of DNA while the zygote possesses the amount of DNA characteristic of the species. *Replication of DNA* is the molecular basis of reproduction and hence continuity of life. Replication faithfully maintains genetic information intact over generations. Continuity of germplasm between all the descendant generations of a species explains many biological similarities that are inherited.

Qualitative and Quantitative Traits

Traits are some biological attributes of individuals within a species for which various heritable differences can be defined. Biological property, characteristics or simply characters are used synonymously for trait. Traits may be measured in terms of color, shape, pattern, molecular structure, or statistical parameters. A *qualitative trait* is one for which a fewer discrete phenotypes can be distinguished by visual observation. A fewer number of genes determine such traits; data can be assorted into a few discrete classes and traits are less affected by environment.

A *quantitative trait* is one for which a range of phenotypes differing by degree exists. Data cannot be separated into a few discrete classes. Various statistical parameters, viz. mean, mode, median, range, variance, standard deviation, coefficient of variation, are used to measure quantitative traits. A large number of genes with small additive effects are generally involved in determination of such traits. Expression of these traits is affected by the environment. Qualitative traits have thus been found more useful for understanding the mechanics of inheritance.

Genotype and Phenotype

Johannsen (1911) evolved the concepts of *genotype* and *phenotype* (cited from Johannsen, 1926). Genotype is the sum total of heredity, i.e. the genetic constitution that an organism receives from its parents. Phenotype is the appearance of the organism, i.e. the sum total of all its characteristics such as color, form, size, behavior, chemical composition, and structure. The phenotype is the external manifestation of a genotype of an organism and is the result of the structure of a genotype in its particular environment. Both are absolutely necessary. The equation phenotype = genotype × environment describes the relationship between genotype and environment. The genotype is relatively stable throughout the life of an individual whereas the phenotype changes continuously during development.

Interplay of the effects of genetic and nongenetic factors on development is called the genotype × environment interaction. Genes determine the potentials and the environment determines whether that potential is to be reached or not.

Genotypic and Environmental Variation

When two or more individuals develop in the same environment and come to possess different phenotypes, the individuals will have different genotypes. This type of variation is known as *genotypic variation*. When individuals with the same genotype develop in different environments, their phenotypes may be quite different, e.g. identical twins raised in different environments. This type of variation is known as *environmental variation*. The variation observed in different individuals in a natural population at the phenotypic level is known as *phenotypic variation*, i.e. the sum total of genotypic variation and environmental variation and their interaction. Mathematically, $V_P = V_G + V_E + V_{GE}$, where V_P = phenotypic variance, V_G = genotypic variance, V_E = environmental variance and

V_{GE} = variance due to interaction between genotypic and environmental factors. Genes, the units of heredity, govern biological properties of an individual. Genotypic variations arise due to *mutations* in genes.

Brief History of Genetics

Genetics began in 1866 when Gregor John Mendel carefully analyzed the mechanism of inheritance. His simple experiments brought forth significant principles that determine how traits are passed from one generation to the next. His hybridization experiments set a stage upon which the mechanism of inheritance could be understood and rules could clearly be formed to detect genes though without knowing their nature or products.

Mendel's experiments were not appreciated for 34 years because nobody believed that genes were discrete objects. In 1900, Mendelism was rediscovered and the birth of genetics took place. In the mid-1900s when biochemistry became highly developed, geneticists began to think about the biochemical nature of genes and the causes of genetic variation. Observations made between the 1920s and 1940s pointed to DNA as the genetic material in almost all organisms, including most viruses. Critical analysis by Watson and Crick (1953a,b) gave the molecular structure of DNA and with this genetics entered the DNA age. This was a very exciting phase in the development of genetics. By the end of the 1960s, it was known how a gene is copied, how a gene is expressed, how a mutation arises, and how genes are turned on and off according to the needs of the cell or organism. It became possible to identify the products of thousands of genes. These developments constituted part of the important branch of genetics called molecular genetics. In the 1970s, genetics underwent its most recent revolution, i.e.

development of *recombinant DNA technology*. This technique, which forms the backbone of genetic engineering, involves procedures to identify, sequence, isolate, clone, and transfer genes from one organism to another bypassing all reproductive barriers at the will of the molecular geneticist. This development had an enormous effect on genetic research, particularly in understanding gene structure, gene expression, and its regulation in plants and animals. The principles of genetics are used not only in basic studies, but also applied in the fields of agriculture, animal husbandry, forestry, medicine, law, gene therapy, and genetic counseling.

Genetics is a *science of potentials*. Human interventions in genetic mechanisms have used these potentials to improve human welfare. It is the genes and not the traits that are inherited. No trait is simple enough to be governed by a single gene. When we talk about the inheritance of a trait, we refer in fact to inheritance of difference(s) existing between two parental lines for a particular trait. A trait is the result of complex interactions involving many genes.

Genetic Analysis

Genetics deals with similarities and differences. To understand how certain differences are inherited, *genetic analysis* in conducted. Genetic analysis explores whether a particular contrasting trait is governed by one or more gene differences and whether the gene in question is nuclear or cytoplasmic. In animals, whether a gene is sex-linked or autosomal is determined. Genetic analysis also includes assignment of a gene to a particular chromosome and ascertaining the specific position of the gene in a chromosome in relation to the other known genes in the nuclear or extranuclear genome. Genetic analysis in certain cases may even be extended to determine the nucleotide sequence of a gene. Several prokaryotes and haploid and diploid eukaryotes are very useful organisms in genetic analysis. The organism used in genetic experiments should fulfill certain conditions: 1) have a number of well-defined contrasting differences, 2) be sexually reproducing and have biparental parentage, 3) undergo inherent recombination, 4) controlled matings should be possible, 5) have a short life cycle, 6) yield a large number of offspring, 7) be convenient in handling, and 8) its culturing should be relatively inexpensive.

MENDELIAN PRINCIPLES

A satisfactory answer to the *mechanics of inheritance* came from the work of an Austrian monk Gregor Johann Mendel (Father of Genetics). He provided the first proof for a theory that explained heredity by transmission of units in the reproductive cells. Thus, he put an end to false notions such as those concerning bloods. Two principles deduced from his work are known as the *principle of "segregation"* and the *principle of "independent assortment"* (now preferably called *"independent segregation"*). For a brief biography of Mendel, readers may see Iltis (1932).

Mendel's Working Method

Mendel avoided complexities that had troubled earlier workers by studying each trait and record of the progeny and parents individually. He focused attention on one trait at a time, e.g. flower color. When the behavior of each single trait was established, he then studied two traits simultaneously, such as flower color and vine length (height). He established *pure lines* of plants for the characters he studied. After that he crossed two plants artificially different in a pair of contrasting traits and observed the appearance of the first hybrid, or "F_1", generation. He then crossed the hybrid plants and raised a second filial generation, "F_2". In the F_2 he counted the number of plants possessing each of the contrasting traits in which the parents differed. From his observations in the F_1 and F_2 generations, he formulated his hypothesis. He tested his hypothesis experimentally by selfing all the F_2 plants to obtain third filial generation, "F_3". He studied seven traits. The contrasting pairs of phenotypes are given in Fig. 10.1. He studied these traits singly through *monohybrid crosses* and two traits together through *dihybrid crosses*.

Symbols

The English alphabet was used by Mendel to represent the factors. An upper case letter, e.g. "A" signified a dominant member and a lower case letter "a" the recessive member. Since a large number of genes are now available, one or more letters of the alphabet are selected to indicate the character they are to represent. The alphabetical letter is chosen on the basis of the *mutant character*, the one which deviates from the *wild type character* (the one most predominant in nature). For instance, the garden pea is normally tall, but the dwarf (affecting stem length) is considered mutant and can be represented by lower-case letter "d" (lower-case italics indicates *recessive* nature). The tall will automatically receive the designation "D" (upper case letters in italics show *dominance*). Each individual receives two doses of factors (genes) for each character, one from the father, the other from the mother. A pair of genes thus represents each character. A true breeding tall plant will be represented by "*DD*" and a true breeding dwarf plant will be represented by "*dd*". The F_1 *heterozygote* will be represented as "*Dd*".

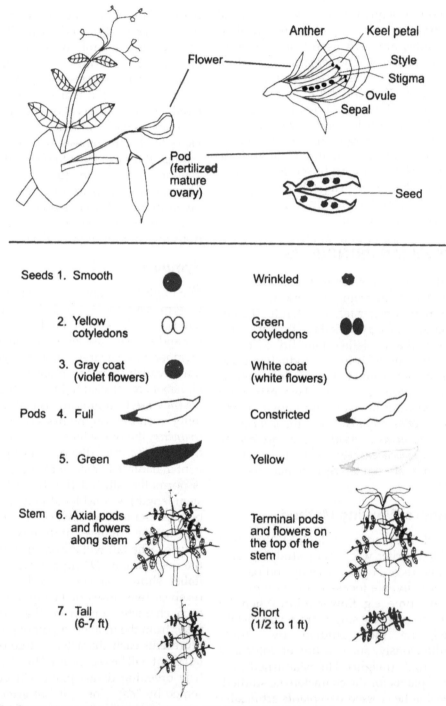

Fig. 10.1 Seven characters in peas studied by Mendel (from Strickberger, 1968)

It is not necessary that the alphabet used should always represent the recessive character since in some cases the mutant may be dominant and the wild type recessive. For example, bar eye, a mutant in *Drosophila melanogaster* is *semidominant* and represented by "B", while the normal wild type eye, which is recessive, is represented as "b".

Another method used is *symbols* to signify the wild type by a plus (+) sign and the mutant by upper-case or lower-case italic letter depending on whether the mutant is dominant or recessive. In the above case, for example, tall would be represented as "++", dwarf as "dd" and F_1 as "$+/d$". The genotypes of gametes are usually written in a circle.

An improvement over the aforesaid method modifies the "+" sign to indicate the character for which it is meant. For instance, in the above case, "$+^d$" or "d^+" might be used to indicate tall and "d" for dwarf. The two methods of assigning symbols to genes are summarized in Table 10.1.

Before attempting a numerical problem wherein one is expected to predict genotypes or phenotypes of an individual, the symbols must be defined.

Monohybrid Crosses

The results of Mendel's experiments involving seven monohybrid crosses in garden pea (*Pisum sativum*) up to the F_2 generation

are given in Table 10.2. The following generalizations were drawn from the results of *monohybrid crosses*: (1) When a pea plant showing one phenotype is crossed with a plant having another contrasting phenotype for the same trait, the F_1 generation consisted of seeds having the same phenotype as one of the parents. The phenotype that appears in the F_1 generation is termed *dominant* while the other that does not appear is termed *recessive*. The F_1 individual is called a *hybrid*; (2) *Reciprocal crosses* between parents produced the same results; (3) When plants of F_1 were allowed to undergo *self-pollination*, the recessive trait, which had remained unexpressed in the F_1 generation, reappeared in the F_2 generation. The F_2 progeny consisted of dominant and recessive phenotypes in the ratio 3:1. This is known as a *monohybrid phenotypic ratio*; (4) When plants of the F_2 generation were self-pollinated, the individuals with recessive phenotypes would always breed true. However, only one-third among the dominants were found to be pure and the remaining two-thirds repeated the ratio of 3 dominant:1 recessive in the F_3 generation.

The gene hypothesis that explains the results of the monohybrid crosses is as follows: Mendel called the determining agents responsible for each trait "*factors*". In a diploid organism, there are two factors for a trait. Each factor is present on each of the two homologous chromosomes at the same

TABLE 10.1 Summary of two systems of assigning symbols to genes

System for symbols	Recessive mutant, a		Dominant mutant, A	
	Symbol for wild type allele	Symbol for mutant allele	Symbol for wild type allele	Symbol for mutant allele
Mendelian system	A	a	A	A
Newer system	a^+	a	A^+	A
	(or $+^a$ or +)	(or a^-)	(or $+^A$ or +)	(or A^-)

Source: Gupta (1994)

TABLE 10.2 Results of Mendel's experiments involving monohybrid crosses in garden pea (*Pisum sativum*)

Contrasting traits of parents	F_1 Generation	F_2 Generation	F_2 Ratio
Tall × Dwarf stem	All Tall	787 Tall 277 Dwarf	2.84:1
Axial × Terminal flowers	All Axial	651 Axial 207 Terminal	3.14:1
Green × Yellow pods	All Green	428 Green 152 Yellow	2.82:1
Inflated × Constricted pods	All Inflated	882 Inflated 299 Constricted	2.95:1
Colored × White flowers	All Colored	705 Colored 224 White	3.15:1
Yellow × Green cotyledons	All Yellow	6022 Yellow 2001 Green	3.01:1
Round × Wrinkled seeds	All Round	5474 Round 1850 Wrinkled	2.96:1

Source: Miglani (2000)

corresponding position. The trait that appears in the F_1 is considered controlled by a dominant factor and the other which remains latent in the F_1 is considered controlled by its recessive form. Thus a factor (later termed *gene*) can exist in more than one alternate form (*alleles*). The regular appearance of a recessive trait in the F_2 was a notable contradiction to the early theory of inheritance as this observation refuted the *blending theory of inheritance*. Figure 10.2 explains the *gene hypothesis* using just one example of a cross between Tall × Dwarf. The *principle of segregation* was deduced from Mendel's work involving monohybrid crosses. Accordingly, the two members of a gene pair separate from each other during meiosis, so that one-half of the gametes carry one member and the other half of the gametes carry the other. *Nondisjunction* (failure of homologous chromosomes to separate during meiosis) is an exception to the principle of segregation (Fig. 10.3). Thus, genes are inferred by observing precise mathematical ratios in the filial generations descending from two parental individuals.

The principle of segregation was also deduced from Mendel's observations from *backcross* experiments. Backcross is defined as crossing the F_1 (or individual with dominant phenotype) with either of the homozygous parents. Backcross is used for transferring genes in conventional plant breeding. Two crosses are illustrated in Fig. 10.4. Cross I is a backcross while cross II is known as a *test cross*. Test cross is defined as crossing the F_1 (or individual with dominant phenotype) with one having a *homozygous recessive* genotype. Test cross is used to determine the genotypes of an individual having a dominant phenotype.

Dihybrid Crosses

What happens when a *dihybrid cross* is made involving two pairs of genes affecting two different characteristics? Mendel selected two varieties of garden pea, one pure for

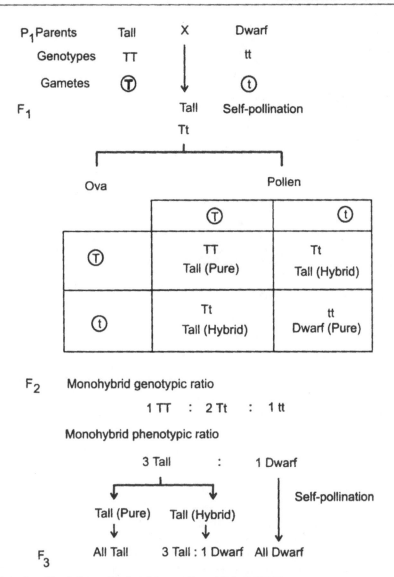

Fig. 10.2 Gene hypothesis for a monohybrid cross (from Miglani, 2000)

round seed and yellow cotyledon and other pure for wrinkled seed and green cotyledon. Since we know from monohybrid crosses that round is dominant over wrinkled and yellow is dominant over green, let the symbol Y denote yellow color and y green color, R round seed and r wrinkled seed. The dihybrid cross is shown in Fig. 10.5. Dihybrid genotypic and phenotypic ratios are given in Table 10. 3.

The *principle of independent assortment* (independent segregation) was deduced from Mendel's observations on dihybrid crosses. Accordingly, different gene pairs segregate independently during meiosis. Mathematically, the principle of indepen-

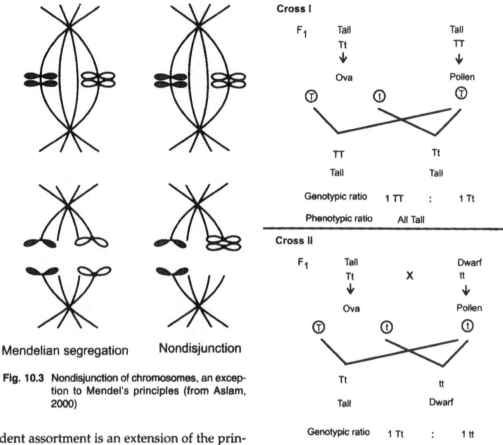

Cross I

F₁	Tall	Tall
	Tt	TT

Ova   Pollen

Ⓣ	①	Ⓣ

TT	Tt
Tall	Tall

Genotypic ratio   1 TT   :   1 Tt

Phenotypic ratio   All Tall

Cross II

F₁	Tall		Dwarf
	Tt	X	tt

Ova   Pollen

Ⓣ	①	①

Tt	tt
Tall	Dwarf

Genotypic ratio   1 Tt   :   1 tt

Phenotypic ratio   1 Tall   :   1 Dwarf

Mendelian segregation   Nondisjunction

Fig. 10.3 Nondisjunction of chromosomes, an exception to Mendel's principles (from Aslam, 2000)

Fig. 10.4 Gene hypothesis for a monohybrid backcross and test cross (from Miglani, 2000)

dent assortment is an extension of the principle of segregation which can be expressed as $(a + b)^n$. This phenomenon produces two recombinant phenotypic classes in addition to the two parental ones in the segregating filial generation (F_2) provided both the gene pairs show *simple* or *complete dominance* and the factors (genes) control different traits and are located in nonhomologous chromosomes. In other words, genes located in the same chromosome (linked) and less than 50 map units apart, are not expected to show *independent segregation*. Thus, *linkage* is an exception to independent segregation and instead of the 9:3:3:1 phenotypic ratio in a dihybrid cross a different ratio will be observed which will depend on the degree of linkage. The principle of independent

segregation can also be understood from backcrosses of dihybrids. Independent segregation is an important means for generating new genotypic and hence phenotypic combinations. The test cross involving two or more genes is used to determine the degree of linkage. Even when different gene pairs show *incomplete dominance* or *codominance*, segregation is independent and the genotypic ratio remains the same while the number of phenotypes increases.

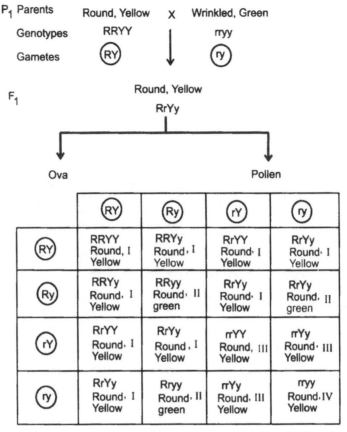

Dihybrid F$_2$ genotypic and phenotypic ratios are summarized in Table 10.3.

Fig. 10.5 Gene hypothesis for a dihybrid cross (from Miglani, 2000)

The types of gametes formed by a hybrid individual, the number of genotypic classes and the number of phenotypic classes in a segregating filial generation depend on the number of segregating gene pairs in the hybrid (Table 10.4).

Figure 10.6 illustrates a similar type of independent assortment of genes for the *dihybrid ratio* 9:3:3:1 in summer squashes. There are two different shapes of fruits, viz. disk and sphere and two different colors, viz. white and yellow.

When white disk (*WWDD*) was crossed with yellow sphere (*wwdd*), the F$_1$ hybrid was white disc (*WwDd*) which is heterozygote for both genes. When the F$_1$ was selfed or intercrossed, there were four types of gametes formed by the male and female parents, viz. *WD*, *wD*, *Wd*, and *wd*. These four gamete types met at random and produced 16 F$_2$ progeny segregating in a ratio of 9 white disk : 3 yellow disk : 3 white sphere : 1 yellow sphere.

P₁ → P_1

P_1 White, Disc   x   Yellow, Sphere
      WWDD   ↓   *wwdd*

F_1 White
             WwDd

             ↓ Selfing

F_2

Male → Female ↓	(WD)	(Wd)	(wD)	(wd)
(WD)	*WWDD* White, Disc	*WWDd* White, Disc	*WwDD* White, Disc	*WwDd* White, Disc
(Wd)	*WWDd* White, Disc	*WWdd* White, Sphere	*WwDd* White, Disc	*Wwdd* White, Sphere
(wD)	*WwDD* White, Disc	*WwDd* White, Disc	*wwDD* Yellow, Disc	*wwDd* Yellow, Disc
(Wd)	*WwDd* White, Disc	*Wwdd* White, Sphere	*wwDd* Yellow, Disc	*wwdd* Yellow Sphere

F_2 segragation: 9 – White, Disc; 3 – Yellow, Disc; 3 White, Sphere; 1 – Yellow, Sphere

Fig. 10.6 Dihybrid ratio (9:3:3:1) inheritance of fruit shape and fruit color in summer squashes (from Sambamurty, 1999)

TABLE 10.3 Genotypic and phenotypic ratios in a dihybrid cross

Genotype	Number(s)	Ratio	Phenotype	Ratio
RRYY	1	1	Round yellow (I)	9
RrYY	3,9	2	(Parental)	
RRYy	2,5	2		
RrYy	4,7,10,13	4		
Rryy	6	1	Round green (II)	3
RrYY	8,14	2	(Recombinant)	
RrYY	11	1	Wrinkled yellow (III)	3
RrYy	12,15	2	(Recombinant)	
Rryy	16	1	Wrinkled green (IV)	1
			(Parental)	

Source: Miglani (2000)

TABLE 10.4 Some important formulae

Segregating gene pairs	n
Number of phenotypic classes	2^n
Phenotypic ratios in case of complete dominance	$(3:1)^n$
Phenotypic ratios in case of incomplete dominance	$(1:2:1)^n$
Genotypic ratios	$(1:2:1)^n$
Number of different types of encounters	4^n
Number of different types of gametes formed by F_1 individual	2^n
Frequency of most common genotypic class	$2^n/4^n$
Frequency of most common phenotypic class in case of complete dominance	$3^n/4^n$

Source: King (1965)

P₁ | Red flower | x | White flower
(layout diagram)

Male → Female ↓	R	r
R	RR Red	Rr Pink
r	Rr Pink	rr White

- Pink flower (incompletely dominant)
- 1 Red : 2 Pink : 1 White

Fig. 10.7 Incomplete dominance in snapdragon for flower color in the F_1 and F_2 generations (from Sambamurty, 1999)

Incomplete Dominance and Codominance

Dominance is not a law since *incomplete dominance* exists. The F_1 progeny is phenotypically an intermediate between the two homozygous parents with contrasting pairs of traits. Flower color in a four o' clock plant, *Mirabilis jalapa*, provides a very good example of incomplete dominance. It could be seen that neither of the alleles was dominant in the hybrid. *Codominance* also showed that dominance is not a law. In the case of codominance, the phenotype of F_1 hybrid revealed characteristics of both parents, e.g. the AB blood group of the ABO system in man.

Incomplete dominance in a monohybrid cross. After the rediscovery of Mendel's principles several investigators started getting different ratios not conforming to the classical 3:1 monohybrid ratio. An example of this kind is the cross of red and white flowered snapdragons (*Antirrhinum majus*). Figure 10.7 gives F_1 hybrids of an intermediate pink color. When this hybrid was selfed, there was segregation of 1 red : 2 pink : 1 white in the F_2 generation. Thus, the R gene was incompletely dominant over its reces-

sive r gene. The validity of this hypothesis can be tested by crossing pink F_1s and the red parent (RR) and the progeny segregation to red and pink in 1:1 ratio. Similarly, when pink (Rr) was crossed with white (rr), the progeny consisted of pink and white in 1:1 ratio. These two experiments verify the assumption that there is incomplete dominance in the red color gene (R).

In the absence of complete dominance, each genotype may have a distinguishable phenotype. Semidominant alleles may produce the same genetic product but in unequal quantity. In the heterozygous condition, the total product is the sum of the separate quantities of the two alleles. Semidominant alleles therefore act additively so that expressions of heterozygotes are intermediate between those of the two homozygotes. The ratio for the monohybrid cross becomes 1:2:1 instead of 3:1.

Incomplete dominance in dihybrid crosses. In snapdragons, intermediate-size leaves are produced by a heterozygous gene combination (BB'). Plants with broad leaves and narrow leaves have the homozygous gene arrangements BB and $B'B'$ respectively. Likewise, pink flowers are controlled by a heterozygous (RR') pair, with red (RR) and white ($R'R'$) attributable to corresponding homozygotes. A cross between plants with broad leaves and red flowers ($BBRR$) and those with narrow leaves and white flowers ($B'B'R'R'$) produced plants with intermediate leaves and pink flowers. The F_2 were classified into nine phenotypic classes corresponding to the genotype combinations as illustrated in Fig. 10.8. The 9:3:3:1 dihybrid ratio was thus replaced by the 1:2:1:2:4:2:1:2:1 ratio, as expected for semidominant involving two pairs of alleles.

Figure 10.9 illustrates incomplete dominance for flower color gene red (RR) vs. white (rr) in snapdragons in a dihybrid cross involving the shape of the flower, i.e.

P	Broad, Red		Narrow, White
	BBRR	x	*B'B'R'R'*

Gametes	*BR*		*B'R'*
		Intermediate, Pink	
F₁	*BB'RR'*x		*BB'RR'*

F₂

Male Female	BR	BR'	B'R	B'R'
BR	*BBRR*	*BBRR'*	*B'BRR*	*BB'RR'*
BR'	*BBRR'*	*BBR'R'*	*BB'RR'*	*BB'R'R'*
B'R	*BB'RR*	*BB'RR'*	*B'B'RR*	*B'B'RR*
B'R'	*BB'RR'*	*BB'R'R'*	*B'B'R'R*	*B'B'R'R'*

Phenotypes	Genotypes	Genotype frequency	Phenotypic ratio
Broad, Red	*BBRR*	1	1
Broad, Pink	*BBRR'*	2	2
Broad, White	*BBR'R'*	1	1
Intermediate, Red	*BB'RR*	2	2
Intermediate, Pink	*BB'RR'*	4	4
Intermediate, White	*BB'R'R'*	2	2
Narrow, Red	*B'B'RR*	1	1
Narrow, Pink	*B'B'RR'*	2	2
Narrow, White	*B'B'R'R'*	1	1

Fig. 10.8 Cross between snapdragons with broad leaves and red flowers and those with narrow leaves and white flowers illustrating semidominant inheritance (from Gardener, 1975)

normal (*NN*) vs. peloric (*nn*). Normal is completely dominant over peloric, which is why the F₁ hybrid was normal pink (*NnRr*). In the F₂ generation, all types of combinations occur, i.e., dominant, incompletely dominant and recessive, and a classical 9:3:3:1 will be replaced by the 6:3:3:2:1:1 ratio.

Trihybrid ratio

In a *trihybrid cross*, three contrasting pairs of traits and hence three gene pairs are considered at one time. In this case also the *Principle of Independent Segregation* is valid according to Mendel's principles.

When a cross is made between a variety with tall, yellow, and round seeds

| P₁ | Red, Normal | x | White, Peloric |

P₁   Red, Normal   x   White, Peloric
            RR NN   *rr nn*

F₁   Pink, Normal
                          Rr Nn
                         ↓ Selfing

F₂

Male → Female ↓	(RN)	(Rn)	(rN)	(rn)
(RN)	*RR NN* Red, Normal	*RR Nn* Red, Normal	*Rr NN* Pink, Normal	*Rr Nn* Pink, Normal
(Rn)	*RR Nn* Red, Normal	*RR nn* Red, Peloric	*Rr NN* Pink, Normal	*RR nn* Red, Peloric
(rN)	*Rr NN* Pink, Normal	*Rr Nn* Pink, Normal	*rrNN* White, Normal	*rr Nn* White, Normal
(rn)	*Rr Nn* Pink, Normal	*Rr nn* Red, Peloric	*rr Nn* White, Normal	*rr nn* White, Peloric

Pink, Normal 6; Red, Normal 3; White, Normal 3; Pink, Peloric 2; Red ,Peloric 1; White, Peloric 1; Total 16

Fig. 10.9 Incomplete dominance in snapdragons (*Antirrhinum majus*). Inheritance of flower color and flower shape (from Sambamurty, 1999).

(*TTYYRR*) and dwarf, green, and wrinkled seeds (*ttyyrr*) of peas, the F_1 hybrid will be tall, yellow and round.(*TtYyRr*) which is heterozygous for all the three genes. The F_2 generations can be represented by a *forking method* or *dichotomous method* to establish all the genotypes and phenotypes. Since there will be 64 boxes if a checkerboard method is used, this becomes complicated and time is required to represent them. The dichotomous method is easy to represent even when there are 4 or 5 genes.

The F_1 hybrid tall, yellow and round (*TtYyRr*) produces eight types of gametes, viz. *TYR, TYr, TyR, Tyr, tYR, tYr, tyR,* and *tyr* (Table 10.5). These eight types of gametes meet at random from both the F_1 parents to produce eight different phenotypes.

Table 10.6 gives the genotypic frequencies in a trihybrid cross. In Table 10.7, all the

genotypes in 64 boxes are given and when counted yield 27 different genotypes.

Genetic Ratios

An important part of genetics today is concerned with predicting the types of progeny that emerge from a cross and calculating their expected frequency, in other words their probability. We have already examined two methods for doing so—*Punnett squares* and *branch diagrams*. Punnett squares can be used to show hereditary patterns based on one gene pair, two gene pairs or more. Such squares are a good graphic device for representing progeny but making them is time consuming. Even the 16-compartment Punnett square takes a long time to write out, but for a trihybrid there are 2^3, or 8 different gamete types, and the Punnett square has 64 compartments. The branch diagram is easier and adaptable for pheno-

TABLE 10.5 Trihybrid phenotypic ratio (27:9:9:9:3:3:3:1) with three genes—tall, yellow, round in pea plant vs dwarf, green, wrinkled

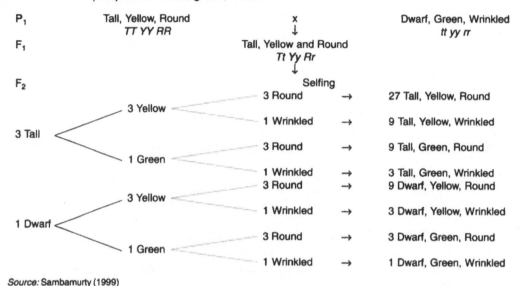

P_1	Tall, Yellow, Round *TT YY RR*	x ↓		Dwarf, Green, Wrinkled *tt yy rr*
F_1		Tall, Yellow and Round *Tt Yy Rr* ↓		

F_2 — Selfing

- 3 Tall
  - 3 Yellow
    - 3 Round → 27 Tall, Yellow, Round
    - 1 Wrinkled → 9 Tall, Yellow, Wrinkled
  - 1 Green
    - 3 Round → 9 Tall, Green, Round
    - 1 Wrinkled → 3 Tall, Green, Wrinkled
- 1 Dwarf
  - 3 Yellow
    - 3 Round → 9 Dwarf, Yellow, Round
    - 1 Wrinkled → 3 Dwarf, Yellow, Wrinkled
  - 1 Green
    - 3 Round → 3 Dwarf, Green, Round
    - 1 Wrinkled → 1 Dwarf, Green, Wrinkled

Source: Sambamurty (1999)

typic, genotypic, or gametic proportions, as illustrated for the dihybrid *A/a; B/b*. (Dash means that the allele can be present in any form, that is, dominant or recessive.)

Notice that the "tree" of branches for genotypes is quite unwieldy even in this case, which uses two gene pairs, because there are $3^2 = 9$ genotypes. For three

gene pairs, there are 3^3, or 27 possible genotypes.

Application of simple *statistical rules* is the third method for calculating the probabilities (expected frequencies) of specific phenotypes or genotypes coming from a cross. The two probability rules needed are the *product rule* and the *sum rule*, considered in that order below.

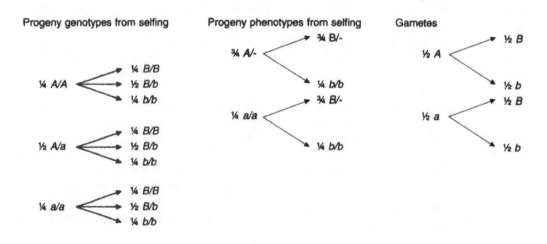

Progeny genotypes from selfing

- ¼ A/A → ¼ B/B, ½ B/b, ¼ b/b
- ½ A/a → ¼ B/B, ½ B/b, ¼ b/b
- ¼ a/a → ¼ B/B, ½ B/b, ¼ b/b

Progeny phenotypes from selfing

- ¾ A/- → ¾ B/-, ¼ b/b
- ¼ a/a → ¾ B/-, ¼ b/b

Gametes

- ½ A → ½ B, ½ b
- ½ a → ½ B, ½ b

TABLE 10.6 Genotypic ratio for the trihybrid cross by forking method

F₁ Tall, Yellow, Round   x   F₁ Tall, Yellow, Round
    Tt Yy Rr   ↓   *Tt Yy Rr*

F₂ generation

			1 RR	→	1 TT YY RR
1 TT	1 YY		2 Rr	→	2 TT YY Rr
			1 rr	→	1 TT YY rr
			1 RR	→	2 TT Yy RR
	2 Yy		2 Rr	→	4 TT Yy Rr
			1 rr	→	2 TT Yy rr
			1 RR	→	1 TT yy RR
	1 yy		2 Rr	→	2 TT yy Rr
			1 rr	→	1 TT yy rr
			1 RR	→	2 Tt YY RR
2 Tt	1 YY		2 Rr	→	4 Tt YY Rr
			1 rr	→	2 Tt YY rr
			1 RR	→	4 Tt Yy RR
	2 Yy		2 Rr	→	8 Tt Yy Rr
			1 rr	→	4 Tt Yy rr
			1 RR	→	2 Tt yy RR
	1 yy		2 Rr	→	4 Tt yy Rr
			1 rr	→	2 Tt yy rr
			1 RR	→	1 tt YY RR
1 tt	1 YY		2 Rr	→	2 tt YY Rr
			1 r	→	1 tt YY rr
			1 RR	→	2 tt Yy RR
	2 Yy		2 Rr	→	4 tt Yy Rr
			1 rr	→	2 tt Yy rr
			1 RR	→	1 tt yy RR
	1 yy		2 Rr	→	2 tt yy Rr
			1 rr	→	1 tt yy rr

Source: Sambamurty (1999)

TABLE 10.7 Genotype frequency of the F₂ generation of the trihybrid ratio tall, yellow, round (F₁) showing different genotypes and phenotypes

Phenotypes	Genotypes	Genotypic frequency	Phenotypic ratio
Tall, Yellow, Round	TT YY RR	1	27
	TT YY Rr	2	
	TT Yy RR	2	
	TT Yy Rr	4	
	Tt YY RR	2	
	Tt YY Rr	4	

(Contd. Table 10.7)

(Contd. Table 10.7)

	Tt Yy RR	4	
	Tt Yy Rr	8	
Tall, Yellow, Wrinkled	TT YY rr	1	9
	TT Yy rr	2	
	Tt YY rr	2	
	Tt Yy rr	4	
Tall, Green, Round	TT yy RR	1	9
	TT yy Rr	2	
	Tt yy RR	2	
	Tt yy Rr	4	
Dwarf, Yellow, Round	tt YY RR	1	9
	tt Yy Rr	2	
	tt Yy RR	2	
	tt Yy Rr	4	
Tall, Green, Wrinkled	TT yy rr	1	3
	Tt yy rr	2	
Dwarf, Yellow, Wrinkled	tt YY rr	1	3
	tt Yy rr	2	
Dwarf, Green, Round	tt yy RR	1	3
	tt yy Rr	2	
Dwarf, Green, Wrinkled	tt yy rr	1	1

Source: Sambamurty (1999)

The product rule states that the probability of *independent events* occurring together is the product of the probabilities of the individual events.

The possible outcomes of rolling dice follow the product rule because the outcome on each separate dice is independent of the others. As an example, let us consider two dice and calculate the probability of rolling a pair of 4s. The probability of a 4 on one dice is 1/6 because a dice has six sides and only one side carries the 4. This probability is written as follows:

$$p \text{ (of a 4)} = 1/6$$

Therefore, with the use of the product rule, the probability of a 4 appearing on both dice is $1/6 \times 1/6 = 1/36$, which is written

$$p \text{ (of two 4s)} = 1/6 \times 1/6 = 1/36$$

The *sum rule* states that the probability of either of two mutually exclusive events occurring is the sum of their individual probabilities.

In the *product rule*, the focus is on outcomes A and B. In the sum rule, the focus is on the concept of outcome A or B. Dice can also be used to illustrate the sum rule. We have already calculated that the probability of two 4s is 1/36 and using the same type of calculation, it is clear that the probability of two 5s will be the same, or 1/36. Now we can calculate the probability of either two 4s or two 5s. Because these outcomes are mutually exclusive, the sum rule can be used to tell us that the answer is 1/36 + 1/36, which is 1/18. This probability can be written as follows:

$$p \text{ (two 4s or two 5s)} = 1/36 + 1/36 = 1/18$$

Now we can consider a genetic example. Assume that we have two plants of genotypes *A/a; b/b; C/c; D/d; E/e* and *A/a; B/b; C/c; d/d; E/e*, and then from a cross between these plants, we want to recover a progeny plant of genotype *a/a; b/b; c/c; d/d; e/e* (perhaps for the purpose of acting as the tester strain in a test cross). To estimate how many progeny plants need to be grown to stand a reasonable chance of obtaining the desired genotype, we need to calculate the proportion of the progeny expected to be of that genotype. If we assume that all the gene pairs assort independently, then we can do this calculation easily by using the product rule. The five different gene pairs are considered individually, similar to five separate crosses, and the appropriate probabilities then multiplied to arrive at the answer.

From *A/a* × *A/a*, one-fourth of the progeny will be *a/a* (see Mendel's crosses); from *b/b* x *B/b*, one-half of the progeny will be *b/b*; from *C/c* × *C/c*, one-fourth of the progeny will be *c/c*, from *Dd/* × *d/d*, one-half of the progeny will be *d/d*; from *E/e* × *E/e*, one-fourth of the progeny will be *e/e*. Therefore, the overall probability (or expected frequency) of progeny of genotype *a/a; b/b; c/c; d/d; e/e* will be ¼ × ½ × ¼ × ½ × ¼ = 1/256. So we learn that hundreds of progeny will need to be isolated to stand a chance of obtaining at least one of the desired genotype. This probability calculation can also be extended to predict *phenotypic frequencies* or *gametic frequencies*.

Binomial Method

It is useful to know the method of expanding the binomial $(a + b)^n$. The *binomial method* is given in Fig. 10.10. (expansion of the binomial a + b through the fifth power). These simple algebraic terms are of great value in studies of *probability*. It is important to learn to obtain coefficients through use of *Pascal's triangle* (Fig. 10.11).

This method can be used to obtain coefficients of *binomial equations*. Note that each number after the first is obtained by adding the numbers above it. The second number in each line indicates the power of the equation. Thus the last line in Fig. 10.10 has 5 as its second number and is used in expanding $(a + b)^5$. This method works nicely for expansions in which the number of events is small, but becomes cumbersome when they become large.

Reverse Application of Laws of Probability

It is possible to apply the principles of probability in reverse to determine the chances of *independent events* when one knows the chances of two such events happening together. For instance, we know that in tossing pennies, the chance of tossing two heads in succession is ¼. Now, suppose that from this figure we want to determine the chance of tossing one head in one toss. We take the square root of ¼ and obtain ½. We can thus say, the chance that either of the two independent events of equal frequency will

$$(a + b)^1 = a + b$$
$$(a + b)^2 = a^2 + 2ab + b^2$$
$$(a + b)^3 = a^3 + 3a^2b + 3ab^2 + b^3$$
$$(a + b)^4 = a^4 + 4a^3b + 10a^2b^2 + 10ab^3 + b^4$$
$$(a + b)^5 = a^5 + 5a^4b + 10a^3b^2 + 10a^2b^3 + 5ab^4 + b^5$$

Fig. 10.10 Expansion of the binomial a + b through the fifth power (from Winchester, 1966)

```
          1       1
       1     2     1
    1     3     3     1
 1     4     6     4     1
1    5    10   10    5    1
```

Fig. 10.11 Pascal triangle (from Winchester, 1966)

occur separately is equal to the square root of the chance that the two will occur together. Please note that the events must be of equal frequency.

Now let us see how this principle could be applied in a genetics problem. Suppose that one baby out of every 100,000 born in the United States dies at birth because of an inherited lung deformity, which results from a *recessive autosomal lethal gene*. How many such lethal genes are there in a living population? Each baby who dies because it is homozygous represents the coincident occurrence of independent events of equal frequency: the union of germ cells, both of which carry the gene. Therefore, the chance that any one germ cell in the population will carry the gene is equal to the square root of 1/100,000 or about 1/316.

Rediscovery of Mendelism

In 1900 three botanists, namely Correns, De Vries, and Tschermak, working on various flowering plants independently drew the conclusions similar to those of Mendel. Later these botanists rediscovered Mendel's research in 1900. Mendel's original paper was republished in Flora 89:364 (1901). Correns was a German botanist and geneticist who in 1900 independently but simultaneously with other biologists conducted research with garden peas and came to the same conclusions drawn by Mendel. Later he worked with four o' clock (*Mirabilis jalapa*) and established the first conclusive example of *extrachromosomal inheritance* or *cytoplasmic inheritance*. De Vries was a Dutch biologist and geneticist. He rediscovered independently but simultaneously with Correns and Tschermak in 1900 Mendel's principles of inheritance. Later, working on *Oenothera larmarckiana*, he coined the term mutation for sudden heritable changes in characters. Tschermak, an Austrian botanist

and geneticist, was one of the codiscoverers of Mendel's papers on garden peas. He found that his results recorded with the findings of Mendel. In the same year, 1900, when Tschermak reported his findings, De Vries and Correns also reported their discoveries of Mendel's papers. Later he applied Mendel's principles of heredity to barley, wheat-rye hybrids, and oats hybrids for development of new plants.

Bateson (1905) confirmed Mendel's work by a series of *hybridization experiments*. *Rediscovery of Mendel's work* went a long way in understanding Mendelism in its real sense. This rediscovery marks the beginning of the science of the formal genetics.

Mendel's research remained dormant and unnoticed by the scientific world until 1900. During these intervening 34 years many developments occurred in biology which prepared the way for the rediscovery of Mendel's work. For instance, during this period Haeckel (1866) recognized the active role of the nucleus in heredity. Weismann (1887), Hertwig (1936), Strasburger (1984), and Kolliker (cited from Verma and Agarwal, 1998) suspected active participation of chromosomes in heredity transmission. Roux (1883) suggested that the chromosomes must contain qualitatively different hereditary determiners arranged in linear orders (cited from Roux, 1905). Weismann (1887) supported the idea of Roux by propounding his germplasm theory. Further, some workers such as Darwin (1868) in England, Vilmorin (1879) in France, Rumpau (1891) in Germany, and Bohlin (1897) in Sweden carried out hybridization experiments very much like those of Mendel on different plants and observed the phenomenon of *dominance*, but failed to provide any conclusive explanation for their findings (cited from Verma and Agarwal, 1998).

TABLE 10.8 Summary of F_2 results obtained by various workers on inheritance of seed color in peas

Investigator	Yellow seeds		Green seeds		Total
	No.	(%)	No.	(%)	
Mendel (1865)	6,022	(75.05)	2,001	(24.95)	8,023
Correns (1900)	1,394	(75.47)	453	(24.53)	1,847
Tschermak (1900)	3,580	(75.05)	1,190	(24.95)	4,770
Hurst (1904)	1,310	(74.64)	445	(25.36)	1,755
Bateson (1905)	11,902	(75.30)	3,903	(24.70)	15,806
Lock (1905)	1,438	(73.63)	514	(26.33)	1,952
Darbishire (1909)	109,060	(75.9)	36,186	(24.91)	145,246

Source: Sinnot et al. (1959)

Later work on peas by other investigators completely confirmed Mendel's results (Table 10. 8). What Mendel had shown and *rediscoverers of Mendelism* verified may be summarized thusly:

1. When parents differ in characters, the offspring resemble one parent but not the other. This is the phenomenon of *dominance*.

2. When the hybrid reproduces, it transmits with equal frequency either the dominant character of one parent or the recessive character of the other, but not both. This is the principle of *segregation*.

3. When parents differ in two or more pairs of characters, each pair shows dominance and segregation independent of the other. As a consequence all possible combinations of the two or more pairs can occur in the reproductive cells of the hybrid, and in their chance frequencies. This is the principle of *recombination*.

Impact of Mendelism and its Rediscovery

An important *consequence of Mendel's work* and its subsequent rediscovery was to replace the blending theory of inheritance with a *particulate theory of inheritance*. Mendel's observation that one of the parental characteristics was absent in F_1 hybrid and reappeared in unchanged form in the F_2 generation was inconsistent with the *blending theory*. It was concluded from Mendel's work that traits from the parental lines were transmitted as two different elements of particulate nature that retained their purity in the F_1 hybrids. This gave rise to the *particulate theory of inheritance*.

CONCEPT OF DOMINANCE

Earlier a gene was considered dominant if it was expressed in the heterozygous condition and recessive if it was masked. Sometimes heterozygous individuals for a particular locus reveal an intermediate phenotype. For example, bar-eyed heterozygote *Drosophila* individuals have a kidney-shaped eye which is different from either of the homozygous parents. Thus heterozygous individuals in this case exhibit *incomplete dominance* rather than a simply dominant or recessive situation. The following example will provide further insight into the *concept of dominance*.

Inheritance of a temperature-sensitive lethal gene. Inheritance of a temperature-sensitive lethal gene at different temperatures shows that the *dominance-recessiveness relationship* is not a property of a gene, but its expression in a particular environment (Fig. 10.12). At 18°C, the heterozygous is identical to both parents and no dominance relationship exists. At 20°C, a^+ is dominant over a and at 27°C, a is dominant over a^+. Thus we see that these genes cannot be classified simply as dominant or recessive without referring to the condition(s) at which the relationships exist.

It we consider alleles from a functional perspective, a clearer understanding of dominant and recessiveness emerges. Assume that a gene a^+ produces an enzyme that converts a substrate "X" to a product "Y". Suppose a mutation changes $a^+ \rightarrow a$; various consequences may be (i) gene a produces a nonfunctional enzyme or none at all. In heterozygous conditions (a^+a), if a^+ can still produce enough normal enzyme, the organism may show a normal phenotype (as done by a^+a^+). In the organism of genotype aa, the reaction "X" → "Y" will not occur and a mutant phenotype will result. Such a mutant gene (a) will be classified as recessive, with a^+ as completely dominant over a. (ii) gene a produces a nonfunctional enzyme or none at all. In heterozygous

conditions (a^+a), a^+ gene may not produce enough normal enzyme. Here reaction "X" to "Y" may proceed slowly. Here an intermediate phenotype may be produced. Thus a^+ will be incompletely dominant over a. (iii) gene a produces an enzyme that has a greater affinity for substrate "X" than does the a^+ gene product. Enzyme produced by gene a converts "X" to a new product "Y_2", which causes a new phenotype to appear in the heterozygote that is identical to that of the homozygous a phenotype. Therefore, gene a may be regarded as dominant. Dominance of a over a^+ may also arise when the a gene product inactivates the normal a^+ gene product.

Scale of Dominance

The *scale of dominance* is represented in Fig. 10.13. Let us consider a locus A with two alleles, A_1 and A_2. Let us assume that the phenotype of homozygote A_1A_1 is P_1 and that of A_2A_2 is P_2. If the phenotype of heterozygote A_1A_2 is (a) same as P_1, A_1 is dominant over A_2; (b) same as P_2, A_2 is

Locus A has two alleles A_1 and A_2 and three genotypes – two homozygotes, A_1A_1 and A_2A_2, and one heterozygote, A_1A_2.

Assume that phenotype of A_1A_1 is P_1 and that of A_2A_2 is P_2.
* A_1 will be dominant over A_2 if phenotype of A_1A_2 is P_1.
* A_2 will be dominant over A_1 if phenotype of A_1A_2 is P_2.
* The two alleles will show incomplete dominance if phenotype of A_1A_2 ranges between P_1 and P_2.
* The two alleles will show codominance if phenotype of A_1A_2 shows expression of both the alleles.
* The two alleles will show overdominance if phenotype of A_1A_2 is beyond the range of P_1 and P_2.
* If phenotype of A_1A_2 falls at mid-parent (M) value, it is termed as no dominance.

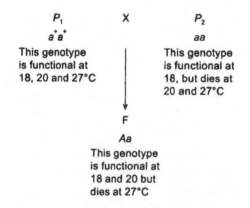

Fig. 10.12 Inheritance of a temperature-sensitive lethal gene (from Miglani, 2002)

Fig. 10.13 Scale of dominance (from Miglani, 2002)

dominant over A_1; (c) ranges between P_1 and P_2, A_1 and A_2 show incomplete dominance; (d) exhibits properties of both P_1 and P_2 phenotypes, codominance is the result; (e) lies beyond the range $P_1 - P_2$, overdominance is said to exist between alleles A_1 and A_2. When the phenotype of F_1 falls at the midparent value, some workers call it *no dominance*.

Dominance, Heterosis, and Hybrid Vigor

The foundation of a comprehensive understanding of *heterosis* was laid by Shull (1908) and East (1936). The concept was further amplified by Shull (1952). The term heterosis has different connotations for different workers. According to Dobzhansky (1952), true heterosis is the mutational *euheterosis* which results from sheltering deleterious recessive mutants by the adaptive superior dominant alleles in the population of sexually reproducing and cross-fertilizing organisms. He also described *"the balanced euheterosis"* which arises out of a balanced type of adaptive mutations and these mutations confer on the heterozygote a better adaptive value than is found in corresponding homozygotes. The condition is sometimes called *overdominance* and was supported by Hull (1952). According to Whaley (1952), heterosis and *hybrid vigor* are often used synonymously. Hybrid vigor precisely refers to the developed superiority of hybrid and heterosis refers to the mechanism by which superiority is developed. Thus, heterosis is the cause and hybrid vigor is the effect. By this definition, hybrid vigor is a manifestation of heterosis. Heterosis is also defined as the decrease or increase for any character of the F_1 hybrid over its parents considered separately. The term heterosis is usually applied to size, rate of growth, or general fitness. The appearance of individuals in the F_2 of a cross that falls beyond the parental range for a quantitative trait is known as *transgressive segregation*. Brewbaker (1964) termed heterosis as the consequence of hybridization and defined it as the deviation of the F_1 or heterozygote from the midparent value, or the relative performance of a hybrid compared with its homozygous parents. When the F_1 value falls in between the midparent value and a parental value, the F_1 is said to show *partial* or *incomplete dominance*. When the F_1 value coincides with the value of one of the parents, the F_1 is said to show *complete dominance* and when it overshoots a parental value a case of overdominance occurs. Thus partial dominance, complete dominance, and overdominance are three different aspects of the phenomenon of heterosis. Trehan and Gill (1987) investigated the biochemical basis of heterosis using acid phosphatase gene-enzyme system of *Drosophila* as a model. They discovered through partial purification of the individual electrophoretic bands that superior performance of the heterozyote compared to the two parents and the F_1 was due to the presence of two homodimeric and one heterodimeric enzyme compared to the two parents which had only one type of homodimeric enzyme.

Reverse Genetics in Mendel's Wrinkled Character

In 1990, scientists cloned the pea gene r (rugosus; old name wrinkled, w) which determines whether the seed is round or wrinkled. It was shown that an isoform of the starch branching enzyme, SBEI, is present in round (RR or WW and Rr or Ww) seeds, but absent in wrinkled (rr or ww) seeds (cited from Verma and Agarwal, 1998). The gene for SBEI is located on the r locus, but a small DNA sequence (0.8 kilobases) called transposon-like insertion, due to which the aberrant SBEI enzyme is produced, interrupts the gene. This defective enzyme leads to metabolic disturbances in the biosynthesis of starch, lipid, and protein. This results in an increase of free sugars due to the failure of starch formation. This probably leads to higher osmotic pressure and hence higher water content and larger cell volume in *rr* (*ww*) seeds. Seeds lose a large proportion of their water on maturation, leading to shrinkage in volume. Since the testa does not shrink with the cotyledons, seeds become wrinkled. Such a follow-up of the molecular basis of a Mendelian trait comes under the purview of the modern branch of genetics called *reverse genetics* (from DNA to phenotype) as opposed to a classical dihybrid cross in *Drosophila*.

MOLECULAR BASIS OF MENDELIAN GENETICS

Let us consider some of Mendel's terms in the context of the cell. First, what is the *molecular nature of alleles*? When alleles such as *A* and *a* are examined at the DNA level by using modern technology, they are generally found to be identical in most of their sequences and differ only at one or a few nucleotides of the thousands of nucleotides that make up the gene. Therefore, we see that the alleles are truly different versions of the same basic gene. Looked at another way, the gene is the generic term and allele specific. (The pea color gene has two alleles coding for yellow and green.) The diagram below represents the DNA of two alleles of one gene; the letter "X" represents a difference in the nucleotide sequence:

Allele 1 ▬▬▬▬▬▬▬▬▬▬
Allele 2 ▬▬▬▬ X ▬▬▬▬

What about *dominance*? We have seen that although the terms dominant and recessive are defined at the level of phenotype, the phenotypes are clearly manifestations of the different actions of alleles. Therefore, we can legitimately use the phrases *dominant allele* and *recessive allele* as determinants of dominant and recessive phenotypes. Several different molecular factors can make an allele either dominant or recessive. One commonly found situation is that the *dominant allele* encodes a functional protein, and the *recessive allele* encodes lack of the protein or a nonfunctional form of it. In the heterozygote, the protein produced by the *functional allele* suffices for the normal needs of the cell; so the functional allele acts as a dominant allele. An example of the dominance of the functional allele in a heterozygote is *albinism*. The general idea can be stated as a formula as follows:

A	plus	*a*		=	*A/a*
(functional protein)	plus	(nonfunctional protein)		=	function

What is the *cellular basis of Mendel's first principle*, the equal segregation of alleles at gamete formation? In a diploid organism such as peas, all the cells of the organism contain two chromosome sets. Gametes, however, are haploid, containing one chromosome set. Gametes are produced by specialized cell divisions in the diploid cells in the germinal tissue (ovaries and anthers). Nuclear divisions called meiosis accompany these specialized cell divisions. The highly *programmed chromosomal movements* in *meiosis* cause equal segregation of alleles into the gametes. In meiosis in a heterozygote *A/a*, the chromosome carrying *A* is pulled in the direction opposite to the chromosome carrying *a*; so half the resultant gametes carry *A* and the other half carry *a*. The situation can be summarized in a simplified from as:

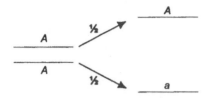

The nuclear spindle, a series of *microtubules* consisting of the protein *tubulin*, generates the force pulling the chromosome to cell poles. Microtubules attach to the *centromeres* of chromosomes by interacting with another specific protein located in that area. Orchestration of these molecular interactions is complex, yet constitutes the basis of the principles of hereditary transmission in eukaryotes.

TEST FOR ALLELISM

Many genetic research programs attempt to understand the genes that contribute to one particular biological process. Such an analysis begins with a collection of related mutant phenotypes centered on the process. For example, if a geneticist were interested in the genes determining flower color, genetic dissection would begin with isolating a set of different flower colors. An important task is to determine how many different genes are represented by the mutations that determine the related phenotypes, because this number defines the set of genes that affect the process under study. Hence it is necessary to have a test for ascertaining whether the mutations are alleles of one gene or different genes. The *allelism test* having the widest application is the *complementation test*, is illustrated in the following example.

Consider a species of harebell (*Campanula*) in which the wild type flower color is blue. Let's assume that by applying mutagenic radiation, three white-petalled mutants could be induced and be available as homozygous pure-breeding strains. We can call the mutant strains $, £, and ¥, using

currency symbols so that we do not prejudice our thinking concerning dominance. When crossed with the wild type, each mutant gives the same results in the F_1 and F_2, as follows:

white $ x blue ⟶ F_1 all blue ⟶ F_2 ¾ blue, ¼ white
white £ x blue ⟶ F_1 all blue ⟶ F_2 ¾ blue, ¼ white
white ¥ x blue ⟶ F_1 all blue ⟶ F_2 ¾ blue, ¼ white

In each case, results show that the mutant condition is determined by the recessive allele of a single gene. However, are they three alleles of one gene or two or three genes? The question can be answered by asking whether the mutants complement each other. *Complementation* is defined as the production of a wild type phenotype when two haploid genomes bearing different recessive mutations are united in the same cell.

Demonstration of the recessive nature of individual mutants is a crucial result that allows us to proceed with a *complementation test*. Dominant mutations cannot be used in a complementation test. In a diploid organism, intercrossing homozygous recessive mutants two at a time performs the complementation test. The next step is to observe whether the progeny have the wild type phenotype.

This unites the two mutations as haploid gametes to form a diploid nucleus in one cell (the zygote). If recessive mutations represent alleles of the same gene, then they will not complement because both mutations represent lost gene function. Such alleles can be thought of generally as a' and a'', using primes to distinguish between two different mutant alleles of a gene whose wild-type allele is a^+. These alleles could have different mutant sites, but they would be functionally identical (that is, both nonfunctional). The heterozygote a'/a'' would be:

However, two recessive mutants in different genes would have a wild-type function provided by the recessive wild alleles. Here we can name the genes a_1 and a_2 after their mutant alleles. We can represent the heterozygotes as follows, depending on whether the genes are on the same or different chromosomes.

Different chromosomes

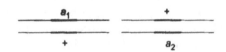

Same chromosome

Let us return to the harebell example and intercross the mutants to unite the mutant alleles to test complementation. We can assume that the results of intercrossing mutants $, £, and ¥ are as follows:

white $ x white £	→	F_1 all white
white $ x white ¥	→	F_1 all blue
white £ x white ¥	→	F_1 all blue

From this set of results, we can conclude that mutants $ and £ must be caused by alleles of one gene (say, $w1$) because they do not complement; but ¥ must be caused by a mutant allele of another gene ($w2$).

The molecular explanation of such results is often in relation to biochemical pathways in the cell. How does complementation work at the molecular level? Although the convention is that it is the *mutants* that complement, in fact the active agents in complementation are the proteins produced by *wild* alleles. A blue pigment called *anthocyanin* causes the normal blue

color of the flower. Pigments are chemicals that absorb certain parts of the visible spectrum; in the harebell, anthocyanin absorbs all wavelengths except blue, which is reflected into the eye of the observer. However, this anthocyanin is made from chemical precursors that are not pigments, that is, they do not absorb light of any specific wavelength and simply reflect back the white light of the sun to the observer, giving a white appearance. The blue pigment is the end product of a series of biochemical conversions of nonpigments. A specific enzyme encoded by a specific gene catalyzes each step as follows:

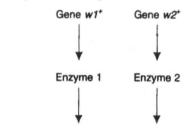

→ Precursor 1 → Precursor 2 → Blue anthocyanin

A mutation in either of the genes in the homozygous condition will lead to accumulation of a precursor that will simply make the plant white. Mutant designations could be written as follows:

$ $w1_s/w1_s$. $w2^+/w2^+$

£ $w1_£/w1_£$. $w2^+/w2^+$

¥ $w1^+/w1^+$. $w2_¥/w2_¥$

However, in practice the subscript symbols would be dropped and the genotypes written as follows:

$ $w1/w1$. $w2^+/w2^+$

£ $w1/w1$. $w2^+/w2^+$

¥ $w1^+/w1^+$. $w2/w2$

Hence, an F_1 from $ × £ will be: $w1/w1.w2^+/w2^+$ which will have two defective alleles for $w1$ and therefore be blocked at step 1. Even though enzyme 2 is fully functional, it has no substrate on which to act. Thus no blue pigment will be produced and the phenotype will be white.

The F_1s from the other crosses however will have wild alleles for both the enzymes needed to make the interconversions to the final blue product. Their genotypes will be:

$w1^+/w1$. $w2^+/w2$. Hence we see why complementation is actually a result of the cooperative interaction of the wild alleles of the two genes. Figure 10.14 summarizes diagramatically the interaction of the complementing and noncomplementing white mutants.

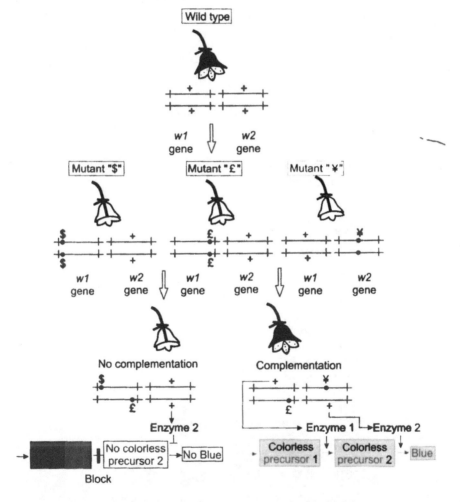

Fig. 10.14 Molecular basis of genetic complementation. Three phenotypically identical white mutants—$, £ and ¥—are intercrossed to form heterozygotes whose phenotypes reveal whether the mutations complement each other. (Only two of the three possible crosses are shown here.) If two mutations are in different genes (such as $ and £), then complementation results in completion of the biochemical pathway (the end product is a blue pigment in this example). If mutations are in the same gene (such as $ and £), no complementation occurs, because the biochemical pathway is blocked at the step controlled by that gene, and the intermediates in the pathway are colorless (white). What would you predict to be the result of crossing $ and ¥? (from Griffiths et al., 2000).

When two independently derived recessive mutant alleles producing similar recessive phenotypes fail to complement, the alleles must be of the same gene.

The terms *dominance, incomplete dominance,* and *codominance* are somewhat arbitrary unless analyzed critically. The type of dominance inferred depends on the phenotypic level at which the observations are being made—organismal, cellular, or molecular. Indeed, the same caution can be applied to many of the categories that scientists use to classify structure and processes; humans for convenience devise these categories for analysis. The type of dominance is determined by the molecular functions of the alleles of a gene and level of analysis.

The leaves of clover plants show several variations on the dominance theme. Clover is the common name for the genus *Trifolium.* There are many species. Some are native to North America, others grow as introduced weeds. Much genetic research has been done with white clover, which shows considerable variation among individuals in the curious V, or chevron, pattern on the leaves. Figure 10.15 shows that the different chevron forms (and absence of chevrons) are determined by *multiple alleles.* In this example, we are dealing with a genetic polymorphism so the wild type/mutant allele symbolism is not used. Study Fig. 10.15 to determine the type of dominance of each allele in various combinations. List the alleles in a way that expresses how they relate to one another in dominance. Are there uncertainties? Does this evidence permit us to say anything about the dominance or recessiveness of allele *v*?

Fig. 10.15 Multiple alleles determine the chevron pattern on the leaves of white clover. The genotype of each plant is shown below it (from Griffiths *et al.*, 2000).

So far, we have mainly been considering diploid organisms—organisms with two homologous chromosome sets in each cell. As seen earlier, the diploid condition is designated $2n$, where n stands for the number of chromosomes in sets of seven chromosomes, so $2n = 14$. The organisms encountered most often in our daily lives (animals and flowering plants) are diploid in most of their tissues.

However, a large part of the earth's biomass comprises organisms that spend most of their life cycles in a haploid condition, in which each cell has only one set of chromosomes. Important examples are most fungi and algae. Bacteria could be considered haploid, but they constitute a special case because they do not have chromosomes of the type considered here. Also important are organisms that are haploid for part of their life cycle and diploid for the other part. Such organisms are said to show *alternation of generations*, referring to the alternation of the $2n$ and n stages. All types of plants show alternation of generations; however, the haploid stage of flowering plants and conifers is an inconspicuous specialized structure dependent on the diploid part of the plant. Other types of plants, such as mosses and ferns, have independent haploid stages.

Do all these various life cycles show Mendelian genetics? The answer is that Mendelian inheritance patterns characterize any species that has meiosis as part of its life cycle, because Mendelian laws are based on the process of meiosis. All the groups of organisms mentioned, except bacteria, undergo meiosis as part of their cycle.

MENDELISM: EXCEPTION OR RULE IN GENETICS

There is also no doubt that Mendelian principles of inheritance hold good for most characters in almost all organisms. But except for the basic principles of inheritance, the whole of modern genetics is an exception to Mendelism. Modern genetics seems to defy if not question the universality of Mendelian inheritance. We shall briefly analyze a few of these aspects.

(a) Mendel believed that one gene governs one character, i.e. advocated the one gene-one character hypothesis. There are many exceptions to this hypothesis. Several genes control the development of one character as in comb shape in fowls, flower color in Indian corn, coat color in mice, etc. (*multigene inheritance*).

(b) Mendel believed that the inheritance of one gene is not influenced by the other and genes assort themselves independently. *Linkage* of genes, wherein characters present in the same chromosome have a tendency to be inherited together. This shows that Mendel's principles are not universal.

(c) In governing a character, one gene will not be influenced by the other according to Mendel. There are several instances of qualitative (flower color in some plants, plumage color in fowls, etc.) and quantitative (skin color in maize, aleurone color in wheat, etc.) *interactions of genes*, again an exception to Mendelism.

(d) Mendel indicated that there are only two alleles for every gene and the characters are always contrasting. This is not true, as proven by the existence of *multiple alleles* and integration of characters as in *quantitative characters*.

(e) According to Mendel, characters are always dominant or recessive. This need not be so. There are instances of *incomplete dominance* or a nonallelic dominant gene suppressing another dominant gene (*epistasis*).

(f) Mendelism is based on *chromosomal inheritance* or *nuclear inheritance*. But there are several instances of inheritance of characters through cytoplasm.

The above facts reveal that every one of the Mendelian hypotheses has been questioned as to its applicability to the entire biological world. Indeed, in modern genetics, Mendelism is something of an exception than a rule. The pattern of inheritance is not as simple as Mendel thought. Mendel was lucky however in choosing the pea plant and characters. Every one of the characters he chose (genes responsible for them) was stated to be present on a different chromosome and no linkage was reported. Had he chosen linked characters, he could not have enunciated the principle of independent assortment. Having said this, it must be conceded that the discoveries of modern genetics have made Mendelism either obsolete or irrelevant. The rules of inheritance propounded by Mendel, though not universally applicable, still hold good for quite a majority of traits and the rules of inheritance remain forever the corner stone on which modern genetics is founded.

Over time the following phenomena have been found to be *exceptions to Mendel's principles*: (1) gene mutations, (2) chromosome mutations, (3) expanded triplet repeats, (4) genome imprinting, (5) complex traits, and (6) multiple gene inheritance. These phenomena are discussed later.

MENDEL'S RESULTS: CHROMOSOME THEORY AND LINKAGE

Mendelian factors, later called *genes*, are located in *chromosomes*. The genes located in the same chromosome, unless separated by long distances, will not exhibit independent assortment. In other words, these genes will show linkage. In light of this fact, we can reexamine the seven pairs of genes studied by Mendel and analyze whether or not they exhibited independent assortment. This is necessary because often a parallel is drawn between the seven pairs of characters used by Mendel and the seven pairs of chromosomes found in pea. This is also used as an argument albeit erroneously, for *independent assortment* observed by Mendel, saying that had he taken more than seven characters, he would have had problems due to linkage. In this case we can examine the relationship between different characters with the standard gene symbols and the chromosomes on which each is located.

In his original paper, Mendel gave results of only two experiments showing independent assortment. One of them involved two characters (*R,r; Ii*) and the other involved three characters (*R,r; I,i; A,a*), such that they represented demonstration of *independent assortment* among only three of the seven characters in three possible combinations (out of a total of 21 possible combinations among seven characters). For other character combinations, experiments were conducted but no results are available in Mendel's original paper. However, he claims that independent assortment was observed in all other combinations as well.

Later, linkage studies conducted by Lamprechi (1961) demonstrated that the seven gene pairs studied by Mendel belonged to only four *linkage groups* (cited from Gupta, 1994). It was also shown that the combinations *I-a*, *v-fa* and *fa-le* despite each belonging to the same linkage group are not linked due to separation by long distances (50 or more map units) and exhibit independent assortment. However, a solitary combination, *v-le*, showed linkage (13%

recombination), suggesting lack of independent assortment. Therefore, it is obvious that either Mendel did not conduct an experiment for independent assortment between *v* or *le* or else overlooked lack of independent assortment in this combination.

CHROMOSOME THEORY OF INHERITANCE

Living cells of most organisms (eukaryotic) are characterized by the presence of a *nucleus*. One component of the nucleus is fine, threadlike strands, the *chromatin*, which consists of *deoxyribonucleic acid* (*DNA*) and basic proteins (*histones* or *protamines*). By the beginning of the twentieth century it was realized that the chromosomes might carry the hereditary material consisting of genes. Chromosomes could be seen under the microscope. During nuclear division the chromatin strands contract and appear thicker as chromosomes under an ordinary light microscope. The number and morphology of the chromosomes are specific, distinct, and ordinarily constant for each species.

Pairing and *disjunction* of *homologous chromosomes* at meiosis explain the mechanism of gene segregation. Independent assortment of genes can likewise be deduced from knowledge of chromosome behavior as observed in microscopic preparations. Diploid cells contain two sets of chromosomes that are derived from two different parents. At meiosis various pairs of chromosomes are assorted and distributed to gametes independent of each other. Gametes may thus contain any mixture of *maternal chromosomes* and *paternal chromosomes*. This is precisely the manner in which the genes behave in inheritance (Fig. 10.16).

Thus chromosomes constitute physical inheritance. In other words, genes are located in the chromosomes. Sutton (1902) and Boveri (1902) independently discovered parallelism between behavior of chromosomes and Mendelian factors (genes) during meiosis and fertilization.

Individuality of chromosomes and genes. Individuality of each chromosome is maintained like the individuality of the genes from generation to generation. It was seen by Boveri (1885) that in the roundworm *Ascaris megalocephala*, a specific number of lobes appear. These lobes are formed during telophase by the free ends of V-shaped chromosomes. The shape and number of lobes are identical in sister cells. In subsequent prophase, chromosomes always appear at the same positions in the lobes as they were at earlier telophase. Similarly, the Mendelian factors (genes) determining different characteristics of an organism are also qualitatively different.

Chromosomes and genes exist in pairs. Chromosomes, like genes, exist in pairs making up the genotype of an individual. Mendel assumed that the hereditary factors exist in pairs. Individuality of each gene is evident from its behavior in F_1 and F_2 generations. Montgomery (1906) observed that the pairing behavior of chromosomes was not random. Homologous chromosomes always showed pairing. Sutton reached a similar conclusion in the grasshopper.

Chromosomal and genetic constitution of a gamete. Each gamete receives one chromosome of each homologous pair. Mendel postulated segregation of the paired genes during gamete formation, each gamete receiving only one factor of each pair.

Independent segregation of chromosomes and genes. The chromosomes of various homologous pairs are randomly assorted in meiosis and segregate independent of the chromosomes of every other pair. Crothers (1913) gave evidence for random orientation of chromosomes from a study of

(a) Segregation of chromosomes and genes

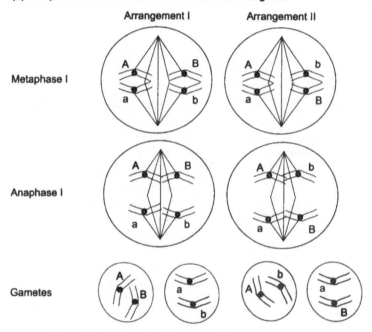

Fig. 10.16 (a) Segregation and (b) independent assortment of chromosomes (from Miglani, 2000)

meiosis in the grasshopper in which $2n = 3$. In one pair of chromosomes, one homolog was larger than the other while another chromosome had no homolog. She was able to observe two metaphase-I arrangements in equal frequency. Mendel held that the factors of each pair are independent of every other pair in their distribution of gametes.

Chromosome and genetic constitution of a zygote. The chromosomes from two parents come together in the zygote, restoring the diploid number of chromosomes for the offspring. Mendel held that the paternal and maternal characters are mixed in the progeny.

Qualitative differences between chromosomes and genes. There are qualitative differences between different chromosome pairs. Every cell must get at least one chromosome of each kind for normal development. Boveri (1904) studied development of the sea urchin ($2n = 36$). It was observed that two cells with identical chromosome number do not necessarily develop in a similar fashion. Every cell must get at least one chromosome of each type for normal development. In a later experiment, Blakslee (1934) observed twelve different types of plants in *Datura*, each having $2n = 25$. If all the chromosomes were qualitatively similar, all plants having one extra chromosome should appear more or less alike. Blakslee observed that shape and size of the capsule differed in each of the twelve types of plants. This indicated the qualitative differences among different chromosomes.

A firm basis for Mendel's laws of heredity exists in the behavior of chromosomes during meiosis and fertilization. This paved the way for the *chromosome theory of inheritance*. Considering the above parallel behavior, Sutton and Boveri independently suggested in 1902 that genes in a linear order are the actual physical hereditary units located in the chromosomes. Since there are two chromosomes of each kind, there must be two genes of each kind, one in each of the homologous chromosomes. Thus chromosomes from parents to offspring carry the genes that actually determine the traits. This led to the formulation of the chromosome theory of inheritance. Chromosome behavior during meiosis is seen under the microscope whereas the existence of genes is inferred from observation of behavior of traits in hybridization experiments. The chromosome theory of inheritance states that chromosomes constitute the physical basis of inheritance and that genes are situated in chromosomes.

Within a decade, a number of experiments provided critical evidence that these relationships were true. Among the first of these experiments were those relating to sexual characteristics. The chromosomal influence on sex, in certain insects, had been shown by McClung (1902) to be associated with a special sex determining "X" chromosome. McClung proposed that a male had one X chromosome per cell (XO) and a female had two (XX), and that the male produced two types of gametes, X and O, in equal proportions. In the insect hemipteran *protenor*, investigated by Wilson (1906), the X chromosome could be clearly seen as a special dark-staining (heteropycnotic) body. As expected, counts of nuclei in protenor tests showed approximately equal amounts of *spermatids* with and without the X chromosome. Since the female gametes all contained X chromosome, the sex of the offspring was determined by the male gamete. Thus, segregation of the X chromosome in the male led to an equal proportion of male and female offspring. This provided an excellent correspondence between segregation of the sex factor and segregation of the sex chromosome.

MULTIPLE ALLELISM

Multiple allelism is a condition wherein a population of two or more alternate alleles represent the same locus in a given pair of chromosomes. Although only two actual alleles of a given gene can exist in a diploid cell, the total number of different allelic forms that might exist in a population is often quite large. The number of possible diploid combinations depends on the number of alleles in a series. If n is the number of alleles in a series, the number of diploid combinations (genotypes) is given by the expression $n(n+1)/2$. For example, if the number of alleles is 5, the number of diploid genotypes in a population will be $\{5(5+1)\}/2 = 15$.

Self-incompatibility alleles in plants. Some plants just do not self-pollinate. A single flower bearing male and female flowers may not produce seed. The same plant may however, cross with certain other plants, so obviously it is not sterile. This phenomenon is known as self-incompatibility. In the case of self-incompatibility, ovum and pollen produced by the same plant do not produce seed. Self-incompatibility has a genetic basis. If a pollen grain bears a self-incompatibility (S) allele that is also present in the maternal parent, then it will not germinate. However, if that allele is not present in the maternal tissue, then the pollen grain produces a pollen tube containing the male nucleus, and this tube mediates fertilization (Table 10.9). A large number of self-incompatibility alleles are known to be present in different plant species. For example, the number of alleles in evening primrose and clover is 50 each while in some species reaches 100.

Self-sterility multiple allelic system in Nicotiana. Multiple allelic systems have also been found in plants, especially for self-sterility genes that prevent fertilization between individuals closely related to each other. As worked out by East and collaborators (1916b) in the tobacco plant, *Nicotiana*, a haploid pollen grain carrying a self-sterility allele, i.e. S^1, will not grow well on a diploid female style carrying the same allele, i.e. S^1S^2, but can successfully fertilize a plant carrying S^2S^3 or S^3S^4, etc. Thus successful fertilizations in these cases, always denote the presence of an allele in the pollen that

differs from the two alleles in the female plant.

Self-sterility systems offer an advantage to new rare alleles and a disadvantage to those that are common. The reason for this is that once an allele becomes common, its frequency is reduced or limited by the many sterile mating combinations to which it is now exposed. Rare alleles, on the other hand, will have successful fertilizations with almost every female plant they meet, until they too become common. In this fashion, self-sterility systems of considerable numbers of alleles can be established, reaching as high as 200 alleles or more in red clover (Bateman, 1947).

Conclusions from Multiple Allelism

The conclusions that can be drawn from studies on multiple alleles are: 1) a gene can exist in several alternate forms; 2) several mutable sites occur within a gene; and 3) members of allelic series can show any type of dominance relationship with one another.

GENE-GENE INTERACTIONS

In characters studied by Mendel, a single gene controlled variation for each trait. Actually, two or more genes may govern variation in many characters. Genes are not merely separate elements producing distinct individual effects but may interact

TABLE 10.9 Genetic control of self-incompatibility

Parents	$S_1 S_2 \times S_1 S_2$	$S_1 S_2 \times S_2 S_3$	$S_1 S_2 \times S_3 S_4$
Pollen	S_1, S_2	S_2, S_3	S_3, S_4
Egg Cells	S_1, S_2	S_1, S_2	S_1, S_2
Progeny	None	$S_1 S_3$	$S_1 S_3 S_1, S_4$
		$S_2 S_3$	$S_2, S_3 S_2 S_4$
Results	Fully incompatible	Semicompatible	Fully compatible

Source: Miglani (2000)

with each other to give completely novel phenotypes. *Gene interactions* may give modified F_2 ratios. This so happens because two or more gene pairs, which interact, influence many characters. This is explained by the few examples here. In all these cases it is assumed that both gene pairs show complete dominance. In the case of independent assortment, we deal with two gene pairs or more, all governing different characters, whereas in *gene-gene interactions* we deal with two or more gene pairs affecting one and the same character. Variations of the latter type show multigene inheritance.

Dominance and Epistasis

In the case of *dominance*, one member of the allelic pair masks the expression of the other member (*interallelic interaction*) whereas *epistasis* is suppression of action of a gene or genes by a gene or genes not allelomorphic to those expressed (*intergenic interaction*). *Epistatic genes* are those that suppress the gene expression while *hypostatic genes* are those whose action is suppressed. Gene interactions are classified on the basis of the manner in which the concerned genes influence or modify the expression of each other. Some of the common interactions are described below.

Recessive epistasis

Complementary gene interaction. Recessive epistasis is exemplified by the work of Bateson and Punnet (1905, 1906, 1908) in the case of flower color in sweet peas. Either gene when homozygous recessive is epistatic to the other. Such genes are called *complementary genes* because two dominant genes complement each other to produce the wild type phenotype. Complementary genes are similar in phenotypic effect when present separately. When together, they interact to produce a wild phenotype. When

two different inbred strains, both with white flowers are crossed, the results obtained are shown in Fig. 10.17. Genes showing this type of interaction are known as complementary genes because two dominant genes interact to produce the wild phenotype (purple in the above example). The mechanism of complementary gene interaction is illustrated in Fig. 10.18.

Another example of complementary gene interaction is provided by HCN content in clovers, in which case two dominant genes, *Ac* and *Li*, are required for production of HCN. Write a biochemical pathway to explain the foregoing gene interaction.

Dominant epistasis—duplicate genes

When two gene pairs seem to be identical in function, either dominant gene or both dominant genes together give the same effect. Such genes are called *duplicate genes* and the type of epistasis is called dominant epistasis. The capsule shape in *Bursa* (shepherd's purse) is a good example (Fig. 10.19). A possible metabolic pathway involved in the aforesaid gene-gene interaction is given in Fig. 10.20. Fruit color in summer squash provides another example of *dominant epistasis* (Fig. 10.21). In this case there is complete dominance but one gene

	Strain 1 White *CCpp*	x		Strain 2 White *ccpp*
F_1	Purple *CcPp*	x	F_1	Purple *CcPp*
F_2	9 *C-P-*			3 *C-pp* 3 *ccP-* 1 *ccpp*
	9 Purple	:		7 White

Fig. 10.17 Complementary gene interaction for flower color in sweet pea. This is one type of recessive epistasis (from Miglani, 2002).

Gene	Product	Representation
C	An enzyme that synthesizes purple pigment	• • • • •
c	No enzyme synthesized, therefore, o pigment formed.	
P	A protein that forms a foundation for the deposition of the pigment	⎡ᴗ ᴗ ᴗ ᴗ ᴗ⎤
p	No protein synthesized, therefore, no foundation formed.	

Genotype (s)	Product(s)	Phenotype	Ratio
C _ P _	• • • • • Foundation + Pigment	Purple	9
C _ p p	• • • • • No foundation + Pigment	White	⎤
c c P _	⎡ᴗ ᴗ ᴗ ᴗ ᴗ⎤ Foundation + No pigment	White	7
c c p p	No foundation + No pigment	White	⎦

Fig. 10.18 Molecular model explaining complementary gene interaction in case of flower color in sweet peas (from Miglani, 2002)

Triangular
$T_1T_1\ T_2T_2$ x   Ovoid
$t_1t_1\ t_2t_2$

F_1 Triangular
$T_1t_1\ T_2t_2$ x F_1 Triangular
$T_1t_1\ T_2t_2$

F_2 9 T_1 - T_2 -
3 T_1 - t_2t_2 1 $t_1t_1\ t_2t_2$
3 $t_1t_1\ T_2$ -

15 Triangular   1 Ovoid

Fig. 10.19 Dominant epistasis exhibited by duplicate genes for fruit shape in Shepherd's purse (from Miglani, 2002)

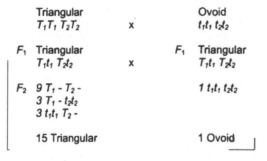

Fig. 10.20 Possible metabolic pathway for duplicate epistasis shown by genes for capsule shape in Bursa (shepherd's purse) (from Miglani, 2002)

when dominant is epistatic to the other. The metabolic pathway given in Fig. 10.22 satisfies the observations made in this cross.

Suppression gene interaction

Suppression of one gene by another is illustrated by inheritance of production of a chemical called "malvidin" in *Primula*.

Dominant gene K is required for production of pigment. However, action of this gene may be suppressed by a nonallelic dominant gene D. Selfing of F_1 dihybrid ($KkDd$) produces in F_2 13 malvadin minus:3 malvidin plus individuals. Confirm the results through suppression gene interaction hypothesis with the help of a cross.

Polymeric effect

Polymeric effect is shown between dominant genes in the case of fruit shape in summer squash (Fig. 10.23). Either dominant gene shows the same effect but interaction between dominant genes produces a novel phenotype. A metabolic pathway for the polymeric effect is given in Fig. 10.24.

The relationship between different types of gene-gene interactions can be understood by studying Table 10.10.

To emphasize once again, all the cases described above hold true when there is *complete dominance* for both gene pairs. The F_2 ratios are modified to give a higher number of phenotypic classes when there is incomplete dominance for two gene pairs (1:2:2:4:1:2:1:2:1) or for one gene pair (3:6:3:1:2:1) (Table 10.11). Try to arrive at these ratios.

Conclusions from Gene-Gene Interactions

The discovery of gene-gene interaction dispelled the notion that each gene produced a single, non-overlapping individual effect and that these effects fitted together like a mosaic to produce the organism. This viewpoint is called *preformationism*. The present view is that development of an organism results from the interaction of gene products with each other and with the environment. This viewpoint is called *epigenesis*.

Multiple genes govern inheritance of a trait. In the final expression of a trait, no single gene acts by itself and the phenotype is an outcome of several integrated reactions. Gene interactions will change the expression of a phenotype beyond the limits of a gene.

White	x	Green
WWYY		wwyy

| F_1 White | x | F_1 | White |
| WwYy | | | WwYy |

F_2 9 W_Y_ 3 W_yy + 3 wwY_ 1 wwyy

12 White   6 Yellow   1 Green

Fig. 10.21 Fruit color in summer squash provides example of dominant epistasis (from Miglani, 2002)

Spherical	x	Spherical
AAbb		aaBB

| F_1 Disc | x | F_1 | Disc |
| AaBb | | | AaBb |

F_2 9 A_B_   3 A_bb + 3 aaB_   1 aabb

9 Disc   6 Spherical   1 Elongate

Fig. 10.23 Polymeric effect in case of fruit shape in summer squash (from Miglani, 2002)

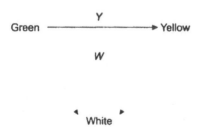

Fig. 10.22 Possible metabolic pathway for dominant epistasis shown by genes for fruit color in summer squash (from Miglani, 2002)

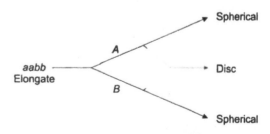

Fig. 10.24 Possible metabolic pathway for polymeric effect shown by genes for fruit shape in summer squash (from Miglani, 2002)

TABLE 10.10 Relationship between different types of gene interactions

Types of epistasis	Genotypes			
	A-B-	A-bb	aaB-	aabb
No epistasis: classical ratio	9	3	3	1
Supplementary interaction Recessive epistasis *aa* epistatic to *B, b*	9	3	4 (3 + 1) Homozygous recessive at one locus produces certain phenotype regardless of allelic condition at other locus	
Dominant epistasis *A* epistatic to *B, b*	12 (9 + 3) Dominant allele at one locus. *A* produces certain phenotype regardless of allelic condition at other locus		3	1
Dominant and recessive epistasis (Inhibitory factors) *A* epistatic to *B, b* *bb* epistatic to *a, A*	13 (9+3+1)		3	
Duplicate recessive epistasis (complementary interaction) *aa* epistatic to *B, b* *bb* epistatic to *A, a*	9	7 (3+3+1) Homozygous recessive at either or both loci produces same phenotype		
Duplicate dominant epistasis (Duplicate factors) *A* epistatic to *B, b* *B* epistatic to *a, A*	15 (9+3+3) Dominant allele at either or both loci produces same phenotype			1
Duplicate genes with cumulative effect (Polymerism) *A-bb* and *aaB-* yield identical phenotype	9	6 (3+3) Dominant allele at either locus produces some phenotype		1

Source: Miglani (2000)

TABLE 10.11 Modifications in F_2 ratios of a dihybrid cross when dominance is lacking for both or one gene pair

Genotype									Dominance lacking in
AABB	AABb	AaBB	AaBb	AAbb	Aabb	aaBB	aaBb	aabb	
1	2	2	4	1	2	1	2	1	Both pairs
3		6		1	2	3		1	One pair

Source: Miglani (2000)

PLEIOTROPY

In general, one gene affects a single character but many cases are known in which a gene influences more than one character. Such genes are known as *pleiotropic genes* and the condition is termed *pleiotropy*. In the pea plant, the gene that affects flower color also influences the color of the seed coat and color of leaf axil. In cotton, the gene that affects the fiber (lint) character also affects plant height, boll size and fertility. Pleiotropy helps in understanding various developmental processes. Several different phenotypic effects result from the action of a single gene.

TOTIPOTENCY

Totipotency is the ability of a cell to proceed through all stages of development and thus produce a normal adult. Each cell contains a complete set of DNA even after differentiation and, hence, each cell of a multicellular eukaryote is totipotent, i.e. it has the capability of developing into a complete individual. Thus all the genes of a cell have the potential to express themselves. Totipotency of the cell is presently used to maintain lines of trees and to produce many ornamental and cultivated plants. In development, a particular gene expresses itself only when the cell requires its product otherwise it remains turned off. Genes that are expressing may do so by performing some primary cellular function. The products of various genes interact to produce a particular phenotypic effect.

HOMEOSTASIS

The developmental pathways of organisms are genetically adjusted to produce the characteristic morphology of the species regardless of variations in internal and external conditions of development. This is termed *genetic homeostasis* or *developmental homeostasis*. Lerner (1954) suggested that more heterozygous individuals should be characterized by increased developmental stability. This view was supported by observations of Eanes (1978) in different organisms that variance of two morphometric characters among heterozygotes was lower than among homozygotes.

Genetic homeostasis is thus a mechanism for promoting the stability of phenotypic expression of a genotype when grown over a wide range of environments. *Homeostatic genotypes* are those, which interact with variable environments in such a way as to maintain relative uniformity of an observable character. Stability of development in variable environments implies physiological adjustments and the capacity of such adjustment is a property of homeostatic genotypes.

Trehan and Gill (1983) assessed the role of each of the three *allozymes* present in the heterozygote in terms of contribution of each component to higher stability of the heterozygote. Stability pattern of 1:1 mixture of flies of two different homozygous genotypes was compared with the heterozygote; the latter showed much better stability. This means that it is the unique *heterodimeric enzyme*, which is responsible for increased stability in the heterozygotes. To support this contention, isolation and partial purification of the three allozymes revealed electrophoretically by the heterozygote was done. The *heterodimeric allozyme* was found to be superior in homeostatic ability to the two *homodimeric allozymes*. Thus it provided evidence that increased stability of acid phosphatase in the heterozygote results from increased stability of the heterodimeric allozyme which constitutes half of the total enzyme present in the heterozygotes.

Homeostasis, as explained above, may operate in cross-pollinated crops. Homeostasis in self-pollinated populations cannot be explained on the basis of heterozygous superiority. Perhaps homeostasis in such populations is brought about by selection in favor of genes that can perform optimally in a range of environments, rather than in a single environment.

MENDELIAN ANALYSIS IN DIPLOID EUKARYOTES

Time of Segregation

Waxy and nonwaxy pollens in corn are used to study the *time of segregation* of members of the same allelic pair. Nonwaxy pollen stains blue with iodine, due to a dominant nuclear gene *Wx*. Waxy pollen stains red with iodine, due to a homozygous recessive condition (*wx wx*). When a cross was conducted between nonwaxy and waxy pollen producing plants and pollen produced by F_1 seeds (*Wx wx*) stained with iodine, about half were red and half blue. Of 6,919 pollen studied by Demerc (1929), 3,437 stained blue and 3,482 red. The result is indicative of the fact that segregation of allelic members is complete before the gametes are formed. To be precise, we say that segregation occurs during meiosis.

Endosperm Genetics

Double fertilization results in formation of endosperm tissue, which partakes, of both paternal and maternal inheritance. In a plant in which the ovary wall and seed coat are thin and transparent, as in a kernel of corn, a direct effect of the male gamete on the character of the endosperm is evident. Thus, if an ear of maize from a type normally bearing white endosperm is pollinated by pollen from a yellow race, the endosperm of the seed produced will be yellow. This direct effect of male gamete on

tissue other than embryonic is known as xenia. Reciprocal crosses between plants producing sugary and starchy pollen (Fig. 10.25) do not show the *xenia effect*. In both hybrids the endosperm is starchy no matter which allele enters through the female line. One *Su* gene is dominant. Other alleles however, may not show the same kind of dominance relations in the endosperm. This is shown in crosses between corn plants producing flinty and floury seeds (Fig. 10.26). The flinty phenotype is due to a dominant gene *F*, the floury due to the homozygous condition *ff*. In these *reciprocal crosses*, the endosperm phenotype differs. Here two recessive alleles (*ff*) suppress the expression of a dominant allele (*F*). Geneticists studying *endosperm traits* can keep one generation ahead in the game of genetic analysis compared to some *sporophytic character*.

Linkage and Genetic Recombination

Genetic recombination is the name given to the redistribution of information inherited from parents in the progeny. Recombination involves the physical exchange of material between chromatids of homologous chromosomes. Nonparental combinations of alleles, obtained in the segregating generations, are called recombinants.

Genetic recombination is a fundamental property of all living systems starting from RNA-containing viruses and ending with higher plants and animals. Although this term applies to the recombination of both linked and nonlinked genes, it is applied in the narrow sense to mean recombination of linked markers (between different genes or alleles). Whereas recombination of the unlinked genes is based on the mechanism of free combination of chromosomes in meiosis, recombination of linked genes

Cross I

Starchy (*Su Su*)	x	Sugary (*su su*)	

egg (*Su*)	pollen (*su*)	Starchy embryo (*Su su*)

Polar nuclei (*Su + Su*)	pollen (*su*)	Starchy endosperm (*Su Su su*)

Cross II

Sugary (*su su*)	x	starchy (*Su Su*)	

egg (*su*)	pollen (*Su*)	Starchy embryo (*Su Su*)

Polar nuclei (*su + su*)	pollen (*Su*)	Starchy endosperm (*Su su su*)

Fig. 10.25 Results of crosses between sugary and starchy pollen producing strains of corn for embryonic and endospermic trait recessiveness over two *su* alleles (from Miglani, 2000)

Cross I

Flinty (*FF*)	x	Floury (*ff*)	

Egg (*F*)	Pollen (*f*)	Flinty embryo (*Ff*)

Polar nuclei (*F + F*)	Pollen (*f*)	Flinty endosperm (*FFf*)

Cross II

Floury (*ff*)	x	Flinty (*FF*)	

Egg (*f*)	Pollen (*F*)	Flinty embryo (*Ff*)

Polar nuclei (*f + f*)	Pollen (*F*)	Floury endosperm (*Fff*)

Fig. 10.26 Results of crosses between flinty and floury seed producing strains of corn for embryonic and endospermic trait (from Miglani, 2000)

takes place through an exchange of segments of homologous chromosomes, a process known as *"crossing over"*. In crossing over, a reciprocal exchange of information takes place between homologous chromosomes, while in gene conversion, the exchange is predominantly nonreciprocal in character.

All the organisms have many more genes than the chromosomes. For example, cultivated wheat has only 21 pairs of chromosomes but over 20,000 genes. This implies that each chromosome must contain many genes. Genes located in the same chromosome would not be expected to assort independently. Such genes are said to show *linkage*.

Terminology

Linkage is the tendency of a parental combination of characters to stay together in the F_2. It is also defined as the tendency of a parental combination of genes to stay together in F_2. Association of genes as a result of their occurrence in the same chromosome is also termed linkage. The tendency of two dominant characters, inherited by F_1 from one parent, to stay together in F_2 is termed *coupling linkage* (Fig. 10.27). The tendency of two dominant characters inherited in an F_1—one from one parent and the other from the other parent, to stay apart in the F_2 generation is known as *repulsion linkage*. Crossing over is the exchange of corresponding segments between *chromatids* of *homologous chromosomes*, by breakage and reunion, following pairing. This process is inferred genetically from the recombination of linked factors in the progeny of heterozygotes and cytologically from the formation of chiasmata between homologous chromosomes. *Chiasma* (plural *chiasmata*) is the X-shaped configuration of the chromosomes in a bivalent in prophase of meiosis I, usually

Coupling phase of linkage

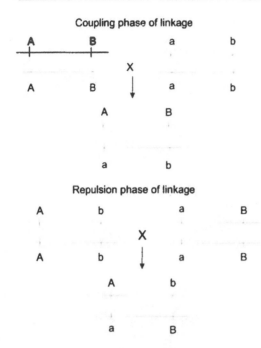

Fig. 10.27 Coupling and repulsion phases of linkage (from Miglani, 2000)

the visible result of prior cytological crossing over. The distance between genes is expressed as *map unit* (*centimorgan*). One map unit equals 1% crossing over. A group of genes showing linkage with one another forms one *linkage group*, i.e. a group of genes located in the same chromosome. A *linkage map* is the arrangement of genes in a linkage group in such a way that the distance between any two of them reflects percent crossing over between them. The *coefficient of coincidence* (CC) refers to the phenomenon which expresses ratio of frequencies of observed and expected double crossovers. It expresses strength of *interference* and is expressed as a ratio of observed frequency of double crossovers divided by expected frequency of double crossovers.

Interference reflects frequency of double crossovers, which fail to be recov-

ered in relation to their expected frequency. Mathematically, interference = 1 – CC.

Detection and Estimation of Linkage

Two-point Crosses. Bateson and Punnet (1905, 1906, 1908) working with sweet pea reported linkage between two gene pairs. They studied flower color (purple and red) and pollen grain shape (long and round). Gene symbols are: R = purple flowers; r = red flowers; R_o = long pollen grain; and r_o = round pollen grain. They crossed purple, long variety with red, round variety. All the F_1s were purple, long. The F_2 progeny were: purple, long – 296, purple, round – 19, red, long – 27, and red, round – 85. In F_2 of a dihybrid cross we except the four phenotypes in the following ratio: 9 purple, long : 3 purple, round : 3 red, long : 1 red, round. Do the aforesaid data fit into the expected ratio? Results of chi-square analysis of the data are given in Table 10.12. The calculated value (χ^2_c) is compared with the theoretical value (χ^2_t). The accepted probability value (p) is 0.05. The degree of freedom (d. f.) for this test would be: number of classes (4) - 1 = 3. χ^2_t, 0.05, 3 = 7.815. Since $\chi^2_c > \chi^2_t$, the observed numbers in different phenotypic classes do not fit into the expected 9:3:3:1 ratio. This means the two gene pairs did not assort independently. Deviation from the Mendelian ratio is not sufficient justification however, for stating that the two gene pairs in question are linked. Linkage between two gene pairs is not the only cause influencing frequencies of certain classes. Abnormal segregation for alleles at a locus can be another reason for departure from the expected ratio. *Chi-square* must first test whether each gene pair segregates and then whether two gene pairs segregate independent of each other again as expected in Mendelian fashion.

TABLE 10.12 Chi-square analysis of two-point cross data on sweet peas

Phenotype	Observed	Expected	$\dfrac{(Obs. - Exp.)^2}{Exp.}$
Purple, long	296	240	$(56)^2/240 = 13.066$
Purple, round	19	80	$(61)^2/80 = 90.262$
Red, long	27	80	$(53)^2/80 = 35.112$
Red, round	85	27	$(58)^2/27 = 124.592$
			$\chi^2_c = 263.032$

Source: Miglani (2000)

If linkage is complete, only two parental classes of offspring are produced. If linkage is incomplete and some recombination occurs, then recombinant classes are also produced but these will not be as frequent as the parental classes. The consequence of incomplete linkage is departure from the frequencies of phenotypes expected according to independent assortment.

Detection of linkage without partitioning chi-square. In cases where segregation is non-Mendelian, a cross $Rr \times Rr$ dose not produce purple and red in the expected 3:1 ratio. Similarly, cross $R_o r_o \times R_o r_o$ dose not produce long and round in the expected 3:1 ratio. For detection of linkage, three chi-square values must be calculated, one each for segregation of the two individual gene pairs and a third for their joint segregation. If the gene pairs singly show Mendelian segregation but jointly fail to show independent assortment, it can be concluded with certainty that the genes in question are linked. This is called detection of linkage. The same principle is applied for detection of linkage in a test cross progeny. Here segregation at individual loci is tested using a 1:1 ratio and independent assortment is tested using a 1:1:1:1 ratio.

Estimation of linkage. Estimation of linkage means measuring the distance between linked genes. For estimation of linkage, percent recombination is calculated as follows: In the F_2 progeny, two classes –

purple, long and red, round—are recombinant types. How do the recombinant types appear in the F_2 progeny? The recombinant types in the F_2 progeny most likely originated from new combinations of the linked genes. The mechanism responsible for origin of new phenotypic combinations is crossing over.

Crossing over is an exception to linkage. Crossing over occurs during *pachytene* of meiosis. Visible consequences of crossing over are seen in the form of X-shaped configurations, called chiasmata, during *diplotene* of prophase I of meiosis. The gene hypothesis of the aforesaid cross is given in Fig. 10.28. *Recombination frequency* is a function of distance between genes in the chromosomes. The greater the distance between two genes, the greater the recombination frequency observed. Let us estimate recombination frequency in the case already discussed. The two parental classes, purple, long and red, round, appear more frequently than expected and the two recombinant classes, purple, round and red, long, appear less frequently than expected because of linkage. The following procedure is followed for estimation of linkage in a dihybrid cross.

Mark the parental pair of *complementary classes* as "P_1" and "P_2". The pair showing the maximum number of flies belongs to the parental types. Mark the pair of complementary classes which arises due to

Parents : Purple, Long   Red, Round

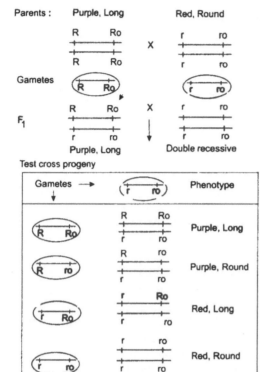

Fig. 10.28 Gene hypothesis showing a two-point test cross in sweet peas (from Miglani, 2000)

Estimation of linkage is done from F_2 data mostly in the case of plants in which crossing over occurs in both sexes in almost equal frequency. More appropriate method of detection of linkage in an F_2 population involves partitioning of chi-square (χ^2).

Detection of linkage through partitioning of chi-square. One has to first test segregation of the two gene pairs separately for their agreement with the respective expected ratios. If two characters controlled by gene pairs Aa and Bb are segregating independently in ratio l_1:1 and l_2:1, respectively, then the joint segregation of the two gene pairs will be tested for the expected ratio of $l_1 l_2$:l_1:l_2:1. In an F_2 population, each of the two gene pairs individually segregate in a ratio 3:1. Simultaneous segregation of these two gene pairs would be expected in the phenotypic ratio of AB 9:Ab 3:aB 3:ab 1 with AB and ab being the parental types and Ab and aB the recombinant. Detection of linkage under such situations is done in a systematic manner by partitioning the total χ^2 for 3 degrees of freedom into three components of 1 degree of freedom each—one for deviation of Aa segregation from 3:1, second for deviation of Bb segregation from 3:1, and third detecting association of the two gene pairs by testing for deviation for joint segregation from 9:3:3:1. The information provided in Table 10.13 form the basis for χ^2 formula to partition the 3 degrees of freedom into three components.

If the observed frequencies in four F_2 phenotypic classes AB, Ab, aB, and ab are a_1, a_2, a_3, and a_4 respectively, for the practical working of the method the following χ^2 formulae can be deduced and used as indicated in Table 10.14.

crossing over as "R_1" and "R_2". The pair showing the least number of flies belongs to this category.

To calculate percent recombination, use the following formula: Number of recombinants/Total progeny × 100. The percent recombination in the above case is 46/427 × 100, i.e. 10.7. The distance between linked genes is measured in map units. One map unit equals 1% crossing over that is equal to one centimorgan (cM). Arrangements of genes in a linkage group in such a way that the distance between any two of them reflects percent crossing over between them is known as a linkage map which uses the symbols of the mutant genes. The linkage map between gene r and r_o is:

If χ^2_A and χ^2_B show goodness of fit but $c^2_{A/B}$ value is above the theoretical χ^2 value, this provides very strong evidence of linkage. Similar formulae can be deduced for detection of linkage in test cross data of a plant population. Seek guidance from your teacher to take up this exercise.

Estimation of linkage in F_2 population. When the existence of linkage has been conclusively proven, then intensity of linkage can be measured. The hypothesis for estimation of linkage is provided by modern genetic theory which regards these deviations from independence as resulting from genes controlling the characters situated on the same chromosome. Intensity of linkage is measured in inverse sense as the fraction of total number of chromosome pairs in which crossing over takes place at gametogenesis. This is known as *recombination fraction*. The smaller this fraction, the more intense the linkage.

In a double heterozygote, two relations in arrangement may exist between genes. Either A - B or a - b may be on the same chromosomes or A - b and a - B may be the arrangement found. In the former case A - b and a - B are the recombinant types. In the latter case recombinants are A - B and a - b. These arrangements are differentiated under the terms coupling and repulsion. Thus coupling is the AB/ab type and repulsion is the Ab/aB type in F_1 individuals.

Several methods of estimating linkage have been proposed from time to time. A good method should fulfill two criteria: the estimation obtained should tend to the theoretical value as the sample is enlarged (consistency) and the estimate should have the lowest possible variance (efficiency). The recombination fraction is denoted by p. It is a measure of frequency of crossing over between the two chromosomes in the region delimited by the two genes under consideration.

Expected frequencies of four phenotypes AB, Ab, aB, and ab can be easily determined. If p and p' are the recombination fractions in males and females respectively, both assumed to be in the coupling phase, the frequencies of four types of gametes would be as given in Table 10.15. From the above gametic frequencies, genotypic frequencies can be obtained which can be grouped into the four phenotypic classes given in Table 10.16.

Since frequencies depend upon θ only, this is the quantity that can be estimated from the observed frequencies for this type of data. Three methods are available for estimation of linkage.

1. *Emerson's Method*

$$\theta = \frac{E - M}{n}$$

where, E is the sum of observed frequencies of parental classes, M the sum of observed frequencies of recombinant classes, and n is $E + M$.

TABLE 10.13 Basis of partitioning chi-square into three components for detection of linkage

Segregation	Phenotypes			
	AB	Ab	aB	ab
A/a	+1	+1	−3	−3
B/b	+1	−3	+1	−3
A/B	+1	−3	−3	+9

Source: Miglani (2000)

TABLE 10.14 Formulae for partitioning of chi-square for detection of linkage

Formula	Used for testing
$\chi^2_A = \dfrac{(a_1 + a_2 - 3a_3 - 3a_4)^2}{3n}$	Segregation for A,a
$\chi^2_B = \dfrac{(a_1 - 3a_2 + a_3 - 3a_4)^2}{3n}$	Segregation for B,b
$\chi^2_{A/B} = \dfrac{(a_1 - 3a_2 - 3a_3 + 9a_4)^2}{9n}$	Independent assortment for A/B

where, $n = (a_1 + a_2 + a_3 + a_4)$

Source: Miglani (2000)

TABLE 10.15 Expected frequencies of male and female gametes F_2 generation

Parent	AB	Ab	aB	ab
Male	$1/2(1 - p)$	$1/2p$	$1/2p$	$1/2(1 - p)$
Female	$1/2(1 - p')$	$1/2p'$	$1/2p'$	$1/2(1 - p')$

Source: Miglani (2000)

TABLE 10.16 Gametic and genotypic frequencies

AB	Ab	aB	ab
$1/4\theta$	$1/4(1 - \theta)$	$1/4(1 - \theta)$	$1/4\theta$

where $\theta = (1 - p)(1 - p')$

Source: Miglani (2000)

This will provide an unbiased estimate of θ. The standard error of the estimate of θ can be obtained from the formula:

$$V = \frac{1 - \theta^2}{n}$$

If $\quad p = p'$, then $\theta = (1 - p)^2$

$$p = 1 - \sqrt{\theta}$$

$$V_p = \frac{V_\theta}{4\theta}$$

S.E. $(p) = \sqrt{V_p}$

2. *Maximum Likelihood Method*

In the case of F_2, it can be shown that

$$\frac{a_1}{2 + \theta} - \frac{a_2}{1 - \theta} - \frac{a_3}{1 - \theta} - \frac{a_4}{\theta} = 0$$

On simplification, this gives the following quadratic equation for θ:

$$2a_4 + \{a_1 - 2(a_2 + a_3) - a_4\}\theta - n\theta^2 = 0$$

Substitute the values of a_1, a_2, a_3 and a_4 and solve the quadratic equation for θ using the formula:

$$\theta = \frac{-b \pm \sqrt{(b^2 - 4ac)}}{2a}$$

where, $a = -n$, $\quad b = (a_1 - 2a_2 - 2a_3 - a_4)$, and $c = 2a_4$.

We know that $= (1 - p)(1 - p')$

If $\quad p = p'$, then $= (1 - p)^2$

$$p = 1 - \sqrt{\theta}$$

In order to calculate standard error for estimate of θ, first calculate V_θ

$$V_\theta = \frac{2(1 - \theta)(2 + \theta)}{n(1 + 2\theta)}$$

$$V_p = \frac{V_\theta}{4\theta}$$

S.E. $(p) = \sqrt{V_p}$

3. *Product-ratio Method*

In this method, we equate the ratio of the product of extreme classes to the product of middle classes, with its theoretical value. This gives us the required equation for estimation of θ.

$$z = \frac{\text{Product of parental classes}}{\text{Product of recombinant classes}}$$

$$= \frac{(2 + \theta)\theta}{1 - 2\theta + \theta^2}$$

This is quadratic in θ.
Solving this equation we get

$$\theta = \frac{1 + z - \sqrt{(1 + 3z)}}{z - 1}$$

The standard error of estimate of p from V_θ:

$$V_\theta = \frac{1 - \theta^2}{n}$$

Then, $V_p = \dfrac{V_\theta}{4\theta}$

where V_p is the variance of p.
Standard error of p is calculated by using the formula

S.E. $(p) = \sqrt{V_p}$.

Three-point test cross. For detection of linkage in test cross progeny of plants and autoso-

mal recessive genes of animals, the following procedure is used. Mating between a heterozygote *AaBbCc* and a triple homozygote *aabbcc* should produce eight classes of offspring: ABC (*AaBbCc*), Abc (*Aabbcc*), aBC (*aaBbCc*), ABc (*AaBbcc*), abC (*aabbCc*), AbC (*AabbCc*), aBc (*aaBbcc*), and abc (*aabbcc*) (Fig. 10.29). If assortment between *Aa*, *Bb* and *Cc* is independent, each of the foregoing eight classes should be in equal ratios of 1/8. If linkage is complete, only two parental classes of offspring are produced. If

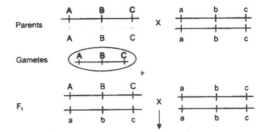

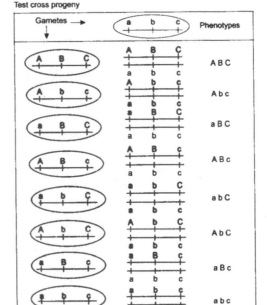

Fig. 10.29 Gene hypothesis of a three-point test cross involving autosomal genes (from Miglani, 2000)

linkage is incomplete and some recombination occurs, then recombinant classes are also produced but these will not be as frequent as the parental classes. The consequence of incomplete linkage is departure from the frequencies of phenotypes expected according to independent assortment.

In case segregation is non-Mendelian, a cross $Aa \times aa$ does not produce expected $1/2$ Aa:$1/2$ aa, cross $Bb \times bb$ does not produce expected $1/2$ Bb:$1/2$ bb, and cross $Cc \times cc$ does not produce expected $1/2$ Cc:$1/2$ cc offspring. Due to non-Mendelian segregation at one or more loci the eight phenotypes in the three-point test cross progeny will not be observed in equal ratios of $1/8$.

Thus for detection of linkage in a three-point test cross, four χ^2 values must be calculated: one each for segregation of the three individual gene pairs and the fourth for their joint segregation. The following steps are followed for estimation of linkage:

The first step is to ascertain the order of the three genes. For this, arrange the eight phenotypic classes in four complementary pairs. Work out the total number of flies in each pair of classes. Mark the parental pair of complementary classes as "P". The pair showing the maximum number of flies belongs to the parental types. Mark the pair of complementary classes which arises due to double crossing over as "DCO". The pair showing the least number of flies belongs to this category. Can you think why? Compare parental and double crossover pairs of complementary classes and determine the gene that changes position. The gene that changes position lies in the middle.

Rearrange the order of genes in the eight phenotypic classes according to the gene order worked out above in all the eight phenotypic classes if the given arrangement is different. Write the genotype of the heterozygous parent of the TC_1 progeny in

proper gene order. By assuming single crossing over between the first two genes, work out the resultant phenotypic classes. Mark these classes as "SCOI". Similarly, assuming single crossing over between the second and third genes, work out the resultant classes and mark them "SCOII". Attempt double crossing over, one event between the first two genes and the other between the second and third genes. What are the resultant phenotypes? Do they match with the classes marked "DCO" in the beginning as those resulting due to double crossing over? If this matching does not occur, you have committed some mistake. Go over the above steps again to correct yourself.

Calculate percent recombination between the first two genes by using the formula: % Recombination = (SCOI + DCO)/N × 100. Calculate percent recombination between second and third genes by using the formula: % Recombination = (SCO II + DCO)/N × 100.

Now construct a linkage map. Show the symbols of the mutant genes indicating distance between them in map units along a straight line.

Calculate the *coefficient of coincidence* (CC) by using the formula: Observed frequency of double crossovers (O_f)/Expected frequency of double crossovers (E_f), where, O_f = DCO/N and E_f = [SCO I + DCO)/N] × [(SCO II + DCO)/N]. See whether the value of CC is less or more than one? What does this value signify?

Calculate the *interference*. Mathematically, interference = $1 - CC$.

Crossing over Occurs at Four-strand Stage

Evidence that crossing over takes place at the four-strand stage comes from *Neurospora* (Fig. 10.30). This experiment is briefly described below:

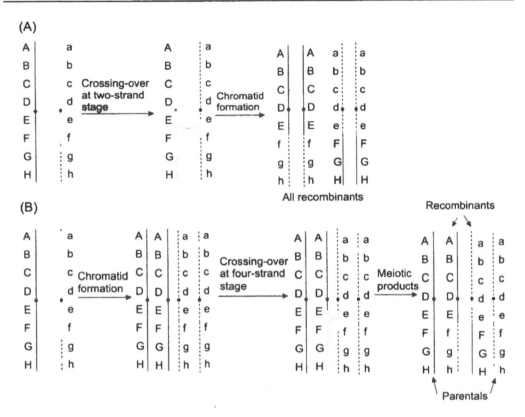

Fig. 10.30 Tetrad analysis of *Neurospora* showing that crossing over takes place at four-strand stage (from Miglani, 2000)

Tetrad Analysis in Neurospora. A cross between a mutant strain carrying linked gene markers and a wild-type strain is expected to produce progeny of recombinant types when crossing over is assumed to take place at the *two-strand stage* (Fig. 10.30A). However, if crossing over is assumed to take place at the four-strand stage, the parental and recombinant type progeny will be expected to appear in 1:1 ratio (Fig. 10.30B). Actual cross yields progeny as expected in the second assumption, thus providing proof that crossing over takes place at the four-strand stage.

Mechanism of Recombination

Theories explaining the *mechanism of recombination* may be divided into three general groups: breakage and copying, complete copy choice, and breakage and reunion.

Breakage and copying hypothesis. This theory (Belling, 1933) proposed that the recombinant chromosome is formed by utilizing the physical section of the parental chromosome and by copying the other. Thus the recombinant chromosome retains part of the parental chromosome due to breakage and at the same time contains part of the new chromosome synthesized by copying the other strand (Fig. 10.31A).

Complete copy choice hypothesis. Lederberg (1955) gave a modified version of Belling's theory. The entire recombinant chromosomes arise from newly synthesized sections which have copied part of their sequence from a section of one parental and the other part from a section of the other parental chromosome (Fig. 10.31B). This hypothesis relates recombination to replication of the hereditary material. This hypothesis requires that homologous chromosomes shall be paired at least at certain points along their length at the time of replication and that new strands may be replicas of one parental strand in one region and of the other in another region. This theory signifies that replication occurs at the chromatid level. It also provides an explanation of how within a tetrad on site may show a 2:2 ratio and a neighboring one a 3:1 ratio (gene conversion).

Breakage and runion hypothesis. According to the theory proposed by Darlington (1937), the two pieces of parental chromosomal information that are combined in the recombinant chromosome arise from the physical breaks in the parental chromosomes with subsequent physical exchange (Fig. 10.31C). The paired parental chromosomes break and rejoin at meiosis, resulting in formation of chiasmata. Recombination does not depend upon the synthesis of new DNA molecules since the information entering the recombinants is merely transferred from the two parents.

Significance of Crossing over

Crossing over is a means of introducing new combinations of genes and hence traits. It thus increases genotypic and phenotypic variability, which is useful for natural selection and for adjustment under the changed environment. Frequency of crossing over depends on the distance between two genes. This serves as a basis for preparing linkage chromosome maps. Linkage maps which use the phenomenon of crossing over as a basis, prove that genes lie in linear fashion in the chromosome. Useful recombinants produced by crossing over are picked by breeders to produce new varieties of crop plants and animals.

MENDELIAN ANALYSIS IN HAPLOID EUKARYOTES

First- and Second-division Segregation

To know the genotype of a particular ascospore arising from crosses between strains of *Neurospora crassa*, ascospores must be isolated and cultured separately. This is called *tetrad analysis*. Because of ordered tetrads in *Neurospora*, the distance between a gene locus and its centromere can be determined. When two members of an allelic pair separate during meiosis I, they are said to show *first-division segregation* (Fig. 10.32A); when two members of an allelic pair do not separate during meiosis I but rather at meiosis II, they are said to show *second-division segregation* (Fig. 10.32B). Since *Neurospora* has ordered tetrads, the tetrads resulting from first- and second-division segregation can be identified.

Different Types of Tetrads and Their Origin

To understand the *types of tetrads*, let us consider a cross $+ + \times a\ a$ in *Neurospora crassa*. In tetrad analysis, if the ascospores have the genotype of one or the other parent (e.g. *a b, a b, + +, + +*), the tetrad is classified as *parental ditype* (PD). If the ascospores have the genotype of neither of the parents, i.e. all ascospores have recombinant types (e.g. *a +, a +, + b, + b*), the tetrad is classified as *nonparental ditype* (NPD), If the ascospores of a tetrad include two parental types and

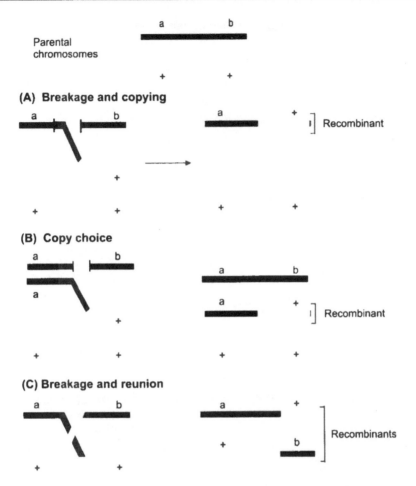

Fig. 10.31 Possible explanations for recombination events: (A) breakage and copying; (B) copy choice; and (C) breakage and reunion (from Miglani, 2000).

two nonparental types (e. g. $a+, +b, a\,b, ++$), the tetrad is classified as *tetratype* (TT). The origin of different types of tetrads is explained in Table 10.17.

Two things must be remembered while mapping genes in *Neurospora crassa*. First, if 10% of the asci in an experiment show crossing over (second division segregation), the percentage of recombination will be only 5 because only two of four chromatids participate in any one crossing over event. Second, if each tetrad (4 chromatids) has a

crossover between centromere and gene, only 50% recombination occurs.

Measuring Distance of a Locus from a Centromere

From the data given in Table 10.18, we shall map the distance of locus t from its centromere. Noncrossover asci (PD) are: $t\,t++$ and $++t\,t$, their total $39+33=72$. Crossover asci (NPD) are: $+t\,t+, t++t, +t++t$, and $t++t$, their total $5+6+9+8=28$. Number of recombinant spores is equal to 28.

TABLE 10.17 Explanation for origin of different kinds of tetrads in the cross *a b* x + + in *Neurospora*

Tetrad type	Event	Products
Parental ditype (PD)	(a) No CO	1/2 + + : 1/2 *a b*
	b) Two-strand DCO between *a-b*	
Nonparental ditype (NPD)	Four-strand DCO between *a-b*	1/2 + *b* : 1/2 *a* +
Tetratype (TT)	a) SCO	1/4 + + : 1/4 + *b* :
	b) Three-strand DCO between *a-b*	1/4*a* + : 1/4 *a b*

Source: Miglani (2000)

TABLE 10.18 Tetrad analysis of a cross + x *t* in *Neurospora crassa*

No. of asci	Spores			
	1 + 2	3 + 4	5 + 6	7 + 8
5	+	*t*	*t*	+
6	*t*	+	*t*	+
9	+	*t*	+	*t*
8	*t*	+	+	*t*
39	*t*	*t*	+	+
33	+	+	*t*	*t*

Source: Miglani (2000)

Percent recombination: $1/2(28) \times 100/100 = 14\%$.

Two-point Cross

Distance of genes in relation to centromere. Let us assume that two genes *a* and *b* in *Neurospora* are segregating. The recorded data are given in Table 10.19. Classify the tetrads as PD, NPD, and TT. We see that PD = (A) + (C) = 58 + 59 = 117 and NPD = (B) + (D) = 57 + 58 = 115. There are no TT tetrads. As a first step, one needs to detect linkage. When there are only two types of tetrads (PD and NPD), check whether PD and NPD are 1:1 ratio or not. This should be done statistically, using chi-square. If PD and NPD are in 1:1 ratio, linkage is absent but if PD and NPD are not 1:1 ratio, linkage exists. In the foregoing example, it can be seen that PD and NPD are in 1:1 ratio and hence linkage is absent which means that the genes *a* and *b*

are located on different chromosomes. Further, the data show no second-division segregation, which means that there is no crossing over between genes and their centromere, i.e. the genes *a* and *b* are very close to their respective centromeres.

Detection and estimation of linkage. Let us consider results of a cross between two strains *pdx* + x + *pan* of *Neurospora* given in Table 10.20. For detection of linkage, classify all the asci [(A) to (G)] as PD, NPD, and TT. In each class, mark the genes showing *first-division segregation* (FDS) and *second-division segregation* (SDS) as in Table 10.20. Compare PD (showing FDS for both loci) with NPD (showing FDS for both loci) for 1:1 ratio, statistically. Compare all NPDs with all TTs. See whether ratio NPD/TT is less or more than 0.25. If PD = NPD and NPD/TT is more than 0.25, the conditions reveal *independent assortment* between *pdx*

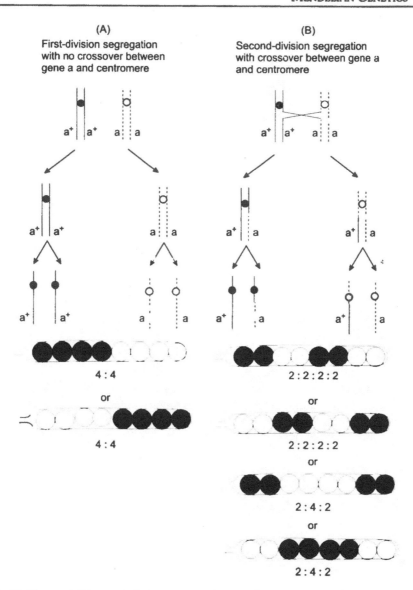

Fig. 10.32 (A) First- and (B) second-division segregation and their products in *Neurospora crassa* (from Miglani, 2000)

and *pan*. If either one or both conditions are not met, this shows the existence of linkage between the genes in question. In the given example, PD (showing FDS for both loci) 15 is statistically not equal to NPD (showing FDS for both loci) 1. Ratio of NPD (all) 1 to TT (all) 20 is 0.05, i.e. less than 0.25. In the given example, according to both criteria, *linkage* is present.

For estimation of linkage, one needs to know first the origin of seven types of tetrads given in the above example. The events leading to these tetrads are shown in Table 10.21. One needs to understand this

TABLE 10.19 Results of tetrad analysis of cross *a b* x + + in *Neurospora crassa*

PD	NPD	PD	NPD
a b	*+ b*	*+ +*	*a +*
a b	*+ b*	*+ +*	*a +*
+ +	*a +*	*a b*	*+ b*
+ +	*a +*	*a b*	*+ b*
FDS 58 FDS	FDS 57 FDS	FDS 59 FDS	FDS 58 FDS

Source: Miglani (2000)

TABLE 10.20 Results of tetrad analysis of cross *pdx* + x + *pan* in *Neurospora crassa*

(A)	(B)	(C)	(D)	(E)	(F)	(G)
pdx +	*pdx pan*	*pdx +*	*pdx +*	*pdx +*	*pdx pan*	*pdx +*
pdx +	*pdx pan*	*pdx pan*	*+ +*	*+ pan*	*+ +*	*+ pan*
+ pan	*+ +*	*+ +*	*pdx pan*	*pdx +*	*pdx pan*	*pdx pan*
+ pan	*+ +*	*+ pan*	*+ pan*	*+ pan*	*+ +*	*+ +*
F 15 F	F 1 F	F 17 S	S 1 F	S 13 S	S 0 S	S 2 S
PD	NPD	TT	TT	PD	NPD	TT

Source: Miglani (2000)

step thoroughly. Are the genes located on the opposite sides or same side of the centromere (c)? Various possibilities that exist are diagrammatically depicted below:

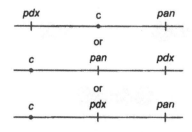

To decide this, compare PD (SDS for both loci) and NPD (SDS for both loci). If one of the classes has a relatively large number of tetrads, the two genes must be located on the same chromosome arm. In the present example, PD = 13 and NPD = 0. This means that the two genes *pdx* and *pan* are located on the same chromosome arm. If PD and NPD would have been approximately equal, the two genes would have been on opposite sides of the centromere. In other examples, one may come across such a situation.

Linkage between genes. Now we need to ascertain the gene which is closer to the centromere. For this, compare tetrads that show SDS for *pdx* with those that show SDS for *pan*. In the present case, classes (D), (E), and (G) show SDS for *pdx* and their total is 16 while classes (C), (E) and (G) show SDS for *pan* and their total is 32. Greater frequency of SDS can be observed for a gene if it lies farther from the centromere than the one which shows lesser frequency. This comparison thus shows that *pdx* is closer to the centromere; *pan* is farther.

Now we come to calculating percent recombination between centromere and gene *pdx* and centromere and gene *pan*. The following formula is used for calculating distance of a gene from the centromere.

TABLE 10.21 Origin of different types of tetrads in a two-point cross *pdx* + x + *pan* in *Neurospora* assuming that the two genes lie in the same arm of a chromosome

Class	Event	Tetrad	Segregation		Tetrad type
			pdx	pan	
(A)	No CO	pdx +	F	F	PD
		pdx +			
		+ pan			
		+ pan			
(B)	Four-strand DCO	pdx pan	F	F	NPD
	a) *pdx-pan* 1 & 3	pdx pan			
	b) *pdx-pan* 2 & 4	+ +			
		+ +			
(C)	SCO	pdx +	F	S	TT
	pdx-pan 2 & 3	pdx pan			
		+ +			
		+ pan			
(D)	Two-strand DCO	pdx +	S	F	TT
	a) cent-*pdx* 2 & 3	+ +			
	b) *pdx-pan* 2 & 3	pdx pan			
		+ pan			
(E)	SCO	pdx +	S	S	PD
	cent-*pdx* 2 & 3	+ pan			
		pdx +			
		+ pan			
(F)*	Four-strand DCO	pdx pan	S	S	NPD
	a) *pdx*-cent 1 & 4	+ +			
	b) cent-*pan* 2 & 3	pdx pan			
		+ +			
(G)	Three-strand DCO	pdx +	S	S	TT
	a) cent-*pdx* 2 & 3	+ pan			
	b) *pdx-pan* 1 & 3	pdx pan			
		+ +			

*Type (F) tetrads are seen only when the two genes in question lie on opposite sides of the centromere
Source: Miglani (2000)

Percent recombination =

$$\frac{\frac{1}{2}(SCO) + DCO}{\text{Total number of tetrads}} \times 100$$

Let us first calculate the distance between centromere (c) and *pan*. Classes (C) and (E) are due to SCO and (B), (D), and (G) due to DCO. Percent recombination between centromere - *pan* is [{½(17+13)} + {1+1+2}]/49 × 100 = 38.7 map units. While calculating distance between centromere

(c) - *pdx* we must keep in mind that *double crossovers* cannot be detected between two markers unless a known third marker (or centromere) lies in between. So double crossovers in this example will be zero; only single crossovers will be found. Classes (D), (E), and (G) are due to single crossing over. Percent recombination is ½(1 + 13 + 2)/49 × 100 = 16.3 map units. The genetic map for *pdx* and *pan* in relation to the centromere will be:

c   pdx   pan
•────────┼──────────────
←─ 16.3 MU ─→   ←── 22.4 MU ──→
←────────── 38.7 MU ──────────→

Once distance between gene *pdx* and centromere and *pan* and centromere is known, distance between genes *pdx* and *pan* can be calculated. Thus the distance between *pdx* and *pan* comes to 22.4 map units.

Interference. Two kinds of information can be obtained from the three-point linkage data. First, what nonsister chromatids are involved in crossing over in the adjacent region? Normally, we expect two-strand double exchanges, three-strand double exchanges and four-strand double exchanges in ratio 1:2:1, if there were no effect of chromatids in one exchange on the chromatids of another exchange. Departure from 1:2:1 ratio would indicate some sort of chromatid interference. This interference could be "positive" or "negative".

A second form of interference also occurs in double exchanges between adjacent regions which arise due to chiasma interference. Accordingly, a crossover (chiasma) in one region interferes in some way with crossing over (chiasma) in an adjacent region no matter what chromatids are involved. In *Neurospora*, negative interference is generally observed.

Gene Mapping in Yeast and *Chlamydomonas*

Yeast and *Chlamydomonas* have unordered tetrads. It is not possible in such cases to distinguish between tetrads resulting from first- and second-division segregation. For detection and estimation of linkage, the tetrads are classified as PD, NPD, and TT type. Detection of linkage in yeast and *Chlamydomonas* is done the same way as in *Neurospora*. In the case of unordered tetrad

data, distances are measured between genes but not between centromere and gene. Percent recombination is (NPD + ½TT)/(PD + NPD + TT). Let us assume data of Table 10.19 giving results of tetrad analysis of cross *pdx* + × + *pan* in *Neurospora crassa* as unordered tetrads. Analysis has already established linkage between *pdx* and *pan* genes. For estimation of linkage, compute total number of PD, NPD, and TT tetrads. Classes (A) and (E) represent PD, class (B) represents NPD, and classes (C), (D), and (G) represent TT tetrads. Thus the number of PD, NPD, and TT tetrads comes to 28, 1, and 20 respectively. Distance between genes *pdx* and *pan* = {1 + ½(20)} /49 × 100 = 22.4% which is the same as observed above.

GENE CONVERSION

Involvement of heteroduplex DNA explains the characteristics of recombination between alleles; indeed, allelic recombination provided the impetus for the development model. When recombination between alleles was discovered, the natural assumption was that it takes place by the same mechanism of *reciprocal recombination* that applies to more distant loci. That is to say, an individual breakage and reunion event occurs within the locus to generate a reciprocal pair of recombinant chromosomes. However, in the close quarters of a single gene, formation of *heteroduplex DNA* itself is usually responsible for the recombination event.

Individual recombination events can be studied in the *Ascomycetes* fungi, because the haploid products of a single meiosis are held together within a large cell, the ascus, in a linear order. Mitosis occurs after production of these four nuclei, giving a linear series of eight haploid nuclei. Figure 10.33 shows that each of these nuclei effectively represents the genetic character of one of the eight strands of the four chromosomes produced by meiosis. Meiosis in a heterozygote should generate four copies of each allele. This is seen in the majority of the spores. But there are some spores with abnormal ratios. They are explained by the formation and correction of heteroduplex DNA in the region in which the alleles differ.

Some asci display ratios of 3:5 or 2:6 in which one or two spores respectively, that should have been of one allelic type are actually of another type. These ratios are a consequence of the process of recombination. An uneven ratio (3:5 or 5:3) can result only from segregation of two mismatched strands in one DNA duplex; only the other heteroduplex must have been corrected.

Even ratios (2:6 or 6:2) could in principle result from independent correction of two heteroduplexes, but are more likely to result from a repair mechanism such as the copying event resulting from a double-strand break (Fig. 10.33). Gene conversion does not depend on crossing over, but correlates with it. Large proportions of aberrant asci show genetic recombination between two markers on either side of a site of interallelic *gene conversion*.

MUTATIONS

Mutation is defined as a change in the DNA at a particular locus in an organism. Mutation is also defined as the process that

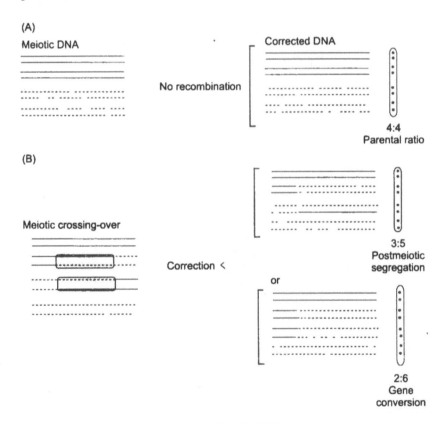

Fig. 10.33 Gene conversion in *Ascomycetes* (from Miglani, 2000)

produces a gene or chromosome set differing from the wild type. The gene or chromosome set that results from such a process is also known as mutation. Mutation is thus a process as well as an effect. Mutation is a change in the genetic material of an organism that is detectable and inherited by offspring and is not caused by recombination or segregation. The term includes nucleotide substitutions, insertions, omissions, duplications, translocations within a gene, and also large chromosomal changes. *Mutagenesis* is the use of physical or chemical agents to increase the rate of mutations. In nature, different alleles of a gene arise through spontaneous mutations and thus provide a basis for genetic variation which serves as raw material for evolution.

Characteristics of Mutations

Mutations are more or less permanent changes in kind, number or sequence of nucleotides in genetic material. The making or breaking of hydrogen bonds in a duplex DNA is a temporary, hence nonmutational change. Mutations are uncoded or unprogrammed changes. Methylation of cytosines or adenines is not a mutation because it is a programmed change since it occurs due to instructions coded by DNA. Mutations are relatively rare and unusual changes. Genetic material tends to be protected against the occurrence of uncoded and unusual changes by the double helical structure and its bonds to proteins. Mutations are regarded as accidental, unintended, undirected, random events as we have no way of knowing when a mutation will occur, which individual, cell, cell organelle, chromosome or gene will mutate. Mutations are unoriented with respect to purpose and result from deviation of a natural process.

Classification of Mutations

Size of mutations

Microlesions. These mutations are also known as point mutations or gene mutations. These are mutations in which only one nucleotide pair is replaced and are not observable even under a microscope. Such mutations can be detected by comparing the nucleotide sequence of wild and mutant DNA/RNA. Due to such a substitution, the number of nucleotide pairs does not change in a gene. Such mutations are of two types. *Transitions* are the change from purine to purine (A → G or G → A) or pyrimidine to pyrimidine (T → C or C → T) while *transversions* are the change from a purine to a pyrimidine (A → T, A → C, G → T, G → C) or vice versa (T → A, C → A, T → G, or C → G).

Intermediate lesions. In this case, either there is addition (insertion) or deletion (removal) of one or a few nucleotide pairs from a gene. This leads to a change in the number of nucleotide pairs in a gene. Thus changes may cause a shift in the translational reading frame and are called *frameshift mutations.* Write the mRNA sequence AUGCCCAAAGGGUUU.... and study the effects on a translational reading frame when one, two or three bases are deleted or added after the initiation codon (AUG). Employ the help of a genetic code dictionary (Table 10.22) for this exercise. Unless a frameshift mutation occurs close to the C-terminus, it diminishes function of the gene.

Macrolesions. Macrolesions are also known as *chromosome mutations* which include changes in chromosome structure and number. These changes at high resolution can be observed with a light microscope.

TABLE 10.22 Genetic code dictionary—nucleotide sequence of RNA codons

First Base ↓	Second Base				Third Base ↓
	U	C	A	G	
U	Phe	Ser	Tyr	Cys	U
	Phe	Ser	Tyr	Cys	C
	Leu	Ser	Nonsense**	Nonsense**	A
	Leu	Ser	Nonsense**	Trp	G
C	Leu	Pro	His	Arg	U
	Leu	Pro	His	Arg	C
	Leu	Pro	Gln	Arg	A
	Leu	Pro	Gln	Arg	G
A	Ileu	Thr	Asn	Ser	U
	Ileu	Thr	Asn	Ser	C
	Ileu	Thr	Lys	Arg	A
	Met*	Thr	Lys	Arg	G
G	Val	Ala	Asp	Gly	U
	Val	Ala	Asp	Gly	C
	Val	Ala	Glu	Gly	A
	Val	Ala	Glu	Gly	G

* initiation codon; ** termination codons
Source: Miglani (2000)

Changes in chromosome structure

These mutations include *duplications, deletions, inversions,* and *translocations.* A deletion or addition may include part of a gene, the whole gene, a group of genes, a complete chromosome, more chromosomes or even a complete set of chromosomes. *Deletions* are usually lethal in a homozygous state. Cytologically, deletions can be detected by failure of a segment of a chromosome to pair properly (Fig. 10.34A). The segment present in a normal chromosome but absent in a mutant homolog loops out during *pachytene synapsis.* This is called the *deletion loop.* The chromosome with a deletion can never revert to a wild-type condition. Deletions may show *pseudodominance* because in this case a recessive gene present in a single dose expresses itself. Some deletions may have phenotypic effects. For example, deletion of part of the short arm of human chromosome 5 leads to *Cri-du-chat syndrome* in which case the individual survives only to age 30.

The reciprocal of a deletion is called *duplication* that at pachytene gives a loop-like structure in which the duplicate part of the chromosome loops out (Fig. 10.34B). There may be a tandem duplication of part or the whole gene. Duplications are very important changes from the evolutionary point of view. In evolution of new genes, the first step is duplication, followed by decreasing the existing function through a nonsense mutation (*pseudogene*), accumulation of more mutations while the gene acquires a new function. Thus, duplications provide additional genetic material potentially capable of giving rise to new genes during the process of evolution. Duplication

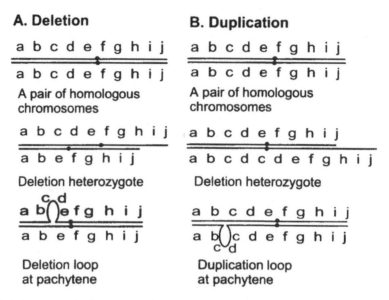

Fig. 10.34 Deletion and duplication loop; each line represents a chromosome (from Miglani, 2002)

of certain genes may produce specific phenotypic effects such as a gene mutation. For example, the semidominant mutations *Bar* in *Drosophila* produces a slitlike eye. Extrasegment present in chromosome carrying a duplication when pairs with its normal homolog, extra segment loops out. This is called a *duplication loop*.

Inversion may involve a portion of a gene or a group of genes, which have been cut out, turned through an angle of 180°, and reattached in reverse order. In the case of inversions, there is no net loss or gain of genetic material and thus heterozygous inversions are perfectly viable. A heterozygous inversion can be detected as an inversion loop. For genes present in the inverted chromosome the linkage relationship is changed. Inversions may involve only one chromosomal arm and not include a centromere (*paracentric inversion*) (Fig. 10.35A) or it may involve both the chromosomal arms and thus include a centromere (*pericentric inversion*) (Fig. 10.35B).

An *intragenic inversion* will affect the amino acid sequence of a polypeptide encoded by that gene. The phenotype may be altered or function may be lost. Inversions have some genetic effects. In a heterozygote for a paracentric inversion, crossing over within the inversion produces dicentric and acentric chromosomes. Individuals producing crossover chromatids do not survive and hence do not produce recombinant gametes. Thus crossovers are not detected. Further, in a heterozygote for a pericentric inversion crossover products are not recovered. Thus, inversions in many genetic experiments are used as suppressors of crossing over.

Translocations involve transfer of part of a chromosome to the same chromosome (*intercalary translocation*) (Fig. 10.36A), or to a nonhomologous chromosome (*simple translocation*) (Fig. 10.36B), or two nonhomologous chromosomes may mutually exchange their parts. In this case, two simple translocations are simultaneously achieved (*reciprocal translocation*)

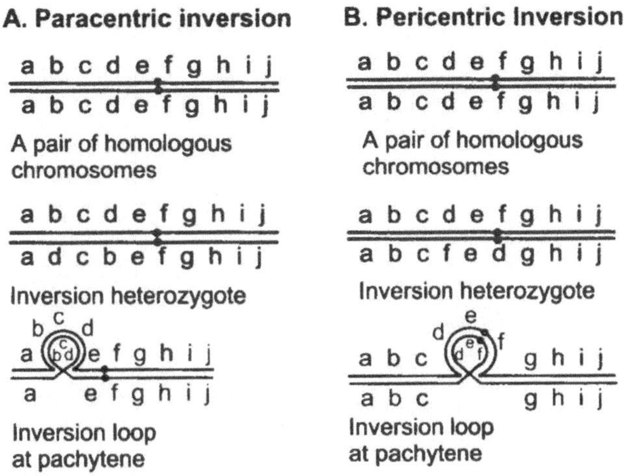

A. Paracentric inversion

a b c d e f g h i j

a b c d e f g h i j

A pair of homologous
chromosomes

a b c d e f g h i j

a d c b e f g h i j

Inversion heterozygote

a e f g h i j

a e f g h i j

Inversion loop
at pachytene

B. Pericentric Inversion

a b c d e f g h i j

a b c d e f g h i j

A pair of homologous
chromosomes

a b c d e f g h i j

a b c f e d g h i j

Inversion heterozygote

a b c g h i j

a b c g h i j

Inversion loop
at pachytene

Fig. 10.35 Paracentric and pericentric inversion loops. Pairing between only one normal and one inverted chromosome shown (from Miglani, 2002)

(Fig. 10.36C). If the break is within a gene, the base sequence of the gene will be changed and its function may be lost. Translocations in homozygous form change linkage relationships. This is the genetic effect. They may drastically alter size of a chromosome as well as the position of its centromere. Translocation is used as an important tool in gene transfers and production of duplications and deficiencies.

Translocations, inversions, and deletions produce semisterility by generating unbalanced meiotic products that may themselves be lethal or result in lethal zygotes.

Changes in chromosome number

Monopholid is a haploid plant produced from a diploid species and contains only a basic number (X) of chromosomes of a species such that each kind of chromosome is represented only once in the nucleus. *Haploids* are individuals with a single set of chromosomes. Thus, monoploid is an individual that contains one chromosome set in a kind of organism usually associated

with a diploid type of life cycle. Monoploids usually arise due to mutations. *Haploids* exist as a single set of chromosomes during almost the entire part of their life and thus monoploids are distinguishable from true haploids. Monoploids may be derived from the products of meiosis. Monoploids have a major role to play in modern approaches to animal and plant breeding.

A change in chromosome number may involve an incomplete set of chromosomes (*aneuploids*) or a complete set of chromosomes (*euploids*).

Aneuploids. Aneuploids may involve, in a disomic ($2n$) individual deletion, of one chromosome ($2n-1$, *monosomic*), deletion of two nonhomologous chromosomes ($2n-1-1$, *double monosomic*), or deletion of a pair of homologous chromosomes ($2n-2$, *nullisomic*). Aneuploids may also arise due to the addition of one chromosome ($2n+1$, *trisomic*), addition of two nonhomologous chromosomes ($2n+1+1$, *double trisomic*), or addition of a pair of homologous chromosomes ($2n+2$, *tetrasomic*). *Nondisjunction* (failure of homologous chromosomes to

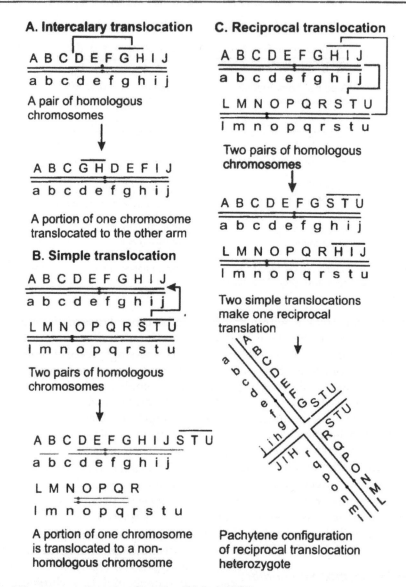

A. Intercalary translocation

A B C D E F G H I J
a b c d e f g h i j

A pair of homologous
chromosomes

↓

A B C G H D E F I J
a b c d e f g h i j

A portion of one chromosome
translocated to the other arm

B. Simple translocation

A B C D E F G H I J
a b c d e f g h i j

L M N O P Q R S T U
l m n o p q r s t u

Two pairs of homologous
chromosomes

↓

A B C D E F G H I J S T U
a b c d e f g h i j

L M N O P Q R
l m n o p q r s t u

A portion of one chromosome
is translocated to a non-
homologous chromosome

C. Reciprocal translocation

A B C D E F G H I J
a b c d e f g h i j

L M N O P Q R S T U
l m n o p q r s t u

Two pairs of homologous
chromosomes

↓

A B C D E F G S T U
a b c d e f g h i j

L M N O P Q R H I J
l m n o p q r s t u

Two simple translocations
make one reciprocal
translation

↓

Pachytene configuration
of reciprocal translocation
heterozygote

Fig. 10.36 Different types of translocations (from Miglani, 2002)

separate during meiosis), as depicted in Fig. 10.3, is the cause of aneuploids. Trisomics play an important role in gene location programs and nullisomics and monosomics are useful in locating newly found genes on specific chromosomes in plants.

Euploids. Euploids may contain more than two complete sets of chromosomes. Individuals having three, four, five, six... sets of chromosomes are called *triploids, tetraploids, pentaploids, hexaploids,* and so on. In animals, *polyploids* are restricted to groups that reproduce asexually or are *hermaphroditic* (an individual with both male and female reproductive organs) and *parthenogenetic* (individual developed from an unfertilized egg). Examples are earth worms, some types of shrimps, and parthenogenetic species of insects and lizards.

Polyploids, which are formed from sets of chromosomes from one single species, are known as *autopolyploids* and those formed from sets of chromosomes of different species are termed allopolyploids or *amphidiploids*. Alfalfa, coffee, potato, and peanut are *autotetraploid* plants. Examples of allopolyploids are *Triticale*, *Brassica*, *Raphnobrassica*, cabbage, and wheat (*Triticum aestivum*). Triploids are characteristically sterile. A practical application of sterility associated with triploidy is the production of seedless varieties of watermelons, bananas, etc. These varieties are more palatable to consumers. Allopolyploidy seems to have played an important force in *speciation* of plants. An alkaloid, colchicine, is used to produce polyploids as it acts by interfering with the spindle apparatus, thereby doubling the number of chromosomes in the cell.

Forward and reverse mutations

Forward mutation. Any mutation away from the standard or wild type.

$$y^+ \rightarrow y \quad \text{or} \quad R \rightarrow r$$

Reverse or back mutation. Any mutation towards the standard or wild type.

$$y \rightarrow y^+ \quad \text{or} \quad r \rightarrow R$$

Same sense, missense, and nonsense mutations

Same sense or neutral mutations. This is a type of mutation in which the changed codon codes for the same amino acid as in the wild-type.

Missense mutation. A point mutation that changes the meaning of a codon such that a new codon codes for a different amino acid.

Nonsense mutation. When a point mutation creates one of the three termination codons, it is called a nonsense mutation. It causes premature termination of protein synthesis.

Write examples of same sense, missense and nonsense mutations using the genetic code dictionary (Table 10.22).

Spontaneous mutations

The operational definition of *spontaneous mutations* is that these mutations are produced in natural or laboratory populations under normal growth conditions. Frequency may thus depend on the growth medium, temperature, pH, etc., which are employed. These are mutations whose cause is not known. They may originate due to errors during DNA replication, environmental mutagens, *tautomerism* or transposable genetic elements. A well-known example of spontaneous mutation was reported in *Oenothera lamarckiana* (evening primrose) De Vries (1901).

Some spontaneous mutations must come about by mistakes in the normal duplication of DNA or RNA. Due to *tautomeric shift* (or ionization) rare base pairs occur spontaneously. *Transitions* may arise due to proton migration within bases and they show unusual pairing. *Tautomers* are rare and short lived. Tautomeric adenine (A^*) pairs with cytosine and this mispairing leads to A.T $\rightarrow$ G.C transition. Tautomeric guanine (G^*) may pair with thymine. This mispair leads to G.C $\rightarrow$ A.T transition. Similarly, T^* G forbidden base pairing leads to T.A $\rightarrow$ C.G transition and C^*=A forbidden base pairing leads to C.G $\rightarrow$ T.A transition.

Proton migration within base pairs can set off a chain of modification that ends in DNA mutation. *Transversions* may arise by mistake pairing between syn-adenine produced by proton migration and rotation and common adenine, and between neoguanine produced by shifting of the glycosidic bond from normal form to extracyclic nitrogen and common guanine. Methylated guanine undergoes tautomeric shifts and pairs with thymine and leads to G.C → T.A transversions. Transversions occur by apposition of two errors—proton migration in one and 180° rotation around the glycosidic bond (syn) in the other. Tautomeric synadenine (A*) may pair with syn A, which leads to A.T → T.A transversion. Similarly, tautomeric neoguanine (G*) may also pair with syn G, which leads to G.C → C.G transversion.

Rare base pairing between two purines or two pyrimidines can cause spontaneous production of transversion. These A.G and T.C pairs use one or two hydrogen bonds. In this case, they have slightly wrong distances or angles for two-sugar (C)-base (N) bonds. Since the DNA backbone is not completely stiff, it can accumulate such distortions at least occasionally, as a mistake.

Apart from these mistakes inherent in the structure and mode of replication of nucleic acids, spontaneous mutations may be caused by a variety of agents made by the cell or present in the medium. Agents of this kind are base analogs, peroxides, nitrous acid, etc., which are produced under some growth conditions. Special enzymes such as catalase, peroxidase, etc., often inactivate these agents. Some of these agents inside the cell act on duplicating DNA while others do so on resting DNA. Processes such as *methylation, depurination*, gross base distortion, and formation of bulky *adducts* are quite common in cellular metabolism.

Induced mutations

Induced mutations are those produced under changed growth conditions, i.e. as a result of exposure of an organism to some agent that does not form a normal growth condition for it. An agent that has the ability to produce mutations is called a *mutagen*. A mutagen can be a *physical agent* (UV light, x-rays, γ-rays, α-particles, β-particles, etc.) or a *chemical agent* (nitrogen mustard, hydrazine, hydrogen peroxide, nitrous acid, base analogs, alkylating agents, dyes, etc.). An individual possessing a mutation is termed a *mutant*. These are those mutations whose cause is known. An excellent account of mechanisms of mutations has been given by Freeze (1963, 1971) and Heslot (1965).

Physical mutagens. Physical agents responsible for induced mutations include UV light, x-rays, γ-rays, α- and β-particles. UV light, x-rays and γ-rays have a very short wavelength and penetrate deep, and hence have a high ionizing capacity. Radiations can change the genetic material. Radiations are used for three purposes: visibility, carrying radio and TV signals, and creating new variability.

The effects of UV light are several. It kills bacteria and also induces mutations among the survivors. The first photochemical lesion found in the DNA of irradiated bacteria was a thymine dimer. UV light produces thymine dimers mainly by linking adjacent thymine bases in the same strand of DNA via C-C bonds. Dimers containing thymine block DNA replication *in vitro* and *in vivo* and are responsible for an unimportant fraction of lethal effects of low doses of UV light in some strains of bacteria. It also causes hydration, i.e. acts on "C" of DNA in water solution, adding a water molecule across a double bond.

UV light can temporarily delay cell divisions. It also produces a chromatid type of aberration and disrupts H-bonding between strands of natural DNA upon exposure of viral suspensions to UV light. This results in inactivation of the virus. In yeast, reversion to *prototrophy* from *auxotrophy* is commonly studied.

Mutations are produced if the UV light-induced lesions are not repaired by the repair mechanism of the cell. Unrepaired *pyrimidine dimers* cause mutations by inducing formation of daughter strand gaps. One gap is formed opposite each lesion. These gaps, lethal if not repaired, cannot be mended by repair replication, because each is located opposite a noncoding lesion. These are reparable only by a mechanism that probably involves some form of genetic recombination between complementary daughter strands. Most UV light-induced mutations are errors introduced into the DNA during inaccurate recombination repair of single-strand gaps.

In many systems, exposure to γ-radiation gives rise to, among other effects, temporary disappearance of mitotic figures from the cell population. This phenomenon is called *mitotic inhibition*. The cause of mitotic inhibition is a block in the cell division cycle immediately prior to prophase I; neither the site of the radiation lesion nor its repair mechanism is known.

X-ray is an electromagnetic radiation emitted in quanta. Its wavelength is (0.05 – 10Å). Two rare chromosomal aberrations induced by x-ray irradiation were recorded on modifying factors in induced mutagenesis. First was the occurrence of ring like metaphase chromosome at mitosis in the root tips from desiccated seeds of *Oryza sativa*. This condition arises due to the failure of separation of the ends. The phenomenon has been explained on the basis of alteration of the cycle of separation of the four regions within a chromosome.

Interlocked chromosomes or a *bridge* at mitosis in the root tips occurs. This has been explained on the basis of translocation following breakage of two chromosomes, leading to formation of a *dicentric chromosome*. The bridge arises due to a full turn of the chromatids over one another at anaphase.

The biological effects of x-rays, γ-rays, α-particles, protons, high energy neutrons, and electrons have been studied extensively. Radiation can have a direct effect on the chromosome, e.g. break it directly or alter one of the DNA bases and an indirect effect, since along the track of each particle a large number of ions or radicals remain which can initiate a chain of chemical reactions. Chromosomes are very sensitive to breakage in the meiotic prophase.

Chemical mutagens. A variety of chemicals capable of inducing mutations are inhibitors of nucleic acid precursors, base analogs, dyes, nitrous acid, hydroxylamine, hydrazine, low pH, heat, hydrogen peroxide, maleic hydrazide, antibiotics, azides, alkylating agents, chemicals used in industries, pesticides, cosmetics, food preservatives, etc.

Chemical agents that induce mutations are classified into two types depending on the physiological state of DNA affected by these agents. In the first group fall those chemicals which affect replicating nucleic acids. Examples of such chemicals are inhibitors of nucleic acid precursors (e.g., 5-aminouracil, azaserine, benzimidazole, caffeine, 8-ethoxycaffeine, 8-mercaptopurine, 5-nitroquinoxaline, paraxanthine, tetramethyl uric acid, theobromine), base analogs (e.g. 5-bromouracil, 5-chlorouracil, 5-iodouracil, 2-aminopurine, 2,6-diaminopurine), and dyes (e.g. proflavin, acridine orange, 2-aminoacridine, methylene blue, toluidine blue, ethidium bromide,

10-methyl acridinium phenanthridine, propidium diiodide). 5-aminouracil and 8-ethoxycaffeine inhibit thymine while azaserine, benzimidazole, caffeine, 8-mercaptopurine, 5-nitroquinoxaline, paraxanthine, tetramethyl uric acid and theobromine inhibit purines. The frequency of mutations increases because the lack of one base either causes a chromosomal break and thus a large alteration or it increases pairing mistakes. It is possible that an inhibitor causes increased formation of another base analog that is actually mutagenic since it is incorporated in place of a normal base. The substances incorporated into DNA without hindering its replication are loosely related to adenine, guanine, cytosine, or thymine.

Base analogs and dyes. Since a base analog differs from its normal form in certain substituents, its electronic structure is modified and occasional errors may occur in the specificity of hydrogen bonds during incorporation into DNA. 5-bromouracil, 5-chlorouracil, and 5-iodouracil replace thymine in DNA and they thus pair with adenine. 5-bromouracil (rare *enol state*) replaces cytosine. It pairs with guanine and thus causes G-C to A-T transitions and vice versa. 2-aminopurine can pair in its normal tautomeric form with two bases, with thymine by two hydrogen bonds and with cytosine by one hydrogen bond. More often, 2-aminopurine pairs with thymine by two hydrogen bonds. Thus, 2-aminopurine is incorporated most frequently in place of adenine which itself would not be mutagenic. When 2-aminopurine pairs with cytosine, it is highly mutagenic. It is expected to induce base pair transitions from A-T to G-C and vice versa. Base analogs, 5-bromouracil and 2-aminopurine do not cause transitions. Since these two base analogs cause transitions in both directions (A-T to G-C and G-C to A-T), the mutations

induced with them can be reversed by use of these substances. The normal base that pairs with base analogs, 2-aminopurine and 2,6-diaminopurine, is not known. Caffeine owes its genetic effect to its interference with one of the steps concerned with the repair of genetic damage. Caffeine-induced chromosome breaks do not join. Acridine dyes attach directly to nucleic acids and thus cause mutations. The presence of acriflavin in the growth medium of *E. coli* removes the *F'* factor. The presence of proflavin prevents assembly of phage T_4. In darkness, acridine dyes induce deletions or additions of bases in DNA. Methylene blue, toluidine blue, and proflavin, when exposed to light oxidize some DNA bases and have a strong lethal effect on phages, provided the dyes are inside the phage. Acridine orange can stack along nucleic acids between two bases of a DNA strand and apparently attaches itself to a phosphate backbone. The two bases are separated by 6.8 Å. Thereafter a base will be missing. X-ray data show that acridine can be sandwiched between purine bases. Acridines induced only deletions or insertions of one base pair in DNA. These dyes thus cause *frameshift mutations*. Ethidium bromide and propidium diiodide cause likewise frameshift mutations.

In the second group fall those chemicals that affect resting nucleic acids. Examples of such chemicals are nitrous acid, hydroxylamine, hydrazine, low pH, heat, hydrogen peroxide, maleic hydrazide, antibiotics, azide, and alkylating agents. The action of some of these mutagens is described below:

Nitrous acid. Nitrous acid (HNO_2) oxidatively deaminates with decreased frequency the bases guanine, cytosine, and adenine in DNA and RNA. In this reaction, the amino group in these bases is replaced by oxygen. HNO_2 induces *point mutations* in T_4 phage and yeast. Its mutagenic effect is not reported for higher plants. Adenine is deaminated to *hypoxanthine*, which then pairs preferentially with cytosine in place of thymine. The result is transition of A-T to G-C. Cytosine is deaminated to uracil, which now pairs with adenine in place of guanine. The result is transition of G-C to A-T. Guanine is deaminated to *xanthine*, which still pairs with cytosine but by only two hydrogen bonds. Thus, deamination of guanine is not mutagenic. Nitrous acid thus causes transitions A-T to G-C in both directions. Thymine, containing no amino group, is not altered by nitrous acid. Nitrous acid also causes *interstrand cross-linking* of DNA. The DNA strands fail to separate and there is no DNA duplication which is lethal for phages. Blockage of replication may result in deletions. One *cross-link* is formed for four deaminations.

Alkylating agents. Alkylating agents carry one, two or more alkyl groups in reactive form which are capable of being transferred to other molecules in which electron density is high. Accordingly, such alkylating agents are called *monofunctional, bifunctional* or *polyfunctional* alkylating agents. For example, in sulfur mustard, one or both alkyl groups may be halogenated. In the case of nitrogen mustard, one, two or three alkyl groups may be halogenated. Although dialkyl sulfates may look bifunctional, they cannot alkylate more than once and so are monofunctional alkylating agents. Similarly, alkylalkane sulfonates are monofunctional alkylating agents. Epoxides and ethylene amines are likewise monofunctional alkylating agents. Some sulfates and sulfonates are monofunctional (dimethyl sulfate, diethyl sulfate, methyl methanesulfonate, ethyl methanesulphonate) while others (e.g. myleran) are bifunctional. Diazomethane, a diazoalkane, is a monofunctional agent. Most nitroso compounds, namely N-methyl-N-nitrosourethane, N-ethyl-N-nitrosourea, 1-methyl-3-nitro-1 nitroso guanidine, are monofunctional alkylating agents.

Alkylating agents may cause crosslinking of DNA and thus lethal effects by inhibiting DNA duplication because DNA strands do not separate due to cross-linking. Alkylating agents may readily react with phosphate groups and thus break the sugar phosphate backbone of DNA. This may induce larger alterations. Most alkylating agents have the same mutagenic effect. Some base of DNA is alkylated. For example, dimethylsulfate upon reaction with DNA may most often produce 7-methyl guanine that may pair with thymine. The result is transition G-C to A-T. Other products of this reaction may be 1-methyl adenine, 3-methyl adenine, 1,3-dimethyladenine or 7-methyladenine. The relative reactivities of various bases for methylation are 100 (7-methyl guanine), 50 (1-methyl adenine), 20 (1,3-dimethyl-adenine), and 2 (7-methyladenine). Diethyl sulfate with DNA produces 7-ethylguanine. N_1 position of cytosine is also alkylated but to a very less degree. This reaction produces 1-methylcytosine, which can pair with adenine. The result is transition G-C to A-T. The alkylation of bases in DNA does not seem to be direct but by *transalkylation* of the alkyl group changing from phosphate to base. The alkylated base might inhibit DNA duplication, causing lethal effects or base pairing mistakes during DNA duplication, resulting in mutagenic effects. Alkylation of thymine may produce 6-O, ethyl thymine which pairs with guanine, resulting in transition T-A to C-G. Alkylation of guanine produces 6-O, ethyl guanine less frequently and 7-alkyl guanine more so. Alkylation of purines in ^{7}N-position gives rise to quaternary nitrogens (N^+) that are unstable. Either alkyl group itself hydrolyzes away from the purine or the alkylated purine separates from deoxyribose leaving it "depurinated". The gap thus left may interfere with DNA duplication. At high pH, *depurinated DNA* may occasionally break. This may induce larger alterations or be lethal but probably does not induce point mutations. From the point of view of inducing mutations, methylating and ethylating agents have been most extensively used. Alkylated guanine in DNA is liberated slowly because of slow hydrolysis at the glycosidic linkage. It may lead to loss of a base pair (deletion) or replacement of a missing base by any available purine or pyrimidine base, thus causing a transition, transversion or return to the original state. Bifunctional agents have toxicity greater than monofunctional agents.

Detection of Mutations in *Neurospora*

Conidia of a particular strain of *Neurospora crassa* are crossed to the wild-type strain of the opposite sex (Fig. 10.37). Haploid spores of this cross are then isolated and grown on a complete medium. Inability of such an isolated strain to grow on a minimal medium indicates a growth defect. Attempts are then made to discover the source of this defect by growing the aberrant strain on a minimal medium supplemented with various additives. In the illustrated example, pantothenic acid added to a minimal medium enables the strain to grow, indicating that the mutant strain is lacking pantothene. Observation of the expected 4 wild type:4 pantothenicless segregation ratio in a cross with a wild-type indicates that mutation is of nuclear origin.

Electrophoresis

A recent method for detection of mutation in different organisms is to screen various enzymes (glucose-6-phosphate dehydrogenase) or proteins (hemoglobin) for slight variation. The technique used is known as *electrophoresis*, which separates DNA and protein molecules in an electric field on the basis of mass/charge ratio. Relative positions occupied by different electrophoretic products (bands) of an individual under test are compared with a wild (standard) individual. Molecules showing variation can be shown to be heritable and their transmission traced between various pedigree generations. Not all variant forms of a protein can be detected by this method.

Applications of Induced Mutations

Mutations are mostly deleterious and recessive and therefore most of them lack practical value. However, a small frequency (one out of 1,000) of induced mutations has been successfully used for production of improved varieties. Gustafson (1947) and Nilsson-Ehle (1948) induced mutations in barley by x-rays. They were able to produce varieties with stiffer straw, denser head, and higher yield than the parents. Mexican wheat varieties though high yielding were red grained. Treatment with γ-rays converted them to amber color. For example, Sharbati Sanora was evolved from Sonora 64 and Pusa Lerma from Lerma Rojo 64A. In wheat, several mutants with lodging resistance, high protein, and lysine content have been obtained and utilized in plant breeding programs. Semidwarf variety of rice was evolved from the tall rice variety with the help of x-rays. Mutants were also obtained for increased protein and lysine content. In some mutants isolated, duration of crop was reduced from 120 to 95 days. Reduction in duration of crop plants has been an important contribution of mutations. For example, in sugar cane, growing period was reduced from 18 months to less than 10 months and in castor, duration reduced from 9 months to 4.5 months. In

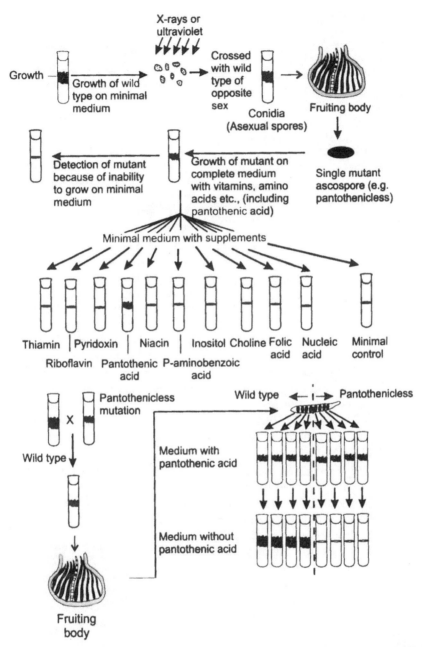

Fig. 10.37 Procedure for induction and detection of nutritional mutations in *Neurospora crassa* (from Miglani, 2002)

California, peanuts with thick shells were produced through induced mutations. Several mutants of *Penicillium* have helped in increasing penicillin yield.

Environmental Mutagenic Hazards

A variety of industrial chemicals are in use and millions of workers regularly exposed to them. Applying fertilizers, insecticides, fungicides, bactericides, etc. is industrializing even our agriculture. Persons not working in industry and agriculture expose themselves to chemicals in other ways, e.g. medicines, food dyes, food preservatives, dyes. Therefore, chemicals have become an important part of our life. A dangerous consequence is their effect on the genetic material. Our technologically oriented society has greatly increased the amounts of various gases, viz. CO, CO_2, SO_2, NO, NO_2, H_2S that can be quite hazardous. For example, sulfurous gases enter the atmosphere through industrial and automobile combustion.

$$SO_2 + H_2O \rightarrow H_2SO_3 \rightarrow H^+ + HSO_3^-$$

Both SO_2 and $H_2SO_3^-$ have been shown to be mutagenic in a number of organisms. (2) NO and NO_2 are produced during combustion of fossil fuels. In the presence of light, $NO_2 + H_2O \rightarrow HNO_2$. HNO_2 is a potent mutagen as such. Also, in the presence of light, $NO_2 \rightarrow NO + O$.

All these compounds are mutagenic. Combustion of fossil fuels in industry and automobiles produces another class of hazardous chemicals known as hydrocarbons. Some of the hydrocarbons are directly mutagenic, while others are metabolically converted into mutagens. Ozone is produced around electric utility plants where sparking is common and in polluted areas exposed to intense sunshine. It enters into free-radical reaction that mimics the effects of ionizing radiations on genetic material.

Now extensive lists of pesticides, drugs, industrial chemicals, food additives are available that have been tested in one or more systems and have been shown to be mutagenic in one or the other, while information on many is lacking. It has been emphasized that all of our environmental chemicals should be thoroughly tested. It has been recommended that "high risk" persons should be offered a routine checkup for any *cytogenetic abnormality*.

Observing a particular chemical to be mutagenic in some test system does not mean it will be mutagenic in all systems. For example, a chemical found mutagenic in the Ames test may not be mutagenic in *Drosophila* or man. Such observations lead to controversy over the safe use of the particular chemicals, e.g. controversy prevailed for a long time over the safe use of DDVP, sachharrine, etc. Before a chemical is declared genetically hazardous, testing must be thoroughly done in various test systems. Basically the genetic material is the same in bacteria, *Drosophila*, wheat, and man. A chemical reacting with DNA in bacteria is also expected to do so in another organism, provided it penetrates the cells. Negative results in one test system do not rule out genetic activity in other organisms.

GENES IN A POPULATION

Laboratory and Natural Populations

Population genetics is a study of the frequencies of genes in a population. Up to now we have been dealing with laboratory populations. From the genetic point of view laboratory and natural populations differ. The relative frequency of alleles at a locus in a laboratory population is usually fixed at some convenient ratio by the design of the experiment, e.g. when two homozygous parents are crossed ($TT \times tt$) to begin an experiment, the alleles are thereby introduced in equal frequency. In natural populations on the other hand, the relative frequency of alleles may vary greatly. In laboratory populations, the system of mating is usually sharply defined, e.g., crossing $F_1 \times F_1$ or F_1 to one of the parents. Thus, the breeder entertains a wide variety of crosses. In natural populations, much more is generally left to chance. Any genotype may mate with any genotype.

Binomial Distribution of Genotype Frequencies

Gene frequencies can be computed in decimals and can be treated and manipulated as probabilities. One chromosome carries one allele and its homologous chromosome carries the other allele. The probabilities of different genotypes in the population can be calculated by combining the chromosomes by chance. In natural populations, the relative frequencies of alleles may vary greatly.

Hardy-Weinberg Law

In 1908, Hardy and Weinberg independently reported the basic principles of population genetics. Their discovery is known as the *Hardy-Weinberg law* which states that in a large population mating at random, after one generation of such mating the frequencies of alleles and of genotypes remain constant generation after generation, provided the following conditions are met: The population must be large enough so that sampling errors can be disregarded. This implies that there should be no random genetic drift. There can be no mutation, either $A \rightarrow a$ or $a \rightarrow A$. If there is the $A \rightarrow a$ mutation must be the same in frequency as the $a \rightarrow A$. If this is not the case, A and a gene frequencies will change over time. Mating must be random, that is all mating combinations must be equally probable. For example, let us consider three genotypes—*AA, Aa,* and *aa.* Various mating

combinations $AA \times AA$; $AA \times Aa$; $AA \times aa$; $Aa \times Aa$; $Aa \times aa$; and $aa \times aa$ should be in the same proportion. All genotypes should be equally viable and equally fertile. No selection should be operative. All gametes must have equal probabilities of forming zygotes. The population must be isolated to avoid individuals, and hence genes, from migrating into and out of it. If these conditions are met, genetic equilibrium will be established; if not, gene frequencies will change. The changing of gene frequencies over time is the entire basis of evolution.

Principle of Genetic Equilibrium

The principle of *genetic equilibrium* states that if the above-mentioned conditions are met, gene frequencies and phenotypic ratio in a population will remain stable. Example: A gene has two alleles, A and a, only in a population so that the total number are $A + a$. Proportion of allele

$$A = \frac{A}{A+a} = p \text{ and that of}$$

$$a = \frac{a}{A+a} = q$$

$$p + q = \frac{A}{A+a} + \frac{a}{A+a} = \frac{A+a}{A+a} = 1$$

Note that $p + q = 1$.

Let us suppose a cross $F_1 (Aa) \times F_1 (Aa)$. Both parents are assumed to form the two types of gametes in the same proportion (A) and (a).

Female → Male ↓	(A)	(a)
(A)	AA $p \times q = p^2$	Aa $p \times q = pq$
(a)	Aa $p \times q = pq$	Aa $q \times q = q^2$

Note that $p^2 + 2pq + q^2 = 1$.

Let us consider a hypothetical population composed of 490 individuals of AA, 420 Aa, and 90 aa genotype. Total number of individuals in the population, 1,000. Frequency of different genotypes: $AA = p^2 = 490/1,000 = 0.49$, $Aa = 2pq = 420/1,000 = 0.42$, and $aa = q^2 = 90/1,000 = 0.90$.

Frequency of allele $A = p = p^2 + pq = 0.49 + 0.21 = 0.7$

Frequency of allele $a = q = q^2 + pq = 0.09 + 0.21 = 0.3$.

Let us now consider whether these frequencies change from one generation to the next if all the conditions of the Hardy-Weinberg law are met.

Let us now calculate frequencies of A and a alleles in the process of these matings.

Note that gene frequencies among the first generation are identical to those among the parents. This is a characteristic of an equilibrium population and would hold all the time, under a defined set of conditions, no matter how many generations were studied.

Genotypic frequencies (Female)	Genotypic frequencies (Male)		
	AA (0.49)	Aa (0.42)	aa (0.09)
AA (0.49)	AA × AA (0.240)	AA × Aa (0.206)	AA × aa (0.044)
Aa (0.42)	Aa × Aa (0.206)	Aa × Aa (0.176)	Aa × aa (0.038)
Aa (0.09)	aa × AA (0.044)	aa × Aa (0.038)	aa × aa (0.008)

Mating	Mating probability	Offspring ratio	Allele ratio in offspring	
			A	a
AA × AA	0.240	1AA	0.240	0.000
AA × Aa	0.206	1AA:1 Aa	0.155	0.051
AA × aa	0.044	1Aa	0.022	0.022
Aa × AA	0.206	1AA:1Aa	0.155	0.051
Aa × Aa	0.176	1Aa:2Aa:1aa	0.088	0.088
Aa × aa	0.038	1Aa:1aa	0.009	0.029
aa × AA	0.044	1Aa	0.022	0.022
aa × Aa	0.038	1Aa:1aa	0.009	0.029
aa × aa	0.008	1aa	0.000	0.008
Total	1.000		0.700	0.300

Equilibrium in Mixed Populations

Let us consider two populations, A and B, that differ in genotypic constitution. Mix populations A and B to get a new population, C. The genotypic makeup of this population would be:

Population A		Population B		Population C	
AA Aa aa	+	AA Aa aa	=	AA Aa aa	
640 320 40	+	250 500 250	=	890 820 290	

Frequency of genotypes = AA 0.445, Aa 0.410, aa 0.145

Frequency of allele A = p = 0.445 + 1/2(0.410) = 0.65

Frequency of allele a = q = 0.145 + 1/2(0.410) = 0.35

Genotype	Expected frequency	Expected individuals
AA	P^2 = (0.65)2 = 0.42	840
aa	2pq = 2(0.65)(0.35) = 0.46	920
Aa	q^2 = (0.35)^2 = 0.12	240

Genotype	Observed	Expected
AA	890	840
Aa	820	920
Aa	290	240

The calculated χ^2 value (24.26) is greater than the theoretical χ^2 value (3.841) at one degree of freedom and 0.05 probability. Observed and expected thus do not agree because population C is not a product of random mating. Genetic equilibrium will however, be reached after one generation of random mating.

QUANTITATIVE INHERITANCE

Quantitative Traits

Quantitative inheritance is the study of inheritance of those traits that exist in a range of phenotypes for a specific character. *Quantitative characters* differ by degree rather than by distinct qualitative differences. The phenotypic range of the quantitative trait appears to be continuous without distinct steps that could be accounted for by distinct genes. Most quantitative traits tend to follow a *normal distribution* curve.

Galton's interest in quantitative traits

Galton (1989) noted that certain physical and mental characteristics, which showed continuous variation, appeared to be inherited, so that taller individuals for example, produced taller children on average. For such traits, although segregation and assortment of individual hereditary factors could not be determined, biometricians were able to demonstrate statistically liklinesses between relatives with respect to numerous continuously distributed traits.

Most of the actual variations among organisms in a natural population are quantitative. Some examples of quantitative traits are yield in crop plants, metabolic activity, reproductive rate and behavior. All these characters are continuous over a range. Continuity of phenotypes in quantitative traits is result of two phenomenon: Each genotype does not have a single phenotype but a wide phenotype range. So phenotypic differences between genotypic classes become blurred and thus it is not possible to assign a particular phenotype to a particular genotype and expression of quantitative traits depends on action and interaction of several genes, each having a small additive effect on the phenotype. Such characters are subject to considerable phenotypic modifications by environment.

Quantitative characters cannot be sharply classified. Quantitative traits are measured in terms of various statistical parameters such as arithmetic mean, mode, median, variance, standard deviation, and coefficient of variation. Data of a quantitative trait fit a normal distribution.

Multiple Factor Hypothesis

Nilsson-Ehle (1909) and East (1916a) postulated the *multiple factor hypothesis*. They explained quantitative inheritance on the basis of action and segregation of a number of allelic pairs having *duplicate* and *accumulative effects* without complete dominance. Studies on ear length in corn, seed weight in beans, and grain color in wheat supported the multiple factor hypothesis.

Ear length in corn. Emerson and East (1913) worked on ear length in corn. They crossed two varieties—Tom Thumb Popcorn (Parent 60) and Black Mexican sweet corn (Parent 54). Details are given in Fig. 10.38. The following conclusions could be drawn from this data: (1) mean of F_1 approximately intermediate between means of the two parental varieties, (2) mean of F_2 equal to mean of F_1, (3) variance in F_2 greater than in F_1; and (4) extreme measurements in the F_2 overlap well in the distribution of parental values. The Emerson and East (1913) experiment thus provided evidence for the multiple factor hypothesis. A hypothetical model to explain the multiple factor hypothesis has been given in Fig. 10.39.

Parent 60		Parent 54
Range : 5-8 cm	x	13-21 cm
Mean : 6.63 cm		16.80 cm
Variance : 0.816	–	1.887
	F_1	
Range	:	9-15 cm
Mean	:	12.116 cm
Variance	:	1.519
F_1	x	F_1
	–	
	F_2	
Range	:	7-19 cm
Mean	:	12.888 cm
Variance	:	2.252

Fig. 10.38 Results of crosses conducted by Emerson and East (1913) for ear length in corn (from Miglani, 2002)

Mendelian genes and quantitative inheritance. After *rediscovery of Mendelism*, the view emerged that genes control inheritance of discontinuous variation. It proved difficult to explain inheritance of quantitative traits on the basis of particulate genes, however. Are the genes controlling quantitative inheritance different from those controlling qualitative inheritance? The answer to this question came from detailed analysis of quantitative characters in beans (*Phaseolus vulgaris*).

Johannsen's technique, based on inbreeding, was to separate the beans into distinctive lines on the basis of seed weight. Since beans are highly self-pollinated, it was easy to establish inbred stocks derived initially from single seeds. He established 19 pure lines. Each pure line produced average seed weights that were individually distinct (Table 10.23). Line 1 was the heaviest (64.2 cg) and Line 19 the lightest (35.1 cg). Genetic differences between lines could be seen on comparing the offspring of different pure lines with the same parental seed weight. A study of Table 10.23 shows that different parental sizes within a particular line produced offspring with uniform mean values. Some degree of variability existed within each pure line that was apparently caused by environment, as can be seen in the variety of seed weights produced by a single sample line, that of number 13 in Table 10.24, for example.

On the basis of *Johannsen's experiment*, a population of individuals that varied with respect to a quantitative characteristic was conceived to consist of a number of genetically different groups (Johannsen, 1903, 1909). Within each group, the quantitative characteristic would have a range of measurements. Some degree of overlapping would occur between the ranges of different groups, so that the separation between them was not distinct. Johannsen's experiment explained that (a) continuous variation noted for quantitative traits was the result of both genotype and environment, and (b) although basically genetic, the distinction between Johannsen's pure lines could not be ascribed to particular genes since none were individually identified. On the basis of these

True-breeding Tall is 78"   x   True breeding short is 60"
XXYYZZ   ↓   *xxyyzz*

F₁: *XxYyZz*
                            69"

F₂:

Frequency	Genotype	Height (inches)
1	*XXYYZZ*	78
2	*XXYYZz*	75
1	*XXYYzz*	72
2	*XXYyZZ*	75
4	*XXYyZz*	72
2	*XXYyzz*	69
1	*XXyyZZ*	72
2	*XXyyZz*	69
1	*XXyyzz*	66
2	*XxYYZZ*	75
4	*XxYYZz*	72
2	*XxYYzz*	69
4	*XxYyZZ*	72
8	*XxYyZz*	69
4	*XxYyzz*	66
2	*XxyyZZ*	69
4	*XxyyZz*	66
2	*Xxyyzz*	63
1	*xxYYZZ*	72
2	*xxYYZz*	69
1	*xxYYzz*	66
2	*xxYyZZ*	69
4	*xxYyZz*	66
2	*xxYyzz*	63
1	*xxyyZZ*	66
2	*xxyyZz*	63
1	*xxyyzz*	60

No. of active alleles	0	1	2	3	4	5	6
Height (inches)	60	63	66	69	72	75	78
Frequency	1	6	15	20	15	6	1

Fig. 10.39 Model explaining multiple-factor hypothesis (from Miglani, 2000)

experiments it was concluded that Mendelian genes also govern quantitative variation.

Yule (1906) suggested that continuous quantitative variation might be produced by a multitude of individual genes, each with a small effect on the measured character.

Grain color in wheat. Nilsson-Ehle (1909) demonstrated an actual segregation and assortment of genes with quantitative effect. According to him, three individual gene

TABLE 10.23 Johannsen's experiments on 19 pure lines. Each pure line produced average seed weights that were individually distinct.

Pure line	Yield of parents (centigrams)						Average weight of offspring
	20 x 20	30 x 30	40 x 40	50 x 50	60 x 60	70 x 70	
1					63.1	64.9	64.2
2			57.2	54.9	56.5	55.5	55.8
3				56.4	56.6	54.4	55.4
4				54.2	53.6	56.6	54.8
5			52.8	49.2		50.2	51.2
6		53.5	50.8		42.5		50.6
7	45.9		49.5		48.2		49.2
8		49.0	49.1	47.5			48.9
9		48.5		47.9			48.2
10		42.1	47.7	46.9			46.5
11		45.2	45.4	46.2			45.4
12	49.2			45.1	44.0		45.5
13		47.5	45.0	45.1	45.8		45.4
14		45.4	46.9		42.9		45.3
15	46.9			44.6	45.0		45.0
16		45.9	44.1	41.0			44.6
17	44.0		42.4				42.8
18	41.0	40.7	40.8				40.8
19		35.8	34.8				35.1
Mean of all lines	47.9						

Source: Miglani (2000)

TABLE 10.24 Number of offspring of different seed weights produced in Johannsen's pure line, number 13

Weight of parents	Weight classes of offspring (centigrams)										Total
	17.5	22.5	27.5	32.5	37.5	42.5	47.5	52.5	57.5	62.5	
27.5			1	5	6	11	4	8	5		40
32.5				1	3	7	16	13	12	1	53
37.5		1	2	6	27	43	45	27	11	2	164
42.5	1		1	7	25	45	46	22	8		155
47.5			5	9	18	28	19	21	3		103
52.5		1	4	3	8	22	23	32	6	3	102
57.5			1	7	17	16	26	17	8	3	95
Total	1	2	14	38	104	172	179	140	53	9	712

Source: Miglani (2000)

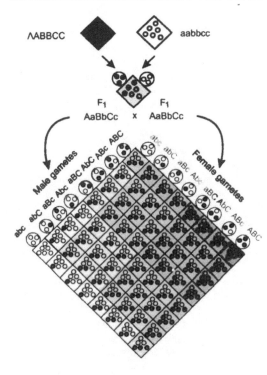

AABBCC   aabbcc

F₁ AaBbCc × F₁ AaBbCc

Male gametes   Female gametes

Fig. 10.40 Inheritance of a quantitative trait (from Miglani, 2000)

pairs were involved in determination of grain color in wheat, i.e., *A,a*; *B,b*; and *C,c* with genes for red (*A, B, C*) dominant over genes for white (*a, b, c*). Each of these three gene pairs segregated in predictable Mendelian fashion, so that the products of heterozygotes for any one pair produced offspring in 3:1 when two gene pair differences segregated simultaneously; the results also followed Mendelian principles, producing 15 red:1 white. Similarly, a cross between heterozygotes for 3 gene pairs produced a close fit to 63 red:1 white. Not all red phenotypes showed identical shading, suggesting that some quantitative effect was involved in determining the degree of redness. A variety of red genotypes occurred containing different numbers of the three red genes, *A, B,* and *C*.

If we imagine that instead of dominance, the action of each red gene is to add a small equivalent degree of redness to the plant, then the range of red phenotypes corresponds to the range of red genotypes. The extreme phenotypes among the F_2 are expected to be quite rare, while the intermediate types should be more frequent. In wheat, three genes determining grain color segregate independently. Consequently, there will be seven classes with respect to number of red genes—6 red, 5 red, 4 red, 3 red, 2 red, 1 red pairs showing a bell-shaped curve red in the ratio 1:6:15:20:15:6:1 respectively (Fig. 10.40). F_2 distribution of color frequencies will be like a normal bell shape (Fig. 10.41).

Polygenes

The term *multiple factor* was replaced by *polygene* by Mather (1943). He defined polygenes as genes with a small effect on a particular character that can supplement each other to produce observable quantitative changes. Some of these quantitative effects can be considered additive if they can be added together to produce phenotypes which are the sum total of the negative and positive effects of individual polygenes. Environmental effects, although

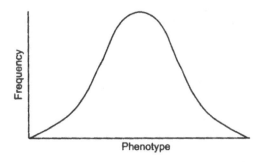

Fig. 10.41 Frequency distribution in F_2 generation of grain color in wheat from the crosses between two strains differing in three genes (from Miglani, 2000)

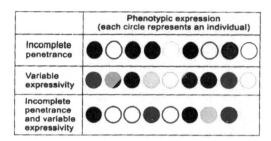

	Phenotypic expression (each circle represents an individual)
Incomplete penetrance	
Variable expressivity	
Incomplete penetrance and variable expressivity	

Fig. 10.42 Effects of penetrance and expressivity on a hypothetical trait "pigment intensity". All individuals in each row have the same genotype (from Miglani, 2002).

not inherited, may also produce modifications in the expected phenotypes of any generation. This factor is also operating to produce the continuity of measurement for quantitative characters.

PENETRANCE AND EXPRESSIVITY

Penetrance represents the proportion of genotypes that shows an expected phenotype. Expressivity deals with the degree to which individuals express a particular effect.

Genes causing lethality may differ in level of penetrance and expressivity. These genes result in a broad range of effects, from a condition in which no survival is observed among affected genotypes to those wherein almost complete viability is observed. Many of these genes which do not cause definite lethality, are termed semilethals or subvitals. It is difficult to draw a line between categories of lethals, but usually only those genes are regarded as lethal which cause the death of the organisms at an early stage. The distinction between penetrance and expressivity is illustrated in Fig. 10.42. Many *developmental traits* not only fail to penetrate sometimes but also show a variable pattern of expression, from very mild to very extreme, when they do penetrate.

References

Aslam, M.S. 2000. *Genetics—Behavioral and Molecular*. Campus Books, New Delhi, India.

Bateman, A.J. 1947. Number of S-alleles in a population. *Nature* 160: 337.

Bateson, W. 1905. In a letter to Sedgewick, A. from Bateson, W. 1928. *Essays and Addresses*. Bateson, B. (ed.). Cambridge Univ. Press, Cambridge, UK.

Bateson, W. and Punnet, R.C. 1905. Experimental studies in the physiology of heredity. *Report in the Evolution Committee of the Royal Society*. II. Harrison and Sons, London, UK.

Bateson, W. and Punnet, R.C. 1906. Experimental studies in the physiology of heredity. *Report in the Evolution Committee of the Royal Society*. III. Harrison and Sons, London, UK.

Bateson, W. and Punnet, R.C. 1908. Experimental studies in the physiology of heredity. *Report in the Evolution Committee of the Royal Society*. VI. Harrison and Sons, London, UK.

Belling, J. 1933. Crossing over and gene rearrangement in flowering plants. *Genetics* 18: 388-413.

Blakslee, A.F. 1934. New jumson weeds from old chromosomes. *J. Hered.* 25: 80-108.

Boveri, T. 1904. *Ergebnisse umber dia Konstitution der chromatisichen Substang des Zelkerns*. G. Fischer, Jena.

Boveri, Th. 1885. Über des Verhalten der centrosomen bei der Bebruchtung des Scrigeleies mebst allgemeimen Bemerkungen uber centrosomen and Vermandtes. *Verh. Phys. Med. Ges. Wurzburg*. 29: 1-30.

Boveri, Th. 1902. Ueber mehroplige Mitosen als Mittel zur Analyse des Zellkerns. *Verh. Phys.-med. Ges. Wurzburg*. 33.

Brewbaker, J.L. 1964. *Agricultural Genetics*. Prentice-Hall, Englewood Cliffs, NJ, USA.

Correns, C. 1900. G. Mendel's regel umber das Verhalten der Nachkommenschaft der assenbastarole. *Deutsch. Bot. Geoell.* 18: 158-168.

Crothers, E.E. 1913. The Mendelian ratio in relation to certain orthopteran chromosomes. *J. Morph.* 24: 487-511.

Darlington, C.D. 1937. *Recent Advances in Cytology*. 2nd ed. Churchill, London, UK.

Darwin, C. 1868. The variation of animals and plants under domestication. Chap. XII. John Murray, London, UK, London, England.

De Lamarck, J.B. 1809. *Philosophie Zoologique.* London, UK. (translator H. Elliot 1914).

De Vries, H. 1900. Sur la loi de disjonction des hybrids. *C.R. Acad. Sci. Paris* 130: 845-847.

De Vries, H. 1901. *Die Mutationheorie.* Veit, Leipzig.

Demerc, M. 1929. Cross sterility in maize. *Zeitschrift induktive* Abstam. Vererbungs. 50: 281-291.

Dobzhansky, Th. 1952. Nature and origin of heterosis. In: *Heterosis*, pp. 218-223. Gowen, J.W. (ed.). Iowa State College Press, Ames, IA, USA.

Eanes, W.F. 1978. Morphological variance of enzyme heterozygosity in the monarch butterfly. *Nature* 276: 263-264.

East, E.M. 1916a. A Mendelian interpretation of variation that is apparently continuous. *Amer. Natur.* 44: 65.

East, E.M. 1916b. Studies on size inheritance in *Nicotiana. Genetics* 1: 164-176.

East, E.M. 1936. Heterosis. *Genetics* 21: 375-397.

Emerson, R.A. and East, E.M. 1913. The inheritance of quantitative characters in maize. *Bull. Agric. Exper. Sta. NB Res. Bull.* 2.

Freese, E. 1963. Molecular mechanism of mutation. In: *Molecular Genetics.* Pt. I, pp. 207-269.Taylor, J.H. (ed.), Acad. Press, London, England.

Freese, E. 1971. Molecular mechanisms of mutations. In: *Chemical Mutagens*, vol. 1, pp. 1-56. Hollaender, A. (ed.). Plenum Press, New York, NY.

Galton, F. 1889. *Natural Inheritance.* MacMillan, New York, NY.

Galton, F. 1909. *Essays in Eugenics.* Eugenics Educ. Soc., London, UK.

Gardener, G.J. 1975. *Principles of Genetics.* John Wiley & Sons, Inc., New York, NY, USA. 622 pp.

Griffiths, A.J.F., Miller, J.H., Suzuki, D.T., Lewontin, R.C., and Gelbart, W.M. 2000. *An Introduction to Genetic Analysis.* W.H. Freeman & Co., New York, NY (7th ed.).

Gupta, P.K. 1994. *Genetics.* Rastogi Publ., Meerut, India.

Gustafson, A. 1947. Mutation in agricultural plants. *Hereditas* 33: 1-100.

Haeckel, E. 1866. *Generelle Morphologie der Organismen.* Reimer, Berlin, Germany.

Hardy, G.H. 1908. Mendelian proportions in a mixed population. *Science* 28: 49-50.

Hertwig, P. 1936. Arbastarde bei Tieren. *Handbuch Vererbungswiss* 21: 1-140.

Heslot, H. 1965. The nature of mutations. In: *The Use of Induced Mutations in Plant Breeding Programmes*, pp. 3-45. Report of the meeting organised by the FAO of the United Nations and the IAEA. Pergamon Press, New York, NY.

Hull, F.H. 1952. Recurrent Selection and overdominance. In: *Heterosis*, pp. 451-473. Gowen, J.W. (ed.). Iowa State College Press, Ames, IA.

Iltis, H. 1932. *Life of Mendel* W.W. Norton, New York, NY. (translations E. Paul and C. Paul).

Johannsen, W. 1903. Uber Erblickeit in Populationen und in reinea Linien. Plenum Press, New York, NY.

Johanssen, W. 1909. Elemente der exakten Erbichkeitslehre, Fischer Publishing, Jena, Germany.

Johanssen, W. 1926. Elemente der exakten Erbichkeitslehre, 3. Jena, Fischer, Germany.

King, R.C. 1965. *Genetics.* Oxford Univ. Press, New York, NY.

Lederberg, J. 1947. Gene recombination and linked segregations in *Escherichia coli. Genetics* 32: 505-525.

Lederberg, J. 1955. Recombination mechanisms in bacteria. *J. Cell. Comp. Physiol.* 45 (suppl. II): 75-107.

Lerner, I.M. 1954. *Genetic Homeostasis.* Oliver and Boyd, Edinburgh, UK.

Mather, K. 1943. Polygenic inheritance and natural selection. *Biol. Rev.* 18: 32-64.

McClung, C.E. 1902. The accessory chromosomes—sex determinants? *Biol. Bull.* 3: 43-84.

Mendel, G. 1866. Experiments in Plant Hybridization. Harvard Univ. Press, London, England (1916) (translator Royal Hort. Soc., London)

Miglani, G.S. 2000. *Basic Genetics.* Narosa Publ. House, New Delhi, India.

Miglani, G.S. 2002. *Advanced Genetics.* Alpha Sci. Intl., Pangbourne, UK.

Montgomery, T.H. 1906. Chromosomes in the spermatogenesis of the Hemiptera: Heteroptera. *Trans. Amer. Phil. Soc.* 27: 97-173.

Nilsson-Ehle, H. 1909. Kreuzungsuntersuchungen an Hafer und Weizen. Lnnds. Univ. *Aarskr, N.F. Afd.* (ser. 2). vol. 5. (2): 1-122.

Nilsson-Ehle, H. 1948. *The Future Possibilities of Swedish Barley Breeding. Svalor Sweden.* pp. 113-126.

Roux, W. 1905. Die Entwicklungsmechanik, ein newer Zweig doe biologischen Wissensahaft. In: *Vortrage and Aufsatze uber Entwicklung—smechanik der Organisemen*, I.W. Rox and Engleman, T.W., Leipzig, Germany.

Sambamurty, A.V.S.S. 1999. *Genetics*. Narosa Publ. House, New Delhi, India.

Shull, G.H. 1908. The composition of a field of maize. *Rept. Amer. Breeders' Assoc.* I4: 296-301.

Shull, G.H. 1952. Beginning of the heterosis concept. In: *Heterosis*, pp. 14-48. J.W. Gowen (ed.). Iowa State College Press, Ames, IA, USA.

Sinnot, E.W., Dunn, L.C., and Dobzhansky, Th. 1959. *Principles of Genetics*. McGraw-Hill Book Co., Inc., New York, NY.

Strasburger, E. 1884. *Neue Untersuchungen über den Befruchtungavorgang bei den Phanerogemen*, Fischer Publishing, Jena, Germany.

Strickberger, M.W. 1968. *Genetics*. MacMillan, New York, NY.

Sutton, W.S. 1902. The chromosomes in heredity. *Biol. Bull.* 4: 213-251.

Trehan, K.S. and Gill, K.S. 1983. Homeostasis at the molecular level. Abstract *XV Intl. Cong. Genetics*, New Delhi, Pt. I; p. 196.

Trehan, K.S. and. Gill, K.S. 1987. Sub-unit interaction; a molecular basis of heterosis. *Biochem. Genet.* 25(11/12): 855-862.

Tschermak, E. 1900. Umber kunstliche Kreuzung bai *Pisun satium*. Zeit. Landev. Versuch. Oest. 3: 465-555.

Verma, P.S. and Agarwal, V.K. 1998. *Cell Biology, Genetics Molecular Evolution and Ecology*. S. Chand & Co. Ltd., New Delhi, India.

Watson, J.D. and Crick, F.H.C. 1953a. A structure for deoxyribose nucleic acids. *Nature* 171: 737-738.

Watson, J.D. and Crick, F.H.C. 1953b. Genetic implications of structure of deoxyribonucleic acid. *Nature* 171: 964-969.

Weinberg, W. 1908. Über den nachweis der verebung beim Menschen. Jahreshefte Verein f. vaterl. *Naturk. in Württemberg* 64: 368-382.

Weismann, A. 1885. Die kontinuitat des keimplasm als grundlage eimer theorie der Verebung. In: *Ausfsatze uber Verebung und verwandte biologische Fragen*. Fischer, Jena, Germany.

Weismann, A. 1887. Über die zahl der richtungskarper and ihre. In: *Bedeutung fur die Vererbung*. Fischer, Jena, Germany.

Whaley, W.G. 1952. Physiology of gene action in hybrids. In: *Heterosis*, pp. 98-113. J.W. Gowen (ed.). Iowa State College Press, Ames, IA, USA.

Wilson, E.B. 1906. Studies on chromosomes. III. The sexual difference of the chromosome groups in *Hemiptera*, with some considerations on the determination and inheritance of sex. *J. Exper. Zool.* 3: 1-40.

Winchester, A.M. 1966. *Genetics*. Houghton Mifflin Company, Boston, MA, USA.

Yule, G.U. 1906. On the theory of inheritance of quantitative compound characters on the basis of Mendel's laws: a preliminary note. *Rept. 3rd Intl. Congr. Genet.* pp. 140-142.

Appendix 1: Questions

1. Define: genetics, trait, preformation, encapsulation, pangenesis, inheritance of acquired traits, blood/blending inheritance, epigenesis, and germplasm theory.

2. Differentiate between: heredity and variation, genotype and phenotype, qualitative and quantitative traits, genotypic and environmental variation.

3. State certain facts about life.

4. Would the science of genetics have been born were there no variation?

5. "Individuals of identical phenotype may have different genotypes and vice versa" State whether this statement is true or false and why?

6. Describe various theories of inheritance put forth from time to time. Which is close to the accepted mechanism of inheritance?

7. What do you understand by continuity of life?

8. Write a brief history of genetics.

9. What do you understand by genetic analysis? Why is genetic analysis considered a prerequisite for crop and animal improvement?

10. What are the necessary conditions that the organism used in genetic experiments should fulfill?

11. Why is Mendel known as the Father of genetics?

12. Give a brief life sketch of Mendel.

13. What do you know about symbols used in genetics while depicting genotypes?

14. Describe the methodology Mendel used to work out the mechanism of inheritance.

15. Define: monohybrid cross, dihybrid cross, incomplete dominance, codominance, principle of segregation, principle of independent assortment (independent segregation).

16. Describe Mendel's observations on the monohybrid crosses.

17. State, giving evidence, whether or not segregation can occur at either of the meiotic division?

18. What generalizations did Mendel draw from his experiments involving dihybrid crosses?

19. Describe inheritance of fruit shape and fruit color in summer squashes by making appropriate dihybrid crosses.

20. Who rediscovered Mendelism? What do you know about the rediscoverers of Mendelism? What was the impact of Mendelism and its rediscovery?

21. Principle of independent assortment (independent segregation) is just an expansion of the principle of segregation. Comment.

22. Do incomplete dominance and codominance provide exceptions to Mendel's generalizations? Give reasons for your answer.

23. Describe incomplete dominance in snapdragon for flower color by showing F_1 and F_2 generations.

24. Describe incomplete dominance in snapdragons (*Antirrhinum majus*) for inheritance of flower color and flower shape. Show F_1 and F_2 generations.

25. What will be the genotypic and phenotypic ratio in a dihybrid cross when there is incomplete dominance for both gene pairs.

26. How are the F_2 phenotypic ratios in a dihybrid cross modified when there is incomplete dominance for (a) both gene pairs? (b) one gene pair only?

27. Work out a trihybrid phenotypic ratio (27:9:9:9:3:3:3:1) with three genes – *Tall, Yellow* and *Round* in pea plant vs. *Dwarf, Green* and *Wrinkled*. Show F_1 and F_2 generations of the crosses.

28. How would you obtain trihybrid genotypic ratio for the cross of Q. No. 27 by the forking method. You may use logical symbols of your choice.

29. The human egg is much larger than the sperm? Does this mean the child inherits more from the mother than the father? Explain fully.

30. How is type of progeny and its frequency predicted from a cross using principles of probability.

31. Describe product rule and sum rule of probability.

32. What is the binomial method. How is this method applied in Mendelian genetics?

33. What is Pascal's triangle? How is it used in determining coefficients of binomial equations?

34. How can principles of probability be applied in reverse to determine the chances of independent events when one knows chances of two such events happening together?

35. Is dominance a law? Give experimental evidence to support your answer.

36. What insight does the experiment on temperature-sensitive lethal mutations provide about dominance?

37. What do you understand by scale of dominance?

38. Define incomplete dominance, complete dominance, no dominance, heterosis, hybrid vigor, and overdominance.

39. Partial dominance, complete dominance and overdominance are three aspects of the phenomenon of heterosis. Comment.

40. What is the molecular basis of Mendelian genetics? Give the answer keeping in view the molecular nature and functional aspect of alleles.

41. How does Mendelian genetics apply to haploid eukaryotic life cycles?

42. Describe diagnostics for alleles using the complementation test. Can we conclusively state whether two mutations are located within a gene or two separate genes?

43. The terms dominance, incomplete dominance, and codominance are somewhat arbitrary unless analyzed critically. Comment.

44. How would you demonstrate that type of dominance is determined by molecular functions of a gene?

45. Is Mendelism an exception or the rule in Genetics?

46. The whole of modern genetics is an exception to Mendelism. Comment on this statement.

47. Modern genetics seems to defy if not question the universality of Mendelian inheritance. Comment on this statement.

48. Discuss some of the deviations from expectations of Mendel's laws. Why didn't Mendel discover these deviations?

49. Describe Mendel's results in the light of chromosome theory and linkage.

50. How do the pairing and disjunction of homologous chromosomes at meiosis explain the mechanism of gene segregation?

51. What is the contribution of the following scientists to development of the chromosome theory of inheritance: Sutton, Boveri, Crothers, Blakslee, and Montgomery?

52. Describe various points of parallelism between behavior of chromosomes and Mendelian factors (genes) during meiosis and fertilization.

53. Describe the chromosomal theory of inheritance.

54. What conclusions do you draw from correspondence between Mendelian factors and chromosomes?

55. What is the relationship between number of alleles at a locus with number of diploid genotypes in a population?

56. Why do some plants not self-pollinate even when the male and female flowers are perfectly normal?

57. What is the genetic basis of self-incompatibility?

58. Describe self-sterility multiple allelic system in *Nicotiana*

59. What conclusions can be drawn from studies on multiple allelism?

60. What has been the contribution of this multiple allelic series in understanding the gene?

61. Differentiate between: dominance and epistasis, and epistatic and hypostatic genes.

62. Describe one example illustrating interaction between two dominant genes. Give a metabolic pathway and a molecular model to explain results of the crosses.

63. Describe the following examples of recessive epistasis: complementary gene interaction and supplementary gene interaction. Give biochemical pathways and molecular models to explain results of the crosses in the above examples.

64. What is dominant epistasis? Explain the following cases of dominant epistasis: duplicate genes and polymeric effect. Give

biochemical pathways and molecular models to explain results of the crosses in the aforesaid examples.

65. Represent the relationships among different types of gene interactions at a glance.

66. What generalizations have been drawn from the study of examples of gene-gene interaction?

67. What is pleiotropy? In what way is this phenomenon important in genetic studies? Give some examples of pleiotropy.

68. What do you understand by totipotency? How has totipotency of a cell been used in crop and animal improvement?

69. Define homeostasis. What do you understand by genetic/developmental homeostasis? Give some examples of homeostasis.

70. What is the biochemical basis of homeostasis in cross-pollinated and self-pollinated crops?

71. How can time of segregation be studied in corn?

72. What is the xenia effect? Is the xenia effect shown for all the endospermic traits? Answer this question with the help of appropriate reciprocal crosses.

73. What is linkage? What is its relationship with independent assortment and crossing over?

74. What is a linkage group?

75. What do you understand by genetic recombination?

76. Differentiate with the help of a diagram between coupling and repulsion phase of linkage.

77. What type of gametes will be produced from an individual of genotypes Ab/aB, with and without crossing over between two genes?

78. What is the procedure for detection and estimation of linkage in a two-point test cross and F_2 progeny in a plant system?

79. What is a linkage map? What is its basis?

80. Outline the procedure for detection and estimation of linkage in a three-point test cross?

81. What do you understand by coefficient of coincidence and interference? How are these concepts related to each other and what do they signify?

82. What is the theoretical basis of detection of linkage through partitioning of chi-square? Describe the procedure along with the formulae used for detection of linkage in an F_2 plant population. Also describe a similar procedure for a test cross plant population.

83. Describe the following methods for estimation of linkage in F_2 plant population: Emerson's method, maximum-likelihood method, and product ratio method. Which of these methods is more efficient and why?

84. Through tetrad analysis in *Neurospora* show that crossing over occurs at the four-strand stage.

85. Explain with the help of suitable diagrams the following hypotheses of the mechanism of recombination: breakage and copying hypothesis, complete copy choice hypothesis, and breakage and reunion hypothesis.

86. What is the significance of crossing over? In what way has it been used during evolution and how has this phenomenon been used for plant and animal improvement?

87. Explain differences between first- and second-division segregation with the help of a diagram.

88. Describe different types of tetrads and their origin.

89. What important things are kept in mind while mapping genes in haploid eukaryotes? Give the answer with special reference to *Neurospora crassa*.

90. Describe the procedure of measuring distance of a locus from a centromere in *Neurospora*?

91. In a two-point cross in *Neurospora*, how are distances of genes calculated in relation to the centromere?

92. What is the procedure of detection and estimation of linkage in a two-point cross in *Neurospora*.

93. What is interference? What different types of interferences can be observed in data showing double exchanges? Give the answer with special reference to *Neurospora*.

94. What is the basic difference between the type of tetrads in *Neurospora* on the one hand and yeast and *Chlamydomonas* on the other? In what way do methods of estimating linkage in these two systems differ?

95. How does gene conversion account for interallelic recombination?

96. What do you understand by the terms mutation, mutagen, and mutant?

97. Describe various characteristics of mutations.

98. Classify mutations on the basis of the following criteria: size of mutations, changes in chromosome structure, changes in chromosome number, direction of mutations, and effects on the sense of genetic code.

99. What are spontaneous mutations? How are the following processes responsible for their origin: spontaneous changes in electronic structure of bases, rare base pairing between two purines or two pyrimidines, chemicals produced under normal growth conditions.

100. What are induced mutations? Give a brief history of induced mutations.

101. What are the different types of physical agents capable of inducing mutations?

102. Describe the mechanism of action of the following radiations: Ultraviolet light, x-rays and γ-rays.

103. What is the target theory of mutagenic action of x-rays?

104. What are the different classes of chemical agents known to induce mutations?

105. Describe the mechanism of action of the following chemical mutagens: inhibitors of nucleic acid precursors, base analogs, dyes, nitrous acid, and alkylating agents.

106. What is the usefulness of various organisms as a model for detection of mutations?

107. Describe the method of detection of nutritional biochemical mutations in *Neurospora crassa*.

108. How is electrophoresis useful in detection of mutations? What type of mutations can be detected through this technique?

109. Describe various applications of induced mutations.

110. What are environmental mutagenic hazards? What classes of agents pose a threat to the genetic health of man. Why is a battery of test systems recommended for testing mutagenicity of an environmental chemical?

111. What are the salient differences between laboratory and natural populations?

112. How and why is binomial distribution used in determination of genotype frequencies?

113. Define as completely as possible the Hardy-Weinberg law describing all the assumptions considered therein.

114. Explain the principle of genetic equilibrium.

115. Do we observe equilibrium in mixed populations in the initial generation? Give reasons for your answer.

116. Describe the basis and method for determination of gene frequency in traits showing: (a) dominant inheritance, (b) codominant inheritance, (c) incompletely dominant inheritance, and (d) multiple allelic traits.

117. Define properties of quantitative traits.

118. What are Galton's observations on inheritance of quantitative traits?

119. What did Yule's work suggest about the inheritance of quantitative traits?

120. In what ways are quantitative traits measured?

121. How have experiments of Emerson and East contributed to our understanding of the inheritance of quantitative traits?

122. What did the experiments of Johannsen on seed weight in beans add to our knowledge of inheritance of quantitative traits?

123. What is the contribution of Nilsson-Ehle's work on grain color in wheat in understanding inheritance of quantitative traits?

124. Describe the multiple factor hypothesis.

125. What do you understand about the inheritance of a quantitative trait?

126. Do the Mendelian genes explain the inheritance of quantitative inheritance?

127. What are polygenes? How do polygenes differ from Mendelian genes?

128. What is the relationship between genotype, phenotype, and environment?

129. Define penetrance and expressivity.

130. What do you understand by incomplete penetrance and variable expressivity?

Appendix 2: Numericals

1. In peas, a tall plant is dominant over a dwarf. (a) If a plant homozygous for tall is crossed with one homozygous for dwarf, what will be the appearance of F_1; of F_2; of the offspring of a cross of F_1 with its tall parent; with its dwarf parent? (b) Let the allele for tall be represented by T and the allele for dwarfness by t. What will be the gametes produced by the parents and the height of the offspring (tall or dwarf) from each of the following crosses: $Tt \times tt$; $TT \times tt$?

2. In four o' clock flowers, red flower color R is incompletely dominant over white r, the heterozygous plant being pink flowered. (a) In the following crosses, in which the genotypes of the parents are given, what are the gametes produced by each parent and what will be the flower color of the offspring from each cross $Rr \times RR$; $rr \times Rr$; $RR \times rr$, $Rr \times Rr$? (b) How would you produce four o' clock seeds, all of which would yield pink flowered plants when sown?

3. Consider three yellow, round peas, labeled A, B, and C. Each was grown into a plant and crossed to a plant grown from a green, wrinkled pea. Exactly 100 peas issuing from each cross were sorted into phenotypic classes as follows:

    A: 51 yellow, round

        49 green, round

    B: 100 yellow, round

    C: 24 yellow, round

        26 yellow, wrinkled

        25 green, round

        25 green, wrinkled

    What were the genotypes of A, B and C? Use gene symbols of your own choosing; be sure to define each one.

4. In nature, the plant (*Plectritis congesta*) is dimorphic for fruit shape; i.e. the individual plant bears either wingless or winged fruits. Plants were collected from nature before flowering and were crossed or selfed with the following results:

	Number of progeny	
Pollination	Winged	Wingless
Winged (selfed)	91	1*
Winged (selfed)	90	30
Wingless (selfed)	4*	80
Winged × Wingless	161	0
Winged × Wingless	29	31
Winged × Wingless	46	0
Winged × Winged	44	0
Winged × Winged	24	0

Interpret these results, and derive the mode of inheritance of these fruit-shaped phenotypes. Use symbols of your choice. The phenotypes of progeny marked with an asterisk probably have a nongenetic explanation. What do you think it is?

5. In maize, one F_1 plant obtained from a cross between tall (*Le Le*) × dwarf (*le le*) was self-pollinated and also backcrossed with both parents. Give the gametes of the two parents, and the F_1; also give the genotypic and phenotypic ratios of progenies due to selfing and backcrossing.

6. A variety of pepper having brown fruit was crossed with a variety having yellow fruit. F_1 plants had red fruits. F_2 consisted of 182 plants with red fruits, 59 with brown fruit, 61 with yellow fruit and 20 with green fruits. What is the genetic basis of these fruit colors in peppers?

7. In weasel two recessive genes, p and i, are known, either or both of which in homozygous form result(s) in platinum coat color (*PPii, ppII, ppii*). The presence of both dominant alleles results in brown color (*P-I-*). Will platinum weasel necessarily breed true? Set up a cross between two platinum weasels which will result in all brown offspring (*PPii* × *ppII*). What would be expected in the F_2 from this cross?

8. A cross was made between a red-flowered plant and a white-flowered plant. The F_1 plants were all red-flowered. In the F_2 the flower colors were 92 red, 30 cream, and 42 white. Explain these results. On crossing

the white-flowered plants of the F_2 among themselves, what proportion of the offspring would be red-flowered? Cream flowered? White?

9. In summer squashes white fruit color is dependent on a dominant gene W, and colored fruit on its recessive allele, w. In the presence of ww, the color may be yellow due to a dominant gene G, or green due to its recessive allele, g. How many different genotypes may be involved in the production of white fruits? If white fruited plant $WWGG$ is crossed with green fruited, what will be the appearance of the F_1? of the F_2?

10. Two white-flowered strains of sweet pea were crossed producing purple-flowered F_1s. In F_2, 53 plants had purple flowers and 43 had white flowers. What is the phenotypic ratio in F_2? What type of interaction is involved? What were the probable genotypes of the parental strains?

11. Red color in wheat kernels is produced by genotype $R\text{-}B\text{-}$, white by double recessive ($rrbb$). The genotypes $R\text{-}bb$ and $rrB\text{-}$ produce brown kernels. A homozygous red variety is crossed to white. What phenotypic results are expected in F_1 and F_2?

12. White fruit color in summer squash is governed by a dominant gene (W) and colored fruit by its recessive allele (w). Yellow fruit is governed by an independently assorting hypostatic gene (G) and green by its recessive allele (g). When dihybrid plants are crossed the offspring appear in the form of 12 white:3 yellow:1 green. What fruit color ratios are expected from the following crosses? (a) $Wwgg \times WwGG$, (b) $WwGg \times wwgg$, (c) $Wwgg \times wwGg$, and (d) $WwGg \times Wwgg$.

13. On chromosome 3 of corn there is a dominant gene (A_1) which together with the dominant gene (A_2) on chromosome

9 produces a colored aleurone. All other genetic combinations produce a colorless aleurone. Two pure colorless strains are crossed to produce all colored F_1. What were the genotypes of the parental strains and the F_1s? What phenotypic proportions are expected in F_2? What genotypic ratios exist among the white F_2?

14. Two pairs of alleles govern the color of onion bulbs. A pure red strain crossed to a pure white strain produces all red F_1. The F_2 was found to consist of 47 white:38 yellow:109 red bulbs. What epistatic ratio is shown by the data? What is the name of this type of gene interaction? If another F_2 is produced by the same kind of cross and 8 bulbs of F_2 have double recessive genotype, how many bulbs would be expected in each phenotypic class?

15. Three fruit shapes are recognized in summer squash – disk, elongated, and spherical. A pure disk- shaped variety was crossed to a pure elongated variety. The F_1 were all disk-shaped. Among 80 F_2 there were 30 spherical, 5 elongated, and 45 disk-shaped. Reduce the F_2 numbers to the lowest ratio. What type of interaction is operative. What is the genotypic ratio of sphere-shaped F_2.

16. In tomatoes, red fruit is dominant to yellow, double loculate fruit dominant over multiloculate fruit, and tall vine dominant over dwarf. A breeder has two pure lines: red, double loculate, dwarf, and yellow, multiloculate, tall. From these two lines he wants to produce a new pure line for trade that is yellow, double loculate and tall. How exactly should he go about this? Show not only which cross to make but also how many progeny should be sampled in each case.

17. In tomatoes, two alleles of one gene determine the character difference of purple (P)

Mating	Parental phenotypes		Number of progeny		
		P, C	P, Po	G, C	G, Po
1.	P, C × G, C	321	101	310	107
2.	P, C × P, Po	219	207	64	71
3.	P, C × G, C	722	231	0	0
4.	P, C × G, Po	404	0	387	0
5.	P, Po × G, C	70	91	86	77

versus green (G) stems, and two alleles of a separate gene determine the character difference of "cut" (C) versus "potato" (Po) leaves. The results for five matings of tomato plant phenotypes were as given in Table on page 344.

Determine which alleles are dominant. What are the most probable genotypes for the parents in each cross?

18. We have dealt mainly with only two genes, but the same principles hold for more than two genes. Consider the following cross:

$A/a; B/b; C/c; D/d; E/e \times a/a; B/b; c/c; D/d; e/e$

a. What proportion of progeny will *phenotypically* resemble (1) the first parent, (2) second parent, (3) either parent, and (4) neither parent?

b. What proportion of progeny will be *genotypically* the same as (1) the first parent, (2) second parent, (3) either parent, and (4) neither parent?

Assume independent assortment.

19. A corn geneticist has three pure lines of genotypes: $a/a; B/B; C/C; A/A; b/b; C/C$, and $A/A; B/B; c/c$. All the phenotypes determined by a, b, and c will increase the market value of the corn, so naturally he wants to combine them all in one pure line of genotype $a/a; b/b; c/c$.

a. Outline an effective crossing program that can be used to obtain the $a/a; b/b; c/c$ pure line.

b. At each stage, state exactly which phenotypes will be selected and give their expected frequencies.

20. Is there more than one way to obtain the desired corn genotype? Which is the best way? Assume independent assortment of the three genes.

21. In tomatoes, round fruit shape (O) is dominant over oblate (o) and smooth fruit skin (P) dominant over peach (p). Plants having round smooth fruit were crossed with those having elongate peach fruit. The F_1 plants were test crossed and the following progeny obtained: Round, Smooth – 360, Round, Peach – 32, Elongate, Smooth – 38, and Elongate, Peach – 370. Test whether the genes o and p are linked? If so, in which phase? Estimate the frequency of recombi-

nation between the genes and prepare their linkage map. Calculate coefficient of coincidence and interference and comment on their nature.

22. In tomato, compound inflorescence (s) is recessive to solitary (S), peach fruit shape (p) recessive to smooth (P), and oblate fruit shape (o) recessive to round (O). When plants heterozygous for genes o, p, and s were test crossed, the following progeny were obtained: Round, Smooth, Solitary – 73, Oblate, Peach, compound – 63, Round, Smooth, compound – 348, Oblate, Peach, Solitary – 306, Round, Peach, Compound – 3, Oblate, Smooth, Solitary – 4, Round, Peach, Solitary – 96, and Oblate, Peach, Compound – 110. Ascertain the genotypes and phenotypes of the two parents and F_1 individuals. What different types of gametes will be produced by the F_1 plants? Calculate frequency of recombination between genes o, p, and s and prepare their linkage map. Calculate coefficient of coincidence and interference and comment on their nature.

23. In *Mirabilis jalapa* (four o' clock), a plant hybrid for red (R) and white flowers (r) is pink (Rr). A plant bearing pink flowers is crossed with one having red flowers and another having white flowers. Give the genotypic and phenotypic ratios expected in the progenies.

24. In pea, tall vine (Le) is dominant over dwarf (le), green pods (Gp) over yellow (gp) and round seed (R) over wrinkled (r). A homozygous dwarf, green, wrinkled pea plant is crossed with a homozygous tall, yellow, round. Employing the forked line method, give the genotypes and phenotypes of the parents, F_1, and F_2 progenies.

25. Genes a and b are linked with 20% crossing over. An $a^+ b^+/a^+ b^+$ individual was mated with an $a b/a b$ individual. (a) Represents the cross on the chromosomes and illustrates the gametes produced in the F_1 individual. (b) If F_1 was crossed with the double recessive, what offspring would be expected and in what proportion? (c) Is this a case of coupling or repulsion linkage?

26. Genes a and b are linked with 20% crossing over. An F_1 individual of genotype $a^+ b/a b^+$

was mated with an *a b/a b* individual. (a) Represent the cross between the original parents on the chromosomes. (b) Illustrate the gametes produced by the F_1 individual. (c) If F_1 was crossed with a double recessive, what offspring would be expected and in what proportion? (d) Is this a case of coupling or repulsion linkage?

27. What phenotypic classes would be expected in F_2 of a dihybrid cross and in what proportion, if crossing over is 20%.

28. Assume that genes *a* and *b* are linked and show 30% recombination. If a wild individual is crossed with a double recessive individual, what will be the genotype of F_1? What gametes will the F_1 produce and in what proportion? If F_1 is crossed with a double recessive, what will be the appearance and genotypes of the offspring.

29. Plants heterozygous for genes *a* and *b* were selfed and gave the following number of plants in four phenotypic classes: *A_B_* = 260, *A_bb* = 40, *aaB_* = 30, and *aabb* = 70. Detect linkage by partitioning χ-square. Estimate linkage by Emerson's, maximum likelihood and product ratio methods. Comment on the efficiency of these methods.

30. A fully heterozygous F_1 corn plant was red with normal seed. This plant was crossed with a green plant (*b*) that had tassel seed (*ts*). The following progeny were obtained: Red, Normal = 129, Red, Tassel = 121, Green, Normal = 127, and Green, Tassel = 123. Do the genes *b* and *ts* show linkage? If so, what is the percentage crossing over? If these genes do not show linkage demonstrate that recombination frequency is 50%. Diagram the cross showing the arrangement of genetic markers on the chromosomes.

31. For each of the following traits in *Neurospora* in which first- and second-division

Trait	Number of Asci	
	First-division segregation	Second-division segregation
Mating type (*A* vs *a*)	331	51
Pale conidial color vs orange	73	36
Fluffy growth form vs normal	42	67

segregation was observed, state the gene-centromere distance (data of Lindergren).

32. The following are the results of ordered tetrad analysis from a cross between a *Neurospora* strain carrying *albino* (*al*) that was also unable to synthesize *inositol* (*inos*) and a wild-type strain (++): (a) Determine whether these two genes are linked and, if so, the linkage distance between them. (b) Which gene has the longest gene-centromere distance?

33. A cross between a *Neurospora* stock bearing the mutant gene snowflake (sn), which had otherwise been wild, to a colonial temperature-sensitive stock (cot), which was otherwise wild, produced a number of regular tetrads showing first- and second-division segregation as given in the following table. Determine whether these genes are linked and calculate their gene-centromere distances.

34. In a cross between a *Neurospora* strain requiring histidine (*hist-2*) and a strain requiring alanine (*al-2*), 646 unordered tetrads were analyzed with the following compositions and numbers.

115	2 *hist al*[+] : 2 *hist*[+] *al*
45	2 *hist al* : 2 *hist*[+] *al*[+]
486	1 *hist al* : 1 *hist al*[+] :1 *hist*[+] *al*:1 *hist*[+] *al*[+]

al inos	al +	al +	al +	al inos	al +	al +
al inos	al +	al inos	+ +	+ +	+ inos	+ inos
+ +	+ inos	+ +	al inos	al inos	al +	+ +
+ +	+ inos	+ inos	+ inos	+ +	+ inos	al inos
4	3	23	36	15	16	22

sn cot	sn +	sn cot	sn cot	+ +	sn +
sn cot	sn +	sn +	sn +	+ cot	sn cot
+ +	+ cot	+ cot	+ +	sn cot	+ +
+ +	+ cot	+ +	+ cot	sn +	+ cot
25	16	11	12	8	8

Determine whether the two genes are linked and, if so, the linkage distance between them.

35. In an experiment with yeast (unordered tetrads) a cross was made between two strains, each different in respect to three different genes, a, b, c, not necessarily linked on the same chromosome. The cross was $+ b c \times a + +$ and produced the following tetrads:

407	$2 + + c : 2 a b +$
396	$2 + b c : 2 a + +$
104	$1 + b c : 1 + b + : 1 a + c : 1 a + +$
92	$1 + + c : 1 + + + : 1 a b c : 1 a b +$
1	$2 + b + : 2 a + c$
1	$2 + + + : 2 a b c$

From these data determine which genes are linked, if any, and the linkage distances.

36. What is the frequency of heterozygote tasters in a random mating population, if the frequency of nontasters is 0.09?

37. What is the frequency of heterozygotes Tt in a random mating population in which the frequency of all the tasters is 0.19?

38. Two full-grown plants of a particular species are given to you which have extreme phenotypes for a quantitative character such as height, e.g. 1 foot tall and 5 feet tall. (a) If you had only a single set of environmental conditions in which to conduct your experiment (e.g. one uniformly lighted greenhouse), how would you determine whether plant height is environmentally or genetically caused? (b) If genetically caused, how would you attempt to determine the number of gene pairs that may be involved in this trait?

39. A cross between two inbred plants that had seeds weighing 20 and 40 centigrams respectively, produced an F_1 with seeds that weighed uniformly 30 centigrams. An $F_1 \times F_1$ cross produced 1,000 plants; four had seeds weighing 20 centigrams, four had seeds weighing 40 centigrams, and the other plants produced seeds with weights varying between these extremes. How many gene pairs would you say are involved in determination of seed weight in these crosses.

40. If skin color is caused by additive genes, give your answers to the following: (a) Can matings between individuals with intermediate colored skins give birth to lighter-skinned offspring? (b) Can such matings produce dark-skinned offspring? (c) Can matings between individuals with light skins produce dark-skinned offspring?

41. Assume that two pairs of genes with two alleles each, Aa and Bb, determine plant height additively in a population. The homozygote $AABB$ is 50 cm high, and the homozygote $aabb$ 30. (a) What will be the F_1 height in a cross between these two homozygous stocks? (b) After an $F_1 \times F_1$ cross, what genotype in the F_2 will show a height of 40 centimeters? (c) What will be the F_2 frequency of these 40-centimeter plants?

42. If three independently segregating genes, each with two alleles, determine height in a particular plant, e.g. Aa, Bb, Cc, so that the presence of each capital-letter allele adds 2 centimeters to a base height of 2 centimeters: (a) Give the heights expected in the F_1 progeny of a cross between homozygous stocks $AABBCC$ (14 centimeters) $\times$ $aabbcc$

(2 centimeters). (b) Give the distribution of heights (phenotypes and frequencies) expected in an $F_1 \times F_1$ cross. (c) What proportion of this F_2 progeny would have heights equal to the parental stocks? (d) What proportion of the F_2 would breed true for the height shown by the F_1?

CHAPTER 11

Protein Synthesis

William V. Dashek

Cytoplasmic protein synthesis has been reviewed many times (Evans and Wood, 1987; Arnstein and Cox, 1992; Nierhaus and Wilson, 2004; see botany textbooks). Thus, this chapter is concerned mainly with the organelle and *in vitro* synthesis of proteins.

BRIEF REVIEW OF *IN VIVO* PROTEIN SYNTHESIS

Transcription

Prior to protein synthesis transcription occurs. One strand of the DNA double helix serves as a template for the synthesis of m-RNA via RNA polymerase (Fig. 11.1). This RNA, which migrates from the nucleus to the cytoplasm, undergoes various maturations including splicing. During this process noncoding sequences are eliminated.

Translation

The primary structure (amino acid sequences) of proteins is synthesized via translation of codons (nucleotide triplets) in r-RNA by ribosomes (Cold Spring Harbor, 2001). Each amino acid is specified by a codon (Table 11.1). Eukaryotic ribosomes, complexes of proteins, and RNA

(Ramakrishnan, 2002; Moore and Steitz, 2003) are larger than their bacterial counterparts (Blaha, 2004). In addition, eukaryotic ribosomes whose synthesis has been reviewed by Scheer and Hock (1999), LaFontaine and Tollervey (2004), LaFontaine (2004) and Spirin (2000), contain many more proteins (Table 11.2). The r-RNA is transiently associated with m-RNA, initiation (Sonenberg and Dever, 2003), elongation (Nierhaus and Wilson, 2004), and termination (Inagaki and Doolittle, 2000) factors.

The process of m-RNA translation is divided into three phases: initiation, elongation, and termination. To accomplish each of these phases, factors are required (Table 11.3, Table 11.4). The initiation phase comprises all the processes requisite for ribosome assembly with an initiator-methionyl t-RNA (met-t-RNA-imet) at the beginning of m-RNA's start codon. The initiator amino acid appears to be coded by AUG. Sonenberg and Dever, 2003 discuss the mechanism and regulation of initiation in eukaryotes.

The elongation phase is responsible for polypeptide synthesis (Nierhaus and Wilson, 2004). Elongation involves pairing of m-RNA and t-RNA as specific aminoacyl

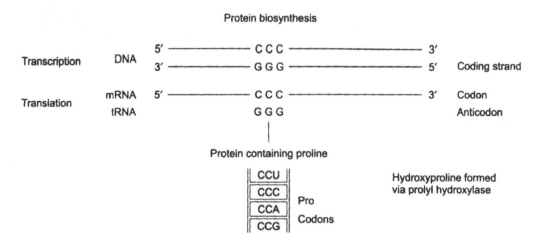

Protein biosynthesis

Fig. 11.1 Diagram of *in vivo* protein synthesis

TABLE 11.1 Triplet condons and corresponding amino acids

	U	C	A	G	
U	UUU / UUC — Phe	UCU / UCC / UCA / UCG — Ser	UAU / UAC — Tyr	UGU / UGC — Cys	
	UUA / UUG — Leu		UAA / UAG — STOP	UGA — STOP / UGG — Trp	
C	CUU / CUC / CUA / CUG — Leu	CCU / CCC / CCA / CCG — Pro	CAU / CAC — His / CAA / CAG — Gln	CGU / CGC / CGA / CGG — Arg	
A	AUU / AUC / AUA — Ile / AUG — Met	ACU / ACC / ACA / ACG — Thr	AAU / AAC — Asn / AAA / AAG — Lys	AGU / AGC — Ser / AGA / AGG — Arg	
G	GUU / GUC / GUA / GUG — Val	GCU / GCC / GCA / GCG — Ala	GAU / GAC — Asp / GAA / GAG — Glu	GGU / GGC / GGA / GGG — Gly	

Source: http://www.colorado.edu/chemistry/bioinfo/AminoAciddata.htm

t-RNAs (Ibba and Soll, 2000) (Fig. 11.2) and brings amino acids to the ribosome (Ribas de Pouplona and Schimmel, 2004). Important in this process are the correct pairing of m-RNA (Fig. 11.3) codons and t-RNA anticodons. Whereas bacteria contain 30-40 t-RNAs possessing different anticodons, approximately 50 different t-RNAs occur in plants. Catalytic centers formed by r-RNA of the ribosomal large subunit function in peptide bond formation. Termination of protein synthesis occurs via stop condons, e.g., UAA, UAG, and UGA. Preiss and Hentze (2003) reviewed additional events for the translation process.

Subsequent to polypeptide release from the ribosome, polypeptides can undergo enzyme-catalyzed folding to yield active three-dimensional forms (Cleland, 1998). The role of molecular chaperones in folding of newly translated proteins has been reviewed by Frydman (2001). In addition, some proteins can be modified by posttranslational reactions (Abraham *et al.*, 1984; Hanes, 1999) including the attachment of carbohydrates (Bill *et al.*, 1999). Finally, certain proteins can be directed to particular cellular loci via targeting mechanisms involving peptide signal sequences (Teasdale and Jackson, 1996; Soll, 1998).

ORGANELLE PROTEIN SYNTHESIS

Both mitochondria and chloroplasts (see Chapter 6) are thought to originate from prokaryotes according to the endosymbiotic

TABLE 11.2 Comparison of eukaryotic and prokaryotic ribosomes

	Subunits	Proteins	Number of bases in rRNA
Prokaryotic	Two, 50s and 30S	31 for Large subunit	Large subunit 23S, 2,904
	Large subunit		5S, 120
	Contains 5S and 23S rRNA	21 for Small subunit	Small subunit 16S, 1,542
	Small subunit contains 16S rRNA		
Eukaryotic	Two, 60S and 40S	50 for Large subunit	Large subunit 28S, 47,00
	Large subunit	32 for Small subunit	5.8S, 154
	Contains 5S, 5.8S, and 28S rRNA		5S, 120
			Small subunit 18S, 1,900
	Small subunit contains 18S rRNA		

TABLE 11.3 Summary of important factors for eukaryotic protein synthesis[a]

Factor	Function
Initiation factors (eIFs)	
eIF1 (general stimulator of initiation)	Prepare m-RNA for correct attachment to the
eIF1A	ribosome
eIF2 (binds to met-t-RNA)	– Ternary complex formation[b]
eIF2B (guanine nucleotide exchange protein)	– GTP/GDP exchange during eIF2 recycling
eIF3 (general role in binding)	– Ribosome subunit antiassociation
eIF4C	
eIF4A (RNA helicase)	– ATPase–dependent RNA helicase
eIF4B (RNA binding protein)	– Stimulates helicase
eIF4F	– Initiation factor complex
eIF5 (stimulates GTPase activity)	Helps dissociate eIF2, eIF3, and eIF4C
eIF6	Helps dissociate 60S from inactive ribosomes
Elongation factors	
eEF1α	Brings the amino-acylated t-RNA to the ribosome
eEF1$\beta\delta$	Guanine nucleotide releases protein required to recycle eEF-1α
eEF2	Required for translocation
Termination factors	
eRF	Release of completed polypeptide chain[c]

[a]For additional factors see http://web.indstate.edu./theme/nwking/protein-synthesis.html. This website also discusses specific steps in initiation, elongation, and termination.
[b]Ternary complex (preinitiation complex) consists of an initiator, GTP, eIF2, and 40S subunits.
[c]Frolova et al. (1994); Drugeon et al. (1997).

TABLE 11.4 Some in-depth reading regarding protein synthesis initiation, elongation, and termination factors

Factor	References
Initiation	Sachs and Varani (2000)
	Pestova *et al.* (2001)
	Schneider *et al.* (2001)
	Sonenberg and Dever (2003)
Elongation	Rodina *et al.* (1999)
	Kisselev and Buckingham (2000)
	Frolova *et al.* (1994)
Termination	Inagaki and Doolittle (2000)
	See numerous references in
	Caraglia *et al.* (2000)

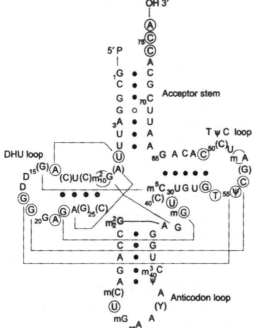

Fig. 11.2 Structure of t-RNA

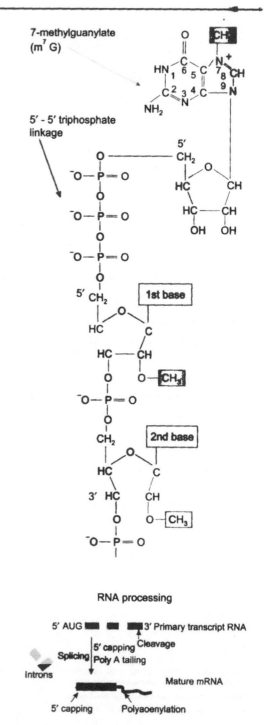

Fig. 11.3 Structure of a portion of m-RNA

theory (Margulis, 1981). Thus, much of the organelle protein synthesizing machinery resembles that of certain prokaryotes. Certain higher plant mitochondria and chloroplasts contain organelle DNA which encodes r-RNAs and t-RNAs (Table 11.5).

TABLE 11.5 Comparison of plant mitochondrial and chloroplast protein synthesizing machinery

Component	Mitochondria[a]	Chloroplast[b]
DNA and RNA	Circular molecule linear forms exist; but mt-DNA is very complex with a molecular weight between 200,000 and ~400,000. The mitochondrial genome ranges from 200 – 2,400 kb and encodes organelle specific r-RNA and t-RNA but few organelle proteins; not all required t-RNAs are encoded by mt-RNAs.	Circular molecule; 120,000 – 160,000 bp; about 120 genes; ~60 genes involved in RNA transcription and translation including genes for r-RNA, t-RNA, and RNA polymerase subunits and ribosomal proteins. Approximately 20 genes encode subunits of chloroplast electron transport complexes and the FoF1, ATPase complex and large subunits of RUBISCO. Most plastid proteins are encoded by the nuclear genome.
Ribosomes	Resemble bacterial ribosomes in structure and sensitivity to chloramphenicol and resistance to cycloheximide; two proteins synthesize inhibitors; have different sedimentation coefficients, different sizes of r-RNA and different sets of ribosomal proteins than cytoplasmic ribosomes.	Component RNAs and proteins are similar to those of eubacteria 70S ribosomes made up of two subunits, 50S and 30S; large subunits possess 5S and 23S r-RNA and the small subunit has 16S r-RNA.

[a]Adapted from: Douce (1985), Bendich *et al.* (1993), Scheffler (1999), Copeland (2002) and Botany on line (http://www.biologie.uni-hamburg.de/b-online/eoo/contents.htm) (2003).

[b]Adapted from Harris *et al.* (1994).

TABLE 11.6 Summary of protein import into mitochondria and chloroplasts[a]

Organelle	Characteristic
Mitochondria and chloroplasts	Import is restricted to sites where the inner and outer organellar membranes contact.
Mitochondria	Cytosolic Hsc 70 and mitochondrial import simulation factor are chaperones which bind cytosolic precursors of mitochondrial proteins, maintaining these in a partially unfolded state which can be translocated into mitochondria; translocation requires ATP. N terminus of proteins imported to mitochondrial destinations other than the matrix often contain a second targeting sequence.
Chloroplast	Import of precursor proteins into the chloroplast stroma involves receptors (import apparatus) in the outer membrane which recognize stromal chaperones; soluble cytosolic components for the import process appear to be required.

[a]Adapted from: http://www.ncbi.nlm.nih.gov/books/bv.fcgi2/cell=bv. Alefsen and Soll J. (1993); Bountry and Chaumont (1993); Meadows *et al.* (1993), Schnell (1998); Whelan (2004).

However, these organelles synthesize few of their proteins. Instead, certain of the organelle proteins are encoded by nuclear genes, synthesized on cytoplasmic ribosomes and imported posttranslation into organelles (Fig. 11.4). Table 11.6 presents some characteristics of protein import.

TABLE 11.7	Components of crude extracts used for *in vitro* protein synthesis

Components

70S or 80S ribosomes

t-RNA

← aminoacyl t-RNAs synthetases, initiation, elongation, and termination factors

amino acids

ATP, GTP[a]

creatine phosphate and creatine phosphokinase[b]

monovalent and divalent cations

[a]Energy source.
[b]Energy regenerating system.

IN-VITRO PROTEIN SYNTHESIS

Three cell-free translation systems are widely employed. These involve extracts form rabbit reticulocytes, wheat germ, and

TABLE 11.8	Comparison of three *in vitro* translation systems

System	Attributes
Rabbit reticulocyte lysate (Fig. 11.5)	"Highly efficient *in vitro* eukaryotic protein synthesis system used for translation of exogenous RNAs."
Wheat germ extract (Fig. 11.5)	"Convenient alternative to rabbit reticuloycte." "Efficiently translates exogenous RNA from a variety of different organisms." "It is recommended for translation of RNA containing small fragments of double stranded RNA...."
Capped or uncapped RNA templates	Useful when employing RNA synthesized *in vitro*

Adapted from: Ambion, TB #187 *in vitro translation basics*. Ambion Corporation (2005).

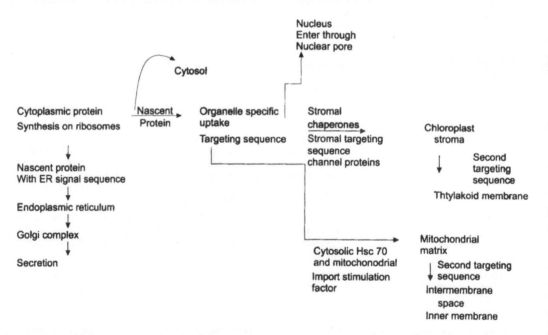

Fig. 11.4 Synthesis and import of organelle proteins

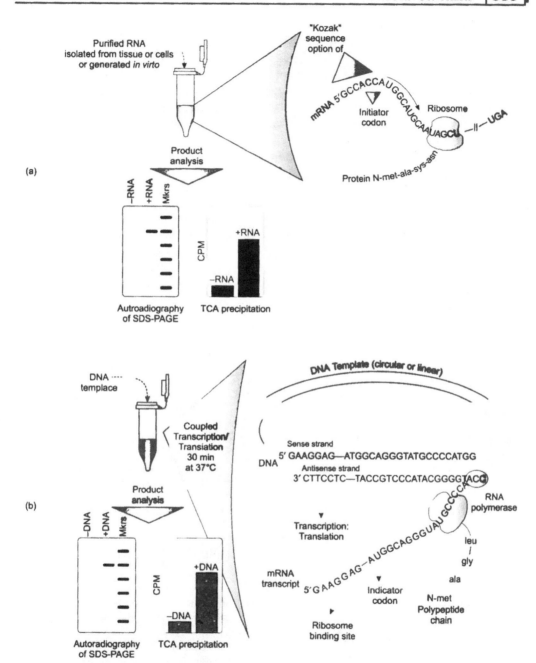

Fig. 11.5 **a** "Standard *in vitro* translation procedure using rabbit reticulocyte lysate or wheat germ extract" (from: www.ambion.com/techlib/tb/tb_187.html with permission)

    b "Coupled *in vitro* transcription translation procedure using *E. coli* extract" (from: www.ambion.com/techlib/tb/tb_187.html with permission)

Escherichia coli, a bacterium. These systems possess an enhanced rate of protein synthesis. Crude extracts of the aforesaid systems can be prepared containing the components required for translation of exogenous RNA. These extracts are supplemented with various chemicals (Table 11.7). The three systems are summarized in Table 11.8.

An overview of an *in vitro* protein synthesizing system is presented in Fig. 11.6 and Table 11.9.

REGULATION OF PROTEIN SYNTHESIS

Plant cells control protein synthesis by a variety of mechanisms. For example, transcriptional and translational events may occur in different cellular compartments. If m-RNA is sequestered in one cellular locale, translation can be prevented. Another regulatory mechanism is m-RNA turnover which can be affected by heat shock, hypoxia, wounding as well as water, and nutrient starvation. Another mecha-

nism affecting protein synthesis is the availability of initiation, elongation and termination factors. In addition, posttranslational events such as glycosylation, phosphorylation, acylation, acetylation, and carboxymethylation can modify protein synthesis. Folding of proteins, a posttranslational event, can alter protein synthesis. Moleuclar chaperones and chaperonins (Ellis, 1996) control folding by preventing the incorrect interaction between parts of other proteins.

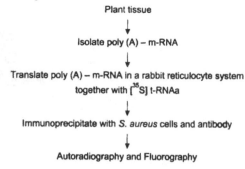

Fig. 11.6 Overview of an *in vitro* protein-synthesizing system (from Dashek, 1997)

TABLE 11.9	Specific for, *in vitro* protein synthesis
Derivation of reticulocyte system	Kits can be obtained from Promega Biotec, Ambion and other companies. These kits employ reticulocyte lysate induced by reticulocytosis in rabbits.
Immunoprecipitation of *in vitro* translation products	This step involves the use of *Staphylococci aureus* cells to precipitate an *in vitro* synthesized protein.
SDS – Page	This step results in the *in vitro* resolution products.
Autoradiography and fluorography	Detects radioactive *in vitro* translation products.

Details of each of these steps are presented in Dashek (1997).

References

Abraham, A.K., Eikhom, T.S., and Pryme, I.F. 1984. *Protein Synthesis Translational and Post-Translational Events*. Humana Press, Totowa, NJ, USA.

Alefsen, H. and Soll, J. 1993. Protein import into chloroplasts: an outline of early events in the translocation process. In: *Plant Mitochondria*, pp. 331-339. Brennicke, A. and Kuck, U. (eds.). VCH, Weinheim, Germany.

Arnstein, H.R.V. and Cox, R.A. 1992. *Protein Biosynthesis*. IRL Press, Oxford, England.

Bendich, A.J., Loretz, C.J., and Monnat, R.J. 1993. The structure of the plant mitochondrial genome. In: *Plant Mitochondria*, pp. 171-186. Brennicke, A. and Kuck, U. (eds.). VCH, Weinheim, Germany.

Bill, R.M., Revers, L., and Wilson, J.B.H. 1999. *Protein Glycosylation* Springer-Verlag, Berlin, Germany.

Biswal, U.C., Biswal, B., and Raval, M.K. 2003. *Chloroplast Biogenesis from Proplastid to Gerantoplast*. Kluwer Acad. Publ., Dordrecht, The Netherlands.

Blaha, G. 2004. Structure of the ribosome. In: *Protein Synthesis and Ribosome Structure*, pp. 53-84. Nierhaus, K.H. and Wilson, D. (eds.). Wiley-VCH, Weinheim, Germany.

Bountry, M. and Chaumont, F. 1993. Protein targeting to plant mitochondria. In: *Plant Mitochondria*, pp. 323-329. Brennicke, A. and Kuck, U. (eds.). VCH, Weinheim, Germany.

Caraglia, M., Budillon, A., Vital, G., Lupoli, G., Tagliaferri, P. and Abbruzzese, A. 2000. Modulation of molecular mechanisms involved in protein synthesis machinery as a new tool for the control of cell proliferation. Eur. J. Biochem. 264: 3919-3936.

Cleland, J.L. 1998. *Protein Folding—In Vivo and In Vitro*. ACS Symposium Series, Oxford Univ. Press, Oxford, England.

Cold Spring Harbor Symposium on Quantitative Biology. 2001. *The Ribosome*. Cold Spring Harbor Laboratory, Cold Spring Harbor, New York, NY.

Copeland, W.C. 2002. *Mitochondrial DNA*. Humana Press, Totowa, NJ, USA.

Dashek, W.V. 1997. *In vitro* synthesis of plant proteins: Polyphenoloxidase. In: *Methods in Plant Biochemistry and Molecular Biology*, pp. 323-334. Dashek, W.V. (ed.). CRC Press, Boca Raton, FL, USA.

Douce, R. 1985. *Mitochondria in Higher Plants, Structure, Function and Biogenesis*. Amer. Soc. Plant Physiol. Monograph Services. Academic Press, Orlando, FL, USA.

Drugeon, G., Jean-Jean, O., Frolova, L., Le Gobb, X., Philippe, M., Kisselev, L., and Haenni, A.L. 1997. Eukaryotic release factor (eRFI) abolishes readthrough and competes with suppressor t-RNAs at all three termination condons in messenger RNA. Nucleic Acids Res. 25: 2254-2258.

Ellis, R.J. 1996. *The Chaperonins*. Academic Press, San Diego, CA, USA.

Evans, E.M. and Wood, E.J. 1987. DNA and protein synthesis (video recording). Biological Society, London, England.

Frolova, L., LeGoff, Y., Rasmussen, H.H., Cheperegin, S., Durgeon, G., Kress, M.,

Armani, I., Haenni, A.L., Celis, J.E. and Philippe, M. 1994. A highly conserved protein family possessing properties of polypeptide chain release factor. Nature 372: 701-703.

Frydman, J. 2001. Folding of newly translated proteins *in vivo*: The role of molecular chaperones. Annu. Rev. Biochem. 70: 603-647.

Hanes, B.D. 1999. *Post-Translational Processing: A Practical Approach*. Oxford Univ. Press, Oxford, England.

Harris, E.H., Boynton, J.E. and Gillham. 1994. Chloroplast ribosomes and protein synthesis. Microbiol. Rev. 58: 700-754.

Hinnebusch, A.G., Dever, T.E., and Soneberg, N. 2004. Mechanism and regulation of initiation in eukaryotes. In: *Protein Synthesis and Ribosome Structure*, pp. 241-322. Nierhaus, K.H. and Wilson, D. (eds.). Wiley-VCH, Weinheim, Germany.

Ibba, M. and Soll, D. 2000. Aminoacyl-tRNA synthesis. Annu. Rev. Biochem. 69: 617-650.

Inagaki, Y. and Doolittle, W.F. 2000. Evolution of the eukaryotic translation system: origins of release factors. Molec. Biol. Evol. 17: 882-889.

Kisselev, L.L. and Buckingham, R.H. 2000. Translation termination comes of age. Trends Biochem. Sci. 25: 561-566.

Ko, K. 1997. Protein synthesis in plant cells. In: *Plant Metabolism*, pp. 17-25. Dennis, D.T., Layzell, D.B., Lefebre, D.D. and Turpin, D.H. (eds.). Longman, Essex, England.

Lafontaine, D.L. and Tollervey, D. 2004. The function and synthesis of ribosomes. Nature Rev. Molec. Cell Biol. 2: 514-520.

Lafontaine, D. 2004. Eukaryotic ribosome synthesis. In: *Protein Synthesis and Ribosome Structure*, pp. 107-144. Nierhaus, K.H. and Wilson, D. (eds.). Wiley-VHC, Weinheim, Germany.

Lodish, H., Berk, A., Zipursky, L., Matsudaira, P., Baltimore, D., and Darnell, J. 2000. *Molecular Cell Biology*. Freeman, New York, NY.

Margulis, L. 1981. *Symbiosis in Cell Evolution*. W.H. Freeman, San Francisco, CA, USA.

Margulis, L. 1998. *Symbiotic Plant: a New Look at Evolution*. Weidenfeld and Nicolson, London, England.

Meadows, J.W., Shackleton, J.B., Bassham, D.C., Mould, R.M., Hulford, A., and Robinson, C. 1993. Transport of proteins into chloroplasts.

In: *Plant Organelles*, pp. 281-292. Tobin, A.K. (ed.). Cambridge Univ. Press, Cambridge, England.

Moore, P.B. and Steitz, T.A. 2003. The structural basis of large ribosomal subunit function. Annu. Rev. Biochem. 72: 813-850.

Nierhaus, K.H. 2004. The elongation cycle. In: *Protein Synthesis and Ribosome Structure*, pp. 323-366. Nierhaus, K.H. and Wilson, D. (eds.). Wiley-VCH, Weinheim, Germany.

Nierhaus, K.H. and Wilson, D.N. 2004. *Protein Synthesis and Ribosome Structure: Translating the Genome*. Wiley-VCH, New York, NY.

Pestova, T.V., Kolupaeva, V.G., Lomakin, I.B., Pilipenko, E.V., Shatsky, I.N., Agol, V.I., and Hellen, C.U. 2001. Molecular mechanisms of translation initiation in eukaryotes. Proc. Natl. Acad. Sci., USA 98: 7029-7036.

Preiss, T. and Hentze, M.W. 2003. Starting the protein synthesis machine: eukaryotic translation initiation. Bioessays. 25: 1201-1211.

Ramakrishnan, V. 2002. Ribosome structure and the mechanism of translation. Cell 108: 557-572.

Ribas de Pouplana, L. and Schimmel, P. 2004. Aminoacylations of tRNAs: Record-Keepers for the genetic code. In: *Protein Synthesis and Ribosome Structure: Translating the Genome*, pp. 169-184. Nierhaus, K. and Wilson, D.N. (eds.). Wiley-VCH, New York, NY.

Rodina, M.V., Savelsberg, A. and Wintermeyer, W. 1999. Dynamics of translation on the ribosome: molecular mechanics of translocation. FEMA Microbiol. Rev. 23: 317-333.

Rodina, M.V. and Wintermeyer, W. 2001. Ribosome fidelity: tRNA discrimination, proofreading and induced fit. Trends Biochem. Sci. 26: 124-130.

Sachs, A.B. and Varani, G. 2000. Eukaryotic translation initiation: there are (at least) two sides to every story. Nature Struct. Biol. 7: 356-361.

Scheer, U. and Hock, R. 1999. Structure and function of the nucleolus. Curr. Opin. Cell Biol. 11: 385-390.

Scheffler, I.E. 1999. *Mitochondria*. Wiley-Liss, New York, NY.

Schneider, R., Agol, V.I., Andino, R., Bayard, R., Cavener, D.R., Chappell, S.A. *et al*. 2001. New ways of initiating translation in eukaryotes? Molec. Cell. Biol. 21: 8238-8246.

Schnell, D.J. 1998. Protein targeting to the thylakoid membrane. Ann. Rev. Pl. Physiol. Pl. Molec. Biol. 49: 97-126.

Soll, P. 1998. *Protein Trafficking in Plant Cells*. Kluwer Acad. Publ., Berlin, Germany.

Sonenberg, N. and Dever, T.E. 2003. Eukaryotic translation initiation factors and regulators. Curr. Opin. Struct. Biol. 13: 56-63.

Sonenberg, N., Hershey, J.W.B. and Matthews, M.B. (eds.). 2000. *Translational Control of Gene Expression*. Cold Spring Harbor Laboratory Press, Cold Spring Harbor, New York, NY.

Spirin, A.S. 2000. *Ribosomes*. Springer, New York, NY.

Teasdale, R.D. and Jackson, N.R. 1996. Signal-mediated sorting of membrane proteins between the endoplasmic reticulum and the Golgi apparatus. Ann. Rev. Cell Devel. Biol. 12: 27-54.

Whelan, J. 2004. Post-mitochondrial protein import: mechanisms and control. Austr. J. Plant Physiol. 26: 725-732.

CHAPTER 12

Plant Metabolism—Respiration

Neil Bowlby

INTRODUCTION

Even though plants meet most of their energy needs through photosynthesis, a number of vital processes require that the glycolytic and respiratory pathways be operational. During periods of darkness, such as at night or underground in the germinating seed, glycolysis, respiration, the pentose phosphate pathway, and the glyoxylate cycle provide the only means for producing the energy and metabolic intermediates used in biosynthetic reactions. This chapter discusses the enzymology, bioenergetics, and regulation of these metabolic pathways.

It is assumed that the reader has a basic knowledge of the operation of glycolysis, the tricarboxylic acid (TCA) cycle, oxidative phosphorylation, and the pentose phosphate pathway from previous study in an introductory biochemistry course. The emphasis here is on the structure and topology of the individual enzymes that have been studied in plant systems. The enzymology and bioenergetics of the different pathways are briefly discussed with special attention paid to areas wherein the pathways in plants differ from those in animals.

Probably the most significant difference from animals occurs due to the presence of plastids in plant cells, which are lacking in animal cells. Many enzymes, especially those of glycolysis and the pentose phosphate pathway, localized either in the cytosol or mitochondria in animals, have analogs localized in the chloroplasts and amyloplasts of plants. Another difference is the physical appearance of plant mitochondria. Figure 12.1 shows an electron micrograph of mitochondria from a spinach leaf homogenate. Unlike the mitochondria from animals, which have a very convoluted inner membrane resulting in an abundance of cristae, the cristae of plant mitochondria are much less conspicuous and the matrix space occupies a much larger percentage of the total volume. The overall size of the mitochondrion in plants is about the same as in animals.

Duplication of enzyme activities in several different compartments of the plant cell requires that coordination of metabolism between the compartments be tightly controlled and regulated. An understanding of the mechanisms involved in this regulation is only beginning to glimmer. The close physical association of mitochon-

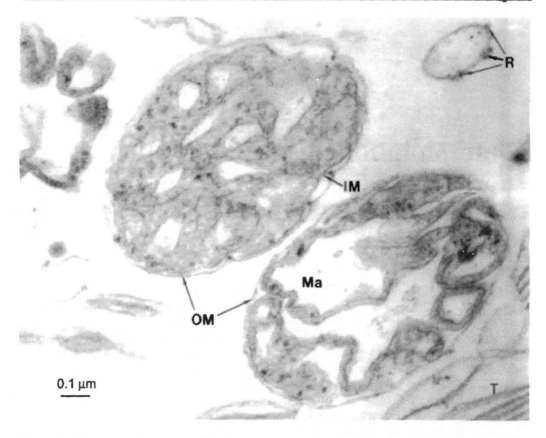

0.1 μm

Fig. 12.1 Electron micrograph of mitochondria from a spinach leaf homogenate. IM, inner membrane; Ma, matrix; OM, outer membrane; R, ribosomes; T, thylakoid membranes (author photo).

drion and chloroplast, and mitochondrion and peroxysome is shown in Figure 12.2. Close proximity allows many metabolic intermediates to be rapidly transported between the two organelles. Some of these transport systems are discussed as they relate to the integration of photosynthesis and respiration in the overall metabolism of the plant cell.

The chapter is divided into sections dealing with glycolysis, the TCA cycle, electron transport and oxidative phosphorylation, the pentose phosphate pathway, the glyoxylate cycle, and photorespiration. Each section begins with a description of the structure and topology of enzymes

involved in the particular pathway, followed by a summary of the bioenergetics and regulation of the metabolic interactions with other pathways. Information about the structure and topology of individual enzymes was obtained from the Comprehensive Enzyme Information System (BRENDA) at Cologne University Bioinformatics Center. Specific reference citations are not given here, but rather the EC number for each enzyme for ready location in the database described at the end of "Further Reading". Information about signature patterns, conserved amino acid sequences that can be used to identify different families of enzymes and proteins,

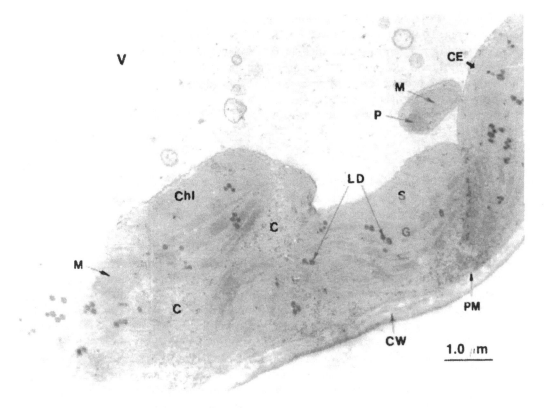

Fig. 12.2   Electron micrograph of intact spinach leaf. C, cytoplasm; CE, chloroplast envelope; Chl, chloroplast; CW, cell wall; G, grana; LD, lipid droplets; M, mitochondria; P, peroxysome; PM, plasma membrane; S, Stroma; V, Vacuole (Photo by the author).

was obtained from the Prosite Database of Protein Families and Domains and the Swiss-Prot and TrEMBL databases hosted by the Expert Protein Analysis System at the Swiss Institute of Bioinformatics. Internet address links (URL) to these sources are also provided at the end of "Further Reading".

GLYCOLYSIS

Structure and Topology of Glycolytic Enzymes

Much of the information regarding the structure and topology of enzymes in plants comes from comparison of protein sequences to well-characterized homologs in bacteria, yeast, and animals. Very few crystal structures of plant enzymes have been published, although protein sequences deduced from an increasing number of gene sequences have allowed a reasonable picture of the structure of the plant enzymes to emerge. For those enzymes in which a crystal structure has been determined, the reader is directed to the Protein Data Bank maintained by the Research Collaboratory for Structural Bioinformatics at the Departments of Chemistry and Chemical Biology and the Center for Molecular Biophysics and Biophysical Chemistry, Rutgers University.

The glycolytic pathway, schematically shown in Fig. 12.3, is often identified as the starting point for respiration since it is involved in the breakdown of glycogen in animals and the production of pyruvic acid for introduction into the TCA cycle. In plants the principal storage molecules for hexoses are starch and sucrose.

Phosphorylase (EC 2.4.1.1)

Starch phosphorylase is an enzyme localized primarily in the plastid, that catalyzes the cleavage of a terminal glucose in starch with condensation of liberated glucose and orthophosphate to form α-D-glucose-1-phosphate. Among plants, the enzyme has been studied from many sources, including *Pisum sativum* (pea), *Solanum tuberosum* (potato), *Spinacea oleracea* (spinach) and *Zea mays* (maize). Enzymes from bacteria (*Escherichia coli*), *Homo sapiens*, and rabbit (*Oryctolagus cuniculus*) have been crystallized; however, no crystal structures exist for any plant source. In addition to starch, the enzyme from the aforesaid plants (pea, potato, spinach, maize) can also hydrolyze and phosphorylate the terminal hexose from a wide variety of sugar polymers, including amylopectin, dextrin, glycogen, maltodextrin and a variety of short-chain maltose polymers. All phosphorylase enzymes require pyridoxal 5'-phosphate as a cofactor. Activators include Mn^{2+} or Mg^{2+} as well as 5'-AMP, β-mercaptoethanol, albumin, and cysteine in potato. Inhibitors include 2,3-diphosphoglycerate, ADP-glucose, UDP-glucose, and divalent metals such as Ag^{2+}, Fe^{2+}, and Hg^{2+}. In the active form starch phasphorylase consists of two identical subunits of approximately 90 kDa, although a tetrameric form (phosphorylase II) of identical 53 kDa subunits has been described from maize. The enzyme has been localized to both the cytoplasm and chloroplast. The pyridoxal phosphate-binding lysine residue is located at position 687 in the *Arabidopsis* sequence in a highly conserved region that identifies members of the phosphorylase family of enzymes (see Table 12.1).

β-fructofuranosidase (EC 3.2.1.26)

β-fructofuranosidase, also called invertase, catalyzes the cleavage of sucrose to produce fructose and glucose. Both these products serve as precursors for entry into glycolysis. Invertases can be classified into three broad categories—acid invertases, neutral invertases, and alkaline invertases, based on pH optimum for activity. Although the cytoplasmic form of the enzyme is thought to be the primary route for monosaccharide entry into the glycolytic pathway, other isozymes have been localized in the vacuole and on the cell wall. The vacuolar isozyme can provide monosaccharides for glycolysis prior to its transport into the cytoplasm. The cell wall isozyme appears to be involved in the response to wounding, as it is normally expressed in very low levels in intact tissues. This enzyme has been extensively studied from numerous plant sources and subcellular locations. In addition to the plant species listed for phosphorylase, the list includes *Avena sativa* (oat), *Beta vulgaris* (sugar beet), *Carica papaya* (papaya), *Citrus sinensis* (sweet orange), *Daucus carota* (carrot), *Glycine max* (soybean), *Hordeum vulgare* (barley), two species of *Lilium* (lily), *Lycopersicon esculentum* (tomato), two species of *Nicotiana* (tobacco), and *Prunus persica* (peach).

Sucrose is the most common disaccharide substrate; however, most enzymes will hydrolyze raffinose, stachyose, and trehalose to varying degrees. In papaya, Ba^{2+}, Mg^{2+}, and Sr^{2+} have been shown to activate the enzyme, while in lily, Co^{2+} has been identified as an activator. Inhibitors of the enzyme include heavy metal ions such as

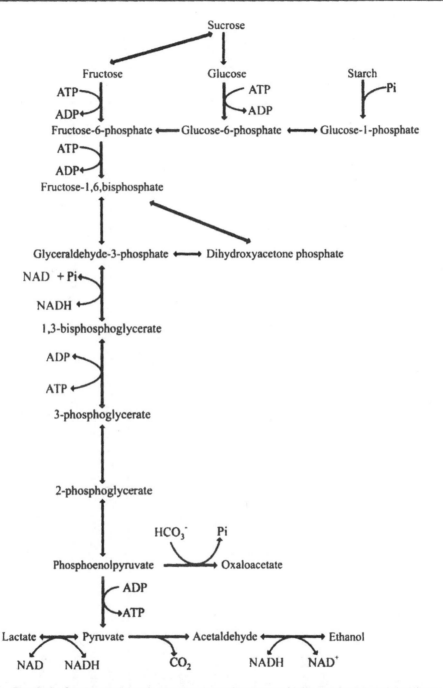

Fig. 12.3 Glycolysis. Sucrose and starch serve as the major sources for hexose input into the pathway. Also shown at the bottom of the figure are the reactions that proceed under anaerobic conditions. All the enzymes of the pathway are located in the cytosol. Double-headed arrows designate reversible reactions.

TABLE 12.1 Consensus signature patterns of enzymes involved in glycolysis[1]

Enzyme Name	Signature pattern from PROSITE database	Position	ID Code[2]
Phosphorylase	E-A-[SC]-G-x-[GS]-x-M-K*-x-x-[LM]-N	679-691	Q9SD76
β-fructofuranosidase	H-x-x-P-x-x-x-x-[LIVM]-N-D*-P-N-G	17-27	Q43857
Hexokinase	[LIVM]-G-F-[TN]-B-S-P-x-x-x-x-[LIVM]-[DNST]-x-x-x-[LIVM]-x-x-W-T-K-x-[LF]	172-197	Q42525
Phosphoglucomutase	[GSA]-[LIVMF]-x-[LIVM]-[ST]-[PGA]-S*-H-[NIC]-P	117-126	O49299
Glucose-6-phosphate isomerase	[DENSA]-x-[LIVM]-[GP]-G-R-[FY]-[ST]-[LIVMFSTAP]-x-[GSTA]-[PSTACMI]-[LIVMSA]-[GSAN]	269-282	P42862
6-phosphofructokinase	[GSA]-x-[LIVMCA]-[LIVMFYWN]-x-x-x-[FY]-[DNT]-Q-x-[GA]-[IV]-[EQS]-x-x-K*	499-516	P16861
Fructose-bisphosphate aldolase	[RK]-x(4)-G-H*-x-Q-[QR]*-G-G-x(5)-D-R	482-500	P46256
Triose-phosphate isomerase	[LIVM]-x-[LIVMFYW]-E-G-x-[LS]-L-K*-P*-[SN]	217-225	P34937
Glyceraldehyde-3-phosphate dehydrogenase	[AV]-Y-E*-P*-[LIVM]-W-[SA]-I-G-T-[GK]	163-173	P25858
	[ASV]-S-C*-[NT]-T-x(2)-[LIM]	154-161	
Phosphoglycerate kinase	[KRHGTCVN]-[VT]-[LIVMF]-[LIVMC]R-x-D-x-N-[SACV]-P	19-29	Q42962
Phosphoglycerate mutase			
Phosphoenolpyruvate hydratase	[LIV]-[LIV]-[LIV]-K-x-N-Q-I-G-[ST]-[LIV]-[ST]-[LIV]-[ST]-[DE]-[STA]	351-364	P26301
Pyruvate kinase	[LIVAC]-x-[LIVM]-[LIVM]-[SAPCV]-K*-[LIV]-E*-[NKRST]-x-[DEQHS]-[GSTA]-[LIVM]	235-247	P22200
L-lactate dehydrogenase	[LIVMA]-G-[EQ]-H*-G-[DN]-[ST]	211-217	P29038
Pyruvate decarboxylase	[LIVMF]-[GSA]-x(5)-P-x(4)-[LIVFMYW]-x-[LIVMF]-x-G-D-[GSA]-[GSAC]	465-484	Q9FVF0
Alcohol dehydrogenase	G-H*-E-x(2)-G-x(5)-[GA]-x(2)-[IVSAC]	66-80	P20306
Phosphoenolpyruvate carboxylase	[VT]-x-T-A-H*-P-T-[EQ]-x(2)-R-[KRH]	168-179	Q02909
	[IV]-M-[LIVM]-G-Y-S-D-S-x-K*-D-[STAG]-G	593-605	

[1] Conserved residues are denoted by bold letters, active site residues by an asterisk.
[2] Sequence identification number in the SwissProt/TrEMBL database from which position information was obtained.

Ag^{2+}, Cu^{2+}, Hg^{2+} and the transition metals Mg^{2+}, Mn^{2+}, and Zn^{2+}. Inhibition by these metals varies among the different classes of invertase with the acid invertases being most susceptible to inhibition. Not surprisingly, glucose and fructose also inhibit the enzyme, both monosaccharides being competitive with respect to sucrose (product inhibition).

By gel filtration chromatography, estimates of the size of the holoenzyme range from 750 kDa and 450 kDa in lily to 28 kDa in sugar beet, suggesting a variety of possible quaternary structures. The subunit molecular weights found in most species are approximately 60-70 kDa, indicating a dimeric or tetrameric arrangement of subunits, although in carrot, invertase was found to be an octomer of 57 kDa subunits. The enzyme is thought to be posttranslationally modified by glycosylation with N-acetylglucosamine.

The only crystal structure of invertase available is that from the bacterium *Thermatoga maritima*. Several structural features have been identified by cross-species comparison of amino acid sequences. Proteins targeted to the cell wall or vacuole contain a signal peptide of approximately 40 amino acids. A conserved aspartic acid in the amino terminal region is thought to participate in catalysis, and several potential glycosylation sites are found throughout the sequence. The consensus signature pattern contains the active site aspartic acid.

Hexokinase (EC 2.7.1.1)

Hexokinase works in parallel with starch phosphorylase in supplying the glycolytic enzymes with hexose phosphates. The major substrates of the enzyme are glucose and ATP and the reaction yields ADP and α-D-glucose-6-phosphate. Although the enzyme has been widely studied in animal tissues, it has only been characterized biochemically in developing potato tubers and cell suspension cultures from *Vinca rosea* (rosy periwinkle), where it has been found to require Mg^{2+} or Mn^{2+} as a cofactor. It has not been crystallized from any plant source but shows sequence similarity to the yeast enzyme, which has been crystallized. The entire protein is about 500 amino acids in length and has a molecular weight of about 50 kDa. It is located on the cytoplasmic side of the chloroplast outer envelope membrane, anchored by a short transmembrane region near the amino terminus at positions 4–24. A phosphonucleotide-binding site (positions 101–106) and hexose-binding site (positions 172–197) have been identified in *Arabidopsis* by sequence comparison to other known binding domains.

Phosphoglucomutase (EC 5.4.2.2)

Phosphoglucomutase catalyzes the intramolecular phosphate group transfer from α-D-glucose-1-phosphate to α-D-glucose-6-phosphate. It has been found in the cytosol and chloroplast and has been studied in pea seeds, potato tubers, spinach leaf, and maize although none of the enzymes from these sources have been crystallized. Similar to the enzyme from rabbit muscle, which has been crystallized, the enzyme is a heterodimer in pea, consisting of 59 and 67 kDa subunits. The enzyme has been isolated by gel filtration as a monomer of about 60 kDa from potato and by sucrose density centrifugation as a 60 kDa monomer from spinach. Activators include α-D-glucose-1,6-diphosphate and EDTA in spinach and potato, cysteine in pea and potato, and imidazole in potato. Inhibitors include Be^{2+}, fructose-1,6-diphosphate and 2,3-bisphosphoglycerate in potato, and fructose-2,6-diphosphate and p-chloromercuribenzoate in pea. In potato and spinach, the enzyme has been isolated as a monomer of about 62 kDa. The deduced amino acid sequence of the corn enzyme has a putative magnesium binding site comprising a potential phosphoserine residue at position 124 in a highly conserved region of the protein that contains the signature sequence, and an aspartic acid triad at positions 300, 302, and 304. By analogy, the enzyme from *Arabidopsis* is also thought to require the Mg^{2+} cofactor as it contains a serine in the signature pattern sequence and the aspartic acid triad in the carboxyl terminal region of the protein.

Glucose-6-phosphate isomerase (EC 5.3.1.9)

Glucose-6-phosphate isomerase catalyzes conversion of α-D-glucose-6-phosphate to α-D-fructose-6-phosphate. Although it has been studied in a number of plant species, including *Brassica oleracea* (cauliflower), *Helianthus annuus* (sunflower), soybean, pea, and spinach, no crystal structure of the enzyme has been produced from a plant source. The enzyme from rabbit muscle has been crystallized. In pea and spinach, the enzyme was identified as a homodimer of approximately 61 kDa subunits. In addition to the cytosolic enzyme involved in glycolysis, a plastidic form was studied in spinach and sunflower. The enzyme from spinach is inhibited by 3-phosphoglycerate, dihydroxyacetone phosphate, glyceraldehyde-3-phosphate (products of later steps in glycolysis), 6-phosphogluconate and erythrose-4-phosphate, intermediates in the pentose phosphate pathway, as well as inorganic phosphate and Zn^{2+}. The putative active site contains a conserved histidine at position 391 and a conserved lysine at position 516 in the sequence from rice.

Phosphofructokinase (EC 2.7.1.11)

Phosphofructokinase (PFK) catalyzes the phosphate group transfer from ATP to α-D-fructose-6-phosphate, producing α-D-fructose-1,6-bisphosphate and ADP. It has been studied in a variety of plant sources including *Cucumis sativus* (cucumber), *Triticum aestivum* (wheat), pea, potato, spinach, and maize. In addition to the cytosolic form of the enzyme, a plastid-localized form has been studied in spinach and cucumber. In potato, the enzyme exists as an oligomer of approximately 800 kDa, most likely an octomer of 80-90 kDa subunits. This is similar to the yeast enzyme, composed of 4 α-subunits and 4 β-subunits of about 100 kDa each. In cucumber, it is thought to be a tetramer composed of 3 different isoforms of 40-50 kDa each. Aggregation of the enzyme in biochemical preparations from plants makes it difficult to determine accurately the quaternary structure of this enzyme. Although ATP is the usual phosphate-group donor, both the cucumber and spinach enzymes can utilize CTP, GTP, ITP, or UTP as the phosphate donor. The bivalent cation Mg^{+2} is required for catalytic action, although this requirement can be met, to some degree, by Mn^{2+} or Co^{2+}. Anions, such as bicarbonate, nitrate, sulfate, and arsenate, and the ammonium cation can stimulate enzymatic activity. Inhibitors include 2-phosphoglycerate, 3-phosphoglycerate, phosphoenol pyruvate, ADP, AMP, pyrophosphate, fructose-1-phosphate and fructose-1,6-bisphosphate. Crystal structures from eukaryotic organisms have not been published, although by analogy to the structure from *E. coli* and the nucleotide sequence from *Saccharomyces cerivisiae* (Baker's yeast) several features have been identified. The active site is composed of a nucleotide binding site defined by residues 225-229, 383-387, and 400-416, an aspartic acid at position 356 that acts as a proton acceptor, and a hexose binding site defined by residues 391, 482, 488, and 500. Residues at positions 482, 488, and 500 are part of the consensus sequence that identifies the phosphofructokinase signature.

Fructose-bisphosphate aldolase (EC 4.1.2.13)

Fructose-bisphosphate aldolase catalyzes the carbon-carbon bond cleavage of α-D-fructose-1,6-bisphosphate to yield D-glyceraldehyde-3-phosphate and glycerone phosphate (dihydroxyacetone phosphate). The enzyme has been studied in *Phaseolus vulgaris* (kidney bean), oat, pea, rice, spinach, wheat, and maize. It is found in the chloroplast as well as the cytosol. As in other higher eukaryotes, aldolase in maize and spinach belongs to Class I aldolase and exists as a homotetramer of 38 kDa subunits. The quaternary structure of pea and wheat enzymes is probably the same based on the holoenzyme molecular weight of about 140 kDa. Although fructose-1,6-bisphosphate is the normal substrate, fructose-1-phosphate has also been shown to bind and be converted to glycerone phosphate and D-glyceraldehyde by enzymes found in spinach, pea, wheat, and maize, and probably in other plants as well. Crystal structures of plant enzymes do not exist; however, the rabbit muscle enzyme provides a homolog for identification of structural features. Referring to the amino acid sequence from pea, the C-1 phosphate group of fructose-1,6-bisphosphate is thought to bind to an arginine at position 52 and a lysine at position 142. The Schiff-base intermediate is carried by the lysine at position 225 and a glutamic acid at position 183 acts as a proton acceptor. The tyrosine at the C-terminus (position 357) is necessary for the substrate preference for F-1,6-bisphosphate over F-1-P mentioned above. The conserved lysine at position 225 is a part of the consensus signature sequence for Class I aldolases.

Triose-phosphate isomerase (EC 5.3.1.1)

Triose-phosphate isomerase catalyzes the interconversion of D-glyceraldehyde-3-phosphate and glycerone phosphate. It is found in the cytosol as well as the plastid and plays an important role in several metabolic processes in addition to glycolysis. It has been studied in *Arabidopsis*, *Latuca sativa* (garden lettuce), *Secale cereale* (rye), rice, petunia, pea, spinach wheat, and maize. In pea, the enzyme was found to be inhibited by 2-phosphoglycerate, 3-phosphoglycerate, fructose-1,6-bisphosphate, fructose-1-phosphate, and phosphoenolpyruvate. Weaker inhibitors of the pea enzyme include 6-phosphogluconate, citrate, ATP, glucose-1-phosphate and glucose-6-phosphate. The rye enzyme was found to be inhibited by arsenate, iodoacetate, *p*-hydroxymercuribenzoate, and sulfate supplied as the copper salt.

Both cytosolic and plastidic forms of the enzyme are present as a dimer of identical 27 kDa subunits. In rye, neither form was found to be glycosylated. There are no crystal structures from plant sources, although several structures are available from other eukaryotes including rabbit and chicken muscle, trypanosome, and yeast. Two active site residues have been identified in highly conserved regions of triosephosphate isomerases: a histidine at position 95 involved in electrophilic catalysis, and glutamic acid at position 165 that is a part of the consensus signature sequence common to all triose-phosphate isomerases.

Glyceraldehyde-3-phosphate dehydrogenase (EC 1.2.1.12)

Glyceraldehyde-3-phosphate dehydrogenase (GAPDH) catalyzes the phosphorylation of D-glygeraldehyde-3-phosphate (G-3-P) to form 3-phospho-

D-glyceroyl phosphate (1,3-bisphosphoglycerate), reducing NAD^+ to NADH in the process. The enzyme first binds G-3-P followed by NAD^+ and inorganic phosphate. In contrast to the enzyme from *Homo sapiens*, the pea enzyme releases 3-phospho-D-glyceroyl phosphate and binds G-3-P before the release of NADH, resulting in a bi-uni-uni-uni-ping-pong mechanism of action. Both cytosolic and two different plastidic forms of the enzyme have been identified in plants, each encoded by separate nuclear genes. It has been studied in *Arabidopsis*, *Marchantia polymorpha* (liverwort), *Phaseolus aureus* (mung bean), *Pinus sylvestris* (Scots pine), soybean, tobacco, rice, petunia, pea, potato, spinach, and maize. In soybean, the enzyme has been found to be inhibited by 2,3-diphosphoglycerate, AMP, and NADH; the pea enzyme is inhibited by 3-phosphoglycerate and glyceraldehyde-3-phosphate as well as NAD^+ and phosphate (substrate inhibition). Contrarily, the spinach enzyme is irreversibly inhibited by pentalenolactone while mung bean GAPDH is inhibited by high concentrations of K^+ and Na^+ (100 mM-1 M). It exists as a homotetramer of 36-37 kDa subunits and again no crystal structures from plant sources found. A conserved cysteine in the middle of the protein is essential for formation of a thioester of the phosphoglycerol intermediate. A histidine at position 183 of the *Arabidopsis* sequence is responsible for activation of the catalytic cysteine residue. The area around the active site cysteine is highly conserved and serves as the signature sequence for all GAPDH enzymes.

Phosphoglycerate kinase (EC 2.7.2.3)

Phosphoglycerate kinase (PGK) catalyzes the interconversion of 3-phospho-D-glyceroyl phosphate and ADP to 3-phospho-D-glycerate and ATP. Although the name "kinase" suggests a phosphate transfer reaction from ATP, the reaction proceeds as a substrate-level phosphorylation of ADP when viewed in the context of glycolysis with pyruvate as an end product. Cytoplasmic and plastidic forms of the enzyme have been found to occur in plants but almost all studies were performed on barley, spinach, and sugar beet. Although ATP is the preferred substrate for the kinase reaction, the spinach enzyme will accept GTP and ITP as phosphate-group donors. All enzymes require Mg^{+2}, although in sugar beet at least, the nucleotide-activating action of Mg^{2+} can be partially replaced by Ca^{2+}, Co^{2+}, Mn^{2+}, or Ni^{2+}, whereas the enzyme from sugar beet is inhibited by Zn^{2+}. The active form of the enzyme is a monomer of about 40 kDa molecular weight. There are no plant-derived crystal structures but several other eukaryotic structures have been solved. The sequence of PGK is highly conserved among all organisms and contains two domains consisting of alpha/beta structural elements. A conserved region near the amino terminus serves as a signature pattern for this enzyme.

Phosphoglycerate mutase (EC 5.4.2.1)

Phosphoglycerate mutase (PGM) catalyzes the intramolecular phosphate group transfer between positions 2 and 3 of phosphoglycerate. Unlike the foregoing enzymes in the glycolytic pathway, which have both cytosolic and plastidic forms, this enzyme is thought to be localized exclusively to the cytosol. The reaction is reversible with the K_M for substrates being about 0.36 mM for 2-phosphoglycerate and 0.32 mM for 3-phosphoglycerate in maize. The enzyme has been studied in maize, wheat, rice, potato, lily, and tobacco. The enzyme found in plants appears to be structurally different from PGM found in yeast and bacteria. That from yeast and bacteria can utilize 2,3-bisphosphoglycerate (BPG) as a substrate, yielding 3-phosphoglycerate as a product, and contains a phosphohistidine residue near the amino terminus. In contrast, the PGM found in higher plants belongs to the family of BPG-independent enzymes that contain a phosphoserine intermediate, and are only distantly related to the PGM found in yeast and bacteria. Whereas in *Homo sapiens* the enzyme is dimeric and in yeast tetrameric, the maize and wheat types exist as a monomer of about 65 kDa, further supporting the notion that the plant enzyme is not structurally similar to PGM found in nonplant species.

Although no metal requirement has been established for plant enzymes, by a comparison of the nuclear genes from several plant species encoding the enzyme, there appear to be binding sites for two Mn^{2+} ions. The metal requirement is also supported by findings that metal chelators such as 1,10- phenanthroline, EDTA, and EGTA inhibit the enzyme from wheat. In wheat and potato, 8-hydroxyquinoline-5-sulfonic acid inhibits enzyme activity, as does *p*-mercuribenzoate in wheat. No plant-derived crystal structures are available, and the several eukaryotic structures that have been described belong to the phosphohistidine-containing family. A consensus signature pattern for these enzymes has not been developed and it is only by sequence comparison of several BPG-independent enzymes from plant sources that structural components of the enzyme have been inferred. By using the *Arabidopsis* sequence as a model, it is thought that one metal ion is bound as a tridentate ligand by an aspartic acid at position 431 and two histidine residues at positions 435 and 502, while the other metal is liganded through two aspartic acid residues at positions 29 and 473, a histidine at position 473, and phosphoserine at position 82.

Phosphoenolpyruvate hydratase (EC 4.2.1.11)

Phosphoenolpyruvate hydratase (enolase) catalyzes the dehydration of 2-phospho-D-glycerate to yield phosphoenolpyruvate and water. The enzyme is located in the cytosol. Only sparse information exists on its enzymology in plants. It has been studied primarily in spinach, although some work has been performed with potato and maize enzymes. It has an absolute requirement for divalent metal cations, Mg^{2+} being the most effective in activation. Other metals showing activation in spinach include Cu^{2+}, Mn^{2+}, Cd^{2+}, and Zn^{2+}, in decreasing order of effectiveness. Fluoride has been found to inhibit the enzyme in spinach. As with enolases from other species, in plants it exists as a homodimer of 50 kDa subunits. The metal ions stabilize the dimeric structure as well as act in a catalytic role. There are no crystal structures from plants, but the structure from yeast serves as a model for eukaryotic enzymes. The active site contains a critical histidine residue at position 164 in the maize sequence as well as metal-binding aspartic acid residues at positions 251 and 329, and glutamic acid at position 302. Free sulfhydryl groups are thought to be essential for activity in the spinach and potato enzymes; four are present in the maize sequence at positions 108, 125, 348, and 410. The signature pattern that identifies enolases is found in the carboxyl terminal portion of the protein.

Pyruvate kinase (EC 2.1.7.40)

Pyruvate kinase (PK) catalyzes the final step in glycolysis under aerobic conditions. The phosphate-group transfer from phosphoenolpyruvate to ADP, the second substrate-level phosphorylation reaction in glycolysis, yields pyruvate and ATP. The enzyme is found exclusively in the cytosol.

Pyruvate kinase from plants has been studied in detail from tomato and spinach, where two isozymes were partially purified. The enzyme requires Mg^{2+} and K^+ for activity. In addition to ADP, the enzyme from spinach will phosphorylate CDP, GDP, and UDP with phosphoenolpyruvate as the donor, although with much lower activity. Citrate was found to inhibit isozyme I but not isozyme II, while glutamate had the opposite effect. Adenine and tryptophan were found to be weak inhibitors. The quaternary structure of a plant enzyme has not been experimentally determined, but by analogy to other eukaryotic enzymes it is thought to be a homotetramer of 55 kDa subunits. The consensus signature sequence contains the putative active site lysine at position 240 (in potato) and one potential Mg^{2+} ligand, a glutamic acid at position 242. Other postulated metal ligands are found at positions 263 (alanine) and 264 (arginine).

L-lactate dehydrogenase (EC 1.1.1.27)

Lactate dehydrogenase catalyzes one of the final steps of glycolysis under anaerobic conditions. It reduces pyruvate to lactate by using NADH as a cofactor. The reaction is reversible, and when aerobic conditions reappear the reverse reaction proceeds, yielding pyruvate and NAD^+. The enzyme is found in the cytosol and has been studied in lettuce, potato, and sweet potato. In potato and lettuce, the enzyme can catalyze the NADH-dependent reduction of glyoxylate to glycolate and the NADPH-dependent reduction of hydroxypyruvate to 2,3-dihydroxypropanoate. Other reactions observed in potato include the NADH-dependent reduction of 2-oxobutyrate to 2-hydroxybutyrate and the reduction of pyruvate by using NADPH as the cofactor. Inhibitors include AMP, ADP, ATP, and GTP as well as fructose-1,6-bisphosphate, glucose-6-phosphate, phosphoenolpyruvate, and the TCA cycle components citrate and isocitrate. In lettuce, 3-phosphoglycerate seems to be inhibitory while in potato, both 3-phosphoglycerate and 2-phosphoglycerate can serve as activators of enzymatic activity. The holoenzyme from sweet potato has a molecular weight of about 150 kDa and by analogy to lactate dehydrogenase enzymes from other organisms is thought to be a homotetramer of 38 kDa subunits. The active site contains a conserved histidine, necessary for catalytic activity, at position 214 in the maize sequence. This histidine is also a part of the signature pattern that defines L-lactate dehydrogenases.

Pyruvate decarboxylase (EC 4.1.1.1)

Pyruvate decarboxylase (PDC) provides another pathway for the removal of pyruvate under anaerobic conditions. It catalyzes the decarboxylation of pyruvate, yielding CO_2 and acetaldehyde. The reaction is not reversible under normal conditions. The enzyme has been studied in *Pastinaca sativa* (parsnip), wheat, maize, sweet potato, pea, rice, and citrus. In wheat, the enzyme can decarboxylate 3-hydroxypyruvate to yield glycolaldehyde and CO_2. It is found in the cytosol and belongs to the thiamine pyrophosphate (TPP, vitamin B_1) family of enzymes with at least three isozymes identified in plants. PDC has a requirement for a divalent metal ion, usually Mg^{2+}, but in sweet potato, and probably other species, Mg^{2+} can be replaced by Zn^{2+}, Mn^{2+}, or Ca^{2+}, in decreasing order of effectiveness. Iodoacetamide, *p*-chloromercuribenzoate, KCl, NH_4Cl, cyanide, lactate, and metal ions such as Li^+, Be^{2+}, Cd^{2+}, Cr^{3+}, Hg^{2+}, and Pb^{2+} are strongly inhibitory in the maize enzyme, while Co^{2+}, Ni^{2+}, and Sr^{2+} show only slight inhibition. Glyoxylate and phosphate have been found to inhibit the enzyme from rice whereas EDTA or phosphate inhibits it from sweet potato. Pyruvamide has been found to activate the enzyme in pea. The holoenzyme has a size of 220 kDa in rice to 390 kDa in wheat, as estimated by gel filtration. It is thought to be a tetramer of identical 65 kDa subunits in pea, wheat and maize; two bands in the 60-65 kDa range have been observed by SDS-PAGE. In sweet potato and rice, only a single band in the 60-64 kDa range has been observed. There has been no report of posttranslational modification. By multiple sequence comparison, a conserved glutamic acid at position 95 in the maize isozyme 1 sequence is thought to be a part of the active site. Although a signature pattern exists for the family of TPP enzymes, several sequences of PDC isozymes from plants do not strictly adhere to the consensus sequence. Most notable is the lack of conservation of proline at position 8, which is replaced by lysine, asparagine, or glutamine depending on the species and isozyme.

Alcohol dehydrogenase (EC 1.1.1.1)

Alcohol dehydrogenase (ADH) catalyzes the NADH-dependent reduction of acetaldehyde to ethanol. The reaction is reversible, converting ethanol to acetaldehyde with reduction of NAD^+. In plants, the enzyme belongs to the zinc-containing family of alcohol dehydrogenases. The enzyme has been extensively studied in many plant species, including *Triticum monococcum* (Einkorn wheat or small spelt), *Triticum turgidum*, (Poulard wheat or Rivet wheat), maize, barley, broad bean, rice, and soybean. There are at least three cytoplasmic isozymes sharing very similar amino acid sequences. Most studies have been conducted on germinating seeds, embryos or seedlings, in which enzymatic activity is especially high. As with ADH from other eukaryotes, the enzyme can utilize a variety of alcohols including propanol, butanol, 3-methylbutanol, allyl alcohol, cyclohexanol, cyclobutanol, *n*-hexanol, *n*-pentanol and D-glucitol. No NADP/NADPH-dependent activity has been reported. As the family name implies, the enzyme requires at least one Zn^{2+} for activity, although a second bound Zn^{2+} may be important for structural integrity. It is inhibited by 1,10-phenanthroline, iodoacetamide, *p*-mercuribenzoate, *n*-ethylmaleimide, Mg^{2+}, and PMSF (Phenylmethylsulphonylfluoride) in maize. Glutathione and β-mercaptoethanol are inhibitory in small spelt and pyrazole was found to inhibit the soybean enzyme. Estimates of the size of the holoenzyme range from 70 kDa in maize to 116 kDa in *Triticum*, leading to the conclusion that the enzyme is a dimer of 35 kDa to 58 kDa subunits. No crystal structures exist for plant enzymes but several ADH structures that have been crystallized share structural similarities that can be used to infer the structure of the plant enzyme. The catalytic Zn^{2+} is liganded by two cysteines at positions 45 and 175, and a histidine at position 67 in the rice isozyme I sequence. The second Zn^{2+} is bound by four closely spaced cysteines at positions 97, 100, 103, and 111. The signature pattern for the zinc family of ADH enzymes includes one of the histidine ligands to the catalytic metal.

Phosphoenolpyruvate carboxylase (EC 4.1.1.31)

Although phosphoenolpyruvate carboxylase (PEPcase) is not considered, strictly speaking, a glycolytic enzyme, it is included here because it provides a route for PEP utilization by the TCA cycle without passing through the pyruvate dehydrogenase pathway. The enzyme catalyzes the irreversible carboxylation of PEP by using bicarbonate as the CO_2 source, releasing inorganic phosphate and oxaloacetate as products. Although this enzyme is a component of the Hatch-Slack pathway of

C4 photosynthesis, the oxaloacetate it produces can be converted to malate, which can move into the mitochondria to be used in the TCA cycle. Because of its importance in C4 photosynthesis, the enzyme has been extensively studied in plant species, including *Amaranthus hypochondriacus* (Prince's feather), *Amaranthus viridis* (Slender amaranth), *Brassica campestris* (field mustard), *Lupinus luteus* (white lupine), *Medicago sativa* (alfalfa), *Sorghum*, soybean, tobacco, pea, spinach, broad bean, and maize. In field mustard and maize, Mg^{2+} is required for activity, while in other species, Mg^{2+} and Mn^{2+} have been shown to activate the enzyme. In broad bean, K^+ was also found to stimulate activity by about 40%. Glucose-6-phosphate can also be an activator of the enzyme in several species while glycerol, glycine, histidine and glucose-1-phosphate activate the enzyme in maize. Substitution of Mg^{2+} with other divalent metals, especially Ca^{2+}, Cd^{2+}, Co^{2+}, Hg^{2+}, Ni^{2+}, or Zn^{2+} results in severe inhibition. Other inhibitors include ATP, citrate, glutamate, L-lactate, malate, oxalate, oxaloacetate, and pyruvate.

The size of the holoenzyme has been reported to be between 400 kDa in *Amaranthus* and 560 kDa in spinach, suggesting that the enzyme is comprised of a tetramer of identical 100 kDa subunits. In maize, the holoenzyme has been isolated as a dimer or trimer. The enzyme contains a serine at position 11 in the soybean sequence that can serve as a site of phosphorylation. In addition, two residues have been identified that participate in the catalytic mechanism. A histidine at position 172 and a lysine at position 602 are completely conserved; their participation in catalysis has been inferred by similarity to other PEPcase enzymes. Two signature patterns have been identified in the regions of the conserved histidine and lysine residues.

Bioenergetics and Regulation of Glycolysis

Glucose destined for the respiratory pathway comes primarily from the breakdown of starch or sucrose; however, it is not the only substrate for catabolic reactions leading to the production of energy in the cell. All the anabolic reaction pathways, such as those leading to the biosynthesis of lipids, carbohydrates, and nucleic acids are driven by intermediates of the respiratory apparatus and can thus serve as substrates for respiration. The dependence of anabolic reactions on the intermediate compounds formed during respiration poses a problem in plants not usually seen in animals, namely that during photosynthesis an excess of ATP is produced by photophosphorylation that would have a tendency to inhibit reactions associated with respiration in the mitochondria. An inhibition or slowdown of respiration, especially the TCA cycle, would lead to subsequent depletion of important biosynthetic intermediates such as alanine and glutamine.

Glycolysis in plants follows the same pathway as in animals and bacteria. Although it occurs primarily in the cytosol and all of the glycolytic enzymes reside there, many of the first steps can also be catalyzed by plastid-specific isozymes. The process can be broken down into two major steps. In the first a hexose phosphate is broken down into two triose phosphates; in the second the triose phosphates are converted to phosphoenolpyruvate and then to pyruvate. The triose phosphates can be used for the synthesis of glycerol, a precursor for lipid synthesis, or for the synthesis of the simple amino acids serine and cysteine used in protein synthesis. Phosphoenolpyruvate is a versatile intermediate, used in the synthesis of phenolic compounds, including the aromatic amino acids, phytohormones, and anthocyanins. Pyruvate has three fates. It can be fermented to lactate or ethanol under anaerobic conditions, be converted to acetyl-CoA and enter the TCA cycle under aerobic conditions, or be converted to alanine for protein synthesis.

As suggested by the reversibility of their reactions, the glycolytic enzymes phosphoglucomutase, glucose-6-phosphate isomerase, glyceraldehyde-3-phosphate isomerase, triose phosphate isomerase, and phosphoglycerate mutase operate at or near thermodynamic equilibrium. In contrast the phosphofructokinase and pyruvate kinase reactions are essentially irreversible and these enzymes are key regulators whose activities are closely tied to the energy level in the cell. When ADP levels are low, the activity of pyruvate kinase decreases, leading to an increase in the level of phosphoenolpyruvate which in turn decreases the activity of phosphofructokinase. The adenylate energy charge however, does not seem to be the only factor regulating the flux of carbon through the glycolytic pathway. Activity of the TCA cycle and the $NAD^+/NADH$ ratio can also have an influence on the reactions of glycolysis.

PENTOSE PHOSPHATE PATHWAY

Structure and Topology of Pentose Phosphate Pathway Enzymes

The pentose phosphate pathway, also known as the hexose monophosphate shunt, provides a means of generating reducing power in the form of NADPH and ribose-5-phosphate for nucleotide and fatty acid biosynthesis. In plants, the pathway occurs in the plastid as well as the cytosol and a set of nuclear genes encode each of the isozymes. The pathway is shown in Fig. 12.4.

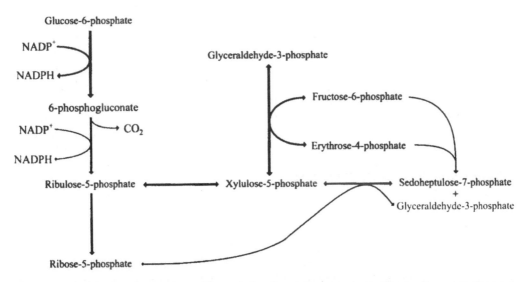

Fig. 12.4 Pentose phosphate pathway. The main function of the pentose phosphate pathway, also known as the hexose monophosphate shunt, is to provide NADPH for anabolic reactions and ribose phosphates for nucleic acid synthesis. Enzymes of the pathway are located in the cytosol.

Glucose-6-phosphate dehydrogenase (EC 1.1.1.49)

Glucose-6-phosphate dehydrogenase (G6PD) catalyzes the irreversible, $NADP^+$-dependent oxidation of D-glucose-6-phosphate to form NADPH and D-gluconolactone-6-phosphate (6-phospho-D-gluconate). The enzyme has only been studied in pea and spinach, but in these two plants both the cytosolic and plastidic forms have been partially characterized. No metal ion requirement or activating compound has been identified in either species. In pea, NADPH has been found to be inhibitory whereas in maize, ATP and ADP have been identified as inhibitors, but NADPH is probably also inhibitory. The quaternary structure determined from gel filtration and SDS-PAGE studies indicates that the enzyme is a homotetramer, although in maize the enzyme may exist as a homodimer in the absence of $NADP^+$. A short lysine-rich sequence in the amino terminal portion of the enzyme at positions 36-45 in the potato cytoplasmic isozyme, is thought to be involved in binding the cofactor. A conserved lysine residue, acting as a nucleophile, is thought to be involved in the catalytic mechanism. The sequence surrounding the catalytic lysine is highly conserved among all organisms and serves as a signature pattern for G6PD enzymes (see Table 12.2).

Phosphogluconate dehydrogenase (EC 1.1.1.44)

Phosphogluconate dehydrogenase catalyzes the irreversible, $NADP^+$-dependent oxidative decarboxylation of 6-phospho-D-gluconate to yield CO_2, NADPH, and D-ribulose-5-phosphate. The enzyme has been studied in *Ricinus communis* (castor bean), *Avena fatua* (wild oat), mung bean, carrot, sugar beet, pea, and soybean. Both cytosolic and plastidic forms have been studied. Although there are no reports of metal ion requirements, in castor bean, $MgCl_2$ was seen to stimulate enzymatic activity. Inhibitors include D-erythrose-4-phosphate, D-fructose-1,6-bisphosphate, or D-ribulose-5-phosphate in sugar beet, and D-fructose-2,6-bisphosphate or D-glucose-1,6-bisphosphate in castor bean. The enzyme from castor bean and sugar beet consists of a homodimer of approximately 50 kDa subunits. The amino terminal portion of the protein contains the cofactor-binding site, while a highly conserved region in the middle of the sequence contains part of the phosphogluconate-binding site, and serves as the signature sequence to the enzymes.

Ribose-5-phosphate isomerase (EC 5.3.1.6)

Ribose-5-phosphate isomerase catalyzes the interconversion of D-ribulose-5-phosphate and D-ribose-5-phosphate. Ribose-5-phosphate can then be used for the synthesis of nucleic acids, coenzymes such as NADH, NADPH, FAD, vitamin B_{12}, and other aromatic compounds. The enzyme from the chloroplast has been studied in spinach and pea. In pea, the enzyme is inhibited by 3-phosphoglycerate, 6-phosphogluconate, erythrose-4-phosphate, glyceraldehyde-3-phosphate, sedoheptulose bisphosphate, and $MgCl_2$. The spinach enzyme is inhibited by 4-phosphoerythronate and 5-phospho-D-ribonic acid. No activators or metal cofactors have been identified. In spinach, it consists of a homodimer of 26 kDa subunits and the same quaternary structure is likely in pea. Although no signature pattern has been identified for the enzyme, comparison of several plant isozymes reveals a consensus pattern containing a conserved pentapeptide in the middle of the protein.

TABLE 12.2 Consensus signature patterns of enzymes involved in the glyoxylate and pentose phosphate pathways[1]

Enzyme name	Signature pattern from PROSITE database	Position	ID Code[2]
Glyoxylate Pathway			
Isocitrate lyase	K-[KR]-C*-G-H-[LMQR]	211-216	P49297
Malate synthase	[KR]-[DNEQ]-H-x(2)-G-L-N-x-G-W-D-Y-[LIVM]-F	276-291	P49081
Phosphoenolpyruvate carboxykinase	L-I-G-D-D-E-H-x-W-X-[DE]-x-G-[IV]-x-N	385-400	P42066
Aspartate transaminase	[GS]-[LIVMFYTAC]-[GSTA]-K*-x(2)-[GSALVN]-[LIVMFA]-x-[GNAR]-x-R-[LIVMA]-[GA]	299-312	O48548
Pentose Phosphate Pathway			
Glucose-6-phosphate dehydrogenase	D-H-Y-L-G-K*-[EQK]	207-213	P37830
Phosphogluconate dehydrogenase	[LIVM]-x-D-x(2)-[GA]-[NQS]-K-G-T-G-x-W	260-271	O22111
Ribose-5-phosphate isomerase	[LIV]-[GD]-[SAY]-[IVF]-D-G-A-D-E-[IV]-[DN]	135-146	Q8RU73
Ribulose-phosphate-3-epimerase	[LIVMF]-H-[LIVMFY]-D[LIVM]-x-D-x(1,2)-[FY]-[LIVM]-x-N-x-[STAV]	91-105	Q43157
Transketolase	[LIVMA]-x-[LIVM]-M-[ST]-[VS]-x-P-x(3)-G-Q-x-[FM]-x(6)-[NKR]-[LIVMC]	195-217	Q43157
	R-x(3)-[LIVMTA]-[DENQSTHKF]-x(5,6)-[GSN]-G-H*-[PLIVMF]-[GSTA]-x(2)-[LIMC]-[GS]	89-109	O20250
	G-[DEQGSA]-[DN]-G-[PAEQ]-[ST]-[HQ]-[PAGM]-[LIVMYAC]-[DEFYW]-x(2)-[STAP]-x(2)-[RGA]	545-562	O20250
Transaldolase	[DG]-[IVSA]-T-[S]-N-P-[STA]-[LIVMF](2)	110-118	O04894
	[LIVM]-x-[LIVM]-K*-[LIVM]-[PAS]-x-[ST]-x-[DENQPAS]-[GC]-[LIVM]-x-[AGVS]-x-[QEKRST]-x-[LIVM]	206-223	O04894
Phosphoenolpyruvate carboxylase	[VT]-x-T-A-H*-P-T-[EQ]-x(2)-R-[KRH]	168-179	Q02909
	[IV]-M-[LIVM]-G-Y-S-D-S-x-K*-D-[STAG]-G	593-605	Q02909

[1] Conserved residues are denoted by **bold** letters, active site residues by an asterisk.
[2] Sequence identification number in the SwissProt/TrEMBL database from which position information was obtained.

Ribulose-phosphate-3-epimerase (EC 5.1.3.1)

Ribulose-phosphate-3-epimerase catalyzes the interconversion of ribulose-5-phosphate and xylulose-5-phosphate. Among plants this enzyme has been studied only in chloroplasts from spinach, in which it was reported that 90% of the total epimerase was associated with the thylakoid membranes. On the basis of gel filtration and SDS-PAGE, the plastidic enzyme was found to be a homooctomer of 23 kDa subunits in spinach, but in potato is thought to exist as a homohexamer. An aspartic acid at position 236 in the spinach chloroplast precursor sequence is believed to act as an electrophile in the catalytic mechanism. There are two conserved sequences within the protein that serve as signature patterns for the ribulose-phosphate-3-epimerase family. The first is near the amino terminus of the mature protein and the second near the active site aspartate.

Transketolase (EC 2.2.1.1)

Transketolase catalyzes the reversible two-carbon keto group transfer from xylulose-5-phosphate to ribose-5-phosphate or erythrose-4-phosphate. If the acceptor sugar is ribose-5-phosphate, the products are glyceraldehyde-3-phosphate and sedoheptulose-7-phosphate. If the acceptor is erythrose-4-phosphate, the products are glyceraldehyde-3-phosphate and fructose-6-phosphate. The enzyme has been studied in wheat and spinach and, as with transketolases from other sources, it requires thiamine diphosphate (vitamin B_1) as a cofactor. Both the cytosolic and plastidic isozymes have been studied. The chloroplast isozyme from both species has an additional requirement for Mg^{2+}, although the cytoplasmic isozyme does not. The spinach enzyme also catalyzes formation of ribulose-5-phosphate from D-glyceraldehyde-3-phosphate and hydroxypyruvate in a decarboxylation reaction. No activators or inhibitors of the enzyme have been described. The enzyme was isolated as a catalytically active homodimer in wheat following reduction with β-mercaptoethanol and gel filtration. The active holoenzyme is a homotetramer in both species but in wheat, dissociates to the dimeric form in the absence of thiamine diphosphate or in dilute solution. The enzyme from maize has been crystallized as a complex with thiamine diphosphate. A histidine residue at position 103 in the spinach chloroplast precursor sequence participates in a proton transfer reaction during catalysis. Two consensus sequence patterns have been identified, the first being in the region of the catalytic histidine, and the second thought to contain conserved acidic residues involved in metal binding.

Transaldolase (EC 2.2.1.2)

Transaldolase performs a reaction similar to transketolase, except that a three-carbon keto group is transferred. The reaction is reversible, utilizing sedoheptulose-7-phosphate as the ketol donor and glyceraldehyde-3-phosphate as the acceptor in one direction, fructose-6-phosphate as the donor, and erythrose-4-phosphate as the acceptor in the other direction. Only the spinach enzyme has been studied from plants. No cofactor or metal requirement was identified. Inhibitors include inorganic phosphate, pyrophosphate, and sulfate. The quaternary structure of a plant transaldolase has not been established, but in potato is assumed to be a homodimer of 48 kDa subunits, based on comparison of the cytoplasmic isozyme sequence to the yeast enzyme. The catalytic mechanism involves a Schiff-base intermediate carried by a conserved lysine residue at position 209 in the potato cytoplasmic isozyme. Two consensus patterns have been described for members of the transaldolase family. The first contains a highly conserved pentapeptide and the second contains the active-site lysine.

Bioenergetics and Regulation of Pentose Phosphate Pathway

Operation of the pentose phosphate pathway is closely tied to operation of the glycolytic pathway. Several intermediates are shared between the two, namely glucose-6-phosphate, fructose-6-phosphate, and phosphoglyceraldehyde. When the pentose phosphate pathway metabolizes glucose-6-phosphate, two molecules of NADPH and one molecule of CO_2 are produced, with no net production of ATP. The NADPH is usually considered as used exclusively for anabolic reactions, but may also be utilized in the mitochondrial electron transport system to produce ATP. This may provide a means to ensure that the pentose phosphate pathway continues to operate efficiently and to provide a continual source of ribose-5-phosphate and erythrose-4-phosphate for biosynthesis of nucleic acids and the indole-containing amino acids.

Phosphoglyceraldehyde and fructose-6-phosphate produced by the pentose phosphate pathway can either enter the glycolytic pathway or, at least in the case of fructose-6-phosphate, be utilized in the biosynthesis of oligosaccharides. Phosphoglyceraldehyde can enter the glycolytic pathway or continue to be used as an intermediate of the pentose phosphate pathway.

The major sites of regulation of the pentose phosphate pathway are actually at points outside the pathway proper. The level of fructose-6-phosphate is largely controlled by the activity of fructose-1,6-bisphosphatase, which effectively reverses the reaction catalyzed by phosphofructokinase. Sedoheptulose-7-phosphate is generated by the action of sedoheptulose-1,7-bisphosphate, a reaction essentially irreversible, and ribulose-5-phosphate is removed from the pathway in another irreversible reaction catalyzed by phosphoribulokinase.

TRICARBOXYLIC ACID CYCLE

Structure and Topology of Tricarboxylic Acid Cycle Enzymes

The tricarboxylic acid (TCA) cycle shown in Fig. 12.5 is also known as the citric acid cycle or Kreb's cycle and provides a means to utilize the pyruvate generated by glycolysis and the compounds generated by the glyoxylate cycle. The intermediates of the TCA cycle are used in the synthesis of a variety of compounds including nucleotides, porphyrins and several amino acids. Enzymes of the TCA cycle are located in the matrix space of the mitochondrion, except for succinate dehydrogenase, which is a part of Complex II of the electron transport chain, an integral membrane protein complex.

Pyruvate dehydrogenase complex

The pyruvate dehydrogenase (PDH) complex consists of three different enzymes that operate in series to catalyze oxidative decarboxylation of pyruvate. The first step is the decarboxylation of pyruvate by enzyme E1, followed by transfer of the hydroxyethyl group to a dihydrolipoamide carrier and the addition of coenzyme A to form acetyl-CoA by enzyme E2. The reduced dihydrolipoamide carrier is then oxidized with NAD^+ to yield NADH by enzyme E3. The complex consists of a core of up to 60 E2

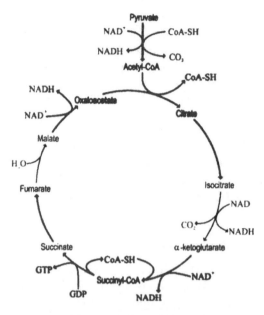

Fig. 12.5 Tricarboxylic acid cycle. The tricarboxylic acid cycle or Kreb's cycle, generates reducing equivalents in the form of NADH that are used during oxidative phosphorylation to produce ATP. It is also the source of most of the CO_2 generated by the various respiratory pathways. Enzymes are located in the matrix space of the mitochondrion.

pea and cauliflower, Ca^{2+} or Mn^{2+} may replace Mg^{2+} and in pea, Ni^{2+} was also able to substitute for Mg^{2+} No metal requirements have been reported for spinach or maize enzymes. The enzyme is inhibited by β-hydroxypyruvate in cauliflower and by glyoxylate in pea and cauliflower. Phosphorylation by a mitochondrial kinase causes reversible inactivation of the enzyme with activity restored by the action of a mitochondrial phosphatase. The enzyme is a tetramer about 150 kDa in size and composed of two 41 kDa α-subunits containing an E1-dehydrogenase domain and two 36 kDa β-subunits containing a transketolase domain. Although PDH contains a TPP cofactor, it does not appear to be related to other TPP-containing enzymes such as pyruvate decarboxylase, and no signature pattern has been developed for this family of dehydrogenases.

The second step of the pyruvate dehydrogenase reaction involves transfer of the hydroxyethyl group from TPP to a lipoic acid carrier, oxidation of the hydroxyethyl group to form an acetyl dihydrolipoamide intermediate, and lastly addition of coenzyme-A to form acetyl CoA. These reactions are catalyzed by E2 or dihydrolipoamide S-acetyltransferase (EC 2.3.1.12) component. The enzyme has been studied in *Arabidopsis*, cauliflower, potato, and maize, although very little information is available on the enzymology of this component of the complex. In potato, treatment of the pyruvate dehydrogenase complex with NaCl allowed for purification of acetyltransferase activity into a fraction containing a protein of 55 kDa, suggesting that the monomer form of the enzyme retains activity. Based on nucleotide sequences from *Arabidopsis* and maize, the predicted size of the mitochondrial precursor protein is about 58 kDa. A lysine residue in a conserved sequence provides the binding site for lipoic acid, and

subunits surrounded by multiple copies of the E1 and E3 enzymes and is thought to be tightly associated with the inner membrane of the mitochondrion. The citrate synthase and malate dehydrogenase enzymes are thought to be in close proximity to the PDH complex.

In the first step of the overall reaction, E1, the pyruvate dehydrogenase subunit E1 (EC 1.2.4.1), decarboxylates pyruvate with the resultant hydroxyethyl group attached to an enzyme-bound thiamine pyrophosphate. The enzyme has been studied in *Brassica oleracea* (cauliflower), pea, spinach, potato, and maize. In addition to the requirement for thiamine pyrophosphate, the enzyme requires a divalent metal ion for activity. The preferred metal is Mg^{2+}, although in

this region in the sequence has provided a signature pattern for lipoyl-binding in E2 enzymes.

In the final step of the reaction, the reduced dihydrolipoamide carrier is oxidized by NAD⁺ on the lipoamide dehydrogenase (EC 1.8.1.4) component (E3) of the complex. The enzyme has been studied in cauliflower, soybean, spinach, and pea. In soybean, reduction of lipoamide by NADH was observed and in spinach, oxidation of NADH could be achieved with either ferricyanide or 2,6-dichlorophenolindophenol as the oxidant. Although oxidation of NADH with free lipoic acid required the presence of EDTA, Mg^{2+}, NAD⁺, and cysteine in spinach, the reverse reaction, reduction of NAD⁺ with dihydrolipoic acid, did not require these supplements. In spinach, p-chloromercuribenzoate caused 50% inhibition of the enzyme when present at 0.1 mM. The active enzyme is a dimer of identical subunits, with each subunit containing a covalently bound FAD cofactor. The enzyme from pea has been crystallized and appears to be the same enzyme that constitutes the L-subunit of glycine decarboxylase. Two conserved cysteine residues are involved in the transfer of reducing equivalents from dihydrolipoamide through FAD and to NAD⁺. A short amino acid sequence surrounding the active-site cysteines serves as a pattern for E3 and other pyridine nucleotide-disulfide oxidoreductases with a class-I active site (see Table 12.3).

Citrate synthase (EC 2.3.3.1)

Citrate synthase catalyzes the irreversible condensation of acetyl-CoA and oxaloacetate to yield citrate and coenzyme A. The enzyme has been studied in *Agave americana* (century plant), *Arabidopsis*, carrot, pea, tomato, and potato. In pea, ATP, ADP, AMP, and $HgCl_2$ were found to be weak to moderate inhibitors while p-hydroxymercuribenzoate caused up to 60% inhibition. In tomato and century plant, ATP in the 5 to 7.5 mM range inhibited the enzyme by about 50%. Exposure of pea leaves to light resulted in activation of the enzyme. No cofactors or metal requirement have been described. The molecular weight of the holoenzyme in tomato was estimated at 104 kDa by gel filtration and SDS-PAGE revealed a single band at 50 kDa, suggesting a dimeric structure for the active enzyme. Two histidine residues at positions 308 and 354 in the *Arabidopsis* mitochondrial precursor protein sequence and an aspartic acid at position 409 are involved in the catalytic mechanism. The second histidine occurs in a highly conserved region that forms a consensus signature pattern for citrate synthases from prokaryotes and eukaryotes.

Aconitate hydratase (EC 4.2.1.3)

Aconitate hydratase (aconitase) catalyzes the interconversion of citrate and its isomer isocitrate. It has been studied in *Cucurbita maxima* (pumpkin), *Sinapis alba* (white mustard), potato, and soybean, with most of the enzymology performed on the enzyme from etiolated pumpkin cotyledons, where several isozymes were identified. One isozyme is found in the mitochondrion while the other is localized to the glyoxysome or cytosol. The apoenzyme contains an iron-sulfur cluster consisting of three iron and four sulfur atoms. An additional iron is bound, converting the enzyme to its active form. No other cofactor or metal requirements have been reported. Competitive inhibitors of citrate include fluorocitrate in pumpkin and oxalomalate in potato. The cytosolic enzyme is a monomer with a molecular weight of 100 kDa in pumpkin. In contrast, based on a nucleotide sequence, the mitochondrial protein in the red alga *Gracilaria verrucosa* is predicted to be only about

TABLE 12.3 Consensus signature patterns of enzymes involved in the TCA Cycle[1]

Enzyme name		Signature pattern from PROSITE database	Position	ID Code[2]
Pyruvate dehydrogenase	E2	[GDN]-x(2)-[LIVF]-x(5)-[LIVMFCA]-x(2)-[LIVFMA]-x(3)-**K***-[GSTAIVW]-[STAIVQDN]-x(2)-[LIVMFS]-x(5)-[GCN]-x-[LIVMFY]	141-170	Q9SWR9
	E3	**G**-**G**-x-**C***-[LIVA]-x(2)-**G**-**C***-[LIVM]-**P**	73-83	P31023
Citrate synthase		**G**-[FYA]-[GA]-**H***-x-[IV]-x(1,2)-[RKT]-x(2)-**D**-[PS]-**R**	351-363	P20115
Aconitate hydratase		[LIVM]-x(2)-[GSACIVM]-x-[LIV]-[GTIV]-[STP]-**C***-x(0,1)-**T**-**N**-[GSTANI]-x(4)-[LIVMA]	375-391	P49609
		G-x(2)-[LIVWPQ]-x(3)-[GAC]-**C***-[GSTAM]-[LIMPTA]-**C***-[LIMV]-[GA]	438-451	P49609
Isocitrate dehydrogenase		[NS]-[LIMYT]-[FYDN]-**G**-[DNGST]-[IIMVY]-x-[STGDN]-[DN]-x(2)-[SGAP]-x(3,4)-**G**-[STG]-[LIVMPA]-**G**-[LIVMF]	255-274	O65852
Succinate-CoA ligase α-subunit		**S**-[KR]-**S**-**G**-[GT]-[LIVM]-[GST]-x-[EQ]-x(8,10)-**G**-x(4)-[LIVM]-[GA]-[LIVM]-**G**-**G**-**D**	189-219	Q8GTQ9
α-subunit		**G**-x(2)-A-x(4,7)-[RQT]-[LIVMF]-**G**-**H***-[AS]-[GH]	274-287	Q8GTQ9
β-subunit		**G**-x-[IV]-x(2)-[LIVMF]-x-[NA]-**G**-[GA]-**G**-[LMA]-[STAVI]-x(4)-**D**-x-[LIVM]-x(3)-**G**-[GREA]	288-312	Q84LB6
Succinate dehydrogenase		**R**-[ST]-**H***-[ST]-x(2)-**A**-x-**G**-**G**	85-94	O82633
		C*-x(2)-**C***-x(2)-**C***-x(3)-**C***-[PEG]	212-223	Q9FJP9
Fumarate hydratase		**G**-**S**-x(2)-**M**-x(2)-**K**-x-**N**	345-354	P93033
Malate dehydrogenase		[LIVM]-**T**-[TRKMN]-**L**-**D***-x(2)-**R***-[STA]-x(3)-[LIVMFY]	178-190	Q9SPB8
Malate dehydrogenase (decarboxylating)		**F**-x-[DV]-**D**-x(2)-**G**-**T**-[GSA]-x-[IV]-[LIVMA]-[GAST](2)-[LIVMF](2)	308-325	P37221

[1] Conserved residues are denoted by **bold** letters, active site residues by an asterisk.

[2] Sequence identification number in the SwissProt/TrEMBL database from which position information was obtained.

80 kDa. Three cysteine residues are ligands to iron atoms in the iron sulfur cluster and form the basis of the signature pattern for aconitase enzymes.

Isocitrate dehydrogenase (EC 1.1.1.41)

Isocitrate dehydrogenase catalyzes the irreversible NAD^+-dependent oxidative decarboxylation of isocitrate to yield α-ketoglutarate, CO_2 and NADH. The enzyme has been studied in pea and potato. The enzyme has a requirement for Mg^{2+}, although Mn^{2+} can substitute for Mg^{2+} in pea. In both species, NADH was an inhibitor, cyanide inhibitory in the pea enzyme, and ATP inhibited the potato enzyme, probably by binding Mg^{2+}. The molecular weight of the holoenzyme was estimated to be 320 kDa, suggesting an octomer of 4 alpha subunits and 4 beta subunits of about 40 kDa each. A conserved serine residue at position 111 in the sequence of the tobacco mitochondrial precursor protein is thought to be involved in binding of isocitrate. A glycine-rich region in the carboxyl terminal portion of the protein provides a consensus sequence that can be used to identify isocitrate dehydrogenases.

Oxoglutarate dehydrogenase complex

The oxoglutarate (α-ketoglutarate) dehydrogenase complex is similar to the pyruvate dehydrogenase complex discussed above. It is comprised of three separate enzymes catalyzing the same types of reactions as pyruvate dehydrogenase. The E1 component catalyzes decarboxylation of α-ketoglutarate to yield a thiamine pyrophosphate-succinyl intermediate. The intermediate is transferred to a dihydrolipoamide carrier for the addition of acetyl-CoA by the action of E2 and following release of succinyl-CoA, the dihydrolipoamide carrier is oxidized by NAD^+ through the action of the dihydrolipoyl dehydrogenase (E3).

The E1 component (EC 1.2.4.2) has been studied in Brassica rapa (turnip), cauliflower, pea, soybean, maize, and potato. It requires thiamine diphosphate as a cofactor. Unlike the E1 component of pyruvate dehydrogenase, comprised of alpha and beta subunits, it appears to be comprised of a single polypeptide of about 105 kDa containing both the E1 dehydrogenase and the transketolase domains. It is present in multiple copies in the active complex. No signature pattern has been defined for this enzyme, although a short sequence (W-E-A-Q-F-G-D-F-x-N-x-A-Q) covering positions 734-746 in the Arabidopsis sequence seems to be highly conserved in the enzymes from both prokaryotes and eukaryotes.

The E2 component of the complex has not been studied in isolation from any plant. The predicted size of the protein from Arabidopsis is about 50 kDa and it is thought to be present in multiple copies, making up the core of the active complex. It contains the consensus lipoyl-binding signature pattern (positions 118-147 in Arabidopsis) described above for pyruvate dehydrogenase.

The catalytic activity of the E3 component of this complex is identical to the E3 component of the pyruvate dehydrogenase complex (EC 1.8.1.4). It requires FAD as a cofactor, and uses NAD^+ to oxidize the dihydrolipoamide component of E2. Several genes encode a lipoyl dehydrogenase in plants but only one sequence, from Arabidopsis, has been identified as encoding the protein from oxoglutarate dehydrogenase. It contains the signature pattern (positions 44-54) containing the active-site cysteine residues.

Succinate-CoA ligase (EC 6.2.1.4)

Succinate-CoA ligase (succinyl-CoA synthetase) catalyzes a substrate-level phosphorylation of GDP to yield GTP. In the process, coenzyme A is released from succinyl-CoA. It has been studied extensively in soybean although some data exist on the enzyme from spinach. In soybean, reverse reaction (formation of succinyl-CoA from succinate coenzyme A and ATP) was found to require Mg^{2+} or Mn^{2+} and was activated by 8-hydroxyquinoline, glutathione or NaCN. The enzyme is completely inhibited by other metals such as Co^{2+}, Cu^{2+}, Hg^{2+}, Pb^{2+}, and Zn^{2+} with moderate inhibition caused by acetate, butyrate, malonate, propionate, EDTA, iodoacetamide, and n-ethylmaleimide. The molecular weight of the purified enzyme from soybean is 160 kDa. Alpha and beta subunits have predicted sizes of 35 kDa and 45 kDa in tomato, suggesting an α/β dimer or α_2/β_2 tetramer as the quaternary structure. A conserved histidine in the active site on the α-subunit acts as a phosphate acceptor during catalysis and is part of one of the three consensus sequences shared by succinyl-CoA ligase, malonyl-CoA ligase, and ATP citrate-lyase. The other two consensus patterns are in glycine-rich regions on each subunit.

Succinate dehydrogenase (EC 1.3.5.1)

Succinate dehydrogenase (SDH), also known as Complex II of the respiratory electron transport chain, catalyzes the two-electron oxidation of succinate to fumarate with the concomitant reduction of ubiquinone (coenzyme Q) to ubiquinol. The complex has been studied in mung bean and soybean. It contains a covalently bound FAD cofactor as well as an iron-sulfur cluster and a b-type cytochrome. No activators have been described but in both species, malonate or oxaloacetate inhibits the enzyme. The holoenzyme consists of 4 different subunits. The flavoprotein subunit (67 kDa) binds FAD at a conserved histidine in the amino-terminal portion of the protein. This region forms a signature pattern for SDH enzymes. The 2Fe/2S cluster is liganded to a 30 kDa subunit by four cysteine residues that make up a signature pattern for iron-sulfur centers. A 15 kDa subunit, containing several transmembrane helices, binds cytochrome b_{560} through a conserved histidine and, along with an additional subunit of about 13 kDa, may be involved in anchoring the complex to the mitochondrial inner membrane.

Fumarate hydratase (EC 4.2.1.2)

Fumarate hydratase (fumarase, fumarate lyase) catalyzes hydration of fumarate to form L-malate. The reaction is reversible, yielding fumarate and water. It has been studied in the unicellular alga *Euglena gracilis*, *Arabidopsis*, pea, and potato. No metal or cofactor requirements have been described. In *Arabidopsis*, the enzyme is inhibited by several TCA cycle intermediates, including α-ketoglutarate, citrate, and succinate, as well as pyruvate, ADP, and AMP. In *Euglena*, ATP and phosphate were found to be inhibitory. The molecular weight of the dimeric holoenzyme from *Euglena* was found to be 120 kDa, and comprising 60 kDa subunits. In pea, the holoenzyme is a tetramer 175-190 kDa in size. Although the enzyme operates primarily in the mitochondrial matrix, a second gene has been identified in *Arabidopsis* that appears to have a signal peptide consistent with targeting to the chloroplast. The active site contains a conserved histidine that participates in the catalytic cycle (position 216 in the mitochondrial precursor protein) and a conserved lysine (position 352) thought to be involved in substrate binding. The lysine residue is contained within a short conserved region that comprises the signature sequence of class II lyases.

Malate dehydrogenase (EC 1.1.1.37)

Malate dehydrogenase (MDH) catalyzes the NAD^+-dependent oxidation of L-malate to form oxaloacetate and NADH. The reaction is reversible, although rapid utilization of oxaloacetate in another turn of the TCA cycle drives the reaction in the direction of malate oxidation. The enzyme has been extensively studied in a variety of plants, including *Lemna minor* (duckweed), *Eucalyptus globulus* (blue gum), *Phaseolus acutifolius* (tepary bean), *Citrus*, *Arabidopsis*, mung bean, broad bean, potato, pea, tomato, wheat, soybean, and maize. No metal or cofactor requirements have been found. In mung bean, ATP inhibits the enzyme, and in soybean, the protein modification reagents dimethylpyrocarbonate, iodoacetamide, and phenylmethanesulfonylfluoride were found to inhibit the enzyme. Although thought of as a TCA cycle enzyme localized to the mitochondrion, at least three other isoforms of the enzyme have been identified, each localized to a different cellular compartment. For instance, in spinach genes have been identified encoding isoforms targeted to the chloroplast, glyoxysome, and peroxysome in addition to the mitochondrion.

The holoenzyme comprises two identical subunits of 38-40 kDa in soybean, tomato, and blue gum, but in spinach, nondenaturing conditions such as treatment with K^+ and/or dithioerythritol result in a variety of forms ranging from 30-40 kDa monomers to 180 kDa oligomers. The presence of Mg^{2+} and NADH stabilized a 127 kDa form in spinach, although the functional significance of this form is not known. The crystal structure of a similar $NADP^+$-dependent enzyme from the chloroplast of *Sorghum* has been determined. Two conserved residues, aspartic acid, and a histidine are involved in a catalytic proton-relay system and a conserved arginine is involved in substrate binding. Aspartate and arginine are within the consensus signature sequence that identifies both the NAD^+-dependent and $NADP^+$-dependent malate dehydrogenases.

Malate dehydrogenase (decarboxylating, EC 1.1.1.39)

Another malate dehydrogenase is present in plant mitochondria, the so-called malic enzyme, which is not found in animal mitochondria. This enzyme is able to perform an oxidative decarboxylation of malate by using NAD^+. The reaction yields pyruvate, which can move into the cytosol, NADH and CO_2. Although it is not a TCA enzyme per se, it is included here because it is found in the mitochondrial matrix and is important for generating pyruvate by a pathway separate from glycolysis. The enzyme has been studied in several members of the Amaranthaceae (amaranth family), *Crassula argentea* (jade plant), *Panicum miliacium* (common millet), *Vigna unguiculata* (cowpea), *Heliocarpus*, cauliflower, and potato. Although the enzyme shows a preference for NAD^+ as a substrate, at least in *Atriplex spongiosa* (an amaranth), jade plant, and potato, $NADP^+$ can be used, although with less than 20% of the activity when NAD^+ is used. A divalent metal cation is necessary for activity. In the Amaranthaceae, Mn^{2+} was an absolute requirement, whereas in cowpea, jade plant, cauliflower and potato, Mg^{2+} or Mn^{2+} were able to activate the enzyme. In cauliflower, Co^{2+} was able to replace Mg^{2+} for activation. Inhibitors include bicarbonate in the amaranths, citrate, Cl^-, NADH, pyruvate, and phosphoenolpyruvate in jade plant, and oxaloacetate in cauliflower and potato.

Fumarate, fructose-1,6-bisphosphate, and AMP activate the enzyme in jade plant, while coenzyme A activates the enzyme in all the species.

The enzyme is active as a heterodimer of similar subunits, each about 60 kDa, but in jade plant and potato, tetramers and octomers have been observed following gel filtration or nondenaturing gel electrophoresis. Two conserved regions involved in cofactor binding have been identified in malic enzymes from prokaryotes and eukaryotes. A third region of unknown function has also been identified and comprises the consensus signature pattern for these enzymes. It is present in both subunits of the enzyme from potato.

Bioenergetics and Regulation of TCA Cycle

The Kreb's cycle, or tricarboxylic acid (TCA) cycle, is the major source of reducing equivalents for the respiratory electron transport chain. For each glucose molecule passed through glycolysis, The TCA cycle and electron transport chain can produce 24 ATP molecules. In addition to its central role in the catabolic reactions that lead to energy production, the TCA cycle plays a central role in anabolic reactions by serving as a source of the building blocks needed for the synthesis of many cell constituents. Regulation of the TCA cycle is therefore an important task of the plant cell. As in animals, the activity of pyruvate kinase is a key point of regulation and is accomplished by the same kind of phosphatase/kinase system. When levels of NADH and acetyl Co-A are elevated, the pyruvate dehydrogenase complex is phosphorylated through the action of pyruvate dehydrogenase kinase and becomes inactive. The activity of the kinase enzyme is decreased when the levels of pyruvate, coenzyme-A, and NAD^+

are elevated. Inactive phosphorylated dehydrogenase is reactivated in the presence of Mg^{2+} or Ca^{2+}, which stimulate the activity of pyruvate dehydrogenase phosphorylase.

Because the biosynthesis of several amino acids uses intermediates of the TCA cycle, especially α-ketoglutarate, fumarate, and oxaloacetate, there must be a way to replace these intermediates. The action of phosphoenolpyruvate carboxylase in the cytosol fulfills this requirement by converting phosphoenolpyruvate to oxaloacetate, which can eventually be imported into the mitochondrion in the form of malate or aspartate. This non-photosynthetic carbon fixation reaction has been called an "anaplerotic" or "filling-up" reaction as it provides those TCA cycle intermediates that have been utilized in anabolic reactions.

GLYOXYLATE PATHWAY

Structure and Topology of Enzymes of Glyoxylate Pathway

The glyoxylate pathway (Fig. 12.6) provides a mechanism by which plants, especially germinating seeds and seedlings, can bypass the decarboxylation steps of the TCA cycle in order to generate pools of TCA cycle intermediates used in anabolic reactions. It also provides a means for utilizing acetyl-CoA produced from the β-oxidation of fatty acids during seed germination. Some of the reactions of the cycle are catalyzed in the glyoxysome by using organelle-specific enzymes from the TCA cycle. Other enzymes are specific to the glyoxylate cycle. The first reaction, formation of citrate from oxaloacetate and acetyl-CoA, and the second reaction, isomerization of citrate to isocitrate, are catalyzed by the glyoxysome-specific citrate synthase and aconitase enzymes

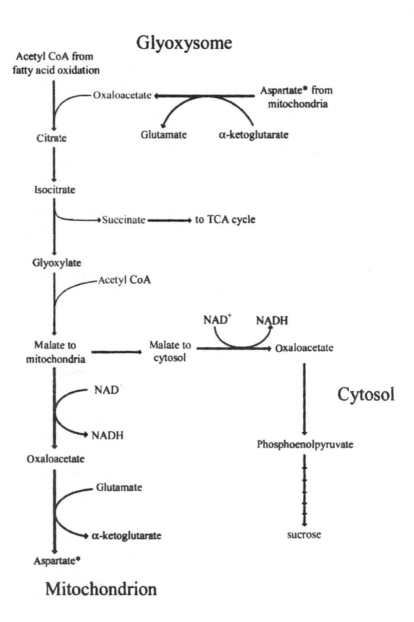

Fig. 12.6 Glyoxylate pathway. The glyoxylate pathway provides a means to utilize the carbon generated by the β-oxidation of fatty acids and bypasses the decarboxylation reactions of the TCA cycle. Some of the reactions are catalyzed by glyoxysome-specific counterparts of TCA cycle enzymes. The malate produced in the glyoxysomes can move into the mitochondria and be utilized in the TCA cycle. Malate in the cytosol can be used in the biosynthesis of sucrose. Aspartate from the mitochondria (denoted by the asterisk) is converted to oxaloacetate in the glyoxysome

respectively. These enzymes are almost identical to those in the mitochondria but are encoded by distinct genes and contain a carboxy-terminal signal sequence that directs them to the glyoxysome.

Isocitrate lyase (EC 4.1.3.1)

Isocitrate in the glyoxysome is cleaved by isocitrate lyase to form glyoxylate and succinate. The enzyme has been studied in several plants, including species of *Pinus* (pine) and *Lupinus* (lupine) genera, *Citrullus vulgaris* (watermelon), castor bean, cucumber, soybean, and maize. It requires a divalent metal cation for activity; the preferred metal is Mg^{2+}, although Mn^{2+} and Ba^{2+} can substitute for it in several species; no other cofactors have been described. Other metals, such as Co^{2+}, Cu^{2+}, Fe^{2+}, and Pb^{2+} are inhibitory in *Pinus densiflora* (Japanese red pine), as are ADP, ATP, and the metabolic intermediates 3-phosphoglycerate, phosphoenolpyruvate, and succinate. Citrate, glycolate, glyoxylate, malate, maleate, malonate, methylsuccinate, oxalate, and tartrate have also been shown to inhibit the enzyme in different species. Dithiothreitol and other sulfhydryl compounds were found to inhibit the Japanese red pine enzyme but stimulated activity in watermelon and lupine. Most reports indicate that the enzyme is a homotetramer with a molecular weight of 200-300 kDa, comprising 64-66 kDa subunits, although based on SDS-PAGE and gel filtration, the subunit size in castor bean was estimated to be 35 kDa. A dimeric form of the enzyme was isolated from soybean, consisting of two 64 kDa subunits. A cysteine residue conserved among bacteria, fungi, and plants is thought to be involved in the catalytic mechanism. A histidine and an aspartate or glutamate are also thought to be involved, although these residues are not as well conserved. The region surrounding the conserved cysteine comprises the signature pattern for isocitrate lyase.

Malate synthase (EC 2.3.3.9)

Malate synthase catalyzes the condensation of glyoxylate and acetyl-CoA to form L-malate and reduced coenzyme A. The enzyme has been studied in *Astasia longa* (a euglenoid alga), *Gossypium hirsutum* (upland cotton), *Pinus taeda* (loblolly pine), Japanese red pine, castor bean, cucumber, maize, pumpkin, rape, and soybean. The enzyme shows a requirement for Mg^{2+} but at least in the case of upland cotton, Ba^{2+}, Co^{2+}, or Mn^{2+} can substitute, although with only about 30% of the activity achieved with Mg^{2+}. Many of the compounds found to inhibit isocitrate lyase are also inhibitory toward malate synthase. An extensive study of inhibitors was performed in upland cotton. Other inhibitors include fluoroacetate, glycolate, and pyruvate in maize, and ATP, ADP, and AMP in castor bean. In Japanese red maple, EDTA at low concentration (10 µM) caused a 10% increase in activity while a higher concentration (1 mM) resulted in a 30% decrease in activity.

The enzyme comprises identical 60 kDa subunits. In cucumber, and probably other species as well, these monomers are synthesized in the cytosol and rapidly processed after moving into the glyoxysome. The exact nature of the processing necessary for oligomerization, is not known; however in castor bean, phosphorylation of a specific serine residue takes place. Many different forms of the holoenzyme have been described, representing different aggregations of the monomer. In cucumber and pumpkin, dimers and octomers were isolated, whereas in castor bean, soybean, pine, and maize, octomers and decamers were observed. A dodecamer, the largest aggregate, was seen in upland cotton. An arginine residue at position 173 in the maize protein acts as a proton acceptor during catalysis, and an aspartate (position 459) acts as a proton donor in the aldol condensation. A highly conserved region in the middle of the protein serves as the consensus sequence that defines malate synthase enzymes.

Phosphoenolpyruvate carboxykinase (EC 4.1.1.49)

Malate produced in the glyoxysome can move to the cytosol and be converted to oxaloacetate by malate dehydrogenase. Phosphoenolpyruvate carboxykinase (PEPCK) then catalyzes the ATP-dependent decarboxylation of oxaloacetate to form phosphoenolpyruvate and ADP. Phosphoenolpyruvate can then be used to synthesize sucrose and other carbohydrates. Very little enzymology has been performed with this enzyme. The only data have been obtained by studying the enzyme from cucumber cotyledons. PEPCK exists as a homotetramer of 74 kDa subunits and is reversibly phosphorylated in response to light. A conserved region in the middle of the protein that is involved in ATP binding has been identified (positions 368-375 in cucumber) and a short sequence nearby has been designated as the signature pattern for these enzymes.

Aspartate aminotransferase (EC 2.6.1.1)

Aspartate aminotransferase, also called aspartate transaminase (AAT) or glutamate-oxaloacetate transaminase (GOT) catalyzes the reversible amino group transfer from aspartate to α-ketoglutarate, or from glutamate to oxaloacetate. This reaction is vital in order to regenerate the oxaloacetate needed in the glyoxysome, since oxaloacetate cannot move between organelles and the cytosol. The enzyme also serves as a link between respiration and biosynthesis of several amino acids. The enzyme has been studied in *Arabidopsis*, alfalfa, carrot, rice, and soybean. Three isozymes have been identified in alfalfa and carrot that probably represent the cytosolic, mitochondrial/peroxysomal, and glyoxysomal enzymes. The enzyme contains a covalently bound pyridoxal phosphate that is required for activity. No metal requirement has been described. In soybean, cyanide, p-chloromercuribenzoate, and sodium mersalyl (a mercury compound) inhibit the enzyme. The active enzyme is comprised of a dimer of identical 40-50 kDa subunits. The pyridoxal phosphate cofactor is bound to a conserved lysine in a region with sufficient similarity among species to generate a consensus pattern for the class I family of aminotransferases.

Bioenergetics and Regulation of Glyoxylate Pathway

The primary function of glyoxysomes is to convert lipids into sucrose through the production of malate and succinate. This is accomplished by compartmentalization of the enzymes involved to glyoxysome. The acetyl-CoA produced by oxidation of fatty acids must be prevented from serving as a substrate for the mitochondrial citrate synthase reaction. This is accomplished effectively by the glyoxysome-specific isozymes of the TCA cycle, citrate synthase, aconitase, isocitrate lyase, and malate synthase. Only a small portion of the acetyl Co-A produced by the glyoxylate cycle actually enters the TCA cycle even after germination and formation of photosynthetically competent cotyledons. Although the activity of these enzymes can be detected in maturing seeds, it is not until germination that the glyoxysome matures and the enzymes become fully active.

Maturation of the glyoxysome and increased accumulation of glyoxysomal enzymes occur as a response to an increase in mRNA encoding the enzymes. Following the emergence of photosynthetically competent cotyledons, the glyoxysome and associated enzymes become less important in meeting the cell's demand for sucrose, being replaced by the enzymes associated with photorespiration. As the plant continues to grow, glyoxysomes in photosynthetic tissues are gradually converted into peroxysomes, whereby photorespiration becomes the predominant metabolic activity in the microbody.

ELECTRON TRANSPORT AND OXIDATIVE PHOSPHORYLATION

Structure and Topology of Electron Transport Chain and ATP Synthetase

Enzymes of the electron transport chain are located on the inner mitochondrial membrane. They are arranged in a way that allows the reducing power of intermediates generated in glycolysis and the TCA cycle to be utilized in generation of a transmembrane proton gradient. The potential energy created by the transmembrane proton gradient is then transduced by the ATP synthase complex to generate ATP as shown in Fig. 12.7. In addition to the cytochrome c oxidase (Complex IV) found in animals and bacteria, plants contain an additional terminal oxidase that is not coupled to proton transport. This "alternative oxidase" does not contribute to energy transduction but is used in some plants, especially aroids, as a source of heat production during the flowering process.

NADH dehydrogenase (ubiquinone) (EC 1.6.5.3)

NADH dehydrogenase, also known as NADH:ubiquinone oxidoreductase or

Complex I, catalyzes the transfer of electrons from NADH to ubiquinone, releasing NAD^+ and ubiquinol. The enzyme has been studied in *Arum maculatum* (cuckoo-pint), sugar beet, and mung bean. In addition to the normal electron acceptor, ubiquinone, the enzyme from *Arum* and sugar beet can use 2,6-dichlorophenolindophenol, ferricyanide, or short tail-length ubiquinones as acceptors. The only metal requirement for the enzyme is iron, used in the iron-sulfur clusters. Inhibitors include dicyclohexylcarbodiimide, mersalyl, and rotenone in sugar beet and mung bean. Four iron-sulfur clusters and a flavin cofactor (FMN) are bound by different subunits. The holoenzyme consists of about 30 different subunits. At least 14 subunits were identified in the sugar beet enzyme, while only five major subunits were identified in the mung bean enzyme. The ubiquinone-binding site is located on subunit 1, a protein of about 36 kDa with 7-8 transmembrane helices. The iron sulfur clusters are bound by several subunits. The 75 kDa subunit participates in binding two 4Fe-4S clusters (N-3 and N-4) along with the 51 kDa subunit. The 51 kDa subunit also contains binding sites for NAD+/NADH and the FMN cofactor. The 49 kDa and 20 kDa subunits are involved in binding another 4Fe-4S cluster and a 24 kDa subunit bind a 2Fe-2S cluster. Several of the subunits of the mitochondrial enzyme seem to have counterparts in the chloroplast. For instance, the 20 kDa subunit has a homolog in the chloroplast that has been identified as being a component of photosystem II (*psbG* gene product).

Succinate dehydrogenase (EC 1.3.5.1)

Succinate dehydrogenase, or Complex II, was described above in the section on the TCA cycle.

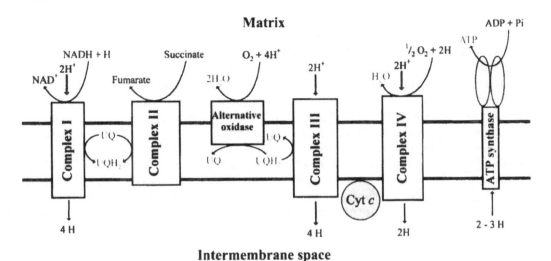

Fig. 12.7 Electron transport pathway and ATP synthesis. Oxidative phosphorylation occurs on the inner mitochondrial membrane (shown as the two horizontal lines), which divides the mitochondrial matrix and intermembrane space. Electrons move from Complex I (NADH dehydrogenase) and Complex II (succinate dehydrogenase), by reducing the mobile carrier, ubiquinone (UQ). Ubiquinol (UQH$_2$) diffuses within the membrane to Complex III (cytochrome b/c$_1$), which transfers electrons to the soluble cytochrome c. Reduced cytochrome c transfers the electrons to Complex IV (cytochrome c oxidase) where they reduce molecular oxygen to water. In the process, protons are translocated from the matrix to the intermembrane space, generating chemiosmotic potential. Accumulated protons flow back to the matrix through the ATP synthase enzyme and drive the phosphorylation of ADP

Ubiquinol-cytochrome c reductase (EC 1.10.2.2)

Ubiquinol-cytochrome c reductase, also known as Complex III or the cytochrome b/c_1 complex, catalyzes the two-electron oxidation of ubiquinol and couples the oxidation to the one-electron reduction of cytochrome c. The enzyme complex is very similar to the cytochrome b_6/f complex found in the thylakoid membranes of chloroplasts. It has been studied in *Euglena*, Jerusalem artichoke, potato, and spinach. It contains two b-type cytochromes, one c-type cytochrome and a 2Fe/2S cluster, known as the Reiske iron sulfur center. The enzyme from spinach is inhibited by a number of compounds, including DBMIB (2,5-dibromo-6-methyl-3-ispropyl-1,4-benzoquinone), 2-heptyl-4-hydroxyquinoline

N-oxide, 2-iodo-6-isopropyl-3-methyl-2′,2, 4′-trinitrophenyl ether, antimycin A, myxothiozol, and UHDBT (undecyl-hydroxyl-dioxobenzoxythiazole). No activating compounds or metal cofactors, other than iron, have been described.

The complex consists of at least 10 polypeptides. The cytochrome b component is a single protein of about 44 kDa in potato. It noncovalently binds 2 cytochromes, cytochrome b_{562} and cytochrome b_{566}, through two pairs of highly conserved histidine residues. A third cytochrome, cytochrome c_1, is covalently bound to a 35 kDa subunit by two conserved cysteine residues and a pair of histidine residues, which act as axial ligands to heme iron. The Reiske iron-sulfur center is bound by a 29 kDa subunit containing the [C-(TK)-H] and [C-P-C-H] binding motifs characteristic of Reiske Fe/S-binding proteins. In addition to the three subunits involved in binding the electron transfer components are two core polypeptides and several smaller polypeptides which remain for the most part uncharacterized with respect to their function. One of the core subunits, a heterodimer of 54 kDa proteins, has been speculated to be involved in the processing of peptides destined for import into the mitochondrion. Both subunits have a high degree of sequence similarity to other peptidases but there does not appear to be a zinc-binding site, which is a common feature of other peptidases.

Cytochrome c

Cytochrome c is a soluble protein reduced by Complex III and oxidized by cytochrome c oxidase (Complex IV). It is found in the intermembrane space and acts as a one-electron carrier between the two complexes. It is a single polypeptide of about 14 kDa and contains a single c-type cytochrome bound by the typical [C-x-x-C-H] binding motif of c-type cytochromes. The two cysteine residues are covalently attached to the porphyrin macrocycle and the histidine residue forms one of the axial ligands to the heme iron. Unlike the proton-coupled electron transport seen in the respiratory chain complexes I – IV, electron transport catalyzed by cytochrome c is not coupled to proton transport across the mitochondrial membrane.

Cytochrome c oxidase (EC 1.9.3.1)

Cytochrome c oxidase (Complex IV) is one of two terminal oxidases found in plant mitochondria. It catalyzes the four-electron reduction of molecular oxygen to form two molecules of water. The reduced form of cytochrome c supplies the electrons and, in the process, four protons are moved across the mitochondrial membrane. Four protons from the matrix side of the membrane are

used in the formation of water, contributing even more to the transmembrane potential that is coupled to ATP synthesis. The oxidase has been purified from sweet potato in an active form. The enzyme as isolated from peas is much less stable. In addition to the well-known inhibition caused by cyanide, the sweet potato enzyme was found to be inhibited by as much as 50% in the presence of 50 mM KCl or 0.3% Triton X-100, a detergent frequently used to isolate membrane protein complexes. It was also found that phosphate in concentrations as low as 15 mM were inhibitory. Activity was enhanced by the addition of phospholipids, especially phosphatidylcholine and phosphatidylethanolamine, which stimulated activity threefold.

The complex consists of at least 10 subunits, although the prosthetic groups are bound by only two of them. The functional core of the complex is formed by subunits I - III. Subunit I, the catalytic subunit, is about 58 kDa in size, binds two hemes (a and a_3) and one copper atom (CuB), and contains 12 transmembrane helices. Electrons passed one at a time from cytochrome c through the CuA center on subunit II are transferred to the binuclear active site by heme a. The binuclear center, consisting of heme a_3 and CuB, accumulates the four reducing equivalents, ultimately using them in reduction of molecular oxygen. The catalytic mechanism of the reduction is an area of intense research. Subunit II contains the CuA center, liganded by a highly conserved tetrad consisting of two cysteine and two histidine residues. It also contains the binding site for cytochrome c consisting of a patch of negatively charged residues facing the intermembrane space. Subunit III is a 29 kDa protein that does not appear to be involved directly with the electron transfer components of the enzyme. It is required for enzymatic activity although the exact nature of its participation is not presently known. It does interact with the two core subunits as well as with several of the smaller subunits that make up the enzyme.

H⁺-transporting two-sector ATPase (EC 3.6.3.14)

This enzyme, commonly known as ATP synthetase, coupling factor or F_0F_1-ATPase, catalyzes formation of ATP from ADP and inorganic phosphate. It uses the electrochemical gradient produced by the electron transport chain to form a high-energy phosphodiester bond. It is almost identical in structure to the chloroplast coupling factor found on thylakoid membranes. Due to the abundance of thylakoid membranes and ease of purification of the enzyme from chloroplasts, most biochemical characterization of the enzyme has been performed on the CF_0CF_1-ATPase. The enzyme has been studied from a variety of sources including *Euglena*, *Oenothera berteriana* (Bertero's evening primrose), maize, rice, wheat, sugar beet, rape, pea, liverwort, and tobacco, but the most common source for the enzyme is from spinach chloroplasts. The reaction is reversible and the enzyme usually assayed in the direction of ATP cleavage. In addition to ATP, other nucleotide triphosphates, such as dATP, GTP, ITP, and UTP can be hydrolyzed. There is a requirement for a divalent metal cation such as Ca^{2+} or Mg^{2+} in the spinach enzyme, but no requirement for Mg^{2+} has been shown in the *Euglena* enzyme, although Ca^{2+} is required for activity. No other metal requirements have been observed. The enzyme from spinach is inhibited by dicyclohexylcarbodiimide (DCCD), oligomycin, Hg^{2+}, permanganate (MnO_4^-) and sulfate. Whereas the membrane-bound CF_1 from spinach requires mild trypsin treat-

ment for activation, similar proteolysis results in inhibition of the *Euglena* enzyme.

The enzyme consists of a multisubunit catalytic domain in the matrix space (F_1) anchored to the inner mitochondrial membrane by the F_0 portion of the protein. The F_1 portion is comprised of a $\alpha_3\beta_3$ hexamer of similar subunits, each of which contains a nucleotide binding site. There is an active site on each β subunit with the α subunits performing regulatory functions. The F_0 domain consists of three different subunits in an $a_1b_2c_{12}$ stoichiometry that are arranged to form a proton-translocating pore through the membrane. Three additional proteins of the F_1 domain, γ, δ, and ε, each present in single copy, connect the catalytic domain to the pore-forming F_0 domain and complete the proton-translocating conduit between the intermembrane and matrix compartments. The α and β subunits are 50 and 55 kDa with the γ, δ, and ε subunits 31, 19, and 14 kDa respectively. The F_0 components are about 30 (a), 17 (b) and 8 (c) kDa, giving rise to a holoenzyme size of approximately 325,000 as determined by equilibrium ultracentrifugation. A crystal structure of the α and β subunits of the F_1 portion of the chloroplast enzyme from spinach is available.

Alternative oxidase (No EC number has been designated)

The alternative oxidase in plant mitochondria catalyzes reduction of molecular oxygen to water directly from ubiquinol without coupling the electron transport to proton translocation across the membrane. The energy released in this reaction is dissipated in the form of heat and thus the enzyme performs a similar function as the "uncoupling protein" found in the brown fat of animals. The enzyme has been studied mostly in members of family *Araceae*, such as voodoo lily and skunk cabbage in which the heat is used to volatilize aromatic compounds in the flower that attract pollinators. The enzyme has also been studied in nonaroid plants, including soybean, potato, and tobacco; here emphasis was given to expression and regulation of the protein levels and activity. It is located in the inner mitochondria membrane with the active site exposed to the matrix side of the membrane, and comprises two identical subunits of about 35 kDa. Several isoforms of the enzyme have been identified. In addition to the normal substrate, ubiquinol$_{10}$, the enzyme can utilize a variety of short-chain ubiquinols as the electron donor. Each subunit of the enzyme is thought to bind a pair of nonheme irons that when the inactive dimer is activated by reduction of a disulfide bond, are thought to form the catalytic site for the four-electron reduction of oxygen. The enzyme is not inhibited by cyanide, as is cytochrome c oxidase, but is inhibited by a number of substituted benzhydroxamic acids, the most commonly used of which is salicylhydroxamic acid (SHAM).

Bioenergetics and Regulation of Oxidative Phosphorylation

The presence of the alternative oxidase in plant mitochondria makes them fundamentally different from animal mitochondria. Any discussion of the bioenergetics and regulation of oxidative phosphorylation in plants must take this into account. Overall, conversion of the potential energy stored in NADH, NADPH, and succinate into ATP occurs in plants just as it does in animals. Electrons are passed between protein complexes that convert the electromotive potential to a chemiosmotic potential in the form of a proton gradient across the inner mitochondrial membrane. The chemiosmotic potential is converted to chemical potential energy in the form of ATP. The

main question to be answered concerns the mechanism of regulation of partitioning of electrons between the two pathways.

Several hypotheses have been postulated regarding the factors that influence the relative activities of the two pathways. In thermogenic tissues such as the spadix of aroids, alternative oxidase is expressed at extremely high levels and thus outcompetes the cytochrome pathway for electrons. The levels of the alternative oxidase are elevated in response to induction of the genes encoding the oxidase in these tissues. For several days in advance of flower opening, mRNA levels are seen to dramatically increase and the oxidase produced in prodigious amounts. After flowering, the levels of mRNA decline, as does expression of the alternative oxidase.

A similar induction of transcripts is seen in potato tuber tissue in response to wounding. This is a nonthermogenic response and probably arises in an effort to keep the TCA cycle operating at high levels in order to supply biosynthetic intermediates involved in the wounding response pathway. The ability to induce the alternative pathway in several root crops, such as turnip and parsnip, in response to wounding may represent a feature of the alternative pathway, namely the ability to keep the TCA cycle operative. When ATP levels are elevated, the TCA cycle is inhibited and the pools of intermediates for anabolic reactions are quickly depleted.

Another possibility for existence of the alternative pathway has to do with what is known as "energy overflow". An increase in carbohydrate catabolism will have a tendency to saturate electron transport through the cytochrome pathway. When the cytochrome pathway becomes saturated, the excess electrons can be rerouted toward the alternative pathway, thus relieving the potential for inhibition of glycolysis and the

TCA cycle. Another hypothesis for induction of the alternative pathway has to do with the availability of ADP. As in animals, the activity of the cytochrome pathway in plants is dependent on the availability of ADP. When ADP levels are low, oxidative phosphorylation decreases, as does the flux of electrons through the alternative pathway. Likewise, when ADP levels are high, the cytochrome pathway becomes more active. If the cytochrome pathway becomes saturated, it has been hypothesized that the alternative pathway becomes engaged to dissipate the excess electrons generated by Complex I and Complex II.

Understanding the role of oxidative phosphorylation in plant mitochondria is further confounded by the presence of chloroplasts and the contribution that photophosphorylation makes to the overall energy budget of the plant cell. Measuring respiration in photosynthetic tissues is extremely difficult due to other reactions that consume oxygen or produce carbon dioxide. While it is generally agreed that regulation of the biochemical reactions that occur in plant mitochondria are intimately tied to the activity of the photosynthetic apparatus, the exact nature of the interaction remains to be elucidated. Whereas past studies have concentrated on describing the biochemistry and bioenergetics of the mitochondrion and chloroplast, attention has focused more recently on regulation of activity in the different organelles of plant cell. What are the signals used to communicate between organelles and the cytoplasm? How are transcription and translation of nuclear and organelle genes coordinated? These and other questions provide a fruitful ground for continuing investigation into the nature of respiration and photosynthesis in plants.

PHOTORESPIRATION

Photorespiration is somewhat of a misnomer because it does not involve oxygen consumption by the mitochondrial electron transport system. It requires the coordinated activity of enzymes from the chloroplast, peroxysome, and mitochondrion as shown in Fig. 12.8. Oxygen is consumed in the chloroplast and peroxysome while CO_2 is liberated in the mitochondrion, but by an enzyme that differs from those associated with the TCA cycle. The process begins in the chloroplast by the oxygenase reaction of ribulose-1,5-bisphosphate carboxylase. This enzyme, normally involved in CO_2 assimilation, possesses an oxygenase activity whereby one molecule of 3-phosphoglycerate and one molecule of 2-phosphoglycolate are produced. Other enzymes in the pathway are discussed below.

Structure and Topology of Photorespiratory Enzymes

Phosphoglycolate phosphatase (EC 3.1.3.18)

Phosphoglycolate phosphatase catalyzes the dephosphorylation of 2-phosphoglycolate to yield glycolate and inorganic phosphate. This reaction occurs in the chloroplast and the glycolate produced by the reaction moves into the peroxysome. The enzyme has been studied in several species of *Panicum*, *Amaranthus caudatus* (tassel flower), tobacco, maize, barley, wheat, soybean, pea, spinach, and mung bean. The enzymes from spinach and tobacco have been found to dephosphorylate methyl phosphate and ethyl phosphate as well as 2-chloroethyl phosphate and 2-hydroxyethyl phosphate. A divalent metal cation is

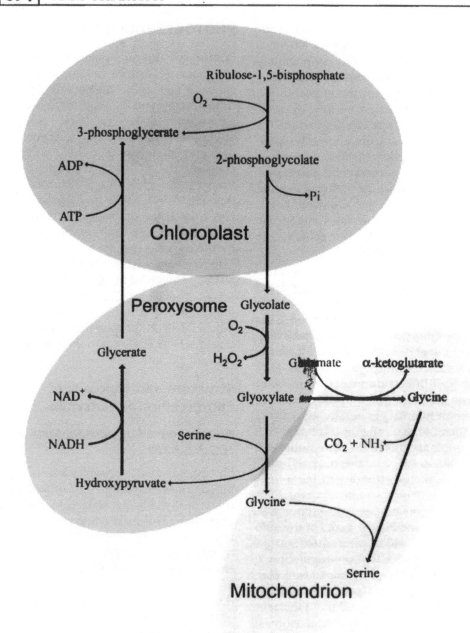

Fig. 12.8 Photorespiration. Photorespiration requires the coordinated activities of the chloroplast, peroxysome, and mitochondrion. It is initiated by the incorporation of molecular oxygen into ribulose-1,5-bisphosphate by ribulose bisphosphate carboxylase/oxygenase (Rubisco). Oxygen is also consumed in the peroxysome during conversion of glycolate to glyoxylate. In the mitochondrion, CO_2 is evolved during synthesis of serine from two molecules of glycine.

required for activity, usually Mg^{2+}, but at least in spinach, pea, and tobacco, Co^{2+}, Mn^{2+}, or Zn^{2+} can substitute for Mg^{2+}, but with lower activity.

At least two isozymes have been identified, each with slightly different sensitivity to inhibitors. Both isozymes appear to be inhibited by 6-phosphogluconate, glycolate, fructose-6-phosphate, p-mercuribenzoate, EDTA, cysteine, and inorganic phosphate. Glyoxylate, glycine, alanine and glutamate also inhibit isozyme II from kidney bean. Estimates of the size of the holoenzyme range from about 100 kDa in pea to about 60 kDa in maize. From SDS-PAGE experiments, the subunit composition is thought to be dimeric in pea and maize, but the tobacco enzyme was observed to be tetrameric, consisting of identical, or nearly identical, 21 kDa subunits.

Glyoxylate reductase (EC 1.1.3.15)

Glyoxylate reductase is located in the peroxysome and catalyzes oxidation of glycolate by using molecular oxygen, producing glyoxylate and hydrogen peroxide. Two molecules of hydrogen peroxide are rapidly decomposed to water and O_2 by the action of catalase. The enzyme has been studied in spinach and requires the flavin cofactor FMN. High concentrations (50 mM) of p-chloromercuribenzoate, iodoacetamide and iodoacetate have been found to be inhibitory. The enzyme belongs to the FMN-dependent alpha-hydroxy acid dehydrogenase family of enzymes, which includes lactate dehydrogenase, discussed above. Unlike lactate dehydrogenase, which contains a cytochrome b_2-binding domain, no metal ion requirements have been discovered for the enzyme. Instead, it appears that electrons from glycolate are transferred to oxygen by the flavin cofactor. The crystal structure of the enzyme from

spinach has been resolved and it appears to exist as a homotetramer or homooctomer of 40 kDa subunits.

Serine-glyoxylate aminotransferase (EC 2.6.1.45)

Glyoxylate is converted to glycine by the action of serine-glyoxylate aminotransferase. The amino group of serine is transferred to glyoxylate, producing glycine and 3-hydroxypyruvate. The enzyme has been studied from several plant sources, including spinach, cucumber, rye, barley, pea, tobacco, and kidney bean. Besides serine, the enzyme from several species can utilize alanine or asparagine as the amino-group donor, resulting in pyruvate or 2-oxosuccinamate as products. Like other aminotransferases, the enzyme requires a covalently bound pyridoxyl-5'-phosphate cofactor for activity. The enzyme from several species is inhibited by glyoxylate, but only in the absence of amino-group donors. In kidney bean, the enzyme is partially inhibited by low concentrations of ammonium (50 μM) and completely inhibited at 50 mM ammonium. Other inhibitors include hydroxylamine, n-ethylmaleimide, and sulfhydryl-group inhibitors. Interestingly, formaldehyde inhibits the enzyme from rye, but the inhibition could be prevented in the presence of glycolate. The holoenzyme consists of a dimer of 40–45 kDa subunits.

Glycine dehydrogenase (decarboxylating) (EC 1.4.4.2)

Glycine dehydrogenase, also called glycine decarboxylase, is a mitochondrial enzyme that catalyzes the first step in the conversion of glycine to serine. The enzyme is a multisubunit complex consisting of P-, H-, T-, and L- proteins. In the first step, glycine is oxidatively decarboxylated by the P-protein using NAD^+ and yields NADH,

CO_2, and the methylamine is transferred to the lipoamide group on the H-protein yielding a protein-bound aminomethyl dihydrolipoyl intermediate. The aminomethyl dihydrolipoyl moiety is transferred to the T-protein where the methylene group is bound to THF, 5,6,7,8-tetrahydropteroyl-polyglutamic acid (tetrahydrofolate) and the amino group is released as ammonia. In the final step, the dihydrolipoic acid is reoxidized by the L-protein and the methylene-THF utilized by serine hydroxymethyltransferase to synthesize serine in the next step. The enzyme has been studied in pea and wheat and requires a pyridoxal-5'-phosphate cofactor on the P-protein. The enzyme is inhibited by serine. The enzyme complex consists of a P-protein dimer of ~100 kDa subunits, H-protein (~14 kDa), T-protein (~40 kDa), and a dimer of ~50 kDa L-protein.

Glycine hydroxymethyltransferase (EC 2.1.2.1)

Glycine hydroxymethyltransferase, also known as serine hydroxymethyltransferase, catalyzes the synthesis of serine from methylene-THF, glycine, and water. The enzyme has been studied in pea, spinach, soybean and mung bean. The enzyme from most sources requires a pyridoxal-5'-phosphate cofactor, although this was not found to be the case with the mung bean enzyme, which had no such requirement. The mung bean enzyme was found to be stimulated in the presence of NADH. Inhibitors of the enzyme include 4-chloro-L-threonine, aminopterin and other tetrahydrofolate derivatives, methotrexate and dichloromethotrexate, KCl, NaCl, and a variety of amino acids. The holoenzyme exists as a homodimer or homotetramer of ~50 kDa subunits.

Hydroxypyruvate reductase (EC 1.1.1.81)

Hydroxypyruvate reductase catalyzes the NADH-dependent reduction of hydroxypyruvate to form D-glycerate and NAD^+. This enzyme is located in the peroxysome and is involved in the regeneration of 3-phosphoglycerate. An isozyme of the reductase, which shows a preference for NADPH as the reductant, is located in the cytosol and is not involved in photorespiration. The glycerate produced in this reaction moves into the chloroplast, where it is phosphorylated by glycerate kinase. The enzyme has been studied in *Cucurbita pepo* (summer squash), cucumber, spinach, barley, watermelon and maize. No cofactors or metal ion requirements have been identified for this enzyme. Inhibitors include phosphohydroxypyruvate, oxalate, and tartronate. The holoenzyme is composed of a dimer of 38 kDa similar or identical subunits.

Glycerate kinase (EC 2.7.1.31)

Glycerate kinase catalyzes the ATP-dependent phosphorylation of glycerate to regenerate 3-phoshoglycerate and ADP. This reaction occurs in the chloroplast using glycerate from the previous step that has moved from the peroxysome to the chloroplast. It generates 3-phosphoglycerate that can be used in the photosynthetic Calvin-Benson cycle. The enzyme has been studied in field mustard, wheat, rye, barley, spinach, and maize. As with many other members of the kinase family of enzymes, glycerate kinase can utilize GTP or UTP as the phosphate donor, although with reduced efficiency compared to ATP. The enzyme from field mustard has an absolute requirement for Mg^{2+}, but in the enzyme from other species, Mg^{2+} can be replaced by

Ca^{2+}, Co^{2+}, Mn^{2+} or Sr^{2+} with no apparent loss of activity. Inhibitors of the enzyme include 3-phosphoglycerate in spinach and maize and pyruvate in maize and field mustard. The enzyme from field mustard has been shown to be inhibited by a wide variety of metabolic intermediates including pyrophosphate and glycolate, and fructose-6-phosphate, glucose-1-phosphate, and glucose-6-phosphate, which only partially inhibit the enzyme. The holoenzyme is a monomer with a size about 40 kDa in spinach to about 47 kDa in maize.

Bioenergetics and Regulation of Photorespiration

Photorespiration inevitably leads to a loss of primary productivity in the plant. Every molecule of oxygen incorporated into rubulose-1,5-bisphosphate as a result of the oxygenase activity of Rubisco decreases the number of CO_2 molecules that can be fixed by the enzyme. There is an energetic cost as well in that close to 10 equivalents of ATP are lost as a result of O_2 fixation. Under experimental conditions, a doubling of biomass was seen when growing plants at low oxygen levels (5%) compared to normal levels (21%). Why would plants allow such a wasteful process to occur? While the process may be wasteful, it is not completely futile. Of all the carbon diverted away from the Calvin-Benson cycle in the form of 2-phosphoglycerate, 75% is recovered and delivered to the chloroplast. It may be that photorespiration provides a way to dissipate photosynthetic energy when CO_2 becomes limiting. Under high light conditions and when CO_2 availability is low, photoinhibition of the photosynthetic apparatus occurs due to a slowdown of carbon fixation. The oxygenase reaction could serve as a means to prevent rubulose-1,5-bisphosphate levels from becoming elevated to the point of inhibition of photo-synthetic electron transfer, a condition that results in photoinhibition.

It may also be that it is an inevitable consequence of the catalytic mechanism of Rubisco. All known plant Rubisco enzymes catalyze the oxygenase reaction and the enzymes of the photorespiratory pathway in plants may have evolved in order to salvage a significant portion of the carbon diverted through the pathway. This idea is reinforced by the observation that cyanobacteria and green algae do not have fully functional photorespiratory pathways.

CONCLUDING REMARKS

The metabolic pathways involved with respiration in plants are much more complex than those in animals are. Animals do not possess the glyoxylate pathway nor do they need to accommodate the results of photorespiration, as do plants. While many of the metabolic pathways of catabolism and anabolism are identical in plants and animals, there are many enzymes and biochemical pathways unique to plants. Indeed, integration of photosynthesis with other metabolic pathways in a cell requires a sophisticated level of communication among the various compartments of the cell. The mitochondria and chloroplast must "talk" to each other about their relative activities and also communicate with the nucleus in order to coordinate gene expression.

The nature of this communication is only just beginning to be understood. Significant progress has been made in recent years in understanding the metabolic consequences of an experimentally induced imbalance in the chemical equilibrium within a cell. The focus of research is now directed toward understanding the signaling pathways involved in the inter- and intracellular regulation of metabolic activity.

Dedication

This chapter is dedicated to the memory of Professor Lee McIntosh. Lee was a mentor, colleague, and friend. He made significant contributions to our understanding of alternative oxidase and was just beginning to unravel the intricacies of communication between energy transducing organelles, and how metabolites are regulated in the plant cell.

Further Reading

Beevers, H. 1961. *Respiratory Metabolism in Plants.* Row, Peterson & Co., Evanston, IL, USA.

Bolton, J., Mataga, N. and McLendon, G. (eds.). 1991. *Electron Transfer in Inorganic, Organic, and Biological Systems.* Amer. Chem. Soc., Washington, DC.

Dennis, D.T. and Turpin, D.H. 1990. *Plant Physiology, Biochemistry and Molecular Biology,* Longman Sci. Tech., Essex, UK.

Devlin, R.M. and Witham, F.H. 1983. *Plant Physiology.* Willard Grant Press, Boston, MA, USA (4th ed.).

Douce, R. 1985. *Mitochondria in Higher Plants. Structure, Function, and Biogenesis.* Acad. Press, Inc., Orlando, FL, USA.

Douce, R. and Day, D.A. (eds.). 1985. *Higher Plant Cell Respiration.* Springer-Verlag, Berlin, Germany.

Forward, D.F. 1965. The respiration of bulky organs. In: *Plant Physiology, A Treatise,* vol. IVA, pp. 311-376. Steward, F.C. (ed.). Acad Press, New York, NY.

Foyer, C.H. and Quick, W.P. (eds.). 1997. *A Molecular Approach to Primary Metabolism in Higher Plants.* Taylor and Francis, London, UK.

Krogmann, D. 1973. *The Biochemistry of Green Plants.* Prentice-Hall, Inc., Englewood Cliffs, NJ, USA.

Kruger, N.J., Hill, S.A., and Ratcliffe, R.G. (eds.). 1999. *Regulation of Primary Metabolic Pathways in Plants.* Kluwer Acad. Publ., Dordrecht, The Netherlands.

Lambers, H. and van der Plas, L.H.W. (eds.). 1992. *Molecular, Biochemical and Physiological Aspects of Plant Respiration.* SPB Acad. Publ., The Hague, The Netherlands.

Öpik, H. 1980. *The Respiration of Higher Plants.* Edward Arnold (Publ.), Ltd., London, UK.

Palmer, J.M. 1984. *The Physiology and Biochemistry of Plant Respiration.* Cambridge Univ. Press, Cambridge, UK.

Stiles, W. and Leach, W. 1932. *Respiration in Plants.* Methuen & Co., Ltd., London, UK.

Yemm, E.W. 1965. The respiration of plants and their organs. In: *Plant Physiology, A Treatise,* vol. IVA, p. 231-310. Steward, F.C. (ed.). Acad. Press, New York, NY.

Database URL:

Comprehensive Enzyme Information System (BRENDA) at the Cologne University Bioinformatics Center: http://www.brenda.uni-koeln.de

Expert Protein Analysis System at the Swiss Institute of Bioinformatics: http://au.expasy.org

CHAPTER 13

Photosynthesis

J. Kenneth Hoober

INTRODUCTION

Sunlight provides a steady, abundant supply to a world increasingly concerned about energy. The cause of this concern is a disconnect between inputs and outputs of energy forms. Forms of energy traditionally used to drive our mechanical devices differ from those that support living systems. Nuclear energy, fraught with waste problems, hydroelectric, geothermal, or wind power, are all forms that can be used for mechanical devices but provide a minor fraction of the total energy input needed in the world economy. The bulk of the available energy has been, and still is, derived from light energy from the sun. Technology has been developed to convert photons to electrons with photovoltaic cells that use absorption of light energy by silicon semiconductor devices to generate electricity as the end product. This approach has great promise as a sustainable energy source but thus far does not compete economically with fossil fuels. Combustion of fossil fuels (coal, oil, and natural gas) is currently the major source of energy used to generate electricity and power machines. Fossil fuels are products derived from **ancient photosynthesis** but, with few exceptions, are not directly usable by living systems. Life is also sustained by converting photons to electrons but through **contemporary photosynthesis**. Except for organisms that live next to thermal vents in the deep oceans and those that can metabolize hydrocarbons, life is sustained by carbohydrates produced by photosynthesis. The major end product of photosynthesis is **glucose 6-phosphate**, the predominant starting material for storage forms of chemical energy and for synthesis of cellular components. When glucose 6-phosphate is metabolized in cells, the energy that was trapped in the molecule by photosynthesis is released to provide the energy required for synthetic reactions and growth. In contrast to animals, **glucose** does not occur in the free form to a significant amount in plants. Of fundamental importance to the biosphere however, is the fact that plants are the primary source of the glucose unit, which they store in polymeric form.

Aerobic metabolism of glucose in all organisms produces CO_2 and H_2O. Persistence of life on earth would not have been possible without a means to convert these "waste products" back to glucose. Otherwise, living organisms would long ago have run out of food. Conversion of CO_2 and H_2O

to 6-carbon molecule glucose requires an input of a substantial amount of energy, as illustrated by the relationship:

$$CO_2 + H_2O + [470 \text{ kJoules per mole of } CO_2] \leftrightarrow 1/6 (C_6H_{12}O_6) + O_2 \quad ...(1)$$

The energy required for complete synthesis of a mole of glucose is six times that shown in eqn (1). The source of energy to drive this reaction is light. The molecular apparatus that performs this process—photosynthesis—is located in the "**thylakoid**" (Greek, "sac-like") **membrane**, which is confined within **chloroplasts** in plants. Light energy is absorbed by molecules of **chlorophyll** (Chl) located in this membrane. Wavelengths of light that contain levels of energy that can be accommodated by the electronic structure of Chl, primarily in the blue and red regions of the visible spectrum (photons with energy levels around 280 and 170 kJoules per mole respectively), are absorbed and generate "excited" states of Chl. From the energy content of these useful wavelengths, it is obvious that more than one photon must be absorbed to fix one molecule of CO_2. The photosynthetic apparatus is designed to collect these packets of light energy and convert them to chemical compounds with sufficiently high levels of energy to drive synthesis of glucose from CO_2 and H_2O. The major end products stored in plants are **starch**, a polymeric form of glucose, or **sucrose**, a disaccharide.

Expanded descriptions of various aspects of photosynthesis can be found in Raghavendra (1998), Ke (2001), Lawlor (2001), Blankenship (2002), and Archer and Barber (2003).

EVOLUTION OF PHOTOSYNTHESIS

Synthesis of Tetrapyrroles

Photosynthesis is nearly as old as life itself. Measurements of the carbon-isotope content in early fossils suggest that a reaction similar to that catalyzed by the enzyme **ribulose 1,5-bisphosphate carboxylase/oxygenase** ("Rubisco") fixed CO_2 into organic material. Because this reaction involves CO_2 as a reactant, CO_2 molecules containing the heavier isotopes of carbon, i.e., ^{13}C and ^{14}C, react slightly more slowly than $^{12}CO_2$. Thus direct incorporation of CO_2 into organic molecules discriminates against the heavier isotopes. A measurable extent of this type of discrimination, resulting in enrichment in ^{12}C, is thought to be possible only with enzymatically catalyzed reactions, and thus isotope ratios have been used as a signature of biological activity. Microfossils with such a signature are present in the oldest rocks on earth, possibly older than 3.8 billion years (Mojzsis et al., 1996).

Tetrapyrroles, required cofactors in photosynthesis, are among the most ancient of biological molecules (see Fig. 13.1 for structures). Availability of cyclic, or rather "macrocyclic", tetrapyrroles that chelate a divalent cation—most commonly a Fe^{2+} or Mg^{2+} ion—allowed development of energy transduction mechanisms, either through electron transport (oxidation-reduction) of the central iron atom in heme or generation of high-energy, excited states of Mg tetrapyrroles, as with the Chls. Phylogenetic evidence indicates that photosynthesis emerged within the purple bacterial lineage somewhat earlier than 3.5 billion years ago (BYA) (Xiong et al., 2000). As deduced from a phylogenetic analysis of genes encoding biosynthetic enzymes, bacteriochlorophyll (BChl) (see Fig. 13.1) was the evolutionary precursor of Chl. Comparison of genomic sequences provided evidence that these genes were subsequently introduced into green sulfur bacteria, green nonsulfur bacteria, and cyanobacteria by lateral gene transfer.

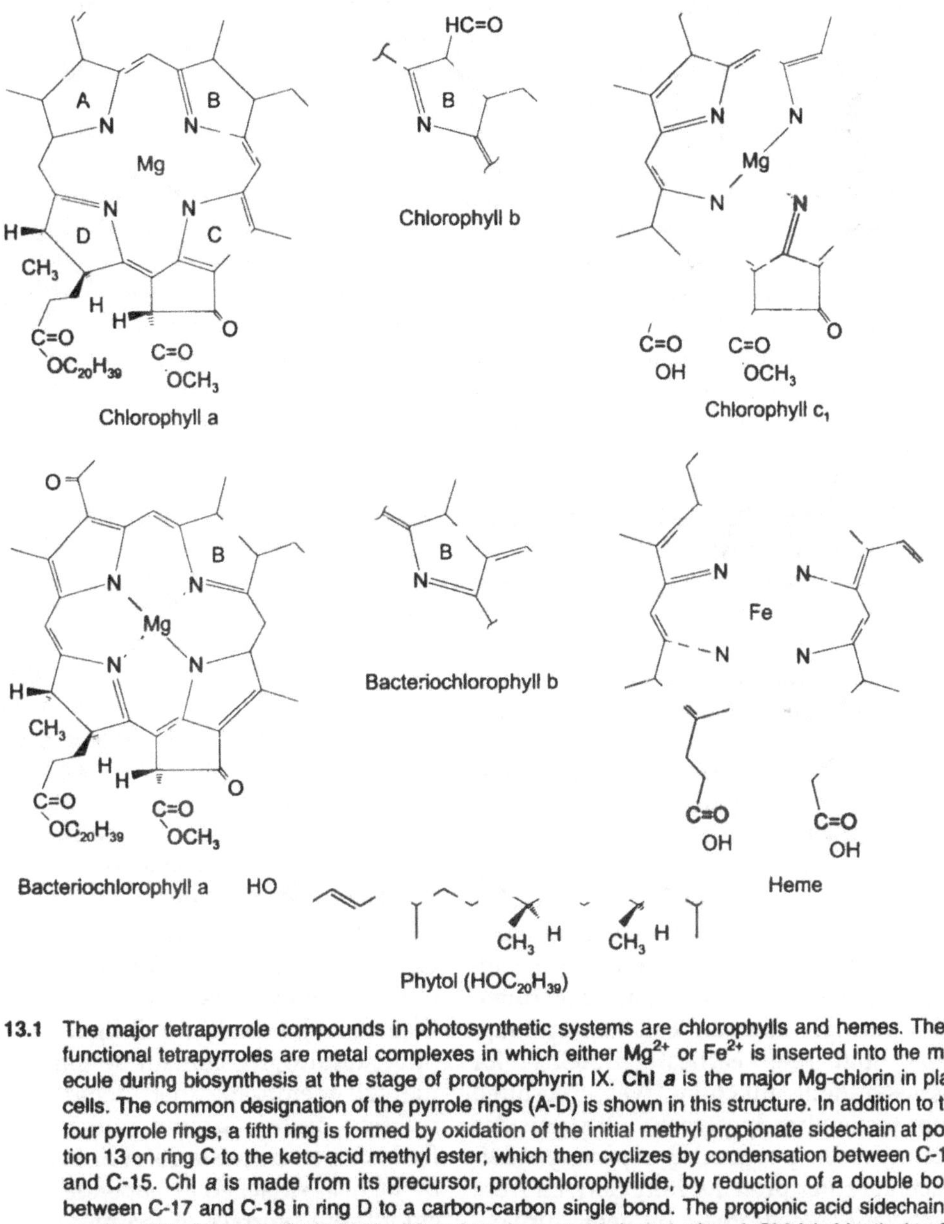

Fig. 13.1 The major tetrapyrrole compounds in photosynthetic systems are chlorophylls and hemes. These functional tetrapyrroles are metal complexes in which either Mg^{2+} or Fe^{2+} is inserted into the molecule during biosynthesis at the stage of protoporphyrin IX. **Chl a** is the major Mg-chlorin in plant cells. The common designation of the pyrrole rings (A-D) is shown in this structure. In addition to the four pyrrole rings, a fifth ring is formed by oxidation of the initial methyl propionate sidechain at position 13 on ring C to the keto-acid methyl ester, which then cyclizes by condensation between $C-13^2$ and C-15. Chl a is made from its precursor, protochlorophyllide, by reduction of a double bond between C-17 and C-18 in ring D to a carbon-carbon single bond. The propionic acid sidechain at position 17 is then esterified with the 20-carbon, isoprenoid alcohol, **phytol**. Chl b is identical to Chl a except that the methyl group at position 7 in ring B is oxidized to an aldehyde group. **Chl c** is similar to protochlorophyllide except that the propionate sidechain at position 17 is oxidized to the *trans*-acrylate form, which remains unesterified. Addition of the double bond extends the conjugated macrocyclic π system to the electronegative carboxyl group. **BChl a** is the major chlorophyll-type pigment in most anoxygenic photosynthetic bacteria. It is similar to Chl a but also has the double bond between C-7 and C-8 in ring B reduced, and the vinyl group at position 3 is oxidized to an acetyl group. The carbon-carbon single bonds in ring B as well as ring D are the primary characteristics of BChl. **Heme** is Fe-protoporphyrin IX. In cytochromes it functions by oxidation and reduction of the central iron ion.

Of the two types of photosynthetic reaction centers, type I, with an iron-sulfur complex as the electron acceptor, emerged among the heliobacterial and green sulfur bacterial lines. Type II, with a quinone as the electron acceptor, emerged among the purple bacteria. The two photosystems were combined for the first time in cyanobacteria by lateral gene transfer. Cyanobacteria were also the first to use Chl rather than BChl as the primary pigment. The redox potential of Chl is higher than that of BChl, and consequently, with the higher energies achieved by excited states of Chl, it became possible to span the difference in redox potential from oxidation of water to oxygen ($E_m' = +0.816V$) to reduction of $NADP^+$ to NADPH ($E_m' = -0.342$ V). This development was achieved about 3.5 BYA. The result was generation of molecular oxygen which, however, did not appear at a significant level in the atmosphere until about 2.3 BYA. Interestingly, the genes for reaction center proteins that *bind* (B)Chl have a deeper lineage than those that encode the enzymes that catalyze *synthesis* of (B)Chl. Lack of congruence of the genes that encode the BChl/Chl biosynthetic enzymes and the reaction center proteins that bind these pigments indicates that the photosynthetic apparatus is a composite structure, with components recruited from multiple sources. This concept suggests that the earliest proteins for photosynthetic functions were recruited from those already present. Of particular interest is the proposal that *photosynthetic* reaction center proteins were derived from *respiratory* cytochrome *b* by gene duplication and subsequent divergence of the genes to encode proteins with new functions, from one that bound Fe-porphyrin cofactors (the heme in cytochromes) to one that bound Mg-chlorin (Chl) cofactors (Xiong and Bauer, 2002).

Evolution of Photosynthetic Membrane and Light-harvesting Complexes

The essential functional structure for photosynthesis is a **membrane** that (a) physically separates two different compartments and (b) contains an energy-transducing apparatus. The membrane allows development of an electrochemical gradient, which is an essential intermediate in photosynthesis. Within the membrane reside the reaction centers and a series of electron transfer components. Photochemistry occurs in the reaction centers, initiated by the absorption of light by Chl. The flux of photons, even in full sunlight, is sufficient for only a few photons to be absorbed by a Chl molecule in a reaction center per second, a rate that is much too slow for productive photosynthesis. Therefore, photosynthetic organisms also developed structures that contain a large number of accessory pigment molecules that harvest light energy, *i.e.*, increase the absorptive cross-section of the system, and funnel the energy into reaction centers. As a consequence, sufficient high-energy states are generated to efficiently drive synthetic reactions.

Different organisms have developed quite different means to increase absorption of light energy. **Green sulfur bacteria** (*e.g.*, *Chlorobium*), which are obligate *anaerobic* photoautotrophs, contain domains within the cell membrane that include a single reaction center, around which electrons flow, driven by absorption of light energy by BChl *a*. Attached as appendages on the cytoplasmic surface of the photosynthetic domains are specialized antenna complexes, which are nearly pure aggregates of up to 10,000 molecules of BChl *c*, *d*, or *e*, called "**chlorosomes**" (Olson, 1998). Light energy absorbed by chlorosomes is transferred to reaction centers through the

FMO complex (after Fenna and Matthews, who first determined its structure, and Olson, who discovered the protein), which is a trimeric arrangement of a protein subunit that binds 7 molecules of BChl *a* (Li *et al.*, 1997). The reaction center in green sulfur bacteria has an iron-sulfur complex as the terminal electron acceptor, similar to photosystem I in plants.

The facultative photoheterotrophic **purple bacteria** (e.g., *Rhodobacter* and *Rhodopseudomonas*) induce synthesis of BChl and expression of genes encoding photosynthetic proteins, and thus grow photosynthetically only in response to low oxygen tension (Drews, 1996). The amount of BChl in these cells is severalfold higher under low, compared to high, light intensity, a reflection of the increased need for light-harvesting capacity. Expansion of the cell membrane through assembly of photosynthetic domains produces extensive invaginations into the interior of the cell. These invaginations form a photosynthetic "intracytoplasmic membrane" system, which usually remains continuous with the respiratory plasma membrane. When cells are broken, the intracytoplasmic membranes fragment into small vesicles called **"chromatophores"**. Expansion of membrane surface area in these organisms dramatically increases the capacity for photosynthetic activities and is required to accommodate the light-harvesting complexes that are integral pigment-protein complexes within the membrane structure itself. **Light-harvesting complex I** (LH-I) surrounds the single type of reaction center, which in this case has a quinone as the electron acceptor similar to photosystem II in plants (Deisenhofer and Michel, 1989) (Fig. 13.2). LH-I is composed of 16 α,β-heterodimer polypeptide subunits, with each polypeptide binding one BChl *a* molecule, and forms an annular structure that

encircles the reaction center (Cogdell *et al.*, 1999). The ring is slightly distorted by a single copy of PufX, the product of the *pufx* gene, which apparently facilitates access of electron carriers such as quinone molecules to the reaction center. In this environment, the association with proteins causes the absorption maximum of BChl *a* in LH-I to shift from its maximum of 773 nm in an organic solvent such as diethyl ether to 875 nm. LH-II, an additional, more peripheral light-harvesting complex, is present at varying amounts. LH-II is also a ring-like structure, composed of 9 α,β-heterodimers of short polypeptides, 53 and 41 amino acid residues long respectively. The N-terminus of each of the 9 α subunits is modified to the N-*carboxy*-methionine residue, which coordinates with a BChl *a* molecule oriented parallel with the membrane near the periplasmic surface. This complex maximally absorbs 800 nm light. The imidazole group of a histidine residue in each α and β peptide coordinates one BChl *a* molecule, which creates a ring of 18 BChl molecules near the cytoplasmic surface of the membrane, oriented perpendicular to the plane of the membrane. These LH-II BChl molecules absorb maximally at 857 nm. Because photons of shorter wavelengths contain more energy than those of longer wavelengths, and excited molecules transfer energy usually to those at a lower energy level, this arrangement of the purple bacterial light-harvesting antenna provides efficient "downhill" transfer of energy, from B800 and B857 in LH-II to B875 in LH-I to the reaction center (Papiz *et al.*, 2003).

Cyanobacteria have an extensive photosynthetic **thylakoid membrane** system that forms concentric layers inside, but separate from, the plasma membrane. Cyanobacteria were the earliest organisms to perform oxygen-producing photosynthesis, with two reaction centers operating in

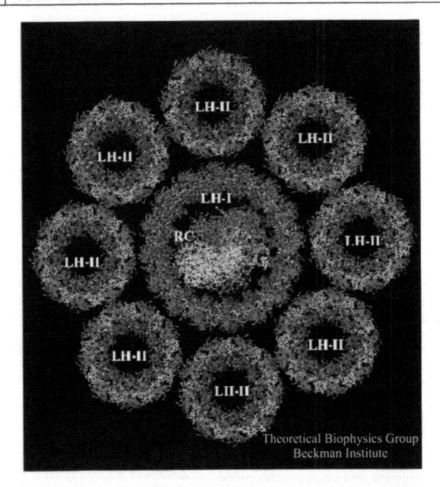

Fig. 13.2 Model of the structure of the photosynthetic apparatus in a nonsulfur purple bacterium. The reaction center (RC) is enclosed by the core light-harvesting complex, LH-I. Surrounding the core complex are the peripheral light-harvesting complexes, LH-II, present in variable amounts depending on growth conditions. Each LH-II complex contains nine BChl *a* molecules that maximally absorb at 800 nm and 18 BChl *a* molecules that absorb at 857 nm. Energy is transferred to the LH-I complex, which contains BChl *a* that absorbs at 875 nm. The reaction center pair of BChl *a* molecules absorb at 870 nm (from Hu *et al.*, 2002).

series. One reaction center, **photosystem (PS) II**, is the quinone type (from the purple bacterial lineage), while the second, **PS I**, is the iron-sulfur type (from the green sulfur bacterial lineage). The major light-harvesting function in cyanobacteria is provided by peripheral, exquisitely designed complexes of proteins that contain covalently bound, linear tetrapyrrole chromophores (phyco-bilins). These complexes, called **"phycobilisomes"**, are assembled from the subunits phycoerythrin (λ_{max} 565 nm), phycocyanin (λ_{max} 620 nm), and allophycocyanin (λ_{max} 650 nm), named according to their reddish or bluish color respectively. The arrangement of these subunits, as illustrated in Fig. 13.3, also provides the "downhill" flow of excitonic

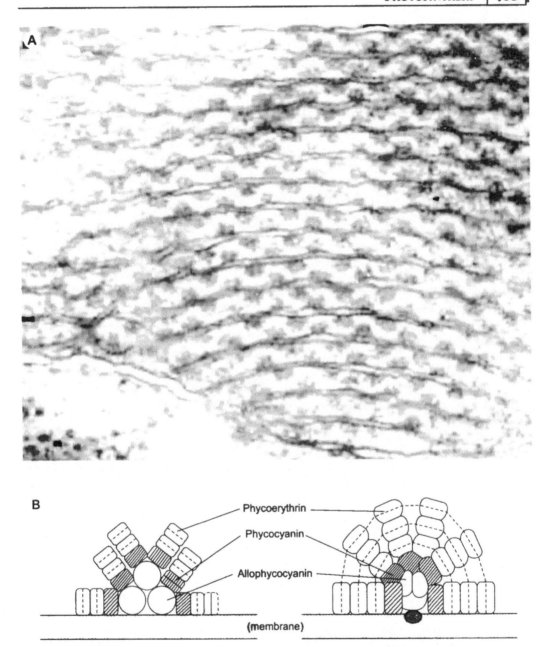

Fig. 13.3 Phycobilisomes are light-harvesting complexes in cyanobacteria and red algae. (**A**) The electron micrograph of a portion of the chloroplast of a red alga shows a parallel array of thylakoid membranes with interdigitating phycobilisomes on the stromal surface. (**B**) Arrangement of phycobiliproteins within the structure of a phycobilisome from a cyanobacterium (*left*) or a red alga (*right*) (adapted from Gantt, 1981: micrograph courtesy of Dr. Elizabeth Gantt).

energy from the highest energy chromophore (shortest wavelength λ_{max}) to the reaction center, which is necessary for efficient trapping of light energy (Gantt, 1981).

Origin of Chloroplast in Eukaryotic Cells

Eukaryotic organisms evolved about 2.7 BYA. An endosymbiotic event, in which an early eukaryotic cell engulfed a prokaryotic cell, occurred about 1.9 BYA that led to development of mitochondria. Extensive evidence supports the appearance of photosynthesis in eukaryotic cells as a second endosymbiotic event, in which a cyanobacterium was productively trapped nearly 1.6 BYA (Yoon et al., 2004; Hedges et al., 2004). This event defined the divergence of animals and plants. It seems that rather quickly thereafter the second endosymbiont evolved into the **chloroplast** and most of the genes in the original cyanobacterial cell were transferred to the nucleus or were discarded. Extensive analysis of cyanobacterial genes in the nucleus of the model plant *Arabidopsis thaliana* indicated that about 18% of the total nuclear genes of this plant are derived from the endosymbiont (Martin et al., 2002). Although chloroplasts contain up to 100 *copies* of the residual **chloroplast genome**, the number of *different* genes that encode proteins in most plant chloroplast genomes is in the range of 70 to 80, only a few percent of that in a modern-day cyanobacterium (Timmis et al., 2004). Analysis of the sequences of the remaining genes, in particular those encoding the large subunit of ribulose 1,5-bisphosphate carboxylase/oxygenase and ribosomal RNA in algae and plants has shown that the plastids in all photosynthetic eukaryotes have a monophyletic lineage (Bhattacharya and Medlin, 1995). It is remarkable that the endosymbiotic event that led to evolution of the plastid seems to have occurred only one

time. This is very close to not happening at all, or it happened within a population of cyanobacterial organisms that had substantial similarity and the ability to take advantage of a unique but temporary environment. The rather high uniformity of the chloroplast genome in nearly all plants, with the exception of some rearrangements, supports the single-event hypothesis rather than multiple events. Consequently, chloroplasts have features similar to cyanobacteria, with a separate, extensive, internal thylakoid membrane system. Of course, because genes encoding most of the plastid proteins are now expressed in the nucleus, an elaborate system is required to import these proteins into the plastid after synthesis in the cytosol.

Many algal cells contain a single chloroplast (Fig. 13.4), but mature plant cells usually contain numerous—in some cases up to as many as 150—plastids (Fig. 13.5). These organelles are surrounded by a two-membrane **envelope**, which contains specific transport systems through which proteins and metabolites pass. The outer membrane contains pores that are less discriminatory than the more specific transport systems that reside in the inner membrane (Ferro et al., 2002). Envelope membranes are tightly appressed, such that transport complexes on the outer and inner membranes provide continuous passageways. Current evidence indicates that the outer membrane of the chloroplast envelope resembles more the outer bacterial membrane than the endocytic membrane that presumably enclosed the engulfed bacterial cell. If the progenitor of the chloroplast were similar to modern cyanobacteria, which are "Gram-negative" organisms that have a structural wall sandwiched between two membranes surrounding the cell, then evolution of the plastid envelope involved loss of the cell wall. Primitive algae, such as

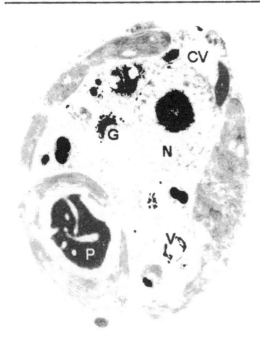

Fig. 13.4 Electron micrograph of a cell of green alga *Chlamydomonas reinhardtii cbn1-113*. The algal cell contains a single cup-shaped chloroplast (c). Within the base of the chloroplast is a structure called the pyrenoid (p), a condensed form of Rubisco, the enzyme that fixes CO_2 into organic compounds. Other organelles, typical of a eukaryotic cell, include the nucleus (n), Golgi apparatus (G), small digestive vacuoles (v) and a contractile vacuole (cv) (from Park *et al.*, 1999; micrograph courtesy of Dr. Hyoungshin Park).

encoding **Chl *a* oxygenase**, the enzyme that catalyzes oxygenation of Chl *a* to make Chl *b*, also has a monophyletic lineage, with the gene in the prokaryotic prochlorophyte organisms such as *Prochloron* and *Prochlorococcus* as the most ancient (Tomitani *et al.*, 1999; Hess *et al.*, 2001). Chl *a* oxygenase activity apparently was present in the early endosymbiont, long before the plant-type **light-harvesting** Chl *a/b*-binding proteins (LHCPs) developed. Tomitani and colleagues (1999) suggested that the ancestral prokaryotic endosymbiont contained both phycobilisomes and a prochlorophyte-type light-harvesting complex containing Chl *b*. Subsequently, ancestral photosynthetic eukaryotes would have diverged into either (a) organisms that lost the ability to make phycobiliproteins but gained the source of the abundant family of Chl *a/b*-binding proteins that provide the antenna in green plants or (b) the ancestors of red algae that retained the ability to make phycobilisomes but lost synthesis of Chl *b*.

Many species of algae are products of "secondary" endosymbiotic events, in which an entire eukaryotic green or red alga was engulfed by another nonphotosynthetic eukaryotic cell (Raven and Allen, 2003). These latter algae contain three or four membranes surrounding the plastid—the two of the original chloroplast envelope, the derivative of the original algal cell membrane, and the endocytic membrane of the secondary host. Most of these species contain Chl *c* instead of Chl *b* (see Fig. 13.1). Such secondary endosymbiosis explains most of algal diversity (Yoon *et al.*, 2002). Even more remarkable are organisms among the dinoflagellates that are products of a third, "tertiary" endosymbiosis (Stoebe and Maier, 2002). Although secondary endosymbiotic events occurred many times, with host organisms with diverse lineages,

Cyanophora, contain chloroplast-like organelles termed **"cyanelles"** that have a prokaryotic-like peptidoglycan layer between the inner and outer membranes of the organelle, which is thought to be a relic of the cell wall of their cyanobacterial ancestor (Pflanzagl *et al.*, 1996).

Cyanobacteria contain only Chl *a*, whereas plants and their algal ancestors in the order Charales, and the eukaryotic green algae belonging to Chlorophyta, contain Chl *b* in addition to Chl *a*. The gene

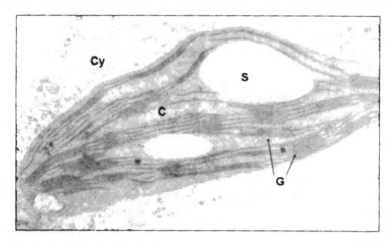

Fig. 13.5 Electron micrograph of a chloroplast in a young leaf of tobacco. Each cell contains numerous chloroplasts, perhaps a hundred or more, distributed around the periphery near the cell membrane. Thylakoid membranes within the chloroplast (C) are arranged in stacks, designated grana (G), connected by stromal lamellae. The chloroplast is surrounded by a double-membrane envelope. Starch grains (S), the final storage form of the products of photosynthesis, accumulate in the stroma. A small portion of the large vacuole that occupies the center of the cell is visible at the bottom of the micrograph. A portion of the cytosol (Cy) is shown at the top of the micrograph. (Micrograph courtesy of Dr. Hyoungshin Park).

the plastids themselves all show a single, i.e., monophyletic, origin. This chapter emphasizes photosynthesis in green plants, including algae in Chlorophyta.

DEVELOPMENT OF CHLOROPLAST

Maturation of Chloroplast

In seed plants, the plastid begins development during seed germination as a simple, double-membraned vesicle called a "**proplastid**". In contrast to most single-celled photosynthetic organisms, which are able to synthesize Chl and the photosynthetic apparatus in darkness, plants require light for conversion of protochlorophyllide (Pchlide) *a* to chlorophyllide (Chlide) *a* (the '-ide' suffix indicates absence of esterified isoprene alcohol). This reaction involves stereochemical reduction of a double bond between carbon-17 and carbon-18 in ring D (see Fig. 13.1) by reduced **nicotinamide adenine dinucleotide phosphate** (NADPH),

catalyzed by Pchlide-NADPH oxidoreductase (POR). In plants, this reaction requires absorption of light energy by the substrate, Pchlide. Three forms of POR occur: PORA preferentially binds Pchlide *b* and PORB has a higher affinity for Pchlide *a* (Reinbothe *et al.*, 2003). PORA is abundant in plastids ("etioplasts") in plants grown in the dark but is rapidly degraded upon exposure of the plants to light. PORB is induced by light and is present during chloroplast development. These forms provide most of the Chl. PORC is a newly discovered form that is present at a lower abundance, whose function is not known (Pattanayak and Tripathy, 2002).

The process by which the thylakoid system forms depends on the conditions under which the plant cell is grown. In seedlings exposed early to light, initial thylakoid membranes form in the proplastid by invagination of the inner membrane (von Wettstein, 2001; Gunning, 2004). Thus, photosynthetic domains are expelled as vesicles from the envelope by a series of accessory proteins involved in vesicle traffic. The vesicles then fuse to form and expand the developing thylakoid system. A mutant of *Arabidopsis thaliana*, deficient in an activity designated vesicle-inducing plastid protein 1 (VIPP1), is unable to induce vesicle formation from the plastid envelope and consequently does not make thylakoid membranes (Kroll *et al.*, 2001). Formation of vesicles has been observed during the initial stage of chloroplast development (a process commonly referred to as greening) of the alga *Chlamydomonas reinhardtii* (Fig. 13.6) and in plant leaves treated with inhibitors of vesicle fusion. These results provide direct demonstrations of the envelope as the source of material for thylakoid membranes. This process has also been demonstrated by application of specific inhibitors of vesicle traffic to isolated chloroplasts (Westphal *et al.*, 2001).

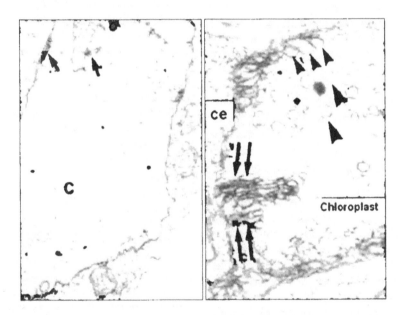

Fig. 13.6 *Left panel:* portion of the chloroplast of a dark-grown, degreened cell of the alga *C. reinhardtii* y1. The chloroplast (C) is depleted of thylakoid membranes, which were diluted as the cell continued to grow and divide. *Right panel:* portion of the chloroplast of a dark-grown, degreened cell of alga *C. reinhardtii* y1 after 15 min of greening in light. *Arrowheads* point to vesicles that appear to be derived from the chloroplast envelope in response to exposure to light. *Arrows* point to an extensive invagination of the chloroplast envelope (ce). The image suggests that development of the thylakoid membrane system occurs with membrane vesicles emanating from the envelope (adapted from Hoober *et al.*, 1991).

Alternatively, when seedlings are grown from germination in darkness, the resultant "**etioplasts**" contain a highly organized, tubular membrane structure called the "**prolamellar body**". PORA is the predominant protein in the prolamellar body and exists in a complex with its substrates, Pchlide and NADPH. Galactosyl diglycerides are the major lipids of this structure. Exposure of these seedlings to light results in reduction of Pchlide to Chlide *a* and dispersal of the prolamellar body into "prothylakoid" membranes. These rudimentary membranes expand by addition of proteins and lipids. Some proteins, such as those in the core complex of photosynthetic units, are made within the plastid, while others, in particular the proteins of the light-harvesting antennae, are imported after synthesis in the cytosol. Lipids are synthesized predominantly on envelope membranes, and the envelope is also the location of the latter steps in Chl and carotenoid synthesis. Thus the envelope is an important interface between the chloroplast and cytosol and serves as the platform for biogenesis of the extensive thylakoid membrane. In plant cells exposed to light, membrane formation occurs over several hours to several days, depending on the organism, to achieve the mature chloroplast.

The 3-dimensional arrangement of membranes in the chloroplast of plants has been established by electron microscopy. The thylakoid membrane, separate from the inner envelope membrane, is differentiated into cylindrical stacks of "appressed" membranes, designated **grana**, that are interconnected with "unappressed" stromal membranes (Figs. 13.5 and 13.7). The highly elaborated, folded membrane system encloses a single, continuous lumen (Mustárdy and Garab, 2003). This arrangement seems to maximize efficiency of the overall process. Thylakoid membranes in algal cells are less differentiated, and in many cases are appressed over much of their surface (see Fig. 13.4). Most important for the process of photosynthesis is formation of the luminal compartment, as will be explained below.

Role of Chlorophyll in Chloroplast Development

Chl is required not only for photosynthesis, but also for assembly of the photosynthetic apparatus and, to a large extent, formation of the thylakoid membrane. The two major forms in green plants are Chl *a* and Chl *b* (see Fig. 13.1). Chls are "macrocyclic," highly conjugated, tetrapyrrole structures with a central Mg^{2+} atom and exist mostly, if not entirely, associated with proteins. Mg^{2+} in Chl usually forms five coordination bonds. Ligands for four coordinate covalent bonds are provided by the pyrrole nitrogens within the Chl molecule. The ligand for the fifth, axial coordination bond is provided by solvent (water) or a functional group on a protein molecule. In the reaction centers and **light-harvesting complexes** (LHCs), the preferred ligand to Chl *a* is the electron-rich, neutral, sidechain imidazole of histidine. Chl *b* differs from Chl *a* only in oxidation of the methyl group on pyrrole ring B to an aldehyde. Although this reaction results in a spectral shift (Fig. 13.8), and consequently expands the spectral range for absorbance of light by the chloroplast, introduction of the additional oxygen atom also influences the coordination chemistry of the central Mg^{2+} atom. The electronegativity of the oxygen atom results in withdrawal of electron density away from the pyrrole nitrogens toward the periphery of the molecule. Consequently, the Mg^{2+} atom in Chl *b* expresses a more positive point charge than that in Chl *a* and is a stronger Lewis acid. As a result, Chl *b* binds more strongly than Chl *a* to the dipolar solvent, water (Ballschmitter *et al.*, 1969). The imidazole sidechain of histidine, the preferred ligand for Chl *a*, is normally not a ligand for Chl *b*, probably because the imidazole sidechain does not have a sufficiently large dipole to displace

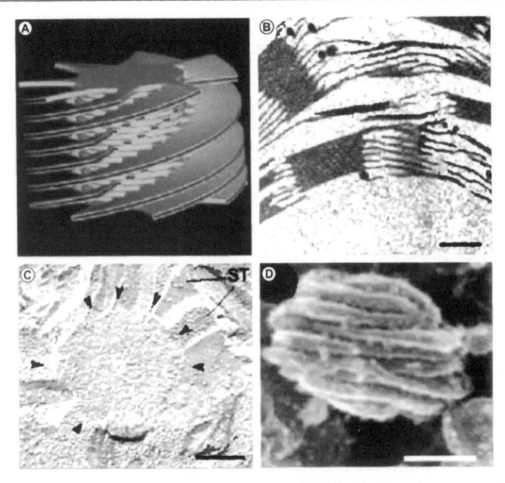

Fig. 13.7 Arrangement of stromal membranes around the grana as depicted by (**A**) computer model constructed from electron micrographs as shown in panels (**B**), (**C**), and (**D**). Reconstruction suggests that each thylakoid in a granum is connected to its neighbor through the stromal lamellae (ST) and that the thylakoid system is a continuous membrane enclosing a single lumen (from Mustárdy and Garab, 2003).

the strongly bound water ligand (Chen *et al.*, 2005). In fact, several molecules of Chl *b* in LHCs retain water as the axial ligand. The recent high-resolution structural determination of the LHCII from spinach defined the specific binding sites for Chl *a* and Chl *b* (Liu *et al.*, 2004). Ligands that favor interaction with Chl *b* usually contain oxygen and thus provide an electrostatic character to the coordination bonds (Fig. 13.9). Most of these

assignments were suggested from results of reconstitution of LHCs with mutant forms of the apoprotein (LHCP) in which an amino acid that provides a ligand was replaced by an amino acid lacking a functional group in the sidechain (Remelli *et al.*, 1999). In mutant strains of plants that lack the ability to make Chl *b*, very few LHCs accumulate. Thus, Chl *a* alone is not sufficient to complete assembly of most of the

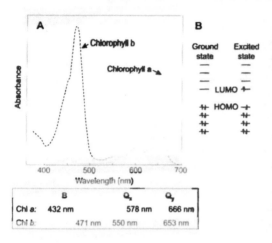

Fig. 13.8 **(A)** Absorption spectra of Chl *a* and Chl *b* in methanol. Indicated below the spectra are the wavelengths of maximal absorption along the major red vector, Q_y, the perpendicular vector Q_x, and the blue absorption band, designated the B or Soret absorption peak. See text for details. **(B)** Diagram describing absorption of light energy by elevation of an electron from the highest occupied molecular orbital (HOMO) to the lowest unoccupied molecular orbital (LUMO).

complexes. It was proposed that the stronger interaction between Chl *b* and the proteins is essential for assembly of stable Chl-protein complexes (Eggink *et al.*, 2001).

About one-third of the Chl *a* and essentially all of the Chl *b* reside in peripheral LHCs. The apoproteins, LHCPs, of the major and minor LHCIIs, associated with PS II, have molecular masses in the range of 25 to 30 kDa. (A Dalton [Da] is a unit of mass that is 1/12 the mass of a carbon atom [atomic weight, 12.011], or approximately the mass of a hydrogen atom.) LHCPs are encoded by a large family of *Lhcb* genes in the nuclear genome and bind between 8 and 14 Chl molecules per protein molecule. The major LHCII in plants exists as a trimer, as determined by electron crystallography of two-dimensional crystals examined at 4.2K

with an electron cryomicroscope (Kühlbrandt *et al.*, 1994). The major LHCPs in LHCII are designated Lhcb1, Lhcb2, and Lhcb3. Each Lhcb1 protein binds 14 Chl molecules, 8 Chl *a*, and 6 Chl *b*, along with 3 xanthophylls, which usually include 2 lutein and 1 neoxanthin molecules (Croce *et al.*, 1999; Liu *et al.*, 2004). (Xanthophylls are carotenoids that contain oxygen atoms.) Minor LHCIIs exist as monomers. Lhcb4 (CP24) binds eight Chls (6 Chl *a* and 2 Chl *b*), Lhcb5 (CP26) binds nine Chls (6 Chl *a* and 3 Chl *b* while Lhcb6 (CP29) binds ten Chls (5 Chl *a* and 5 Chl *b*). The minor LHCs also contain fewer carotenoids, usually with 1 lutein, 0.5 neoxanthin and 0.5 to 1 violaxanthin (Croce *et al.*, 2002; Pascal *et al.*, 2002).

In plants, four apoproteins—Lhca1, Lhca2, Lhca3, and Lhca4—are involved in the antenna for PS I. Because of sequence homology, the structure of LHCI monomers is probably quite similar to that of LHCII. Apoproteins of LHCI coordinate 6 to 9 Chl *a*, 3 Chl *b*, 1 lutein, 0.5 violaxanthin and bind substoichiometric amounts of β-carotene instead of neoxanthin (Schmid *et al.*, 2002). LHCI occurs as dimers, generally with heterodimers of Lhca1 with Lhca4, which comprise LHCI-730 that absorbs maximally at 730 nm, and Lhca2 with Lhca3, which provides LHCI-680, with maximal absorption at 680 nm.

Assembly of Light-harvesting Complexes during Chloroplast Development

Experiments with a model organism, the alga *Chlamydomonas reinhardtii*, have been instructive with regard to the pathway for insertion of the major LHCPs into the thylakoid membrane. The chloroplast of cells grown in light is filled with thylakoid membranes. When these green cells are

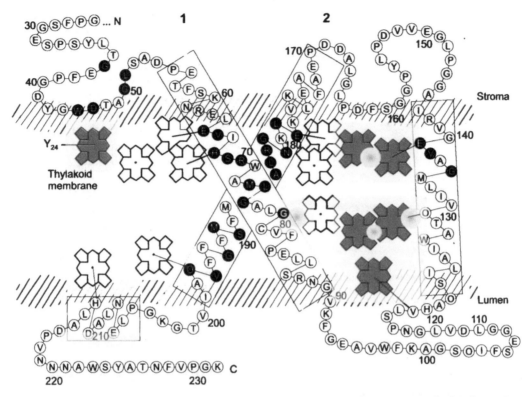

Fig. 13.9 Model of distribution of Chl *a* (*open symbols*) and Chl *b* (*green symbols*) in the major light-harvesting complex (LHCII) of green plants. Coordination bonds are indicated by solid lines from amino acids to Chls. Five Chl molecules have water as a ligand (blue dot) (adapted from Green and Durnford, 1996, with assignments as described by Liu *et al.*, 2004).

subsequently grown in darkness, the membranes are diluted among the progeny, leaving the chloroplast nearly depleted of membranes (Fig. 13.6). Exposure of the dark-grown cells to light initiates Chl synthesis and membrane assembly. Assembly of LHCs can be monitored by Förster resonance energy transfer from Chl *b* to Chl *a* as these molecules are brought sufficiently close (10 Å or less) as a result of incorporation into LHCs (Fig. 13.10). With the technique of immunoelectron microscopy, the initially assembled LHCs were detected in the chloroplast envelope and associated invaginations, as shown in Fig. 13.11 (White *et al.*, 1996; Eggink *et al.*, 2001). LHCPs made

in excess accumulated outside of the chloroplast in vacuoles. Furthermore, when cells were grown in darkness, conditions under which Chl synthesis did not occur, the proteins were not retained by the chloroplast and instead accumulated in the cytosol and vacuoles (Park and Hoober, 1997). A model that illustrates the current concepts in LHCII assembly is shown in Fig. 13.12.

Other than the requirement for Chl, and in particular Chl *b*, little is known about the mechanism of incorporation of these proteins into membranes. Chl *a* oxygenase, the enzyme that catalyzes synthesis of Chl *b*, is localized on chloroplast envelope membranes (Eggink *et al.*, 2004). Eventual

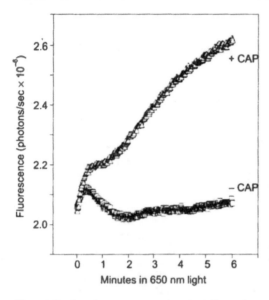

Fig. 13.10 Förster resonance energy transfer analysis of assembly of LHCs during greening of the alga *C. reinhardtii* y1. Cells were exposed to light of 650 nm, which was sufficient for photoconversion of Pchlide to Chlide *a* and also for excitation of Chl *b*. Fluorescence of Chl *a* was continuously monitored at 680 nm. After the first minute of illumination in untreated cells (-CAP), energy absorbed by newly assembled LHCs was trapped as the result of connection of the complexes with reaction centers. When chloramphenicol was added, which inhibited synthesis of reaction center core proteins on chloroplast ribosomes (+CAP, 50 μg ml⁻¹), energy absorbed by "free" LHCs was reemitted as fluorescence (adapted from White *et al.*, 1996).

accumulation of these proteins in thylakoid membranes is determined by molecular interactions that occur initially within the envelope, even before the proteins are completely imported into the plastid. In mutants unable to make Chl *b*, few if any of the LHCPs are retained in the chloroplast, although the plants still produce Chl *a*. Other factors are also likely involved in facilitating assembly of LHCs. For example,

mutants that lack the ability to synthesize the major carotenoid in LHCs, the xanthophyll lutein, assemble LHCs much slower than normal (Park *et al.*, 2002). Further, mutants deficient in subunits of a stromal complex designated the "chloroplast signal-recognition particle" are deficient in LHCs (Hutin *et al.*, 2002), which suggests that these proteins may be involved in assembly of the complex. Plants lacking a membrane protein, ALB3, are deficient in LHC assembly (Bellafiore *et al.*, 2002), although the specific action of this protein is not clear. ALB3 seems to be involved in facilitating insertion of LHCPs into the membrane. Kinetic measurements of the development of photosynthetic activities indicate that connection between photosystems and LHCs occurs immediately upon assembly of the antennal complexes, most likely in the chloroplast envelope.

ABSORPTION OF LIGHT ENERGY

Excitation of Chlorophyll

Chlorophylls, as cyclic tetrapyrroles, are conjugated, organic molecules that efficiently absorb light. Within the electronic structure of the molecules are orbitals that have distinct energy levels. Wavelengths of light with energy levels that match the allowed electronic transitions within the molecule are described by the absorption spectrum (see Fig. 13.8). Photons with more or less energy pass through the molecule without being absorbed. Absorption spectra are interpreted in terms of electrons lifted from the energetically highest *occupied* molecular orbitals (HOMO) to the energetically lowest *unoccupied* molecular orbitals (LUMO). For these transitions to occur, the molecule must absorb a photon that contains precisely the energy by which the orbitals differ (Gouterman, 1961). The

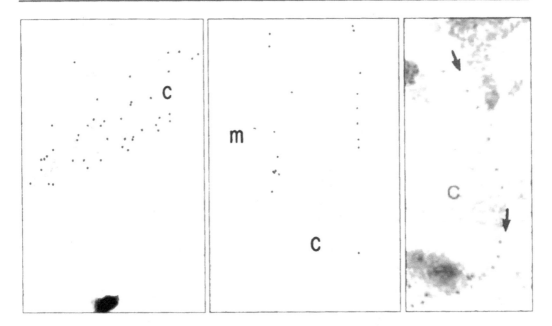

Fig. 13.11 Immunoelectron microscopic localization of LHCPs in the developing chloroplast of *C. reinhardtii* y1. (*Left panel*) LHCPs in a chloroplast developing normally during 6 hours of exposure to light at the normal growth temperature of 25°C were localized to thylakoid membranes in the interior of the chloroplast (c). Thin sections of cells were treated with IgG antibodies raised against LHCP, followed by incubation with protein-A, which binds tightly to IgG molecules. Protein-A was conjugated with gold particles that were visible in the electron microscope. (*Middle panel*) Companion cells those in the *left panel* were exposed to light in the presence of a high concentration of chloramphenicol (200 μg ml^{-1}), which suppressed synthesis of Chl. LHCPs were detected in the chloroplast (c) only within the envelope. The micrograph also includes a mitochondrion (m) (from Eggink *et al.*, 2001.) (*Right panel*) Dark-grown cells of the alga were exposed to light at 38°C, which provided conditions that enabled detection of the first LHCPs to be incorporated into membranes during chloroplast development. After 15 min exposure to light, gold particles were observed in the chloroplast (c) only over the chloroplast envelope, designated by arrows (adapted from White *et al.*, 1996).

ability of the Chl molecule to absorb light is also a function of the direction of the interacting waveform, with stronger absorption occurring when the directional vector is parallel with the *y* axis of the molecule, which lies across pyrrole rings A and C of the macrocycle (see Fig. 13.1). Consequently, this generates the Q_y absorption band, which usually has the lowest energy (longest wavelength). Absorption of light with the vector along the shorter axis, across pyrrole rings B and D, generates the Q_x absorption band. The shorter wavelength of absorbed photons indicated by the Q_x band therefore reflects a larger energy gap that must be crossed, but these electrons reach the same LUMO when excited as those absorbed in the Q_y band. These transitions occur with photons with wavelengths between 550 and 700 nm. Another set of allowed transitions is described by the higher energies required to reach the LUMO+1 orbital. Absorption of photons with the appropriate energy for this transition is described by the B (or Soret) bands, with wavelengths in the 420 to 470 nm range.

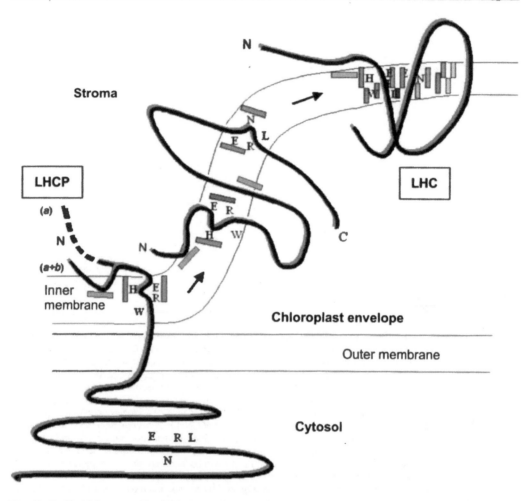

Fig. 13.12 Model for assembly of light-harvesting complexes based on experimental results obtained with the model alga *C. reinhardtii*. LHCPs are imported into the chloroplast, N-terminus (N) first, directed by the "transit sequence". The transit sequence enters the stroma and is removed by a specific protease while most of the protein still extends into the cytosol. Chl *a* possibly binds to a motif with the generic sequence -ExxHxR- in the first membrane-spanning region of the protein. Extensive evidence suggests that Chl *b* is required to retain the protein in the chloroplast, which possibly binds to the backbone peptide bond at tyrosine-24 (see Fig. 13.9). In the absence of Chl *b* (dashed N-terminal region [a]), the protein is not held sufficiently strongly to prevent retraction into the cytosol (Park and Hoober, 1997). Binding of the initial Chls to N-terminal sites allows the remainder of the LHC to assemble (a + b). Upon assembly, LHCs associate with newly assembled PS II and the resultant expansion of the inner envelope membrane causes formation of vesicles (see Fig. 13.6) (adapted from Hoober and Eggink, 2001).

The energy content of a photon is related to its frequency, or wavelength, as described by the following relationship.

$$E = hv = hc/\lambda \qquad \ldots(2)$$

where E is energy in Joules, h Plank's constant (6.63×10^{-34} Joules S), λ frequency or the inverse of λ, the wavelength of light, and c the speed of light (3.0×10^{10} cm s^{-1}). The

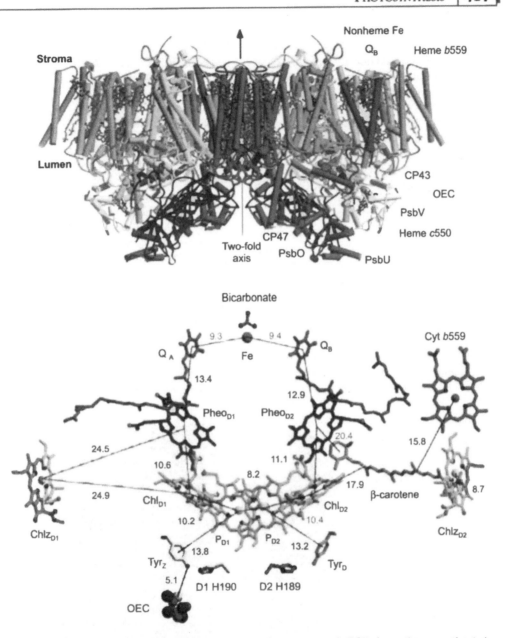

Fig. 13.13 Structure of the core complex and reaction center of PSII from the cyanobacterium *Thermosynechococcus elongatus*. (*Upper*) View of the PSII dimer from the side. Extending from the luminal side of the membrane is the oxygen evolving complex (purple). PsbO on the luminal side of the complex is the 33 kDa protein that stabilizes the Mn cluster. (*Lower*) Arrangement and distances of the cofactors within the reaction center shown with proteins removed. P_{D1} and P_{D2} are the Chl molecules of the special pair, P680. Electrons flow from the Mn cluster (oxygen evolving complex, OEC) on the luminal surface to P680 through a tyrosine radical, Tyr_Z, and then in sequence to Chl_{D1}, $Pheo_{D1}$, Q_A and Q_B on the stromal surface (see text for details) (from Ferreira *et al.*, 2004).

excited state of the molecule can return to ground state by several means, either by release of vibrational energy as heat, release of a photon of slightly lesser energy (**fluorescence**), or transfer of the packet of energy—an "**exciton**"—to another molecule. Rapid decay from the LUMO+1 to the LUMO orbital occurs by an internal, vibrational mode. Energy leaves the molecule either by radiationless energy transfer, fluorescence or heat by decay from the LUMO to the ground state. Most of the energy entering photosynthetic systems is absorbed by light-harvesting complexes, which contain the bulk of the Chl in thylakoid membranes. Excitons are transferred extremely rapidly, on the order of hundreds of femtoseconds (10^{-15} s), between Chl molecules within an antennal complex. Excitons exit an LHC through a specific Chl *a* molecule and may travel through many other Chl-protein complexes before eventually reaching a reaction center. Trapping of energy from LHCI by PS I occurs in about 25 picoseconds (25×10^{-12} s), whereas transfer of energy from LHCII to the reaction center of PS II occurs in about 130 picoseconds (130×10^{-12} s) (van Amerongen and van Grondelle, 2001).

Reaction Centers

The heart of each "reaction center" is a pair of Chl *a* molecules in close, nearly parallel juxtaposition (in PS I, one of the pair is Chl *a'*, a stereo-isomer around C-13^2). This pair of Chl *a* molecules in PS I has been designated P700 because of its absorption maximum. In PS II, the special pair is designated P680. When an exciton reaches one of the Chl molecules in the reaction center, an electron is lifted to a higher energy level. The "excited" molecule consequently becomes a strong reducing agent and donates an electron to a nearby electron acceptor. The acceptor thus achieves a negative charge while the donor has a positive charge. This "**charge separation**" is the key

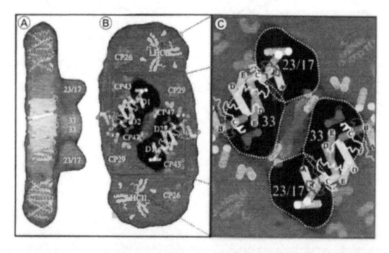

Fig. 13.14 Structural model of the subunits within the PSII supercomplex. (A) and (B) show representations from the side and luminal surface respectively. Proteins of the Mn cluster involved in oxygen evolution protrude from the luminal surface, as shown in (A). The reaction centers, composed of proteins D1 and D2, occur as a dimer, each with its associated Chl-protein complex. Higher magnification of the luminal surface shown in (C), with the positions 17 kDa, 23 kDa, and 33 kDa proteins of the Mn cluster indicated in outline over the reaction center subunits (from Nield *et al.*, 2000).

photochemical event in photosynthesis (Allen and Williams, 1998).

The reaction center of PS II (Fig. 13.13) is composed of two very similar proteins, designated D1 (38.0 kDa) and D2 (39.4 kDa), which are encoded by genes *psbA* and *psbD* respectively in the chloroplast genome (Hankamer *et al.*, 1997). These two proteins together bind six molecules of Chl *a*. The PS II reaction center also contains two molecules of **pheophytin** *a* (Chl *a* molecules without the central Mg^{2+} atom), a tightly bound **plastoquinone** molecule, and an iron atom. The biochemically purified PSII reaction center also includes the α (9 kDa) and β (4 kDa) subunits of cytochrome b_{559}

and a small peptide product (4 kDa) of the *psbI* gene. Each reaction center is associated with two Chl *a*-protein complexes that act as "core" antenna, CP43 (51 kDa) and CP47 (56 kDa), encoded by the *psbC* and *psbB* genes respectively, which were initially designated according to their molecular masses as estimated by electrophoresis. CP43 and CP47 bind 12 and 14 Chl *a* molecules respectively (Zouni *et al.*, 2001). A cluster of Mn^{2+} ions on the electron donor side of the reaction center is involved in oxygen evolution. This complex is stabilized by three proteins, 26.5, 20.2, and 16.5 kDa in mass. The complete "**core complex**" contains approximately 25 different proteins. The core

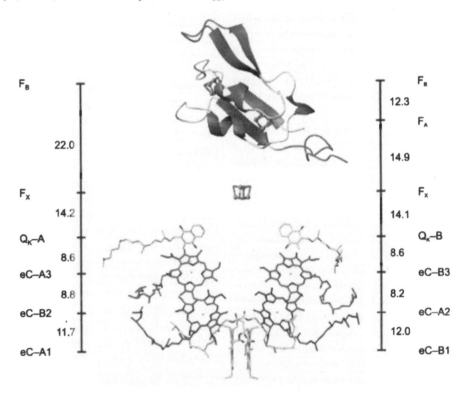

Fig. 13.15 Positions of cofactors in the reaction center of PSI from *Synechococcus elongates*, with the center-to-center distances between them in Angstroms. Pairs of Chls and phylloquinones arranged in two branches, A and B, from P700 Chls (eC-A1 and eC-B1) to the iron-sulfur center, F_X. Iron-sulfur clusters F_A and F_B are shown within protein PsaC. The phylloquinones are indicated by Q_K and one or both may be the electron acceptor A_1 (see Fig. 13.18) (from Jordan *et al.*, 2001).

complex was purified by sucrose gradient centrifugation and cryoelectron microscopy revealed that PS II exists as a dimeric structure that also includes the minor LHCs CP26 and CP29 and a single, tightly bound LHCII trimer per reaction center (Fig. 13.14). This overall structure was designated the "**supercomplex**" (Nield *et al.*, 2000). In thylakoid membranes, approximately four to six additional, more loosely bound, LHCII trimers are associated with each PSII reaction center and thereby increase the absorptive cross-sectional area of the photosynthetic unit. The minor LHCIIs connect the major, peripheral antenna to the reaction center. The slightly more blue-shifted spectra of the major LHCs relative to the absorption maximum of the reaction center ensures that energy will flow toward, and be trapped by, the lower energy levels of P680. Together with the functional antenna, a PS II unit contains a total of about 280 Chl molecules per reaction center.

In PS II, electrons are transferred from P680 to a Chl *a* and then to a pheophytin *a* molecule on protein D1. Experimental data suggest that the initial charge separation occurs predominantly between the Chl *a* and pheophytin *a* to form the radical pair state $Chl^+ Phe^-$, followed by electron transfer from the P680 Chl *a* on D1 to produce the terminal radical pair state $P680^+ Phe^-$ (Barter *et al.*, 2003). The electron is then transferred to a tightly bound plastoquinone molecule (Q_A) on subunit D2 and next to a loosely held plastoquinone (Q_B) on D1 to generate the semiquinone form. After a second electron is transferred to the latter acceptor, along with two protons, the reduced quinol (PQH_2) leaves the reaction center and is replaced by another plastoquinone molecule from the pool in the membrane. Reduction of plastoquinone to the quinol is a two-electron reaction, as shown in eqn (3), and involves uptake of two protons (H^+) from the stroma. In oxygenic photosynthesis, the electron hole in P680 is then filled by an electron abstracted from water (see below, "Generation of end products").

$$R = C_{45} \text{ isoprenoid sidechain} \quad \quad ...(3)$$

The reaction center of PS I is also composed of two principle, similar proteins, PsaA (84 kDa) and PsaB (83 kDa), encoded by genes *psaA* and *psaB* in the chloroplast genome (Jordan *et al.*, 2001). These proteins bind 6 Chl *a* molecules, 2 **phylloquinone** molecules and an **iron-sulfur center** (Fig. 13.15). The structure of the PS I core complex was resolved to 2.5 Å, which revealed that the complete complex contains a total of 12 protein subunits, 90 additional Chl *a* molecules, a total of three Fe_4S_4 centers (designated F_A, F_B, and F_X) and 22 various isomers of β-carotene. Seventy-nine of the Chls are bound to the large proteins of the PS I reaction center, PsaA and PsaB, mostly to the N-terminal domains that serve as the core antenna. In the excited state, P700 donates an electron to an adjacent Chl *a* molecule (A_0). The electron is then transferred to a phylloquinone molecule (A_1) and subsequently to the series of iron-sulfur centers. From F_B, electrons are transferred to the Fe_4S_4 cluster of the electron-carrying protein, **ferredoxin**. Whereas the reaction center of PS II exists as a dimeric structure, PS I in plants is monomeric (Kargul *et al.*, 2003). (However, in cyanobacteria and another photosynthetic prokaryote *Prochlorococcus*, PS I is trimeric.) Estimates indicate that about five dimeric LHCIs are associated with each PSI complex. Along with the light-harvesting antenna, LHCI, PS I contains a total of about 215 molecules of Chl.

Both reaction centers have essentially a symmetrical structure. The two subunits in each reaction center differ slightly but create two potential branches for electron transfer. Interestingly, electron flow occurs preferentially across one branch. A substantial amount of work has been devoted to determining whether any electrons flow across the alternate branch. Although evidence exists for functional ability of both branches, electrons normally flow over only one branch in PS II. In PS I, electron transfer is about 10-fold faster (~13 ns) on the PsaB side than on the PsaA side (~140 ns) (Guergova-Kuras et al., 2001). With one branch kinetically much more rapid, activity of the alternate branch can only be observed when the primary pathway is impaired by mutation.

Fluorescence Induction Curves

A useful technique has been refined over the past two decades to noninvasively monitor energy flow through PS II. When the quinone acceptors of electrons from the reaction center are oxidized, light energy is nearly quantitatively trapped by energy transduction to produce electrons that reduce the quinones first to a semiquinone when one electron is added and then the fully reduced quinol when the second electron is added. However, when photons are absorbed by the light-harvesting complexes more rapidly than the rate at which the reduced quinol at the Q_B site can exchange with an oxidized plastoquinone molecule, Q_A remains reduced and electron flow through PS II is blocked. Subsequent excitons produced from absorption of photons by LHCII are deactivated by nonradiative decay with release of heat or, in a minor fraction of events, by emission of photons of

slightly lower energy, the process of **fluorescence**. Consequently, transients of fluorescence intensity provide information on the state of the reaction center. Figure 13.16 describes the fluorescence transient when cells of the alga *Chlamydomonas reinhardtii* were exposed first to a low intensity red light, which yields a low level of fluorescence of Chl *a* molecules when the reaction center remains oxidized (F_o). Subsequent irradiation causes a rise first to a plateau level, indicative of PS II reaction centers that are not directly connected to electron transport chains. Without a means for electrons to exit these reaction centers, they are rapidly reduced. Fluorescence then continues to increase, but more slowly, to a peak value (F_p) as Q_A becomes reduced in the bulk of the PSII centers. The fluorescence transient therefore indicates that electrons initially arrive at Q_A, and then Q_B, at a higher rate than reduced Q_B (quinol) can be oxidized. As PS I and the carbon fixation pathway become activated, the kinetically more rapid PS I drains electrons from PSII through the electron transport chain, Q_B is oxidized, the backup in PSII is relieved, and the fluorescence decreases to a steady-state level (F_s). However, when electron transport from Q_A is blocked by an herbicide that binds to the Q_B site, the efflux of energy from the reaction center is blocked and fluorescence rapidly rises to a maximal level (F_m). The same maximal level can be achieved with a brief, intense flash of white light that floods the reaction center with excitons. The quantum yield, Φ, an indication of the integrity of the reaction center, is described by the simple relationship, $\Phi = (F_m - F_o)/F_m = F_v/F_m$. This value for chloroplasts in most organisms grown under normal conditions is 0.6 to 0.8, as shown in Fig. 13.16.

GENERATION OF END PRODUCTS

Evolution of Oxygen

Oxygen is produced only by photosynthetic systems that use Chl. In these organisms, two photosystems, PS I and PS II, operate in series and thus span a wider redox range than possible with the single reaction center in photosynthetic bacteria. Oxidation of water requires an oxidant with a redox potential more positive than E_m = +0.816 V,

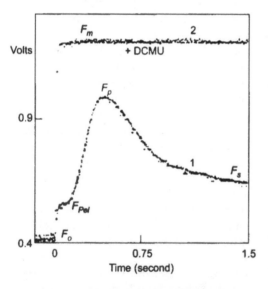

Fig. 13.16 Fluorescence induction curve obtained with green cells of alga *C. reinhardtii*. Cell suspensions were dark adapted before exposure to modulated measuring light of low intensity (2.5 µmol photons m^{-2}s^{-1}) for determination of F_o and then to actinic light (80 µmol photons m^{-2}s^{-1}) to obtain fluorescence transients. Fluorescence was measured at room temperature in the absence (curve 1) or presence (curve 2) of 10 µM 3-(3,4-dichlorophenyl)-1,1-dimethylurea (DCMU), a herbicide that binds to the Q_B site and blocks electron transfer to plastoquinone (adapted from White and Hoober, 1994).

the midpoint potential of the O_2/H_2O couple. BChl in reaction centers of green and purple bacteria achieves redox potentials of about E_m = 0.5 V, because the energy of a photon at 870 nm, the absorption maximum of the purple bacterial special-pair BChl, is considerably less than that at 680 nm, the absorption maximum of P680 in PS II (from eqn 2, 138 kJoules *vs* 180 kJoules per mole photons). After donating an electron to pheophytin, the oxidized reaction center of PSII, P680$^+$, with a redox potential E_m = +1.2 V, abstracts an electron from a tyrosine residue in D1. The tyrosine radical, designated Z, has a redox potential of about +0.93 V, which is sufficient to abstract electrons from water. Z connects P680 with the oxygen-evolving complex located on the luminal surface of the membrane (see Figs. 13.13 and 13.14). The core of this complex is a structure formed by four manganese atoms coordinated with several oxygen atoms and one each of Cl$^-$ and Ca^{2+} ions (Ferreira *et al.*, 2004). The Mn cluster is stabilized by three proteins, 26.5, 20.2, and 16.5 kDa in mass. P680$^+$ abstracts one electron at a time from tyrosine Z, which in turn pulls an electron from the Mn cluster. Consequently, for an oxygen molecule to be released by the complex, four electrons from two water molecules must be consecutively collected by the Mn cluster and transferred to P680. Electron collection from water and delivery to P680 occur by changes in the valence states of Mn. When chloroplasts are exposed to a series of very brief flashes of light, such that only a single turnover of P680 occurs per flash, the result shown in Fig. 13.17 is obtained, with a burst of oxygen produced every fourth flash. For each oxygen molecule produced, four protons from water are released into the *luminal* environment.

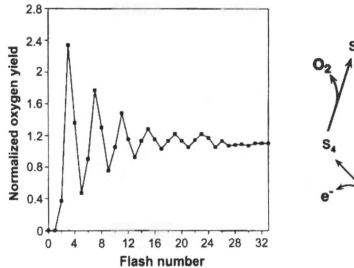

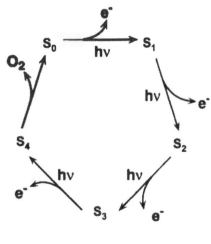

Fig. 13.17 Oxygen evolution as a function of light flashes. **(A)** Pierre Joliot and Bessel Kok discovered this periodic pattern when chloroplasts were exposed to a series of flashes sufficiently short for only a single turnover of P680 with each flash. This pattern indicated that four photons were required to abstract four electrons from water to generate one molecule of oxygen. Studies of this process led to the scheme shown in **(B)**, which suggests the Mn cluster of the oxygen-evolving complex is progressively driven through four states. Because oxygen is released spontaneously from S_4 and the first burst of oxygen occurs after the third flash, these results indicate that the resting state is S_1. Because not every center was hit with each flash, periodicity gradually damped as the experiment proceeded.

Proton Motive Force and ATP

The electron acceptor of PS II, Q_B, is reduced on the stromal surface of the membrane. Reduction of each plastoquinone molecule to the quinol (PQH_2) involves uptake of two protons from the *stromal* environment (eqn 3). PQH_2 is oxidized by a complex of membrane proteins, intermediate between PS II and PS I, that contains cytochrome b_6, cytochrome f, and a Rieske iron-sulfur (Fe_2S_2) center (Kurisu *et al.*, 2003; Stroebel *et al.*, 2003). The iron atoms in the heme cofactors in the **cytochrome b_6/f complex** accept only electrons and since electrons are transferred to a site on the luminal side of cytochrome b_6, the protons from PQH_2 are released into the *luminal* compartment (Fig. 13.18). If this complex operated as a simple, forward transfer of electrons, one H^+ would be transported across the membrane per each electron transferred from PS II to the cytochrome b_6/f complex. However, measurements indicate that *two* H^+ are transported per each electron. To explain this observation, a mechanism called the "Q-cycle" has been proposed. In this cycle, one electron from the quinol, PQH_2, in the "Q_o" site of the cytochrome b_6/f complex is transferred to one of the two heme groups, b_L, in cytochrome b_6 and the second to the Rieske iron-sulfur center, which can accept only one electron. The electron in cytochrome b_6 is then transferred from the b_L to the b_H heme, which reduces a plastoquinone tightly bound in the "Q_i" site to the semiquinone form. An electron from oxidation of a second PQH_2 in the Q_o site is also

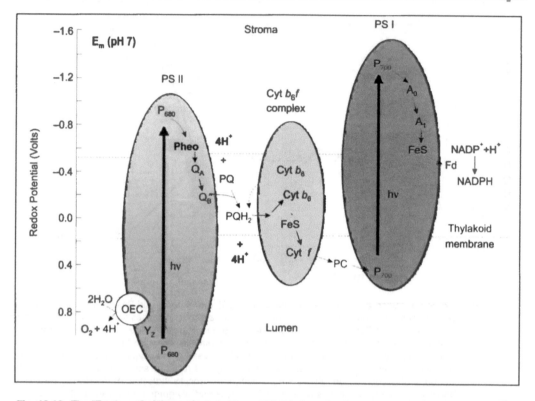

Fig. 13.18 The "Z-scheme" of linear electron transport in photosynthesis is generated when the sequence of electron carriers is plotted according to their midpoint redox potentials. Thick arrows indicate changes in redox potential of the electron donors in each reaction center, the Chl pairs P680 and P700, when they receive excitation energy from the antenna. The physical positions of the electron carriers Pheo, Q_A, and Q_B in PS II and A_0, A_1, and the Fe_4S_4 clusters in PS I are shown in Figures 13.13 and 13.15. Plastoquinone (PQ) is reduced by PS II on the stromal side to PQH_2 and reoxidized by the cytochrome b_6/f complex on the luminal side of the membrane, which contains a Rieske Fe_2S_2 center. Operation of the "Q cycle", which transfers electrons through cytochrome b_6, results in transfer of 4 H^+ from the stroma to the lumen. Plastocyanin (PC) shuttles electrons from cytochrome f to PS I. Ferredoxin (Fd), a small protein containing an Fe_4S_4 cluster, is the first soluble, stromal electron carrier to accept electrons from PS I. Ferredoxin reduces $NADP^+$ to NADPH in a reaction catalyzed by the flavin-containing enzyme, ferredoxin-$NADP^+$ reductase. Arrows trace the path of electrons through the complexes.

transferred through b_L and b_H to fully reduce the semiquinone (PQ^-) in Q_i to the quinol ($PQ^=$), while the other electron is transferred to the Rieske center. In the process, two additional protons (H^+) are pulled from the stroma during reduction of the plastoquinone to produce PQH_2 in the Q_i site. The PQH_2 is then released from the Q_i site and binds to the Q_o site. Cycling of electrons through cytochrome b_6, and oxidation of the second PQH_2 by the complex, results in transfer of a second H^+ per each electron, or two H^+ per each electron passing through the cytochrome b_6/f complex, from the stroma to the lumen (Sacksteder *et al.*, 2000). Each electron that reaches the Rieske center is transferred to cytochrome *f* and then to a small luminal, copper-containing protein called **plastocyanin**.

Oxidation of water releases one H^+ per electron. When the uptake of protons accompanying reduction of plastoquinone (Q_B) on the acceptor side of PS II is combined with the subsequent release of protons into the lumen by the Q-cycle, a net gain of three H^+ is achieved in the thylakoid lumen, relative to the stroma, during transfer of one electron from water to NADPH. Therefore, reduction of two $NADP^+$ by four electrons from two water molecules results in a total gain of 12 H^+ in the thylakoid lumen. The resultant proton gradient across the membrane generates a "**proton motive force**", which is roughly equally divided between the proton concentration gradient and the accompanying voltage potential ($\Delta\psi$) caused by the imbalance of positive charges across the membrane. This force drives synthesis of **adenosine triphosphate** (ATP). The experimentally measured ratio of H^+/ATP is approximately 4, which indicates that three ATP molecules are synthesized for each two NADPH molecules, the stoichiometry required for CO_2 fixation by the **reductive pentose-phosphate cycle** (see discussion below). Whereas the stromal pH is 7.5 to 8.0 when the chloroplast is actively engaged in photosynthesis, the luminal pH is maintained between 5.8 and 6.5, which provides a transmembrane gradient of approximately a 100-fold difference in proton concentration. The ability to develop the proton motive force requires a physical boundary between these compartments, which is provided by the relatively impermeable thylakoid membrane. Maintaining a small volume of the lumen is also required to achieve a relatively high proton concentration. Activity of ATP synthase is regulated by the concentration of inorganic phosphate in the stroma

(eqn 4), and only when the phosphate level is reduced to a very low level, such that the consequently low activity of ATP synthase slows H^+ flow across the membrane, does the luminal pH drop below 5.8 (Kramer *et al.*, 1999).

ATP synthase is a bipartite protein with a membrane component, CF_o, attached to a large, peripheral domain, CF_1, which extends from the membrane on the stromal surface (Fig. 13.19). This enzyme was initially designated the "chloroplast coupling factor (CF)" because of its function in coupling electron transfer during photosynthesis to ATP synthesis. (The designation "CF" distinguished the chloroplast ATP synthase from the similar enzyme in mitochondria, designated "F".) The energy in the proton motive force causes CF_o, an integral, annular structure of 12 to 14 "c"

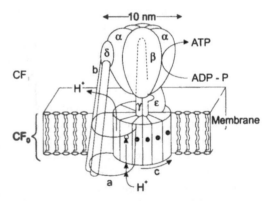

Fig. 13.19 Structure of CF_oCF_1 ATP synthase. The membrane-spanning complex, CF_o, includes approximately 12 copies of the "c" subunit, which are associated with the γ and ϵ subunits. This complex rotates within the membrane, driven by the flux of protons through the membrane. The "α", "β", and δ subunits extend from the membrane and attach to, and serve as a 'stator' for, the CF_1 complex composed of three each of the α and β subunits. Rotation of the γ subunit causes consecutive conformational changes in CF_1 (from Elston *et al.*, 1998).

subunits, to rotate within the plane of the membrane. The stromal domain, CF_1, is a "segmented sphere" of six subunits, three α subunits and three β subunits in alternating order, and is held stationary by the "a", "b" and δ subunits that extend from the membrane and contact the CF_1 domain. The γ subunit is attached to CF_0 and extends up through the central core of CF_1. As CF_0 rotates, it drives rotation of the γ subunit, which causes conformational changes in the catalytic sites on the β subunits. The reaction

$$ADP + \text{inorganic phosphate (Pi)}$$
$$\leftrightarrow ATP + H_2O \qquad ...(4)$$

is thermodynamically reversible, although the hydrolysis (reverse) reaction is often considered practically irreversible in an aqueous environment. (Many biosynthetic pathways couple the energy released by ATP hydrolysis to push reactions forward that are of themselves thermodynamically unfavorable). In one conformation, one of the β subunits binds ADP and Pi extremely tightly and the reversible reaction (eqn 4) occurs out of contact with the environment. When the γ subunit rotates 120°, the next conformation causes affinity for ATP in this site to be essentially lost and the product ATP diffuses from the enzyme. At the same time, the catalytic site in the adjacent β subunit changes from one that loosely binds ADP and Pi to the tight-binding form that makes ATP. Again, when rotation of the γ subunit continues, ATP dissociates from this site. Thus, ATP synthase operates as a molecular rotary motor by undergoing conformational changes in which each of the three catalytic sites is changed, in sequence, from one that loosely binds ADP and Pi to the tight-binding site that generates ATP to the low affinity site from which ATP dissociates (Boyer, 1998). When the CF_1 domain

is disconnected from CF_0, the enzyme operates in reverse as an ATP hydrolase.

Production of NADPH

Electrons are transferred from the copper ion in plastocyanin to PS I on the luminal surface. A specific subunit of PS I, the protein PsaF, is required in addition to a surface loop of PsaB, one of the PS I reaction center proteins, to provide a docking site for plastocyanin that brings the reduced copper atom (Cu^+) near P700 (Sommer et al., 2002). As described above, transfer of excitons from LHCI to P700 powers transfer of electrons from P700 across the membrane, through a Chl a molecule (A_0) and the phylloquinone (A_1) to the iron-sulfur centers on the stromal surface of PS I. The soluble iron-sulfur protein, **ferredoxin**, is then reduced on the stromal surface. Ferredoxin is a one-electron carrier, whereas the ultimate electron acceptor $NADP^+$ is reduced to NADPH in a two-electron reaction. This reaction is catalyzed by **ferredoxin-NADP$^+$ reductase**, an enzyme that carries a flavin adenine dinucleotide (FAD) prosthetic group (Deng et al., 1999). The FAD cofactor of the reductase accepts two electrons, one at a time, to fully reduce the flavin, which then reduces $NADP^+$ in a two-electron reaction to generate NADPH (eqn 5), the major reductant for CO_2 fixation in the chloroplast. Reduced plastocyanin fills the electron hole in P700 as electrons flow into NADPH

$$FADH_2 + NADP^+ \leftrightarrow FAD +$$
$$NADPH + H^+ \qquad ...(5)$$

Distribution of Photosystems in Thylakoid Membranes

Operation of the photosystems in oxygenic photosynthesis is usually diagrammed as shown in Fig. 13.18. The complete system, with two reaction centers acting in series, provides a linear flow of electrons from water to $NADP^+$. This process has been described as the Z-scheme, based on the flow of electrons through components that are arranged according to the midpoint of their redox potentials. The system can be short-circuited, depending on environmental conditions, when reduced ferredoxin adds electrons back into the chain at the level of plastoquinone or the cytochrome b_6f complex. The net effect of this cyclic process is a transfer of protons from the stroma to the thylakoid lumen, resulting in a proton gradient from which ATP is formed but NADPH is not produced. Cyclic electron flow is required to maintain balance in the energy requirements within the chloroplast (Munekage *et al.*, 2004).

Juxtaposition of the photosystems, which is implied by Fig. 13.18, exists only in a small part of the thylakoid membrane. In reality, the photosystems are mostly quite segregated from each other. In a typical chloroplast of a higher plant, such as spinach, tobacco or barley, about 80% of the membrane lies within the grana, which are cylindrical stacks 10 ± 5 lamellae high and about 0.4 to 0.5 μm in diameter (see Fig. 13.7). The remaining 20% of the membrane exists as "stromal lamellae" that interconnect the grana. PS I, which requires access to ferredoxin and $NADP^+$ in the stroma, and ATP synthase are localized predominantly on stromal lamellae and at the edges and surfaces of grana. About 85% of PS II, with its full antenna (designated PS IIα), is within the grana, while a minor portion, with a smaller antenna (PS IIβ) is in the stromal lamellae. This distribution of photosystems within the membrane system was established biochemically, with purified fragments of granal membranes enriched in PS II and fragments of stromal lamellae containing most of the PS I (Albertsson, 1995).

A direct demonstration of the density of PS II units in granal membranes was provided by freeze-fracture images of the membrane interior (Fig. 13.20). In this technique, membranes are embedded in ice. When the ice block is fractured, the fracture plane travels along the internal bilayer interface, which exposes the interior of the membrane. A high density of 14- to 20-nm diameter particles, similar in size to the PS II dimeric "supercomplex" (about 16 nm without LHCII antenna), was revealed on the interior face of the luminal half of stacked, granal membranes (EF_s face), whereas the facture face of the luminal half of unstacked, stromal membranes (EF_u face) contained particles about half this size and at a much reduced density. The complementary interior face of the stromal half of the unstacked membrane, the PF_u face, contained a very high density of 10- to 12-nm diameter particles, a size expected for monomeric PS I. The abundant LHCII particles were retained by the outer half of the granal membrane (PF_s face) when the membrane was fractured. The PS II particles on the EF_s face interdigitate between these PF_s particles, which reflects the close contact of the antenna with the core complexes.

Long-range segregation of most of the PS II and PS I units presents an interesting problem. It is generally accepted that PS II and PS I function in series, although cyclic electron flow can occur around PS I, mediated by return of electrons from ferredoxin back to plastoquinone and cytochrome b_6/f. Mutants deficient in electron transfer from plastocyanin to P700 are light sensitive, which indicates that the major dissipation of energy absorbed by PS II is via PS I. Thus the flow of electrons from water to $NADP^+$ must occur rapidly, mediated presumably by highly mobile electron carriers. The rapid flow of electrons is reflected in the

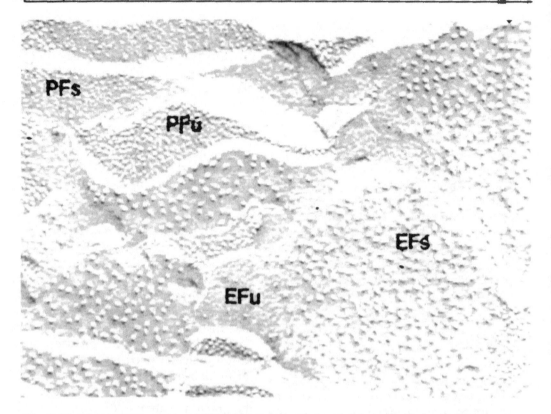

Fig. 13.20 A freeze-fracture image of thylakoid membranes from spinach. An ice block containing the membranes was fractured, which revealed the inner faces of the lipid bilayer membrane. Protein complexes that spanned the membrane were retained on one leaflet or the other depending on which surface of the membrane the proteins were most strongly anchored in the surrounding ice. Interior faces of the stacked, granal membranes provided the surfaces designated EF_S for the luminal leaflet and PF_S for the stromal leaflet. Corresponding faces for stromal thylakoid are designated EF_U and PF_U. Large particles on the EF surfaces represent PS II core complexes, which are more abundant in granal membranes. A high density of smaller particles on the PF_U face represent PS I complexes. LHCII complexes were retained by the PF_S leaflet. This image shows that PS II and PS I particles do not occur to any significant extent in each other's territory (adapted from Staehelin, 2003; micrograph courtesy of Dr. Andrew Staehlin).

fluorescence transient illustrated in Figure 13.16. The major intermediate between the two photosystems, the cytochrome b_6/f complex, is distributed on granal and stromal membranes. This complex has been isolated and crystallized as a large, dimeric structure that is expected to have, because of its size, limited mobility (Kurisu *et al.*, 2003; Stroebel *et al.*, 2003). PQH_2, the electron carrier from PS II to the cytochrome b_6/f complex, is a small, mobile membrane lipid, but the rate of long-range molecular diffusion in thylakoid membranes is two orders of magnitude slower than that expected for a fluid lipid membrane, possibly because of the high density of protein complexes in the membrane. At this rate, the $t_{1/2}$ of electron transfer from PS II would be of the order of many minutes instead of the measured rate of several hundred milliseconds. Studies of electron flow in thylakoid membranes indicated that only a few plastoquinone molecules (about six) are associated with each PS II reaction center, which shuttle electrons over short distances to nearby cytochrome b_6/f complexes (Kirchhoff *et al.*, 2000). Reduction of the cytochrome b_6/f complex is kinetically the slowest redox step in electron flow.

The short-range diffusion limit of plastoquinone in thylakoid membranes leaves the task of electron transport from PS II-cytochrome b_6/f microdomains to PS I up to plastocyanin. Diffusion of plastocyanin, the small protein that transfers electrons from the cytochrome b_6/f complex to PS I, is localized in the thylakoid lumen, a narrow compartment with many proteins extending from the membrane into this space. Diffusion may also be restricted by lack of bulk water within the lumen. Thus it is clear that the rate of transfer of electrons from PS II to ferredoxin is much slower than the rate of conversion of excitons to electrons in the reaction centers, which occur on a time scale of pico- to nanoseconds. The conclusion emerges that the rate of photosynthesis is limited at saturating intensities of light, by the rate of diffusion of these electron carriers. Although activation of the CO_2-fixation cycle and transfer of electrons out of PS I maintain the PS II reaction center in a relatively oxidized state, at high light intensities an energy "backup" in PS II is readily detected by the release of some of the absorbed energy as fluorescence.

The elaborately folded granal structure raises an interesting question about the purpose of this arrangement, because structure and function are usually complementary. One possible explanation for membrane differentiation, with PS II in grana and PS I in stromal lamellae, follows from their kinetic differences. PS I has much faster trapping kinetics than PS II (Trissl and Wilhelm,

1993), and were PS I sufficiently near to the major antenna, LHCII, most of the absorbed energy would be drained off by PS I. Because the proton gradient required for ATP synthesis is generated by electron flow from water, through PS II, to the cytochrome b_6/f complex, segregation of the photosystems is essential to maintain a high level of ATP synthesis, via the proton gradient, relative to NADPH production. To some extent, this problem has been solved in cyanobacteria, which do not form grana, by connecting the light-harvesting phycobilisomes only to PS II. In some eukaryotic algae, thylakoid membranes are also not appressed, and this problem is partially solved since the number of PS II units is severalfold higher than that of PS I. Moreover, many eukaryotic algae have thylakoid membranes that are appressed over most of their surface, without discrete distinction between granal and stromal membranes.

PHOTOINHIBITION: DAMAGE AND REPAIR OF PS II REACTION CENTER

When LHCII absorbs light energy at levels in excess of the rate at which electron flow can dissipate energy in the PS II reaction center, photooxidative damage to the D1 subunit (PsbA) can occur. Such conditions are caused when light absorption exceeds chloroplast capacity for CO_2 fixation and is referred to as **"photoinhibition"**. The resultant inactive PS II reaction center can only be repaired by replacement of the damaged D1 subunit with a newly synthesized protein, which occurs at a surprisingly rapid rate. Repair requires partial dissociation of the reaction center, removal of the damaged D1 and its replacement. A compensatory, or protective, mechanism in most plants and algae is the ability for LHCII monomers to dissociate from PS II at high light intensities and move toward PS I, when PS II is running faster than PS I and the plastoquinone pool consequently becomes highly reduced. Dissociation of LHCII from PS II requires phosphorylation of the protein, a reaction that is regulated by binding of PQH_2 to cytochrome b_6. The phosphorylation of LHCII is catalyzed by a specific protein kinase, Stt7 (state transition, thylakoid) (Depège et al., 2003). As a consequence, more energy is transferred from LHCII to PS I, a condition referred to as "state 2". When light intensities return to lower levels, the kinase activity is reduced, and phosphatase enzymes remove the phosphate group from LHCII. Reconnection of LHCII with—and transfer of energy predominantly to—PS II restores the default condition referred to as "state 1" (Haldrup et al., 2001; Allen, 2003).

PROTECTION OF PS II BY CAROTENOIDS

Although repair of photodamaged PS II reaction centers can occur over a period of minutes, plants have developed another adaptive mechanism to protect D1 by releasing absorbed energy as heat in the antenna, before it is transferred to the reaction center. This process is called "nonphotochemical quenching" of Chl fluorescence (NPQ) and is triggered by two factors, an excessive proton gradient across the thylakoid membrane and synthesis of zeaxanthin. (Zeaxanthin is a member of a class of carotenoids, the "xanthophylls", that contain oxygen atoms.) At high light intensities, the deepoxidation reaction by which the xanthophyll violaxanthin is enzymatically converted to zeaxanthin, is stimulated (Fig. 13.21). The energy level of the lowest excited singlet state (denoted S_1)

of zeaxanthin is 14,550 cm^{-1}, which is lower than S_1 state of Chl a, about 14,700 cm^{-1}. The S_1 energy level of violaxanthin, at 14,880 cm^{-1}, is too high for energy transfer from Chl (Frank et al., 2000). Elevated amounts of zeaxanthin will quench excited states of Chl and allow dissipation of absorbed light energy by thermal decay within the antenna. Interconversion of violaxanthin and zeaxanthin in response to light intensity has been described as the "xanthophyll cycle" (Demmig-Adams and Adams, 1992).

High levels of zeaxanthin alone do not suffice for nonphotochemical quenching, as shown by npq2 mutants that lack the enzyme that catalyzes epoxidation of zeaxanthin to violaxanthin and therefore accumulate zeaxanthin. A large pH gradient across the thylakoid membrane, which is a factor in feedback regulation of electron transport, is also required. The effects of both factors are mediated by the PsbS protein, which is structurally similar to apoproteins of LHCII. Mutants lacking PsbS are unable to perform the quenching required for protection of the reaction center (Li et al., 2000). When two pH-responsive glutamate residues in loops of the protein exposed on the luminal side of the membrane were changed to glutamine residues, the resulting PsbS protein was unable to perform efficient quenching of energy in the antenna (Li et al., 2004).

Additional carotenoids, including the xanthophyll lutein in the core of LHCII, and the α- and β-carotenes and xanthophylls within the membrane environment, provide general protection against photooxidative damage. When the energy in excited Chl molecules is not transferred to the reaction center or quenched by protective mechanisms, the initial singlet state, which is short-lived, can convert to the triplet state.

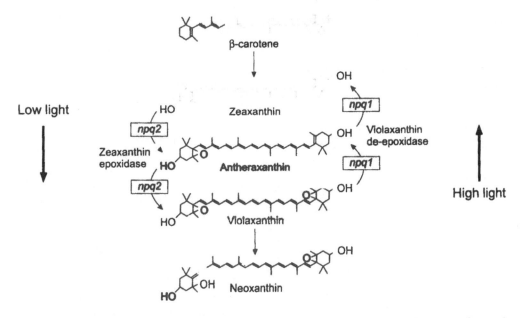

Fig. 13.21 The xanthophyll cycle. Xanthophylls are oxygenation products of α- and β-carotene. The major product from α-carotene is lutein, a dihydroxy carotenoid in light-harvesting complexes. Oxidation of β-carotene produces zeaxanthin, an isomer of lutein, which is normally converted through antheraxanthin to violaxanthin, a major xanthophyll in chloroplasts. Violaxanthin is the precursor of neoxanthin, another major xanthophyll. In plants exposed to high light intensity, which results in extensive reduction of the plastoquinone pool and generation of a large pH gradient across the thylakoid membrane, violaxanthin is converted back to zeaxanthin by deepoxidation. Zeaxanthin is able to quench excited states of Chl and prevents build-up of excitation pressure on PSII. Mutant strains defective in the ability to "nonphotochemically quench" excited states are designated *npq1* and *npq2* and lack the ability to interconvert violaxanthin and zeaxanthin (adapted from Niyogi, 1999).

Because this **"intersystem crossing"** involves reversal of the orientation of the spin of an electron, the Chl molecule now has two unpaired electrons with parallel spin, which restrains its decay to the ground state. Consequently, triplet states are relatively long-lived and decay only when an electron returns to antiparallel orientation. Triplet states can however, react with oxygen, and the transfer of energy to oxygen allows one of the unpaired, parallel-spin electrons in the *ground-state, triplet* oxygen molecule to flip. The resultant antiparallel orientation of the two electrons allows pairing to produce *singlet* oxygen, which has an unfilled orbital. Singlet oxygen is extremely reactive, and particularly in a membrane environment containing fatty acids with many unsaturated double bonds, will cause damage of the membrane by two-electron oxidation reactions. The abundance of oxygen molecules produced by photosynthesis therefore requires a mechanism to prevent this outcome. Carotenoids effectively quench the triplet state of Chls and thus provide a major protective "barrier" to damage of the photosynthetic apparatus.

INCORPORATION OF CARBON AS CO_2 INTO CARBOHYDRATE

Reductive Pentose-phosphate or "C_3" Cycle

The energy-laden products, ATP and NADPH, of *photo*-synthesis, referred to as "light reactions", are produced in very small quantities. These compounds are immediately consumed in **anabolic** (synthetic) reactions and remade. As such, they are not designed as energy-storage products. Photosynthesis absorbs much more energy over a day than the plant needs to sustain itself moment to moment. The excess is converted to storage forms such as starch or sucrose for use during the night. From many years of selection of desired traits by plant breeders, the vegetables we currently find in the grocery store have an ability to accumulate storage products in extraordinary large quantities.

Three molecules of ATP and two molecules of NADPH, the stoichiometry produced by linear electron transport in the light reactions, drive the **reductive pentose-phosphate cycle**, with each turn of the cycle resulting in incorporation of one molecule of CO_2 into carbohydrate. Thus the cycle must operate six times for synthesis of one molecule of glucose (Fig. 13.22). ATP is used to synthesize **ribulose 1,5-bisphosphate** from ribulose 5-phosphate in the reaction catalyzed by **ribulose 5-phosphate kinase**. Ribulose 1,5-bisphosphate then reacts with CO_2 in a reaction catalyzed by the most abundant protein in the biosphere, **ribulose 1,5-phosphate carboxylase/oxygenase**, or "Rubisco". The enzyme is so named because it cannot completely discriminate between CO_2 and O_2 (Fig. 13.23). The products of carboxylation reaction are two molecules of **3-phosphoglycerate**. The reaction with oxygen results in one molecule of 3-phos-phoglycerate and one of **2-phosphoglycolate** (Hartman and Harpel, 1994). The latter reaction is not anabolic and subsequent metabolism of 2-phosphoglycolate results in production of CO_2 in the pathway known as "**photorespiration**".

Rubisco is a large enzyme, occurring in eukaryotic photosynthetic organisms as a protein with 8 large subunits of about 52 kDa in mass and 8 small subunits between 12 and 16 kDa in mass. The total mass of the enzyme is about 520 kDa. Active sites on the enzyme are formed at the interface of the large subunits of alternating orientation, which provides 8 active sites for each protein molecule. The enzyme requires Mg^{2+} for activity, which can only be bound to the enzyme after a sidechain ε-amino group of lysine-201 reacts with a molecule of CO_2 to form a carbamate group,

$$\text{(-lysine- N} \overset{\overset{\displaystyle H}{|}}{} - \overset{\overset{\displaystyle O}{\|}}{C} - O^-)$$

This interesting reaction, in which CO_2 is required not as a substrate but to *activate* the enzyme, is catalyzed by **Rubisco activase**, a protein that undergoes conformational changes driven by hydrolysis of ATP (Spreitzer and Salvucci, 2002). The carbamate negative charge, along with sidechain carboxyl groups of glutamate and aspartate residues, coordinate the Mg^{2+} in the active site of Rubisco. The Mg^{2+} ion consequently positions the "substrate" CO_2 to achieve reaction with the other substrate, ribulose 1,5-bisphosphate. The enzyme is very "sluggish", catalyzing only about 3 reactions per second per protein molecule. The concentration of CO_2 that provides half-maximal activity of the enzyme, designated the K_M, is 10 μM, or approximately the concentration of CO_2 in water at 30°C and ambient atmospheric concentrations of CO_2. Thus the enzyme normally operates at only

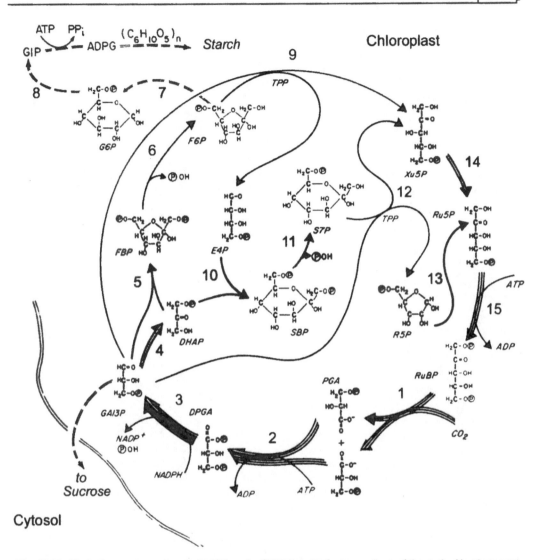

Fig. 13.22 Reductive pentose-phosphate (C₃) cycle. Thick lines indicate reactions of the cycle. Numbers next to each reaction refer to reactions listed in Table 13.1.

half its maximal rate, which suggests that plants should respond well to an elevated concentration of CO_2 in the atmosphere. The enzyme evolved in the early biosphere under conditions in which CO_2 was 10-fold or more higher than that in the current atmosphere while O_2 levels were very low. Nature has not been able to improve much

on this reaction over the eons and has apparently compensated for the seemingly inefficient catalytic activity by making large quantities of the enzyme. Nevertheless, the activity of Rubisco remains one of the limiting factors in photosynthesis.

The product of the carboxylation reaction, **3-phosphoglycerate**, is a 3-carbon

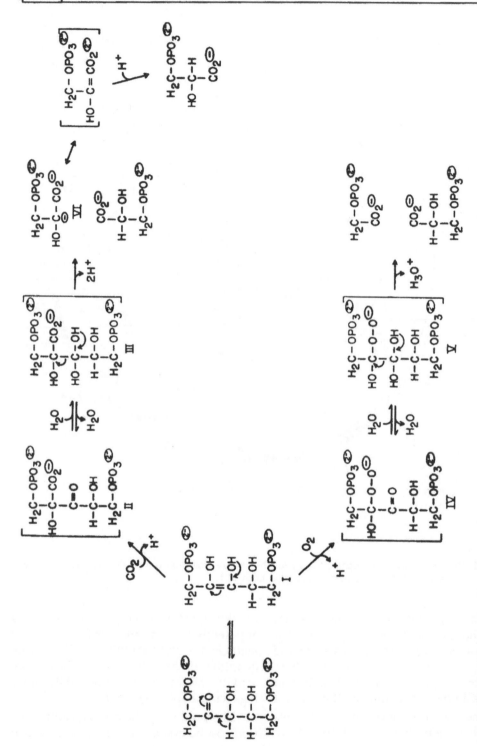

Fig. 13.23 Reactions catalyzed by Rubisco. The reaction of ribulose 1,5-bisphosphate with CO_2 results in two molecules of 3-phosphoglycerate (upper pathway). The reaction of O_2 results in one molecule of 3-phosphoglycerate and one of 2-phosphoglycolate (lower pathway) (from Hartman and Harpel, 1994).

compound, which led to designation of this pathway as the C_3 cycle. As described below, this is one of the three major pathways for assimilation of carbon from the atmosphere. (In the other two major pathways for assimilation of carbon from the atmosphere, a 4-carbon compound is the initial product of carbon assimilation [see below]). 3-Phosphoglycerate is converted to **1,3-bisphosphoglycerate** in a reaction with ATP. Because two molecules of 3-phosphoglycerate are now in play, this step in the cycle uses two molecules of ATP. These two, along with the ATP molecule involved in synthesis of ribulose 1,5-bisphosphate, account for the three required for each CO_2 molecule fixed. The product, 1,3-bisphosphoglycerate, has an "activated" carboxyl group, which can be reduced by NADPH to **glyceraldehyde 3-phosphate**. This reaction uses the two NADPH molecules produced in light reactions. The enzyme is named for the reverse reaction, **glyceraldehyde 3-phosphate dehydrogenase**. The remaining steps of the cycle are rearrangements of the molecules to arrive at a molecule of glucose. The reactions involved in assimilation of CO_2 are listed in Table 13.1.

"C_4" Pathway

Whereas over 90% of land plants contain *only* the reductive pentose-phosphate cycle, the remainder develop additional pathways for assimilation of carbon (Ku *et al.*, 1996). These alternate pathways are adaptations that provide greater efficiencies in more hostile environments, such as areas with higher temperatures and a more arid climate. In these pathways, HCO_3^- is the initial carbon source instead of CO_2. Bicarbonate is formed by dissociation of carbonic acid, which is formed when CO_2 in solution is hydrated (eqn 6).

$$CO_2 + H_2O \leftrightarrow H_2CO_3 \leftrightarrow HCO_3^- + H^+ \quad ...(6)$$

Hydration of CO_2 is catalyzed very rapidly by the enzyme **carbonic anhydrase**. Because the solubility of CO_2 in water decreases with increasing temperature, the concentration of bicarbonate can therefore reach much higher levels than that of CO_2 under these conditions. Consequently, incorporation of carbon into organic molecules is more efficient. The initial reaction in this pathway is catalyzed by **phospho-*enol*pyruvate carboxylase**, which adds the one-carbon unit to phospho*enol*pyruvate (eqn 7). In contrast to the reaction catalyzed by Rubisco, O_2 is not a substrate or inhibitor of this reaction. The product, **oxaloacetate**, is unstable and rapidly reduced to **malate** or converted to the amino acid **aspartate** by a transamination reaction (Fig. 13.24). The initial product of carbon assimilation is a 4-carbon intermediate, and thus the pathway has come to be known as the "C_4 **pathway**".

$$\underset{\text{Phosphoenolpyruvate}}{H_2C = \overset{\overset{\displaystyle PO_3^-}{|}}{\overset{\displaystyle |}{C}} - \overset{\overset{\displaystyle O}{\|}}{C} - O^-} + HCO_3^- \rightarrow$$

$$\underset{\text{Oxaloacetate}}{^-O - \overset{\overset{\displaystyle O}{\|}}{C} - CH_2 - \overset{\overset{\displaystyle O}{\|}}{C} - \overset{\overset{\displaystyle O}{\|}}{C} - O^-} + Pi \qquad ...(7)$$

The presence of the C_4 pathway is usually accompanied by a leaf morphology referred to as "Kranz" (crown) anatomy (Fig. 13.25). The vascular tissue in these plants is surrounded by a layer of "**bundle sheath**" cells that perform the C_3 pathway. Chloroplasts in bundle sheath cells contain thylakoid membranes that are not differentiated into grana or tightly appressed. Surrounding the bundle sheath cells are layers of "**mesophyll**" cells that initiate the C_4 pathway. Thylakoid membranes in the chloroplasts in these cells are differentiated into the typical granal and stromal

TABLE 13.1 Reactions in reductive pentose-phosphate cycle

A. Fixation of CO_2

(1) Ribulose 1,5-bisphosphate + CO_2 $\xrightarrow{\text{Rubisco}}$ 2 3-phosphoglycerate

(2) 3-Phosphoglycerate + ATP $\xrightarrow{\text{Phosphoglycerate kinase}}$ 1,3-bisphosphoglycerate + ADP

(3) 1,3-Bisphosphoglycerate + NADPH + H^+ $\xrightarrow{\text{Glyceraldehyde 3-phosphate dehydrogenase}}$ Glyceraldehyde 3-phosphatase + $NADP^+$ + Pi

B. Synthesis of glucose

(4) Glyceraldehyde 3-phosphate $\xleftrightarrow{\text{Triose phosphate isomerase}}$ Dihydroxyacetone phosphate

(5) Glyceraldehyde 3-phosphate + Dihydroxyacetone phosphate $\xrightarrow{\text{Fructose bisphosphate aldolase}}$ Fructose 1,6-bisphosphate

(6) Fructose 1,6-bisphosphate $\xrightarrow{\text{Fructose bisphosphate}}$ Fructose 6-phosphate + Pi

(7) Fructose 6-phosphate $\xrightarrow{\text{Glucose phosphate isomerase}}$ Glucose 6-phosphate

   (Glucose 6-phosphate $\xrightarrow{\text{Glucose 6-phosphatase}}$ Glucose + Pi)

(8) Glucose 6-phosphate $\xrightarrow{\text{Phosphoglucomutase}}$ Glucose 1-phosphate (substrate for starch synthesis)

C. Regeneration of ribulose 1,5-bisphosphate

(9) Fructose 6-phosphate + Glyceraldehyde 3-phosphate $\xrightarrow{\text{Transketolase}}$
   Xylulose 5-phosphate + Erythrose 4-phosphate

(10) Erythrose 4-phosphate + Dihydroxyacetone phosphate $\xrightarrow{\text{Fructose bisphosphate aldolase}}$
   Sedoheptulose 1,7-bisphosphate

(11) Sedoheptulose 1,7-bisphosphate $\xrightarrow{\text{Sedoheptulose bisphosphatase}}$ Sedoheptulose 7-phosphate + Pi

(12) Sedoheptulose 7-phosphate + Glyceraldehyde 3-phosphate $\xrightarrow{\text{Transketolase}}$
   Ribose 5-phosphate + Xylulose 5-phosphate

(13) Ribose 5-phosphate $\xrightarrow{\text{Ribose phosphate isomerase}}$ Ribulose 5-phosphate

(14) Xylulose 5-phosphate $\xrightarrow{\text{Ribose phosphate 3-epimerase}}$ Ribulose 5-phosphate

(15) Ribulose 5-phosphate + ATP $\xrightarrow{\text{Ribose 5-phosphate kinase}}$ Ribulose 1,5-bisphosphate + ADP

membranes. Interestingly, some species, such as in family Chenopodiaceae, contain both pathways *within the same cell*, with chloroplasts differentiated into the two types typical of bundle sheath and mesophyll cells (Voznesenskaya *et al.*, 2002).

Phospho*enol*pyruvate carboxylase is located predominantly in mesophyll cells. The initial products of carbon assimilation, malate or aspartate, are transported to bundle sheath cells, where they are oxidized or deaminated and then decarboxylated to generate CO_2. The CO_2 thus produced is used by Rubisco in chloroplasts of the bundle sheath cells to initiate the typical reductive pentose-phosphate cycle (Fig. 13.22). The other product of the decarboxylation reaction is pyruvate, which in plants that use malate as the "carbon carrier" is returned to the mesophyll cells. In those plants in which aspartate is the initial product, pyruvate is transaminated

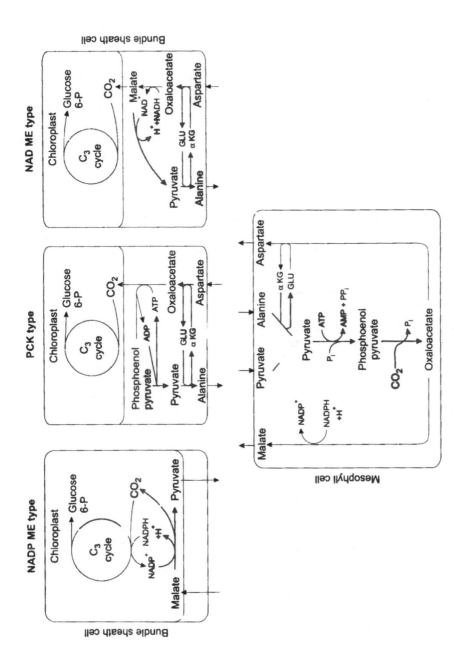

Fig. 13.24 Fluxes of carbon in the C_4 pathways of photosynthesis. In mesophyll cells, CO_2 is rapidly hydrated to form the bicarbonate ion, HCO_3^-. Bicarbonate reacts with phosphoeno*l*pyruvate to form oxaloacetate, catalyzed by phosphoeno*l*pyruvate carboxylase. Oxaloacetate is either reduced to malate or transaminated to aspartate, depending on the specific plant. These intermediates are transported to adjacent bundle sheath cells, where a reversal of these reactions provides a source of CO_2. This CO_2 is fixed in the bundle sheath cells by the reductive pentose-phosphate (C_3) pathway. The differing pathways in the variety of C_4 plants are illustrated in the upper three schemes.

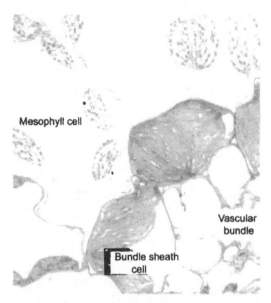

Mesophyll cell

Vascular bundle

Bundle sheath cell

Fig. 13.25 Morphological differentiation between the bundle sheath cells that surround the vascular tissue and the outer layer of mesophyll cells in a C_4 plant. Thylakoid membranes in the chloroplasts of bundle sheath cells are not organized into grana, in contrast to the extensive granal structures in mesophyll cells. Rubisco and the C_3 pathway are localized almost entirely in bundle sheath cells, while phospho*enol*pyruvate carboxylase and the C_4 pathway are localized in mesophyll cells (micrograph courtesy of Dr. C. C. Black, Jr.).

to alanine, which is then returned to the mesophyll cells. In these latter plants, the carrier must transport nitrogen as well as carbon. The advantage of the interplay between these cells is significantly higher concentration of CO_2 in the chloroplast of bundle sheath cells, which results from its production *in situ* by decarboxylation of the carbon carrier, than could be achieved under normal atmospheric conditions. Moreover, bundle sheath cells have a much lower PS II to PS I ratio than mesophyll cells, which reduces the amount of oxygen produced by PS II during light reactions.

Bundle sheath cells thus seem to perform more cyclic photosynthetic electron transport than mesophyll cells. The higher concentration of CO_2, combined with a lower oxygen level, increases the efficiency of "carboxylation" reaction of Rubisco over "oxygenation" reaction, which allows these plants to thrive under less favorable conditions. However, this adaptation has an energy cost of two additional ATP molecules consumed per CO_2 assimilated. The additional energy is required to convert pyruvate to phospho*enol*pyruvate in mesophyll cells, which expends two equivalents of ATP to convert AMP, the product of this reaction catalyzed by **pyruvate orthophosphate dikinase**, to ADP and then to ATP (Fig. 13.24).

Development of the C_4 pathway probably occurred just over 7 MYA, when the carbon isotope ratio in fossil organic material abruptly changed from one that showed strong discrimination against ^{13}C, a characteristic of CO_2 fixation by Rubisco, to a less discriminatory ratio characteristic of bicarbonate incorporation by phospho*enol*pyruvate carboxylase. The change in ratio indicated a rapid expansion of C_4 plants across the earth. C_4 plants have high rates of photosynthesis and growth, and 11 of the 12 most productive species are C_4 (see Raghavendra, 1998). Although only about 3% of the approximately 250,000 current species of plants contain the C_4 pathway, the agricultural importance of several of these species (e.g., maize, sugar cane, and sorghum) results in about 30% of the primary productivity of plants provided by C_4 plants.

An interesting variation on the C_4 pathway occurs in plants that live in unusually hot, dry climates. These plants keep their stomates—openings in the leaves through which O_2, CO_2, and H_2O exchange with the atmosphere—closed during the day but

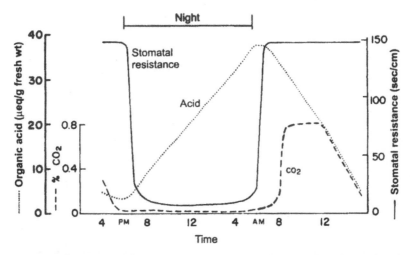

Fig. 13.26 Changes in internal gas phase CO_2 concentration (——), acid content of vacuole, primarily malate (·····), and stomatal resistance to diffusion of water vapor (——) in a typical CAM plant. The drop in stomatal resistance at nightfall marked the opening of these leaf structures. CO_2 entering through the stomata at night was incorporated into malate by the CAM pathway (Fig. 13.27). At dawn, the stomata closed and subsequent decarboxylation of malate markedly increased the CO_2 levels within the plant during the day. Over the course of the day, the released CO_2 was fixed into carbohydrate (starch) by the reductive pentose-phosphate cycle.

open during the night (Fig. 13.26). Photosynthesis during the day generates starch, which is stored in chloroplasts. At night, starch is metabolized through glycolysis to phospho*enol*pyruvate, which is carboxylated to oxaloacetate by phospho*enol*pyruvate carboxylase. Oxaloacetate is reduced to malate and stored in the large vacuole in the cells. During the day, malate is decarboxylated and the resulting CO_2 is fixed by Rubisco and the C_3 cycle, driven by photosynthetic electron transport (Fig. 13.27). This pathway, in which light reactions and CO_2 assimilation reactions are separated *temporally*, is a remarkable adaptation to a hostile environment and referred to as "**crassulacean acid metabolism**". Plants containing this pathway are commonly known as "CAM plants". Leaves of these plants are generally thick and fleshy and store large quantities of water. The most common examples of such plants are cacti and agave plants that survive surprisingly well in the hot, arid deserts of southwestern US and Mexico.

Regulation of CO_2 Assimilation by Light

The activity of a large number of soluble enzymes in the chloroplast stroma is regulated to prevent "futile" reactions that would dissipate the energy captured by the reactions on the thylakoid membrane. A general factor controlling enzymatic activity is the change in environment in the stroma between day and night. During light-driven photophosphorylation, transfer of protons from the stroma to the thylakoid lumen for generation of the proton gradient results in pH of the stroma rising to nearly 8, while pH is near 7 during the night. For an enzyme with a pH optimum near 8, this change in pH is sufficient to dramatically affect activity. The

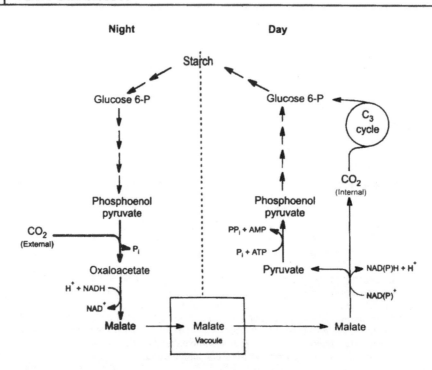

Fig. 13.27 Pathway for the flux of carbon in crassulacean acid metabolism in CAM plants. During the night starch is degraded through a glycolytic pathway to phospho*enol*pyruvate. CO_2 entering through the stomata from the atmosphere at night is hydrated to carbonic acid and fixed into oxaloacetate by phospho*enol*pyruvate carboxylase (see Fig. 13.24). Oxaloacetate is reduced by cytosolic NAD^+-malate dehydrogenase to malate, which is transported into and accumulates in the vacuole of the cell. During the day malate is released from the vacuole and converted to pyruvate through a decarboxylation reaction catalyzed by $NADP^+$-dependent malate dehydrogenase. Pyruvate is converted to phospho*enol*pyruvate by an unusual reaction catalyzed by pyruvate phosphate dikinase, which provides the substrate for glucose and starch synthesis by a reversal of glycolysis (gluconeogenesis). The CO_2 released by the decarboxylation reaction of malate is refixed photosynthetically by the reductive pentose-phosphate cycle. Loss of the CO_2 produced in this reaction is prevented by closure of the stomata during the day (see Fig. 13.26).

lower concentration of H^+ in the stroma during the day is compensated for by an increase in concentration of Mg^{2+}. Several enzymes, including Rubisco, require Mg^{2+} for activity. The higher pH and concentration of Mg^{2+} support significantly higher activity of these enzymes.

Light energy is required to drive the photosynthetic reactions, and these reactions in turn regulate activity of several key enzymes in CO_2 assimilation. Chloroplasts regulate many of their enzymes by covalent

modification. A common mechanism is disulfide-sulfhydryl interchange (-S—S- ↔ 2 -SH) between cysteine residues in the protein, in which light-driven reduction of the disulfide bond to sulfhydryl groups occurs during the day and reoxidation to the disulfide form in darkness (Buchanan, 1991). For example, at the end of the day, the photosystems stop driving the electron transport required to generate the proton gradient. Left on its own, ATP synthase would then run in reverse, as an ATP hydrolase, using the remaining ATP made during the day for a retrograde transfer of protons from stroma to the lumen. This hydrolysis of ATP is inhibited by oxidation of a pair of sulfhydryl groups in the γ subunit, cysteine-199 and cysteine-205, to a disulfide bond, which locks the rotary motion of the γ subunit of ATP synthase (see Fig. 13.19). Upon activation of electron transport at dawn, ferredoxin is reduced by PS I, which in turn reduces a small (12 kDa) protein called **thioredoxin**. Thioredoxin contains cysteine residues that undergo reversible oxidation and reduction. Production of reduced thioredoxin leads to reduction of the disulfide bond in the γ subunit, and the activity of ATP synthase is restored.

Regulation of the reductive pentose-phosphate cycle is critical because the cycle must run in the forward direction during the day, using ATP and NADPH to drive synthesis of glucose 6-phosphate from CO_2. Glucose 6-phosphate is the substrate for storage of carbohydrate as starch. In darkness, the cycle operates in reverse and is then called the *oxidative* pentose-phosphate cycle. Glucose 6-phosphate, produced by breakdown of starch, is metabolized in the oxidative pathway to 6-phosphogluconate and then decarboxylated to ribulose 5-phosphate, with $NADP^+$ reduced to NADPH in each step. These reactions serve as the primary source of NADPH and ribose 5-phosphate in darkness. To prevent "futile" cycling, or to ensure that the system does not work against itself, the cycle is regulated. **Glucose 6-phosphate dehydrogenase**, which oxidizes glucose 6-phosphate to 6-phosphogluconate—the first step in the oxidative cycle, is inhibited by the product of the reaction, NADPH. At pH 8, pH of the stroma during active photosynthesis, the ratio of concentrations of NADPH over $NADP^+$ is sufficiently high to completely inhibit the enzyme. In addition, the enzyme is also inhibited by ribulose 1,5-bisphosphate, which is the substrate for Rubisco (Lendzian and Bassham, 1975). Thus by reducing $NADP^+$ to NADPH and synthesizing ATP from ADP and phosphate, the light reactions of photosynthesis ensure that the oxidative pentose-phosphate pathway is blocked during the day. Furthermore, just to make sure, glucose 6-phosphate dehydrogenase is inactivated in light through reduction by thioredoxin.

Several enzymes in the cycle, including fructose 1,6-bisphosphatase, sedoheptulose 1,7-bisphosphatase, ribulose 5-phosphate kinase, and NADP-dependent glyceraldehyde dehydrogenase, are activated in light by reduced thioredoxin. Each of these enzymes of the carbon assimilation cycle is inactivated when two cysteine-SH groups in the enzyme are oxidized to a disulfide (-S—S-). Restoration of activity requires light-activated PS I to reduce ferredoxin, which then reduces thioredoxin. Although the overall redox state of the chloroplast remains highly reducing during the night, the redox potentials of sulfhydryl groups on these regulated enzymes are unusually low and are oxidized to the disulfide form when the status of the stroma becomes only slightly more oxidative.

A particularly well-studied enzyme that is regulated by sulfhydryl-disulfide interchange is **NADP-dependent malate dehydrogenase**, which is involved in the C_4 pathway of carbon fixation but also present in C_3 chloroplasts. As shown in Fig. 13.28, the enzyme is inactive in the oxidized form and fully activated by treatment with reduced thioredoxin. In the oxidized form, a disulfide bond occurs near the C-terminus of the protein, between cysteine-361 and cysteine-373 (Ashton *et al.*, 2000). This bond causes the otherwise flexible C-terminal extension to be constrained into a sharp turn and fold into the active site (Fig. 13.29). Consequently, binding of malate to the active site is blocked and no conversion to oxaloacetate is measured. Addition of a sulfhydryl reducing agent such as dithiothreitol is relatively ineffective in activating the enzyme as compared with reduced thioredoxin, which suggests that thioredoxin has an additional activity, perhaps through pro-

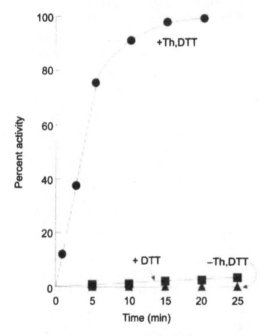

Fig. 13.28 Activation of the chloroplastic NADP$^+$-dependent malate dehydrogenase. The oxidized form of the enzyme, in which two cysteine residues are linked with a disulfide (-S—S-) bond, is inactive. Dithiothreitol, a 4-carbon molecule containing a sulfhydryl group (-SH) on each terminal carbon, was ineffective in activating the enzyme but was capable of reducing thioredoxin to its sulfhydryl form. Addition of reduced thioredoxin rapidly activated the enzyme by reducing the disulfide bond to sulfhydryl groups (adapted from Ferte *et al.*, 1982).

tein-protein interactions (Ashton *et al.*, 2000).

Rubisco is also a highly regulated enzyme. In darkness, when the carbamate group of Rubisco dissociates, the enzyme binds one of its substrates, ribulose 1,5-bisphosphate, very tightly (K_d of 20 nM) in a nonproductive complex. Without the carbamate group, the enzyme cannot bind the Mg^{2+} ion required for binding the other substrate, CO$_2$. Formation of the carbamate group at the active site is thus effectively blocked. Rubisco activase, which requires ATP for activity, facilitates removal of the ribulose 1,5-bisphosphate and thus allows activation of Rubisco (Hartman and Harpel, 1994). To further ensure inactivity of Rubisco in darkness, any of the enzyme remaining in the carbamylated form binds very tightly (K_d of 32 nM) to a reaction-intermediate analogue, 2-carboxyarabinitol 1-phosphate, which inhibits enzymatic activity. When reductants are generated in light, a phosphatase is activated that hydrolyzes the phosphoester bond, which causes concentration of free 2-carboxyarabinitol 1-phosphate to drop below the K_d and allows dissociation of the inhibitor from the active site (Holbrook *et al.*, 1991). Moreover, the kinase reaction that phosphorylates ribulose 5-phosphate to produce ribulose 1,5-bisphosphate, the substrate for Rubisco, is activated by thioredoxin. Consequently, the reaction catalyzed by Rubisco can only proceed when conditions for CO$_2$ fixation are suitable. Rubisco activase is also inactive in darkness and is activated by thioredoxin. The remarkable regulation of these enzymes ensures that ATP is not dissipated in darkness needlessly, and reactions that use ATP as a substrate occur only in light under which it can be replenished (Geiger and Servaites, 1994).

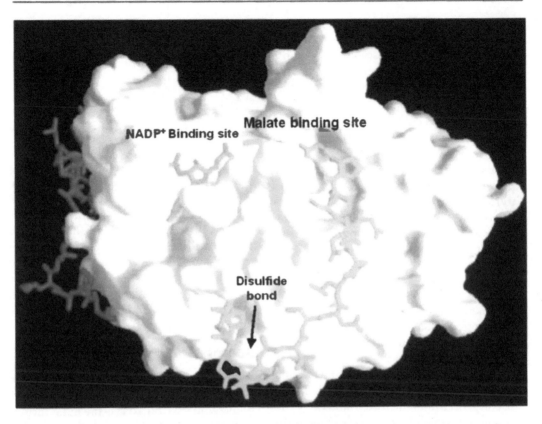

Fig. 13.29 Model of structure of the oxidized, inactive NADP⁺-dependent malate dehydrogenase. Figure shows surface view of the enzyme monomer with stick view of the C-terminal extension (cyan) that extends into the active site near the bound NADP⁺ (green), which is partially obscured by the active site loops. The presence of the negatively charged C-terminus in the active site stabilizes the binding of positively charged NADP⁺ but not NADPH. Thioredoxin reduces the disulfide bond between cysteine-361 and cysteine-373 (the enzyme contains 385 amino acids) and promotes dissociation of the C-terminal extension from the active site (adapted from Ashton *et al.*, 2000).

END PRODUCTS OF CARBON AS-SIMILATION

Starch

Similar to storage of excess calories by animals in the form of lipids and a polymer of glucose, plants also use these materials to store energy. Lipids, the long-term storage form of carbon with highest caloric content per unit weight, accumulates primarily as fat in adipose tissue in animals and in oil bodies in the seeds of plants. In animals, glucose is polymerized into a highly branched molecule, glycogen. In plants, glucose is polymerized into long, mostly linear polymers, **starch** or **amylose**, which has roughly 2,000 glucose units per polymer. Starch is slightly branched, much less so than another, more abundant, highly branched polymer, amylopectin. Up to 10,000 glucose units occur in amylopectin, which has an average chain length between branches of about 21 (Nelson and Pan,

1995). Chains are linked by α-1,4 glycosidic bonds (Fig. 13.30), with branches extending from the carbon-6 hydroxymethyl group in α-1,6 bonds. Starch accumulates as large aggregates or "grains" in the stroma of chloroplasts (see Figs. 13.4 and 13.5)

In the stroma of the chloroplast, glucose 6-phosphate, the ultimate product of the reductive pentose-phosphate cycle, is converted to glucose 1-phosphate by **phosphoglucomutase**. **ADP glucose pyrophosphorylase** catalyzes the reaction between ATP and glucose 1-phosphate to generate ADP-glucose and pyrophosphate (eqn 8).

$$\text{ATP + glucose 1-P} \leftrightarrow$$
$$\text{ADP-glucose + PPi} \qquad ...(8)$$

ADP serves as the "leaving group" when the oxygen in the 4-OH of the glucose residue at the end of a polymer attacks carbon-1 of ADP-glucose (eqn 9), the reaction catalyzed by **starch synthase**. Thus the polymer grows at the "non-reducing" end, which is also the end from which degradation occurs to release glucose 1-phosphate as the result of **starch phosphorylase** activity (eqn 10).

$$\text{HO-Glucose-(glucose)}_n +$$
$$\text{ADP-glucose} \leftrightarrow \text{glucose}(\alpha 1 \to 4)\text{-}$$
$$\text{O-glucose-(glucose)}_n + \text{ADP} \qquad ...(9)$$

$$\text{Glucose-(glucose)}_n +$$
$$\text{Pi} \leftrightarrow \text{glucose 1-phosphate} +$$
$$\text{(glucose)}_n \qquad ...(10)$$

ADP-glucose pyrophosphorylase is the rate-limiting activity in starch synthesis and is allosterically activated by 3-phosphoglyceric acid, the initial product of CO_2 fixation (Stark *et al.*, 1992). Hence the enzyme is most active during periods of maximal rates of photosynthesis. The enzyme is inhibited by phosphate (Pi) whose concentration increases in darkness. The higher concentration of Pi in the chloroplast promotes breakdown of starch in darkness, catalyzed by starch phosphorylase.

Sucrose

Whereas starch forms insoluble granules, **sucrose** is a highly soluble end product of photosynthesis. Sucrose is the main form of carbon transport in plants and plays a key role in reproduction and propagation, as

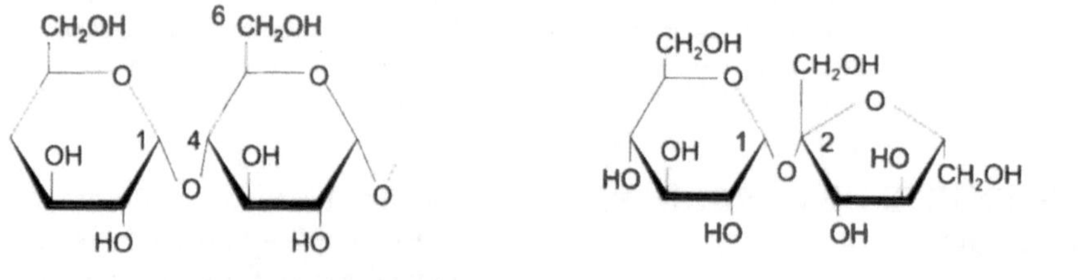

A. α(1 → 4) linkage in strach

B. Sucrose, D-glucose (α, 1 → β,2)-D-fructose

Fig. 13.30 Structures of the linkages in starch and sucrose. The linear chain of starch consists of glucose molecules linked α(1 → 4). Branches occur by additional α(1 → 6) linkages, in which a glucose molecule at a branch point is linked to other molecules through the 6-hydroxyl as well as the 4-hydroxyl groups. In sucrose, the glycosidic bond links the anomeric carbon-1 of glucose, with the hydroxyl in the α configuration, with the anomeric carbon-2 of fructose, with the hydroxyl in the β configuration, to form an α1 → β2 linkage. The net result of formation of a glycosidic bond is elimination of a molecule of water.

seen in flower nectar. As a readily available source of energy, sucrose sustains the initial stages of growth after dormant periods in temperate plants. Sucrose is found in all cells of a plant and is enzymatically hydrolyzed to fructose and glucose, the ultimate building blocks for all other organic compounds in plants.

Unlike starch, sucrose is synthesized in the cytosol. Carbon leaves the chloroplast primarily as the triose-phosphates, glyceraldehyde 3-phosphate or dihydroxy-acetone-phosphate, in exchange for inorganic phosphate. This exchange is facilitated by a specific transporter on the chloroplast envelope. In the cytosol, reversal of the glycolytic pathway converts triose-phosphates to fructose 1,6-bisphosphate, fructose 6-phosphate and then glucose 6-phosphate. Glucose 1-phosphate is made from glucose 6-phosphate in the reaction catalyzed by phosphoglucomutase and is then used to produce UDP-glucose by **UDP-glucose pyrophosphorylase**. This reaction is mechanistically similar to the reaction in the plastid (eqn 8) but uses UTP rather than ATP. UDP-glucose reacts with fructose 6-phosphate to form sucrose 6-phosphate in the reaction catalyzed by **sucrose-phosphate synthase**. This enzyme is inhibited by high concentrations of Pi, as would occur in darkness, and also by phosphorylation of a serine residue (Huber and Huber, 1996). Sucrose 6-phosphate is hydrolyzed to sucrose by **sucrose-phosphate phosphatase**. Glucose and fructose are linked in sucrose between the "reducing" groups of each, *i.e.*, (α1 → β2) (Fig. 13.30), which produces a much more stable disaccharide than the isomers maltose or lactose. Excess sucrose is transported from the cytosol into the vacuole by a specific transport system on the tonoplast, the membrane that surrounds the vacuole.

CONCLUSIONS FOR REACTIONS OF PHOTOSYNTHESIS

ATP and NADPH are the chemical currencies that photosynthetic organisms use for synthesis of glucose from CO_2 and H_2O. Glucose is stored as starch in the chloroplast and used to support metabolism of the plant in darkness or is consumed by animal cells to provide energy for their metabolism and growth. Intermediates in the pathway to glucose, at the level of triose-phosphates, are also exported from the chloroplast and used for synthesis of sucrose in the cytosol, the other major storage form of energy as carbohydrates in plants. Although less in bulk, sucrose is highly desired as a source of energy by animals. Metabolism of starch and sucrose generates CO_2 and H_2O and the cycle repeats.

References

Albertsson, P.-A. 1995. The structure and function of the chloroplast photosynthetic membrane—a model for the domain organization. Photosyn. Res. 46: 141-149.

Allen, J.F. 2003. State transitions—a question of balance. Science 299: 1530-1532.

Allen, J.P. and Williams, J.C. 1998. Photosynthetic reaction centers. FEBS Lett. 438: 5-9.

Archer, M.D. and Barber, J. 2003. *Molecular to Global Photosynthesis*. World Scientific, Singapore.

Ashton, A.R., Trevanion, S.J., Carr, P.D., Verger, D., and Ollis, D.L. 2000. Structural basis for the light regulation of chloroplast NADP malate dehydrogenase. Physiol. Plant. 110: 314-321.

Ballschmitter, K., Cotton, T.M., and Katz J.J. 1969. Chlorophyll-water interactions. Hydration, dehydration and hydrates of chlorophyll. Biochim. Biophys. Acta 180: 347-359.

Barter, L.M.C., Durrant, J.R., and Klug, D.R. 2003. A quantitative structure-function relationship for the photosystem II reaction center: supermolecular behavior in natural photosynthesis. Proc. Natl. Acad. Sci. USA 100: 946-951.

Bellafiore, S., Ferris, P., Naver, H., Göhre, V., and Rochaix, J.D. 2002. Loss of Albino3 leads to the specific depletion of the light-harvesting system. Plant Cell 14: 2303-2314.

Bhattacharya, D., Yoon, H.S., and Hackett, J.D. 2003. Photosynthetic eukaryotes unite: endosymbiosis connects the dots. Bio Essays 26: 50-60.

Blankenship, R.E. 2002. *Molecular Mechanisms of Photosynthesis*. Blackwell Science, Ltd., Oxford, UK.

Boyer, P.D. 1998. ATP synthase—past and future. Biochim. Biophy. Acta 1365: 3-9.

Buchanan, B.B. 1991. Regulation of CO_2 assimilation in oxygenic photosynthesis: the ferredoxin/thioredoxin system. Arch. Biochem. Biophys. 288: 1-9.

Chen, M., Eggink, L.L., Hoober, J.K., and Larkum, A.W.D. 2005. Influence of structure on binding of chlorophylls to peptide ligands. J. Amer. Chem. Soc. 127: 2052-2053.

Cogdell, R.J., Isaacs, N.W., Howard, T.D., McLuskey, K., Fraser, N.J., and Prince, S.M. 1999. How photosynthetic bacteria harvest solar energy. J. Bacteriol. 181: 3869-3879.

Croce, R., Weiss, S., and Bassi, R. 1999. Carotenoid-binding sites of the major light-harvesting complex II of higher plants. J. Biol. Chem. 274: 29613-29623.

Croce, R., Canino, G., Ros, F., and Bassi, R. 2002. Chromophore organization in the higher-plant photosystem II antenna protein CP26. Biochemistry 41: 7334-7343.

Deisenhofer, J. and Michel, H. 1989. The photosynthetic reaction center from the purple bacterium *Rhodopseudomonas viridis*. Science 245: 1463-1473.

Demmig-Adams, B. and Adams, W.W. III. 1992. Photoprotection and other responses of plants to high light stress. Ann. Rev. Plant Physiol. Plant Molec. Biol. 43: 599-626.

Deng, Z., Aliverti, A., Zanetti, G., Arakaki, A.K., Ottado, J., Orellano, E.G. *et al.* 1999. A productive NADP⁺ binding mode of ferredoxin-NADP⁺ reductase revealed by protein engineering and crystallographic studies. Nature Struct. Biol. 6: 847-853.

Depège, N., Bellafiore, S., and Rochaix, J.D. 2003. Role of chloroplast protein kinase Stt7 in LHCII phosphorylation and state transition in *Chlamydomonas*. Science 299: 1572-1575.

Drews, G. 1996. Forty-five years of developmental biology of photosynthetic bacteria. Photosyn. Res. 48: 325-352.

Eggink, L.L., Park, H.S., and Hoober, J.K. 2001. The role of chlorophyll *b* in photosynthesis: hypothesis. BMC Plant Biol. 1: 2.

Eggink, L.L., LoBrutto, R., Brune, D.C., Brusslan, J., Yamasato, A., Tanaka, A., and Hoober, J.K. 2004. Synthesis of chlorophyll *b*: location of chlorophyllide *a* oxygenase and discovery of a stable radical in the catalytic subunit. BMC Plant Biol. 4: 5.

Elston, T., Wang, H., and Oster, G. 1998 Energy transduction in ATP synthase. Nature 391: 510-513.

Ferreira, K.N., Iverson, T.M., Maghlaoui, K., Barber, J., and Iwata, S. 2004. Architecture of the photosynthetic oxygen-evolving center. Science 303: 1831-1838.

Ferro, M., Salvi, D., Rivière-Rolland, H., Vermat, T., Seigneurin-Berny, D., Grunwald, D., Garlin, J., Joyard, J., and Rolland, N. 2002. Integral membrane proteins of the chloroplast envelope: identification and subcellular localization of new transporters. Proc. Natl. Acad. Sci. USA 99: 11487-11492.

Ferte, N., Meunier, J.-C., Ricard, J., Buc, J., and Sauve, P. 1982. Molecular properties and thioredoxin-mediated activation of spinach chloroplastic NADP-malate dehydrogenase. FEBS Lett. 146: 133-138.

Frank, H.A., Bautista, J.A., Josue, J.S., and Young, A.J. 2000. Mechanism of nonphotochemical quenching in green plants: energies of the lowest excited singlet states of violaxanthin and zeaxanthin. Biochemistry 39: 2831-2837.

Gantt, E. 1981. Phycobilosomes. Ann. Rev. Plant Physiol. 32: 327-347.

Geiger, D.R. and Servaites, J.C. 1994. Diurnal regulation of photosynthetic carbon metabolism in C_3 plants. Ann. Rev Plant Physiol. Plant Molec. Biol. 45: 235-236.

Gouterman, M. 1961 Spectra of porphyrins. J. Molec. Spectrosc. 6: 138-163.

Green, B.R. and Durnford, D.G. 1996. The chlorophyll-carotenoid proteins of oxygenic photosynthesis. Ann. Rev. Plant Physiol. Plant Molec. Biol. 47: 685-714.

Guergova-Kuras, M., Boudreaux, B., Joliot, A., Joliot, P., and Redding, K. 2001. Evidence for two active branches for electron transfer in photosystem I. Proc. Natl. Acad. Sci. USA 98: 4437-4442.

Gunning, B.E.S. 2004. www.rsbs.anu.edu.au/profiles/gunning (accessed February 28, 2004).

Haldrup, A., Jensen, P.E., Lunde, C., and Scheller, H.V. 2001. Balance of power: a view of the mechanism of photosynthetic state transitions. Trends Plant Sci. 6: 301-305.

Hankamer, B., Barber, J., and Boekema, E.J. 1997. Structure and membrane organization of photosystem II in green plants. Ann. Rev. Plant Physiol. Plant Molec. Biol. 48: 641-671.

Hartman, F.C. and Harpel, M.R. 1994 Structure, function, regulation, and assembly of D-ribulose-1,5-bisphosphate carboxylase/oxygenase. Ann. Rev. Biochem. 63: 197-234.

Hess, W.R., Rocap, G., Ting, C.S., Larimer, F., Stilwagen, S., Lamerdin, J., and Chisholm, S.W. 2001. The photosynthetic apparatus of *Prochlorococcus*: insights through comparative genomics. Photosyn. Res. 70: 53-71.

Holbrook, G.P., Galasinski, S.C., and Salvucci, M.E. 1991. Regulation of 2-carboxyarabinitol 1-phosphatase. Plant Physiol. 97: 894-899.

Hoober, J.K., Boyd, C.O., and Paavola, L.G. 1991. Origin of thylakoid membranes in *Chlamydomonas reinhardtii y*-1 at 38°C. Plant Physiol. 96: 1321-1328.

Hoober, J.K. and Eggink, L.L. 2001. A potential role for chlorophylls *b* and *c* in assembly of light-harvesting complexes. FEBS lett. 489: 1-3.

Hu, X., Ritz, T., Damjanović, A., Autenrieth, F., and Schulten, K. 2002. Photosynthetic apparatus of purple bacteria. Quart. Rev. Biophys. 35: 1-62.

Huber, S.C. and Huber, J.L. 1996. Role and regulation of sucrose-phosphate synthase in higher plants. Ann. Rev. Plant Physiol. Plant Molec. Biol. 47: 431-444.

Hutin, C., Havaux, M., Carde, J.P., Kloppstech, K., Meiherhoff, K., Hoffman, N., and Nussaume, L. 2002. Double mutation cpSRP43⁻/cpSRP54⁻ is necessary to abolish the cpSRP pathway required for thylakoid targeting of the light-harvesting chlorophyll proteins. Plant J. 29: 531-543.

Jordan, P., Fromme, P., Witt, H.T., Klukas, O., Saenger, W., and Krauss, N. 2001. Three-dimensional structure of cyanobacterial photosystem I at 2.5 Å resolution. *Nature* 411: 909-917.

Kargul, J., Nield, J., and Barber, J. 2003. Three-dimensional reconstruction of a light-harvesting complex I-photosystem I (LHCI-PSI) supercomplex from the green algal *Chlamydomonas reinhardtii*. J. Biol. Chem. 278: 16135-16141.

Ke, B. 2001. *Photosynthesis: Photobiochemistry and Photobiophysics*. Kluwer Acad. Publ., Dordrecht, The Netherlands.

Kirchhoff, H., Horstmann, S., and Weis, E. 2000. Control of the photosynthetic electron transport by PQ diffusion microdomains in thylakoids of higher plants. Biochim. Biophys. Acta 1459: 148-168.

Kramer, D.M., Sacksteder, D.A., and Cruz, J.A. 1999. How acid is the lumen? Photosyn. Res. 60: 151-163.

Kroll, D., Meierhoff, K., Bechtold, N., Kinoshita, M., Westphal, S., Vothknecht, U.C., Soll, J., and Westhoff, P. 2001. *VIPP1*, a nuclear gene of *Arabidopsis thaliana* essential for thylakoid membrane formation. Proc. Natl. Acad. Sci. USA 98: 4238-4242.

Ku, M.S.B., Kano-Murakami, Y., and Matsuoka, M. 1996. Evolution and expression of C₄ photosynthesis genes. Plant Physiol. 111: 949-957.

Kühlbrandt, W., Wang, D.N., and Fujiyoshi, Y. 1994. Atomic model of plant light-harvesting complex by electron crystallography. Nature 367: 614-621.

Kurisu, G., Zhang, H., Smith, J.L., and Cramer, W.A. 2003. Structure of the cytochrome *b₆f* complex of oxygenic photosynthesis: tuning the cavity. Science 302: 1009-1014.

Lawlor, D.W. 2001. *Photosynthesis*. Bios Sci. Publ., Oxford, United Kingdom (3rd ed.).

Lendzian, K. and Bassham, J.A. 1975. Regulation of glucose-6-phosphate dehydrogenase in spinach chloroplasts by ribulose 1,5-diphosphate and NADPH/NADP⁺ ratios. Biochim. Biophys. Acta 396: 260-275.

Li, X.P., Björkman, O., Shih, C., Grossman, A.R., Rosenquist, M., Jansson, S., and Niyogi, K.K. 2000. A pigment-binding protein essential for regulation of photosynthetic light harvesting. Nature 403: 391-395.

Li, Y.F., Zhou, W.L., Blankenship, R.E., and Allen, J.P. 1997. Crystal structure of the bacteriochlorophyll *a* protein from *Chlorobium tepidum*. J. Molec. Biol. 271: 456-471.

Li, X.P., Gilmore, A.M., Caffarri, S., Bassi, R., Golan, T., Kramer, D., and Niyogi, K.K. 2004. Regulation of photosynthetic light harvesting involves intrathylakoid lumen pH sensing by the PsbS protein. J. Biol. Chem. 279: 22866-22874.

Liu, Z., Yan, H., Wang, K., Kuang, T., Zhang, J., Gui, L., An, X., and Chang, W. 2004. Crystal structure of spinach major light-harvesting complex at 2.72 Å resolution. Nature 428: 287-292.

Martin, W., Rujan, T., Richly, E., Hansen, A., Cornelsen, S., Lins, T. *et al.* 2002. Evolutionary analysis of *Arabidopsis*, cyanobacteria, and chloroplast genomes reveals plastid phylogeny and thousands of cyanobacterial genes in the nucleus. Proc. Natl. Acad. Sci. USA 99: 12246-12251.

Mojzsis, S.J., Arrhenius, G., McKeegan, K.D., Harrison, T.M., Nutman, A.P., and Friend, C.R.L. 1996. Evidence for life on earth before 3,800 million years ago. Nature 384: 55-59.

Munekage, Y., Hashimoto, M., Miyake, C., Tomizawa, K.-I., Endo, T., Tasaka, M., and Shikanai, T. 2004. Cyclic electron flow around photosystem I is essential for photosynthesis. Nature 429: 579-582.

Mustárdy, L. and Garab, G. 2003. Granum revisited. A three-dimensional model—where things fall into place. Trends Plant Sci. 8: 117-122.

Nelson, O. and Pan, D. 1995. Starch synthesis in maize endosperms. Ann. Rev. Plant Physiol. Plant Molec. Biol. 46: 475-496.

Nield, J., Orlova, E.V., Morris, E.P., Gowen, B., van Heel, M. and Barber, J. 2000. 3D map of the plant photosystem II supercomplex obtained by cryoelectron microscopy and single particle analysis. Nature Struct. Biol. 7: 44-47.

Niyogi, K.K. 1999. Photoprotection revisited: genetic and molecular approaches. Ann. Rev. Plant Physiol. Plant Molec. Biol. 50: 333-359.

Olson, J.M. 1998. Chlorophyll organization and function in green photosynthetic bacteria. Photochem. Photobiol. 67: 61-75.

Papiz, M.Z., Prince, S.M., Howard, T., Cogdell, R.J., and Isaacs, N.W. 2003. The structure and thermal motion of the B800-850 LH2 complex from *Rps. Acidophila* at 2.0 Å resolution and 100K: new structural features and functionally relevant motions. J. Molec. Biol. 326: 1523-1538.

Park, H. and Hoober, J.K. 1997. Chlorophyll synthesis modulates retention of apoproteins of light-harvesting complex II by the chloroplast in *Chlamydomonas reinhardtii*. Physiol. Plant. 101: 135-142.

Park, H., Eggink, L.L., Roberson, R.W., and Hoober, J.K. 1999. Transfer of proteins from the chloroplast to vacuoles in *Chlamydomonas reinhardtii* (Chlorophyta): a pathway for degradation. J. Phycol. 35: 528-538.

Park, H., Kreunen, S.S., Cuttriss, A.J., DellaPenna, D., and Pogson, B.J. 2002. Identification of the carotenoid isomerase provides insight into carotenoid biosynthesis, prolamellar body formation, and photomorphogenesis. Plant Cell 14: 321-332.

Pascal, A., Caffarri, S., Croce, R., Sandonà, D., Bassi, R., and Robert, B. 2002. A structural investigation of the central chlorophyll *a* binding sites in the minor photosystem II antenna protein, Lhcb4. Biochemistry 41: 2305-2310.

Peterson, R.B. and Havir, E.A. 2001. Photosynthetic properties of an *Arabidopsis thaliana* mutant possessing a defective PsbS gene. Planta 214: 142-152.

Pflanzagl, B., Zenker, A., Pittenauer, E., Allmaier, G., Martinez-Torrecuadrada, J., Schmid, E.R., De Pedro, M.A., and Löffelhardt, W. 1996. Primary structure of cyanelle peptidoglycan of *Canophora paradeoxa*: a prokaryotic cell wall as part of an organelle envelope. J. Bacteriol. 178: 332-339

Raghavendra, A.S. 1998. *Photosynthesis: A Comprehensive Treatise*. Cambridge Univ. Press, Cambridge, UK.

Raven, J.A. and Allen, J.F. 2003. Genomics and chloroplast evolution: what did cyanobacteria do for plants? Genome Biol. 4: 209.

Reinbothe, C., Buhr, F., Pollmann, S., and Reinbothe, S. 2003. *In vitro* reconstitution of light-harvesting POR-protochlorophyllide complex with protochlorophyllides *a* and *b*. J. Biol. Chem. 278: 807-815.

Remelli, R., Varotto, C., Sandonà, D., Croce, R., and Bassi, R. 1999. Chlorophyll binding to monomeric light-harvesting complex: a mutational analysis of chromophore-binding residues. J. Biol. Chem. 274: 33510-33521.

Sacksteder, C.A., Kanazawa, A., Jacoby, M.E., and Kramer, D.M. 2000. The proton to electron stoichiometry of steady-state photosynthesis in living plants: a proton-pumping Q cycle is continuously engaged. Proc. Natl. Acad. Sci. USA 97: 14283-14288.

Schmid, V.H.R., Potthast, S., Wiener, M., Bergauer, V., Paulsen, H., and Storf, S. 2002. Pigment binding of photosystem I light-harvesting proteins. J. Biol. Chem. 277: 37307-37314.

Sommer, F., Drepper, F., and Hippler, M. 2002. The luminal helix l of PsaB is essential for recognition of plastocyanin or cytochrome c_6 and fast electron transfer to photosystem I in Chlamydomonas reinhardtii. J. Biol. Chem. 277: 6573-6581.

Spreitzer, R.J. and Salvucci, M.E. 2002. Rubisco: structure, regulatory interactions, and possibilities for a better enzyme. Ann. Rev. Plant Biol. 53: 449-475.

Staehelin, L.A. 2003. Chloroplast structure: from chlorophyll granules to supra-molecular architecture of thylakoid membranes. Photosyn. Res. 76: 185-196.

Stark, D.M., Timmerman, K.P., Barry, G.F., Preiss, J., and Kishore, G.M. 1992. Regulation of the amount of starch in plant tissues by ADP glucose pyrophosphorylase. Science 258: 287-292.

Stoebe, B. and Maier, U.G. 2002. One, two, three: nature's tool box for building plastids. Protoplasma 219: 123-130.

Stroebel, D., Choquet,Y., Popot, J.L., and Picot, D. 2003. An atypical haem in the cytochrome b_6f complex. Nature 426: 413-418.

Tomitani, A., Okada, K., Miyashita, H., Matthijs, H.C.P., Ohno, T., and Tanaka, A. 1999. Chlorophyll b and phycobilins in the common ancestor of cyanobacteria and chloroplasts. Nature 400: 159-162.

Trissl, H.W. and Wilhelm, C. 1993. Why do thylakoid membranes from higher plants form grana stacks? Trends Biochem. Sci. 18: 415-419.

van Amerongen, H. and van Grondelle, R. 2001. Understanding the energy transfer function of LHCII, the major light-harvesting complex of green plants. J. Phys. Chem. 105: 604-617.

von Wettstein, D. 2001. Discovery of a protein required for photosynthetic membrane assembly. Proc. Natl. Acad. Sci. USA 98: 3633-3635.

Voznesenskaya, E.V., Franceschi, W.R., Kiirats, O., Artyusheva, E.G., Freitag, H., and Edwards, G.E. 2002. Proof of C_4 photosynthesis without Kranz anatomy in Bienertia cycloptera (Chenopodiaceae). Plant J. 31: 649-662.

Westphal, S., Soll, J., and Vothknecht, U.C. 2001. A vesicle transport system inside of chloroplasts. FEBS Lett. 506: 257-261.

Westphal, S., Soll, J., and Vothknecht, U.C. 2003. Evolution of chloroplast vesicle transport. Plant Cell Physiol. 44: 217-222.

White, R.A. and Hoober, J.K. 1994. Biogenesis of thylakoid membranes in Chlamydomonas reinhardtii y1: a kinetic study of initial greening. Plant Physiol. 106: 583-590.

White, R.A., Wolfe, G.R., Komine, Y., and Hoober, J.K. 1996. Localization of light-harvesting complex apoproteins in the chloroplast and cytoplasm during greening of Chlamydomonas reinhardtii at 38°C. Photosyn. Res. 47: 267-280.

Xiong, J. and Bauer, C.E. 2002. A cytochrome b origin of photosynthetic reaction centers: an evolutionary link between respiration and photosynthesis. J. Molec. Biol. 322: 1025-1037.

Xiong, J., Fischer, W.M., Inoue, K., Nakahara, M., and Bauer, C.E. 2000. Molecular evidence for the early evolution of photosynthesis. Science 289: 1724-1730.

Yoon, H.S., Hackett, J.D., Ciniglia, C., Pinto, G., and Bhattacharya, D. 2004. A molecular timeline for the origin of photosynthetic eukaryotes. Molec. Biol. Evol. 21: 809-818.

Yoon, H.S., Hackett, J.D., Pinto, G., and Battacharya, D. 2002. The single, ancient origin of chromist plastids. Proc. Natl. Acad. Sci. USA 99: 15507-15512.

Zouni, A., Witt, H.T., Kern, J., Fromme, P., Krauss, N., Saenger, W., and Orth, P. 2001. Crystal structure of photosystem II from Synechococcus elongatus at 3.8 Å resolution. Nature 409: 739-743.

CHAPTER 14

Plant Hormones and Signal Transduction

Marcia Harrison

PLANT SIGNALS

A signal is defined as a chemical or environmental change that induces a cellular response. As sessile organisms, plants have developed a broad series of responses to numerous signals from the environment, symbiotic partners, and pathogen attack, in addition to internal signaling that controls growth and development. Environmental signals range from changes in temperature, water availability, and light (daylength, unidirectional light, or light wavelength), to salt or metal exposure, changes in orientation to gravity, and wind or touch. Chemical signals produced by the plant coordinate developmental and growth changes between tissues, between cells, and between subcellular compartments. Major plant signaling compounds include internally derived compounds such as hormones, polypeptides, and sugar, as well as factors associated with cell-to-cell signaling such as those involved in responses to pathogens.

Plants Respond to Diversity of External Signals

Plants respond to environmental signals through coordinated growth changes, resulting in a distinct response for each particular signal (Table 14.1 and Fig. 14.1). Light, heat, cold, and drought stresses initiate numerous changes that may allow a plant to acclimate to changing environment. Environmental signals often act by altering conformation and activity of existing cellular proteins as part of the response. The change in conformation then initiates a series of cellular reactions. For example, a sudden elevation in temperature causes release of heat-shock factors from the cellular proteins to which they are otherwise bound. The released factors act as transcription activators that bind to regulatory regions of heat-shock protein genes, inducing transcription of those genes (Taiz and Zeiger, 2002). Relative to lighting changes, plants not only have photosynthetic pigments, but also possess photosensory systems that respond to different wavelengths of light and regulate numerous developmental responses. For example, blue-light receptors regulate events such as stomatal opening, stem elongation, and phototropism (Fig. 14.1a). Phytochromes represent a family of red light-absorbing protein pigments that regulate events such as seed germination, flowering (Fig. 14.1b),

TABLE 14.1 Examples of plant responses to environmental changes and other external factors

Signal type	Cellular and physiological responses
Environmental Signals	
Drought	Reduces cell elongation; reduces leaf size; causes leaf curling; causes stomatal closing. These responses lower the surface area available for desiccation.
Gravity	Causes upward stem curvature; causes downward root curvature; involved in orientation of lateral organs. These responses help photosynthetic organs to orient toward the light, and roots to orient toward water and nutrients in soil.
Light	Causes curvature toward light; regulates stomatal opening and closing; regulates seed germination and seed dormancy; reduces stem growth and induces chlorophyll production when provided to dark-grown seedlings; regulates timing of flowering. These responses allow the plant to best use available light or to proceed developmentally when adequate light is present.
Temperature	Induces acclimation to heat or cold shock. These responses help plants live through temperature stresses.
Wind/touch	Reduces stem growth; alters direction of growth of an organ. These responses keep plants short when taller plants would be injured by wind.
External Chemical Signals	
Salts and metals	High levels trigger stress responses such as reduced root growth, lowered photosynthesis, and stomatal closing. Acclimation responses include reduction of ion uptake by the roots and altered ion channel activity in order to prevent accumulation of toxic levels in the cell.
Nod factors from bacteria	Induce root hair curling, the initial step in forming nitrogen-fixing nodules. This response allows plants to live symbiotically with bacteria in nitrogen-poor soil.
Pathogen elicitors	Induce ion movement across the plasma membrane; induce cell death around the infection site. These responses allow the plant to react and isolate the area around a pathogen attack.

and the series of changes associated with transition from dark- to light-grown seedlings (Fig. 14.1c and d). A seedling grown in darkness lacks chlorophyll and has an elongated stem with an apical hook containing the unexpanded leaves and stem apex. Light irradiation of dark-grown seedlings induces chlorophyll synthesis and leaf expansion (Fig. 14.1d). In gravity-induced signaling, change in orientation of a plant causes downward movement of heavy starch-filled organelles called amyloplasts as gravity acts upon them. This is one component that triggers a series of cellular events that lead to growth differences between the top and bottom of the plant organ

(Fig. 14.1c). In horizontal stems, cell growth is greater on the lower side of the stem compared to cells on the upper side, causing the stem to curve upward. In roots, the change in orientation causes more growth on the upper side compared to the lower, resulting in the downward curvature of roots in response to gravity (Fig. 14.1c).

External chemicals can also trigger signaling events in plants. These chemicals can be in the form of metals, salt, or organic compounds produced by a pathogenic or symbiotic organism. For example, a chemical signal from a symbiotic bacterium induces cellular growth and formation of a root nodule in the group of plants com-

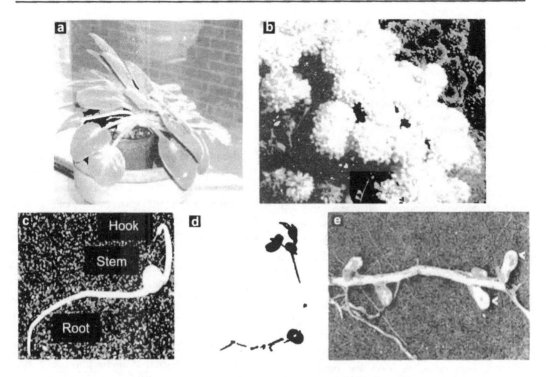

Fig. 14.1 Examples of plant responses to environmental and external signals.

(a) Unidirectional light causes differential growth between the shaded and illuminated side of the plant organ. In this African violet both the stem and leaf petioles exhibited increased growth on the shaded side causing the plant to orient toward the light.

(b) Shortened daylength during late summer induced a change from leaf production to flower produc·tion in *Chrysanthemum*.

(c) Because it was grown in the absence of light, this pea seedling lacked chlorophyll and had an elongated stem with an apical hook that included unexpanded leaves and stem apex. The seedling also demonstrates gravitropism, the effect of a change in orientation to gravity. Changing the orientation of a plant organ results in a series of cellular events, leading to greater elongation of the cells on one side of the organ compared to cells on the opposite side. The resultant differential cell growth causes the stem to orient upward and roots to curve downward.

(d) Light irradiation to dark-grown seedlings signals numerous growth and developmental changes, including chlorophyll synthesis and leaf expansion.

(e) A chemical signal from a symbiotic bacterium causes formation of a root nodule (indicated by arrowheads). The root nodule provides protection and nutrients for the symbiotic bacteria, while the bacteria collect atmospheric nitrogen and alter it to a form usable by the plant (Reproduced courtesy Dr. Jim Deacon, University of Edinburgh).

monly known as legumes (pea, clover, etc.) (Fig. 14.1e). The root nodule provides physical protection and nutrients for the symbiotic bacteria while the bacteria collect atmospheric nitrogen and alter it into a form usable by the plant. Thus legumes with symbiotic bacteria can grow in relatively nitrogen-poor soils. In contrast, a chemical released by a pathogen can be recognized by a plant protein, which then causes an immediate ion release from the cell, triggering death of· the plant cells immediately around the infection site (Taiz and Zeiger, 2002). Although this response results in a small zone of dead cells, it prevents further infection by the pathogen.

Plant-produced Chemical Signals

Plant hormones coordinate most of the growth and development events in plants (see Chapter 5 to review plant hormone structures and actions) (Table 14.2). The five classical plant hormones—abscisic acid (ABA), auxin, cytokinin, ethylene, and gibberellic acid (GA)—and their responses were studied extensively throughout the tweitieth century, beginning in the 1920s with characterization of auxin as a plant growth regulator by Went and others (Arteca, 1996 includes an historical overview of auxin and the discovery of other plant hormones). The role of plant growth regulators seemed to be analogous to animal hormones, defined by Bayliss and

Starling (1904) as substances produced in one part of an organism and transported to another part to induce a response there. However, the concept that plant growth-regulating substances function in the same manner as animal hormones was not experimentally demonstrated despite numerous studies on their biosynthesis and the biochemical changes that occur in response to changing levels of these hormones. Understanding the hormonal mechanisms of action of plant growth-regulating substances began with genetic and molecular studies later in the twentieth century. It is now known that plant hormones are comparable to animal hormones in that they bind to proteins called receptors, and induce a series of interconnected biochemical steps that direct and amplify the response, causing a change in cellular physiology. These steps are collectively known as a "signal transduction cascade".

TABLE 14.2 Major plant hormones and signaling molecules, and some responses under their regulation

Signal type	Cellular and physiological responses
Hormones	
Abscisic acid	Induces leaf and fruit abscission; stimulates stomatal closing; induces seed dormancy; promotes leaf senescence; synthesis increases with drought conditions.
Auxin	Increases cell elongation and division; reduces apical dominance; regulates root and shoot gravitropism; regulates shoot phototropism; promotes lateral root growth.
Brassinosteroid	Stimulates stem elongation; stimulates cell division; inhibits root growth; slows abscission; stimulates seed germination.
Cytokinin	Regulates cell division; delays senescence; releases apical dominance; induces shoot formation in tissue cultures.
Ethylene	Induces fruit ripening; inhibits cell growth; synthesis increases as a result of wounding and stress conditions.
Gibberellic acid	Stimulates stem elongation and leaf growth; stimulates fruit development; stimulates seed germination.
Cell-to-Cell Signaling Factors	
Jasmonic acid	Involved in defensive responses to stress and pathogens.
Sugar	Regulates photosynthesis; controls cell division and elongation; involved in regulation of germination.
Polypeptide	Involved in wounding responses; regulates expression of defense genes; controls meristem cell fate; regulates cell division.

In addition to the classical plant hormones, other signaling compounds play important roles in regulating defense against stress conditions and coordinating events for normal plant growth and development. Brassinosteroids, characterized as the first true plant steroid hormones, inhibit root growth, stimulate cell elongation and division, and induce seed germination (Bishop and Koncz, 2002). Jasmonates are a group of compounds produced from the breakdown of plasma membrane fatty acids in response to the biotic stress of pathogen attack, and to abiotic stresses such as drought conditions and wounding (Turner et al., 2002). Jasmonates either act locally or may travel throughout the plant in a defense response. Until recently, polypeptide hormones such as insulin, which are well characterized in animals, had not been identified in plants. However, plant polypeptides are now known to be important signals in pathogen responses and as regulators of plant growth and development (Ryan et al., 2002). The small polypeptide CLAVATA3 controls the fate of the dividing cells in the shoot apical meristem, maintaining a precise number of stem cells as new cells are produced to differentiate into leaves or flowers. Sugars also have distinct hormone-like characteristics in the regulation of plant growth and development. The ability to sense sugar availability has been well documented in prokaryotic organisms, and the role of sugar as a modulator of growth and development events is now receiving attention in plant and animal studies. In plants, sugars regulate the rate of photosynthesis and critical development events such as cell division, seed germination, and cell elongation (Rolland et al., 2002).

Plant Hormones and other Signals Coordinate Cellular Responses

Plant hormones are not only involved in responses to environmental changes, but are also critical in cell-to-cell communication (Table 14.3). This overlap occurs because hormones are key components of environmental responses and their synthesis and activation is interconnected in complex regulatory networks. Signal transduction cascades for individual signals have been researched extensively over the last 20 years and numerous pathways associated with hormone function and responses to the environment characterized. Networks connecting the signaling events between these pathways (commonly referred to as "cross talk") are now beginning to be understood. For example, herbivory triggers a wounding response that quickly activates the polypeptide hormone systemin (Ryan et al., 2002).

TABLE 14.3 Plant signaling events and major signals that interact in each response

Signaling event	Signals
Whole Plant Signaling	
Apical dominance	Auxin; cytokinin
Wounding; herbivory	Systemin; ethylene; jasmonic acid
Flowering	Light (daylength); gibberellic acid
Seed germination	Abscisic acid; brassinosteroid; ethylene; gibberellic acid; light; sugar
Phototropism	Light (unidirectional); auxin
Cell-to-Cell Signaling	
Meristem organization	CLVATA3; gibberellic acid

Systemin stimulates the cellular production of jasmonic acid, which in turn activates signal transduction events that result in increased expression of stress-response genes (Turner *et al.*, 2002). The wounding caused by herbivory also increases ethylene production, which itself induces a signal transduction cascade leading to cellular responses including reduced growth and leaf abscission. In regulation of seed germination in *Arabidopsis*, environmental signals such as cold and light, and the hormones GA, ethylene, and brassinosteroid interact to control initiation of germination, while ABA serves as the major regulator to maintain seed dormancy (Bentsink and Koornneef, 2002).

This chapter focuses on recent models of plant signal transduction pathways and networks to demonstrate the diversity of plant signaling responses. To illustrate the complexity of plant signaling, the chapter concludes with an examination of networks discussed within the framework of understanding the interaction of the environment and hormones in the stimulation or inhibition of seed germination.

SIGNAL TRANSDUCTION OVERVIEW

The mechanism for cell physiological response to a signal can be broken down into three major processes: perception, transduction, and response.

- *Perception* is the ability to recognize or perceive a signal. It can involve physical binding of a signal to its receptor, or conformational change of a cytoplasmic component in response to environmental signals such as light or change in orientation to gravity. Because the signal initiates the signaling event, it is often called the *primary messenger*.

- *Transduction* is a series of biochemical events induced after the cell perceives the signal. It involves induction of soluble cellular molecules, *secondary messengers*, which initiate a cascade of biochemical reactions that serve to amplify the signal, ultimately altering production or activity of the molecules will directly affect cellular function.

- *Response* to a signal is manifested as a change in cellular function caused by specifically altering protein function, inducing protein degradation, or changing gene expression. Specific response is characterized by the available targets (enzymes, proteins, ion channels, genes) of the signal cascade within a cell type. In that way, different cell types can each respond in a unique manner to the same signal. Various cell types may thus respond to the same signal by activating or inactivating a different set of cellular functions.

An overview of signal transduction pathway components is presented in Fig. 14.2. In this model, a membrane-bound receptor is composed of two identical proteins, which become functional when they bind to the signal and dimerize. The signal-bound receptor activates a cytoplasmic relay protein that initiates a series of transduction events leading to specific cellular responses. In plants, the major targets of signal transduction pathways include changes in enzyme activation in metabolic pathways, ion channel activity, ubiquitin-regulated protein degradation by the proteasome, and gene regulation. Changes in signal transduction involve the combination of activation and repression of various genes and metabolic activities. In the diagrams of signal transduction

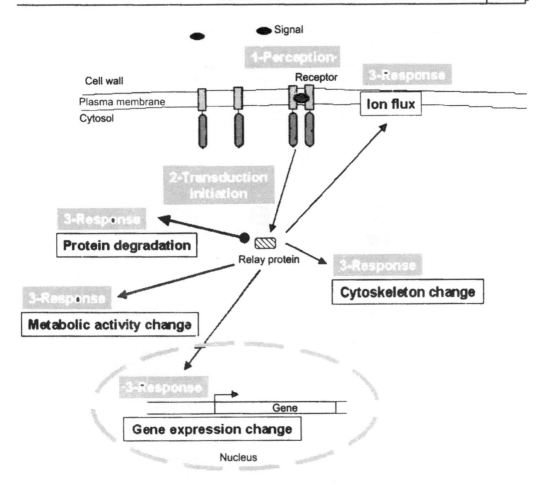

Fig. 14.2 Overview of perception, transduction, and response events for a chemical signal. 1) *Perception* is achieved when the signal binds with the membrane receptor. 2) Signal binding activates a cytoplasmic relay protein, initiating a series of biochemical reactions in a *transduction* cascade initiating from the relay protein and continuing to the cellular responses (indicated by →). 3) Cellular *responses* can be the result of altered metabolic activity in the cytoplasm, protein degradation (indicated by ←→), ion flux due to changes in channel activity, cytoskeletal changes, or activation of transcription factors in the nucleus that regulate target genes (shown as increased transcription by ↱).

pathways that follow, cytoplasmic relay proteins are connected by arrows to the proteins they activate. In diagrams of signal repression, when a relay protein inactivates the next protein in the pathway, the two components are connected by a dead arrow (⊥), while protein repression that occurs by degradation is indicated by a double-stemmed arrow (←→).

Figure 14.3 provides three signal transduction models that demonstrate different types of mechanisms involved in plant signaling. Ethylene regulation of transcription is an example of how the combination of activation and repression steps transduces or communicates the hormone signal (Fig. 14.3a). In this model, the unbound ethylene receptor (shown as a homodimer

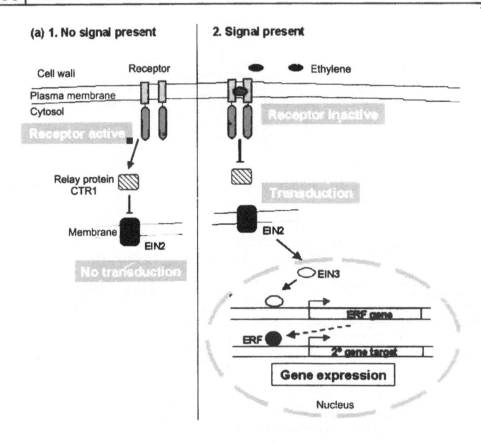

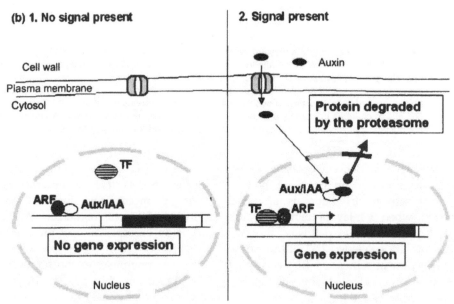

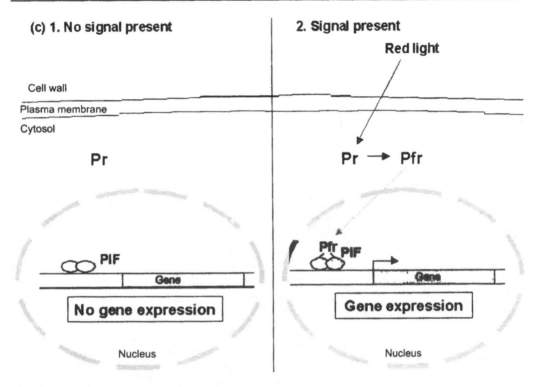

(c) 1. No signal present
2. Signal present

Red light

Cell wall

Plasma membrane

Cytosol

Pr

Pr → Pfr

PIF

Pfr PIF

Gene

Gene

No gene expression

Gene expression

Nucleus

Nucleus

Fig. 14.3 Three models of signal transduction regulation.

(a) Transduction of ethylene involves the combination of activation and repression steps. 1. *No signal present:* The unbound ethylene receptor shown on the plasma membrane is constitutively active as a homodimer, activating a relay protein (CTR1) in the cytoplasm. The activated relay protein inactivates (⊥) the next component (EIN2) in the transduction cascade. 2. *Signal present:* Ethylene binding results in inactivation of the receptor, thus preventing activation of the relay protein. This allows a cascade of events from EIN2 (located on an unidentified membrane), leading to activation of transcription factor EIN3. EIN3 increases (↑) transcription of a primary gene target that encodes the ethylene response factor (ERF). After transcription, translation in the cytoplasm, and import of ERF into the nucleus (indicated by dash arrow), ERF regulates expression of secondary genes

(b) Auxin regulates gene expression by removing an inhibitory transcription factor from a primary response gene. 1. *No signal present:* Auxin-regulated genes are repressed by the binding of AUX/IAA to the auxin response factor (ARF) factor; both are bound as a heterodimer to the promoter proximal area of an auxin-regulated gene. 2. *Signal present:* Auxin enters the cell via an auxin carrier at the plasma membrane, and binds to AUX/IAA, tagging it for degradation by the proteasome. Without AUX/IAA, another transcription factor (TF) can bind to ARF at the promoter proximal area to induce transcription (↑) of the auxin response genes

(c) In phytochrome-regulated transcription, the activated-form of phytochrome Pfr, moves into the nucleus to interact with transcription factors. 1. *No signal present:* In the absence of red-light stimulation, phytochrome remains in its inactive, Pr form. 2. *Signal present:* Red light induces conversion of Pf to Pfr, phytochrome's active state. Pfr is transported into the nucleus, where it complexes with the phytochrome interacting factor (PIF), bound as a dimer on the promoter proximal region of a primary gene. The Pfr-PIF complex induces transcription (↑) of genes that encode additional transcription factors, ultimately regulating genes involved in the red-light response

located on the plasma membrane but may also reside on the ER membrane) is constitutively active and functions by activating a relay protein, CTR1. The activated relay protein *inactivates* the membrane-localized component (EIN2), which prevents further transduction of the signal (Wang *et al.*, 2002). However, when ethylene is present, it binds to and *inactivates* the receptor, thus preventing activation of the relay protein. As a consequence, the EIN2 transduction component, which is usually inactivated by the relay protein, will become active. This allows a cascade of events, leading to activation of the primary gene target, which encodes an ethylene response factor (ERF). ERF is a transcription factor that regulates numerous secondary gene targets that encode proteins responsible for the physiological changes involved in ethylene-regulated stress signaling and growth changes. While ethylene often induces gene expression as shown in Figure 14.3a, it may also downregulate expression by inducing formation of transcriptional repressors.

The transduction pathway of the plant hormone auxin is an example of regulation via protein degradation by the proteasome (Fig. 14.3b). In a current model proposed by Kepinski and Leyser (2002), auxin stimulates gene expression by directly targeting specific transcription factors for degradation. Thus, auxin-regulated genes are repressed under low auxin concentrations and induced whenever auxin is present in higher concentration. In this scenario, repression of an auxin-regulated gene occurs when the transcription factor AUX/IAA forms a dimer with another DNA-bound transcription factor, auxin response factor (ARF), located on the promoter proximal area of an auxin-regulated gene. The AUX/IAA-ARF dimer prevents transcription of the target gene. An increased concentration of auxin within the cell induces proteasomal degradation of the AUX/IAA transcription factor, perhaps by binding to and modifying the AUX/IAA, thus tagging it for degradation. With AUX/IAA degradation, other transcription factors bind to ARF, inducing transcription of the auxin-response genes. Protein degradation by the proteasome has also been found to be important in phytochrome degradation and in the signal transduction pathways for jasmonates, gibberellin, brassinosteroids, and cytokinin (Hellmann and Estelle, 2002).

In a third type of signal transduction mechanism, phytochrome acts as a receptor for red light-induced expression of numerous genes (Fig. 14.3c). Phytochrome is composed of a polypeptide with an attached chromophore. When the inactive state of phytochrome (red light absorbing form, Pr) is exposed to red light, the chromophore changes from *cis* to *trans* conformation, activating phytochrome to its far red absorbing form, Pfr. In the model shown in Figure 14.3c, red light causes the conformation change to Pfr, which also activates phytochrome transport into the nucleus (Schäfer and Bowler, 2002). In the nucleus, phytochrome complexes with phytochrome interacting factors (PIFs) bound to a response element in the promoter proximal region of a light-regulated gene. Pfr binding with the PIF dimer activates gene expression, inducing transcription of additional factors responsible for regulating red light-induced genes.

EXPERIMENTAL APPROACHES TO UNDERSTANDING PLANT SIGNAL TRANSDUCTION

Signal transduction components and their placement within metabolic pathways were primarily identified through a genetic approach. In this method, plants were screened for aberrant phenotypes relative to normal functioning of the signaling process. Once a mutated gene was identified, its characterization was elucidated through a combination of biochemical and molecular techniques to determine gene expression changes, tissue localization of the protein product, and interaction of the protein product with other proteins (Table 14.4). Recent microarray methods provide gene expression analyses of thousands of genes in a single experiment, allowing a rapid assay for changes in the expression of genes involved in environmental and hormonal signaling.

Genetic Approach

The functions of individual plant signal transduction components have been most extensively identified through studies of mutations of specific genes. In this approach, mutagenized plants are screened for an altered response to a specific signal. Since the screens are based on observable developmental or physiological/growth responses, genes involved in pathways with well-characterized responses were the first and most easily identified. Numerous mutants for plant hormones and growth regulators have been identified by screening for insensitive or overly sensitive responses to the addition of a hormone. In some cases, the technique also identifies mutants lacking biosynthesis of specific hormones such as ethylene or GA. However, no deficiency mutants of homones essential for growth and development, such

as cytokinin and auxin, have been identified, presumably because such mutations are likely to be lethal.

In genetic screening for ethylene response mutants, the effect of high levels of ethylene that induce a characteristic "triple response" (exaggerated apical hook curvature, shortened and thickened stem) in dark-grown seedlings was used as the normal ethylene response phenotype (Fig. 14.4a, wild-type plant grown with ethylene exhibits the triple response). Plants that do not respond to the presence of higher than normal levels of ethylene were characterized as "ethylene-insensitive" (*ein*) mutants (Fig. 14.4b; see Box 14.1 for nomenclatural conventions for *Arabidopsis*). These ethylene insensitive mutants are known as "loss-of-function" mutants since even in the presence of the hormone there is no response because the mutated genes do not produce a functional protein (McCourt, 1999). Gain-of-function mutants that exhibit the triple response phenotype without additional exogenous ethylene are called "constitutive triple response" (*ctr*) mutants (Fig. 14.4c). Plants containing mutations for both loss-of-function and gain-of-function genes (called double mutants) are used to determine the relative positions of their proteins within a pathway. For example, when the gain-of-function protein in a pathway follows a loss-of-function gene product, the double mutant plant will express the gain-of-function phenotype (McCourt, 1999) (Fig. 14.4d). Conversely, if the gain-of-function protein precedes the loss-of-function protein, the loss-of-function phenotype will be apparent.

Let's examine the genetic analysis approach used to identify ethylene insensitive mutants. By 1995, seven *ein* mutants were identified in the ethylene response pathway in *Arabidopsis* (Roman *et al.*, 1995). For the seven *EIN* genes, cellular

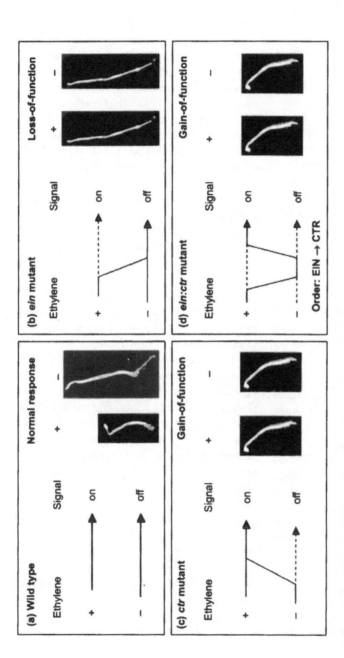

Fig. 14.4 Comparison of gain-of-function and loss-of-function ethylene mutations

(a) Under normal conditions in air, dark-grown wild-type *Arabidopsis* seedlings have a highly elongated hypocotyl with an apical hook. Presence of the hormone ethylene induces a signal transduction pathway resulting in the characteristic triple response wherein the seedlings have exaggerated apical hook curvature, and short and thickened stems.

(b) In "loss-of-function" mutants, signal transduction switches permanently OFF, resulting in no response even in the presence of ethylene. The *ein* mutants do not respond to ethylene and remain highly elongated even in the presence of high ethylene levels

(c) In "gain-of-function" mutants, signal transduction switches permanently ON, resulting in a constitutive response. Thus *ctr* mutants exhibit the triple response phenotype even without the addition of exogenous ethylene

(d) In double mutants carrying both loss-of-function and gain-of-function genes, the signal switches OFF and then ON if the loss-of-function gene product is upstream of the gain-of-function gene product in the signal transduction cascade. Because the *ein:ctr* mutants exhibit the *ctr* phenotype , it is surmised that EIN is positioned upstream of CTR. Graphics adapted from P. McCourt, 1999: Genetic analysis of hormone signaling from Annual Reviews from the Annual Review of Plant Physiology and Plant Molecular Biology, vol. 50, 1999; this Figure appeared on page 229 as Figure 2. Images reproduced by kind permission of Dr. Randy Scholl, Director of the Arabidopsis Biological Resource Center, Ohio State University.

Nomenclatural conventions for *Arabidopsis* genes and gene products

According to nomenclatural conventions for *Arabidopsis*, a three-letter designation based on a description of the mutant phenotype is used for the gene and gene product abbreviation (Council of Biology Editors, Style Manual Committee, 1994; and at http://www.arabidopsis.org). For example, EIN is used as an abbreviation of the ethylene-insensitive phenotype. This three-letter designation is followed by a number that represents the gene designation. By convention the wild-type gene abbreviation is upper case and italicized (e.g., *EIN1*), while the mutant form of the gene is presented as italicized lower case (*ein1*). The protein product is upper case and nonitalicized (e.g. EIN1). Genes can also be named based on their biochemical function if the function has been experimentally characterized. However, phenotypic mutants have been used to identify the bulk of the signal transduction components, so phenotypic nomenclature is the most prevalent form.

localization and pathway sequence for EIN1, EIN2, EIN3, and EIN4 gene products have been identified. Since the *ctr1* mutant could reverse the phenotype *ein1* and *ein4* mutants, the EIN1 and EIN4 proteins were positioned earlier in the ethylene response pathway than CTR1 (Roman *et al.*, 1995). It is now known that EIN1 and EIN4 are ethylene receptors, which supports data that they are early in the pathway (Wang *et al.*, 2002). Mutants *ein2* and *ein3* were not reversed in double mutants with *ctr1*. Therefore, their gene products, EIN2 and EIN3, were placed downstream of CTR1. The current model of ethylene signal transduction places EIN2 under the negative control of CTR1, whereas EIN2 itself is a positive regulator of EIN3, a transcription factor in this pathway (Fig. 14.3a) (Wang *et al.*, 2002).

UV light or a mutagenic chemical treatment usually produces a loss-of-function mutant with a single point mutation that reduces or abolishes the function of the gene's protein product. Loss-of-function mutants (also known as "knockout mutants") are also induced by random insertion of T-DNA, a specific portion of the *Agrobacterium* Ti plasmid. Because this plasmid is responsible for insertion of *Agrobacterium* DNA into the plant genome, T-DNA insertions cause loss of function when the T-DNA inserts within the coding region of a gene. Since the base sequence of the Ti plasmid is known, a probe for the inserted sequence allows the affected gene to be located within the plant genome.

Screening plants for mutagen-induced phenotypic changes can include selection for gain-of-function as well as loss-of-function changes in gene expression. In gain-of-function mutants, mutation causes increased gene expression by altering regulation of the gene so that the protein is produced in larger than normal quantities. On occasion, T-DNA insertions may disrupt promoter sequences in a manner that causes increased transcription of the gene, resulting in a gain-of-function mutation. More commonly a variation of T-DNA insertion is used wherever a reporter gene (a gene encoding a product readily identified by histochemical or microscopic analysis) with a minimal promoter and transcription start site is included in the inserted DNA sequence. Thus, when the T-DNA inserts upstream of a gene that is normally expressed, the gene's regulatory machinery will turn on transcription of the reporter. This "enhancer trap" method has the advantage of allowing the researcher to screen for altered gene expression by looking for an increase in the reporter gene product, which is often easier to detect than changes in the affected plant gene (Geisler *et al.*, 2002). Thus, the enhancer trap insertions can be used to detect gene expression changes even if they do not cause conspicuous phenotypic changes. Enhancer trap mutants are also useful for identifying tissue-specific gene expression for the targeted genes. Figure 14.5 (a, b, and c) shows examples wherein enhancer-trap mutants were used to locate the specific tissue types in which genes of interest were expressed.

Molecular and Biochemical Approaches

Once a specific signal transduction protein is identified and its DNA and protein sequences acquired, further experiments can reveal information about the signaling component. Since the *Arabidopsis* genome has been fully sequenced, complete genetic information is available for study. While many gene sequences have also been identified for other plant species, the wealth of information pertains to *Arabidopsis*; hence many examples of signal transduction pathways in this chapter have been taken from that system.

With the sequence information known, manipulation of gene sequences to produce transgenic plants has become a widespread technique for studying tissue expression levels. Manipulations can include combining promoter sequences with reporter genes, and altering sequences to study the effects of the specific nucleotide changes affecting either the protein product or the regulator region of the gene. For example, a plant carrying a gene construct made of the promoter proximal region for a signal transduction component attached to a reporter gene can be analyzed for tissues with the reporter activity. Cellular localization of a chimeric protein carrying the reporter green fluorescent protein (GFP) can be observed by fluorescence microscopy (Fig. 14.5d and e). Alternatively, the sequence could be altered in a specific region and the effect of this site-directed mutagenesis could be observed in the transgenic plant. This technique is useful in identifying important regulatory components of the gene.

Other molecular techniques have been used to identify the role of transcription factor binding, protein phosphorylation, and protein-protein binding between signal proteins (Table 14.4). One common method for determining protein-protein interactions is by yeast two-hybrid screening (Fields and Song, 1989). In this system, yeast cells are transformed with a plasmid carrying components of the regulatory portion of a reporter gene along with genes for the potential interacting proteins. The transgenic yeast cells produce a chimeric protein containing the putative protein-interacting fragment (known as the "bait") and the DNA binding domain upstream of the reporter gene. When properly expressed, one end of the chimeric protein will be bound upstream to the reporter gene while the other end is free to interact. The chimeric "prey" protein contains the putative interacting protein domain and an enhancer-binding domain, which is necessary to activate reporter gene expression. If the "prey" binds to the "bait" in a protein-protein interaction, transcription of the reporter gene will increase. In this manner, yeast carrying the interacting bait and prey can be screened based on their reporter gene activity.

Genomic Approach

With completion of the entire DNA sequence for *Arabidopsis*, the ability to study the expression of large numbers of genes became possible through GeneChip microarrays (Zhu, 2003). GeneChips are

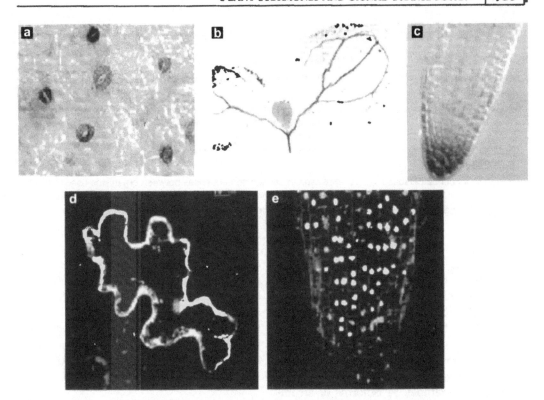

Fig. 14.5 Examples of use of reporter genes in understanding tissue and cellular location of specific gene products. Images a-c show histochemical analysis for reporter-protein expression in *Arabidopsis* enhancer trap mutants. Images d and e demonstrate cellular localization of specific protein-GFP constructs.

(a) Reporter gene expression (dark color) localized in guard cells of stomata.

(b) Reporter gene expression (dark color) localized in vascular tissue of cotyledons.

(c) Reporter gene expression (dark color) localized in apex of root tip.

(d) Fluorescent reporter, GFP (bright areas), illustrating cytoplasmic localization of GFP-protein construct in an epidermal leaf cell.

(e) Fluorescent reporter, GFP (bright areas), illustrating nuclear localization of GFP-protein construct in the root tip cells.

All images copyrighted by the American Society of Plant Biologists and reprinted by kind permission. Images a to c from Geisler, Jablonska, and Springer, 2002. Images appeared on page 1749 as Figure 1C and H.

Image d from Holk *et al.* 2002: Image appeared on page 95 as Figure 4F.

Image e from Zhao *et al.* 2002. Image appeared on page 1226 as Figure 6C.

small glass slides spotted with thousands of DNA probes to detect individual RNA transcripts from control and signal-treated plants. Microarray analyses reveal that the signal-induced shifts in the expression of hundreds of genes affect not only their own signaling pathways, but also reflect interaction with the transduction elements of other pathways. Thus, microarray experiments provide what is commonly referred to as a "global" perspective of the complexity of signaling pathways and the interactions that form signal transduction networks (Zhu, 2003).

TABLE 14.4 Experimental approaches to understanding signal transduction

Experimental Approach	Contribution
Genetic systems	Identification of signaling components; order of components in a pathway
Molecular	
Gene constructs/transgenic plants	Subcellular localization of gene products; tissue localization of gene products; identification of important sequences (using site-directed mutagenesis)
Yeast two-hybrid screen	Identification of protein-protein interactions
Genomics	
Genomic sequence database	Comparison of conserved sequences between organisms used to predict structure and function of gene products
Microarray analysis	Global changes in gene expression for various signals; identification of interacting pathways

Sequence information for known signal transduction components can be used to identify homologous forms of these components within the same species and between species. This has led to the realization that many gene products have numerous isoforms (proteins that perform the same function but whose sequences differ to some degree). For example, 34 isoforms of an important relay protein, calcium-dependent protein kinase (discussed later in the chapter), have been identified in *Arabidopsis* (White and Broadley, 2003). Results of published microarray analyses (available in public-access databases such as http://www.arabidopsis.org) can be useful to screen for changes in expression of the various isoforms. In this way, microarray data can be a tool for preliminary identification of key isoforms involved in a specific signaling event. Additionally, sequence comparisons can be useful in revealing relationships between different proteins. For example, sequence comparisons have revealed a homology between cytokinin receptors, ethylene receptors, phytochrome, and bacterial signaling systems (Hwang *et al.*, 2002). This approach allows a researcher to hypothesize a function in one organism based on sequence similarities and conserved domains present in other

organisms. However, confirmation of the protein's cellular functions needs to be demonstrated in the laboratory by functional analysis experiments. As will be see through examples in this chapter, structural similarities based on sequence comparison do not always indicate the primary cellular function of a protein.

MAJOR TYPES OF PLANT SIGNAL PERCEPTION

For chemical signals, perception involves binding of the signal with a receptor that is usually on a membrane. Alternatively, an environmental signal such as light may induce conformational change in its receptor protein, thus initiating a transduction cascade. Perception can be regulated by several mechanisms. Presence of a signaling molecule and presence of receptor molecules on specific cell types determine whether the cell will respond to a specific signal. The level of perception can be regulated by the nature of the receptors, such as their ability to bind to the signal and the number of receptors available to bind with the signal. Additionally, receptors for a specific molecule may have different isoforms that alter the responsiveness of the tissue to a signal. For example, five types of ethylene receptors have been identified (Wang *et al.*, 2002). It is hypothesized that the different dimer combinations of the ethylene receptors may act to regulate the level of response within a tissue.

The majority of known plant receptors contain kinase domains that are responsible for phosphorylating relay proteins in the initiation of the signal transduction cascade. The examples presented in this chapter are 1) receptors within a two-component transduction system, 2) receptor-like kinases, and 3) Lys-M receptors. Figure 14.6 provides examples of receptor structures.

Two-component System-like Receptors

A major group of plant signal receptors is similar to two-component signaling systems found in the environmental response pathways in bacteria (Hwang *et al.*, 2002). In prokaryotic two-component signaling systems, the signal (environmental stimulus or signal binding) initiates autophosphorylation of the histidine residue in the cytoplasmic kinase domain (the first component) of a membrane receptor. A relay protein transfers the phosphate to a receiver domain on a protein called a response regulator (the second component). In plants, the signal-binding element, the histidine kinase domain, and the receiver domain are often found on a single protein (Fig. 14.6a and b) (Hwang *et al.*, 2002). Signal binding occurs on an extracellular N-terminus or within the transmembrane area. The C-terminus is the site of phosphorylation by a protein kinase, usually to histidine or serine/threonine residues (infrequently to Try, as found in animal cells). The N-terminal signal-binding region is variable, while the cytoplasmic kinase domain is highly conserved with a conserved histidine residue (Hutchinson and Kieber, 2002,

Hwang *et al.*, 2002). In a model two-component system using cytokinin perception as an example, cytokinin binds to the extracellular domain of the receptors, facilitating dimerization (Fig. 14.7) (Hutchinson and Kieber, 2002). The dimerized receptors autophosphorylate at the histidine residues within the cytoplasmic kinase domain. The phosphate is then transferred from the histidine residue to an aspartate residue in the receiver domain near the C-terminus of the cytokinin receptor. The phosphate on the receiver domain is relayed to a response regulator by a histidine phosphotransfer protein, initiating a signal transduction cascade within the cell.

In plants, ethylene, cytokinin, and phytochrome receptors are classified as histidine kinase protein receptors. Overall, there have been 54 histidine protein kinases and related proteins identified in *Arabidopsis* (Hwang *et al.*, 2002). The ethylene receptor system consists of five ethylene receptors, which have been identified in *Arabidopsis* and tomato (Wang *et al.*, 2002). The ethylene receptors are categorized into two subfamilies (Hwang *et al.*, 2002) based on the presence or absence of a complete histidine kinase domain. ETR1 (ethylene resistant) and ERS1 (ethylene response sensor) have complete histidine kinase domains and function as homodimers. The other three receptors (ETR2, EIN4, and ERS2) have degenerate histidine protein kinase domains. Additionally, ETR1, ETR2, and EIN4 have receiver domains while ERS1 and ERS2 do not. Hence, it is speculated that formation of heterodimers (composed of two different receptors) may be needed in order to produce a functional unit that contains both kinase and receiver domains (Hwang *et al.*, 2002). Current evidence supports ethylene acting as a negative regulator when it binds the receptor (see Fig. 14.3a). This implies that the

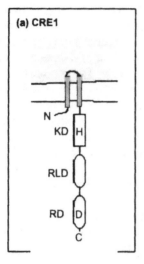
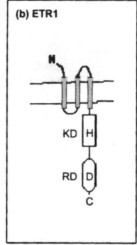
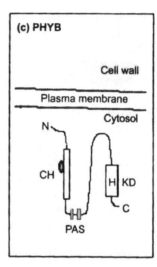

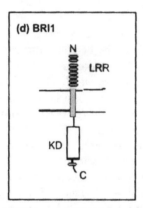
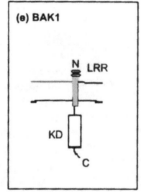
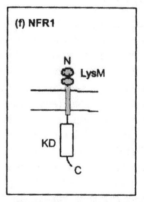

Fig. 14.6 Examples of plant receptor structures

(a) Histidine kinase protein receptors CRE1 (cytokinin response 1) showing cytoplasmic N-terminus, two transmembrane domains, and cytoplasmic region containing kinase domain (KD) with histidine (H) residue, receiver-like domain (RLD), and receiver domain (RD) with asparagine (D) residue adjacent to C-terminus

(b) Histidine kinase protein receptor ETR1 (ethylene resistant 1) showing an extracellular N-terminus, three transmembrane domains, and cytoplasmic region containing kinase domain with histidine residue, and receiver domain with asparagine residue adjacent to C-terminus

(c) Cytoplasmic histidine kinase protein receptor PHYB (phytochrome B) showing attached chromophore (CH), PAS protein-protein interaction domains, and kinase domain containing histidine residue adjacent to C-terminus

(d) Leucine-rich repeat receptor-like kinase BRI1 (brassinosteroid insensitive) showing numerous extracellular leucine-rich repeats (LRR), transmembrane domain, and cytoplasmic serine/threonine kinase domain adjacent to C-terminus

(e) Leucine-rich repeat receptor-like kinase BAK1 (brassinosteroid insensitive 1-associated receptor kinase 1), which has few extracellular LRR, a transmembrane domain, and cytoplasmic serine/threonine kinase domain adjacent to C-terminus

(f) Nod-factor receptor (NFR1) showing LysM domain repeats in N-terminal extracellular region, a transmembrane domain, and a cytoplasmic serine/threonine kinase domain adjacent to C-terminus

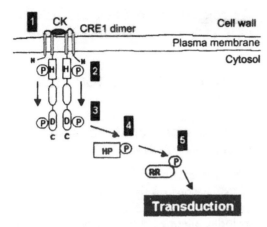

Fig. 14.7 Cytokinin perception mechanism as simplified two-component signaling model. 1) Cytokinin (CK) facilitates dimerizaton of two receptors. 2) Dimerization initiates autophosphorylation at the histidine (H) residues in the cytoplasmic kinase domain. 3) Phosphate transfer to an aspartate (D) within the receiver domain on the receptor protein. 4) Phosphate transfer to a relay protein, histidine phosphotransfer protein (HP). 5) HP then transfers a phosphate to response regulator (RR), which initiates transduction cascade within the cell.

kinase domain is active when in the unbound state. The mechanism of kinase inactivation caused by ethylene binding has not yet been demonstrated (Hwang et al., 2002). While experimental evidence of ETR1 activation of histidine phosphotransfer protein and response relay systems has been demonstrated *in vitro*, ethylene signaling is best characterized for its interaction with CTR1 protein, a serine/threonine protein kinase (Wang et al., 2002). Thus, ethylene transduction pathways that use the histidine kinase may represent an alternative pathway for ethylene-mediated responses.

Phytochromes are a group of cytoplasmic proteins (5 forms in *Arabidopsis*), which perceive a light signal by the chromophore attached on the N-terminal region (Hwang et al., 2002) (Fig. 14.6c). The C-terminus contains a kinase domain homologous with histidine kinase domains as well as two unique PAS domains that are involved in protein-protein interactions. Thus, phytochromes may function as kinases that can phosphoryate other proteins as well as interact with other proteins (such as transcription factors) as part of their signaling mechanism. However, it should be noted that although phytochromes have been classified as histidine-protein kinases based on sequence similarities between bacterial photoreceptor-histidine kinases and phytochromes, there is no experimental evidence that these receptors possess histidine kinase activity (Hwang et al., 2002). Thus, sequence homology is useful in the chemical characterization of proteins, but additional functional analysis will be necessary to provide the real picture of the cellular roles of these receptors.

Leucine Repeat Rich-receptor-like Kinases

Leucine repeat rich-receptor-like kinases (LRR-RLK) compose a large family of proteins (610 genes in *Arabidopsis*) characterized by numerous extracellular leucine-rich areas, a single transmembrane domain, and a cytoplasmic serine/threonine kinase domain (Fig. 14.6d and e). More than half the *Arabidopsis* RLK gene products have structural features that suggest they function as receptors (Diévart and Clark, 2003). In plants, LRR-RLKs have been identified as receptors involved in brassinosteroid signaling, symbiosis recognition, meristem maintenance, wounding response, stomatal development, and responses to pathogen attack (Diévart and Clark, 2003). LRR-RLKs are receptors for polypeptide hormones such as CLAVATA3 (CLV3) and the wounding response hormone, systemin. In meristem maintenance, CLV3 binds to an LRR-LRK receptor complex composed of CLAVATA1 and CLAVATA2 located on the plasma membrane (Ryan et al., 2002). CLAVATA1 contains a cytoplasmic kinase domain, which is phosphorylated after CLV3 binding, thus initiating phosphorylation relay events in a signal transduction cascade.

Two plant LRR-RLKs, BRI1 (brassinosteroid insensitive 1) and BAK1 (BRI1-associated receptor kinase 1), have been identified in brassinosteroid perception in plants (Li, 2003) (Fig. 14.6d and e). In animal steroid signaling, the steroid typically binds to a cytoplasmic receptor, and the hormone-receptor complex acts as a transcription factor to regulate gene expression changes. On the other hand, the BRI1 receptor and its interacting protein, BAK1, are located on the plasma membrane (Bishop and Koncz, 2002). Binding of brassinosteroid initiates BRI1-BAK1 dimerization that results in phosphorylation of the heterodimer (Li, 2003). The heterodimer initiates phosphorylation events that induce a signal transduction cascade. This type of receptor dimerization is thought to be a central event in LRR-RLK perception and activation (Diévart and Clark, 2003).

LysM-receptor Kinases

The plant Nod factor receptors (NFR) recognize Nod factors from symbiotic bacteria and are classified as LysM-receptor kinases (Radutolu et al., 2003). LysM receptor kinases have an extracellular N-terminus containing two LysM domains (lysin motif, short amino acid sequences with peptidoglycan binding, originally identified in bacterial lysin proteins), a single transmembrane domain, and a cytoplasmic serine/threonine kinase domain (Fig. 14.6f). In a current model presented by Radutolu and colleagues (2003), Lotus NFR1 and NFR5 dimerize when bound to a Nod factor. The

dimer is the functional unit that activates a relay protein associated with the cytoplasmic domain of NFR1.

SIGNAL TRANSDUCTION CASCADE MECHANISMS

Overview of Signal Transduction Cascades in Plants

After a signal binds to its receptor, a series of cytoplasmic events occur to communicate the signal within the cell. These events may include altering the activity of proteins through phosphorylation and dephosphorylation of specific amino acids within the polypeptide chain, activating ion transporters, and activating enzymes that cleave membrane phospholipids to produce secondary messengers. Transduction cascades may target specific genes, resulting in production of signal-induced proteins, or may regulate cytoskeletal arrangement, an important component of events such as movement of vesicles, cell division, and cell growth.

Small GTPases

Small GTPase proteins are highly conserved GTP-binding proteins that act as signal transduction switches in eukaryotic cells. Numerous small GTPases exist in plants and are classified into five families based on their conserved sequences (Yang, 2002). The subfamily ROP (Rho family-related GTPases from plants) comprises an important group of small GTPase signaling molecules involved in ABA, pathogen, and stress-mediated signaling. The ROP protein regulation of actin organization controls the tip-directed growth of organs such as root hairs and pollen tubes (Jones et al., 2002), and development of the unique lobe-shape leaf epidermal cells known as pavement cells (Fu et al., 2002).

Small GTPases are approximately 200 amino acids long and contain four GTP-binding domains, a single effector domain, and a plasma membrane attachment site (Yang, 2002). Small GTPases are found in the cytoplasm in their inactive form, but migrate to the plasma membrane when they become activated. After perception of a signal, the guanine-binding domain is activated by a guanine nucleotide exchange factor (GEF) causing exchange of the GTP for the GDP that was bound to the GTPase (Fig. 14.8). The effector domain of the active GTPase then binds to and activates an effector molecule, continuing the transduction cascade to the specific target.

In root hairs, the small GTPase ROP is normally localized in the hair tip, thus directing growth to that area (Fig. 14.9a and b). In a genetic approach to understanding the role of ROP in directional growth, domi-

nant negative (DN) and constitutive active (CA) mutants for ROP are used to analyze the effect of this GTPase. In *Arabidopsis CA-rop1* mutants, ROP is distributed throughout the root hair, resulting in growth in all directions rather than the tip-directed elongation seen in wild-type root hairs (Fig. 14.9c and d) (Molendijk *et al.*, 2001). In another example, epidermal cell shape is regulated by ROP GTPase signal transduction to the cytoskeleton (Fu *et al.*, 2002). Epidermal pavement cells have distinctive lobes and neck regions that are defined by the cytoskeleton as the cells develop. The neck region is formed by microtubules, whereas the lobes have diffuse F-actin associated with them (Fu *et al.*, 2002). Changes in expression of ROP2 alter the shape and size of epidermal pavement cells (Fu *et al.*, 2002). Pavement cells of the *CA-rop2* mutants exhibit multiple direc-

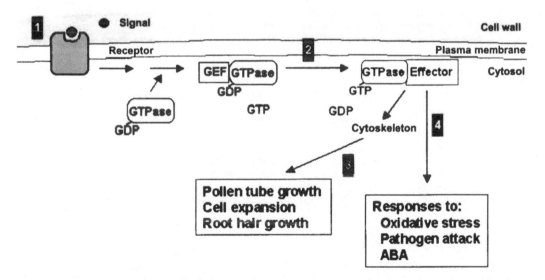

Fig. 14.8 Model of small GTPase function in signal transduction. 1) Binding of signal to receptor initiates signaling to activate binding of guanine nucleotide exchange factor (GEF) to GTPase, located in the cytoplasm in its inactive form. 2) GEF causes exchange of GDP for GTP to produce active GTPase, which then binds to an effector molecule. 3) Activated effector may regulate distribution of cytoskeleton filaments in directional growth of root hairs, pollen tubes, and epidermal cells. 4) Activated effector also involved in signal transduction to specific pathways involved in regulation of oxidative stress, resistance to pathogen response, and inhibition of ABA response.

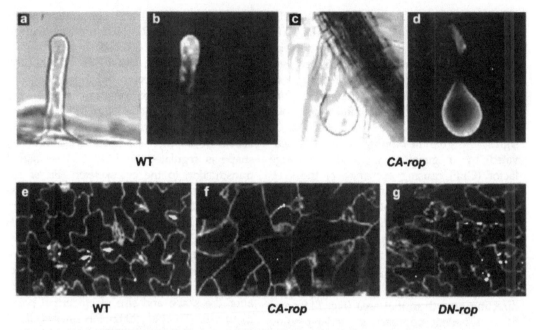

Fig. 14.9 Role of small GTPase ROP in directional growth of root hairs and epidermal pavement cells

(a) Light image of wild type (WT) root hair tip

(b) Tip-localization of GFP-ROP in wild-type root hair, using fluorescence microscopy

(c) Light microscopy image of root hair from *Arabidopsis CA-rop* mutant

(d) *Arabidopsis CA-rop* mutant showing localization of GFP-ROP distributed along edges of swollen root hair, using fluorescence microscopy

(e) Epidermal cells with distinct lobes (indicated by arrowheads) in wild-type plant. GFP-tubulin construct used to highlight cell shape for fluorescence microscopy

(f) Epidermal cells exhibiting multiple directions of growth, causing overall swelling and few lobes in *CA-rop* mutant expressing GFP-tubulin construct

(g) Epidermal cells with reduced cell size and lobe development in *DN-rop* mutant expressing GFP-tubulin construct

Images a, b, e, f, and g copyrighted by American Society of Plant Biologists and reprinted by kind permission. Images a and b from Jones *et al.* 2002. Images appeared on page 770 as Figures 5G and H. Images e, f, and g from Fu *et al.* 2002. Images appeared on page 783 as Figures 4C, F, and I. Images c and d copyrighted by European Molecular Biology Organization and reprinted by kind permission. Images from Molendijk *et al.* 2001. Images appeared on page 2783 as Figures 7G and H.

tions of growth and swollen cells with few lobes, while *DN-rop2* mutants have inhibited actin organization and smaller cells with fewer lobes than wild-type pavement cells (Fig. 14.9e, f, and g).

G Proteins

Other GTP-binding proteins known as G proteins (guanine-nucleotide binding pro-

teins) act as relay proteins during signal transduction. G proteins are composed of three heterodimeric subunits (Gα, Gβ, and Gγ), although monomeric G proteins composed of only the Gα subunit also exist. G protein regulation has been propsed as a signal transduction mechanism for responses to auxin, gibberellin, light, and pathogens (Assmann, 2002). However, in plants, few G protein components have been identified relative to the numerous G protein isoforms found in animals (Jones, 2002). While potential G protein-linked receptors have been proposed, true G protein-linked receptors have yet to be identified (Assmann, 2002). Hence it is generally assumed that in plants G proteins are activated by an upstream relay protein, which has been activated after signal perception. Activation of an upstream relay protein causes GTP binding to the Gα subunit, displacing GDP (Fig. 14.10). The GTP-bound form is the active state of the complex and separates from the Gβ and Gγ subunits upon activation. The free activated Gα subunit may activate or inhibit target proteins such as membrane-associated enzymes, Ca^{2+} channels, and GTPases.

Common enzyme targets include phospholipases, which catalyze cleavage of phospholipids in the membrane to produce secondary messengers. In one mechanism, G proteins activate phospholipase D (PLD), producing the secondary messenger phosphatidic acid (PA) from phospholipid hydrolysis in the plasma membrane (Fig. 14.10a). PA has been shown to activate small GTPases and protein kinases in signal transduction pathways associated with ABA, GA, and stress responses (Munnik and Musgrave, 2001). The free active Ga subunit has been found to directly activate small GTPases in transduction of disease resistance responses in rice (Suharsono et al., 2002), and increase influx of Ca^{2+} into the cytoplasm via activation of Ca^{2+} channel activity in Arabidopsis (Jones, 2002)

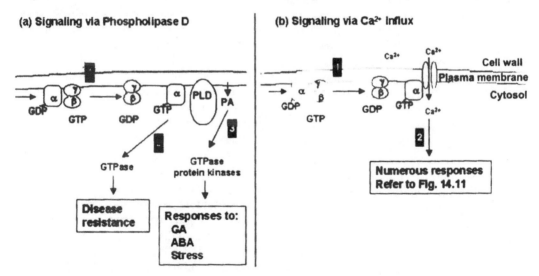

Fig. 14.10 Overview of G protein signal transduction cascades.
(a) G protein signaling via phospholipase D (PLD). 1) Activation of G protein-linked receptor or an upstream relay protein causes GTP binding to Ga subunit, displacing GDP. 2) Free activated Ga subunit may directly activate small GTPase in transduction of disease resistance responses. 3) Free activated Ga subunit may also activate PLD, causing release of second messenger, phosphatidic acid (PA), from plasma membrane. PA activates small GTPases and/or protein kinases in signal transduction pathways associated with disease resistance, ABA, GA, and stress signaling
(b) G protein signaling via Ca^{2+} influx. 1) Activation of heterotrimeric G protein produces free activated Ga subunit. 2) Free activated Ga subunit activates Ca^{2+} channel activity, increasing influx of second messenger, Ca^{2+}, into cell (see Fig. 14.11 for details of Ca^{2+} signal transduction).

(Fig. 14.10b). The intermediates in these GTPase-regulated events have not been experimentally demonstrated.

Calcium Ions as a Secondary Messenger

Influx of calcium ions (Ca^{2+}) into the cytosol is a well-characterized response to environmental, hormonal, symbiotic, and pathogenic signals (a long list of examples in plants is provided by White and Broadley, 2003). Ca^{2+} is maintained in relatively low concentrations in the cytoplasm and sequestered extracellularly or internally in the ER, vacuole, mitochondria, or plastids (Cheng et al., 2002; White and Broadley, 2003). Regulation of Ca^{2+} influx is often controlled by activation of phospholipase C (PLC) by a G-protein or upstream activator (Fig. 14.11). The activated PLC cleaves the membrane phospholipid phosphatidylinositol 4,5-bisphosphate (PIP_2) to produce secondary messengers, inositol 1,4,5-trisphosphate ($InsP_3$) and diacylglycerol (DAG), which can be converted to PA. Free $InsP_3$ binds to an $InsP_3$ receptor/Ca^{2+} channel located on the plasma membrane or intercellular membrane. Activation of calcium channels by $InsP_3$ causes transient influx of Ca^{2+} into the cytosol. The Ca^{2+} influx is typically manifested as a pulse within the cell, and the ion acts as a second messenger in continuing the transduction cascade by direct binding to specific Ca^{2+}-binding proteins, calmodulin (CaM), calcineurin B-like proteins (CBL), or calcium-dependant protein kinases (CDPK) (White and Broadley, 2003). These Ca^{2+}-binding proteins regulate target proteins in transducing the signal.

Ca^{2+} influx pulses are generally rapid. They have been shown experimentally to occur less than a minute after red-light irradiation of dark-grown seedlings, and in roots after the addition of a Nod factor (White and Broadley, 2003). These pulses create Ca^{2+} gradients within the cell, which are thought to be important for establishing cell polarity. In root hair initiation, a high concentration of Ca^{2+} occurs at the apical end of the epidermal cell where the root hair initiates, and high Ca^{2+} pulses are maintained in the root hair tip as it elongates (Fig. 14.12). Changes in Ca^{2+} distribution alter the direction of tip-directed growth in root hairs. In the CA-rop mutants discussed earlier, high Ca^{2+} levels occur at multiple points in the swelling root hair, indicating Ca^{2+} involvement in ROP GTPase regulation of tip-oriented growth (Molendijk et al., 2001).

Calmodulins and calcineurin B-like proteins

Calmodulins (CaMs) and calcineurin B-like proteins (CBLs) are small proteins that bind multiple Ca^{2+} ions. Active Ca^{2+}-bound forms of CaMs and CBLs function by directly binding to and activating protein targets, often protein kinases (Fig. 14.11). CBLs are associated with the plasma membrane and interact with only one class of kinases, called CBL-interacting protein kinases. These kinases are involved in the signal transduction of specific environmental signals, such as drought, cold, salt stress, and wounding (White and Broadley, 2003). CaM is a highly conserved eukaryotic cytosolic protein with many different protein targets, including calcium/calmodulin-dependent protein kinases (CaMKs). CaM and CaM-like proteins have been found to be involved in many responses including those associated with touch, hormones, oxidative stress, and temperature change (White and Broadley, 2003). One interesting example of calcium signaling concerns plant ability to respond to touch, a response thought to have evolved through mechanical stimulation by wind. The touch signal

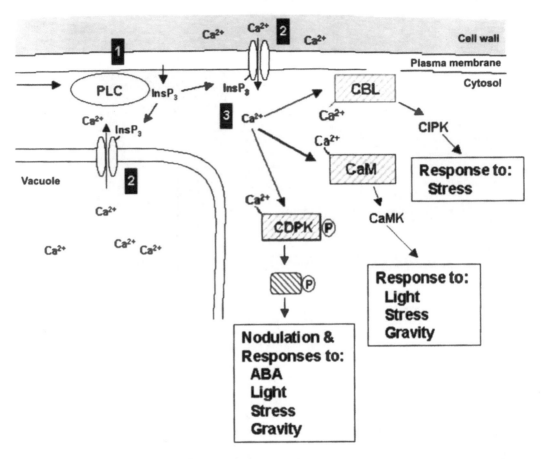

Fig. 14.11 Overview of calcium-regulated signal transduction cascade. 1) Upstream regulator activates phospholipase C (PLC), causing cleavage of inositol 1,4,5-trisphosphate (InsP₃) from the membrane. 2) Free InsP₃ binds to InsP₃ receptor/Ca²⁺ channel located on plasma membrane or an intercellular membrane. Activation of calcium channels by InsP₃ causes transient influx of Ca²⁺ into the cytosol. 3) Ca²⁺ binds directly to specific Ca²⁺-binding proteins, calcineurin B-like proteins (CBL), calmodulin (CaM), or calcium-dependent protein kinases (CDPK). CBL and CaM regulate target proteins (CBL-interacting protein kinases, CIPK; calcium/calmodulin-dependent protein kinases, CaMK) in transducing the signal. CDPKs become phosphorylated when bound to Ca²⁺, and the activated CDPKs phosphorylate target proteins that continue the signal transduction pathway.

causes reduced cellular growth, resulting in short, thick stems. Mechanical stimulation by touch greatly increases the expression of several genes (touch-related genes, *TCH*) (Braam *et al.*, 1997). *TCH1* encodes a CaM, and *TCH2* and *TCH3* encode CaM-related calcium-binding proteins. *TCH4* encodes the enzyme involved in xyloglucan crosslinking of cellulose microfibrils and may play a role in modifying the cell wall in response to environmental stresses (Braam *et al.*, 1997; Khan *et al.*, 1997). *TCH* gene expression is stimulated by many other environmental factors (e.g. cold, heat) as

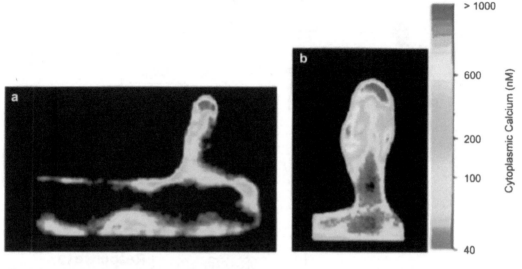

> 1000

600

200

100

40

Cytoplasmic Calcium (nM)

Fig. 14.12 Calcium localization in *Arabidopsis* root hairs.
(a) Wild-type root hair has the highest (red) Ca^{2+} concentration in the rapidly growing tip region.
(b) Root hair from *CA-rop* mutant has high (red) Ca^{2+} levels at multiple points in the swelling (rather than elongating) root hair. Images a and b copyrighted by European Molecular Biology Organization and reprinted by kind permission. Images from Molendijk *et al.* 2001. Images appeared on page 2784 as Figures 9D and L.

well as plant hormones such as brassinosteroids; hence TCH proteins are thought to be important components in response to environmental signals.

Calcium-dependent protein kinases

CDPKs are considered the major type of calcium-binding proteins, with 34 isoforms identified in *Arabidopsis* (White and Broadley, 2003). CDPKs are involved in regulation of events such as nodulation, stomatal movement, phytochrome responses, stress responses, and gravitropism. Many CDPKs contain myristoylation sites that serve as binding sites to membranes (Cheng *et al.*, 2002). CDPKs become phosphorylated when bound to Ca^{2+}, and the activated CDPKs then phosphorylate target proteins (Fig. 14.11). Dephosphorylation by phosphatase is considered a regulatory mechanism for activation and deactivation of CDPKs (Cheng *et al.*, 2002).

Phosphorylation Events: Mitogen-activated Protein Kinases and 14-3-3 Proteins

Mitogen-activated protein kinase (MAPK) cascades have been identified in regulation of stress response pathways, ethylene, brassinosteroid, ABA, and auxin signaling (DeLong *et al.*, 2002; Peck, 2003). MAPK cascades include a series of activations that move from MAPK kinase kinase to MAPK kinase to MAPK, often leading to transcription of target genes. In auxin signaling, MAPK cascades activate binding of transcription factors to auxin-regulated genes (DeLong *et al.*, 2002) (Fig. 14.13). Small GTPases have also been found to activate specific MAPK cascades in signaling events (Molendijk *et al.*, 2001). Interestingly, MAPK signal transduction has been found to be a key component in developmental regulation of stomatal distribution in the

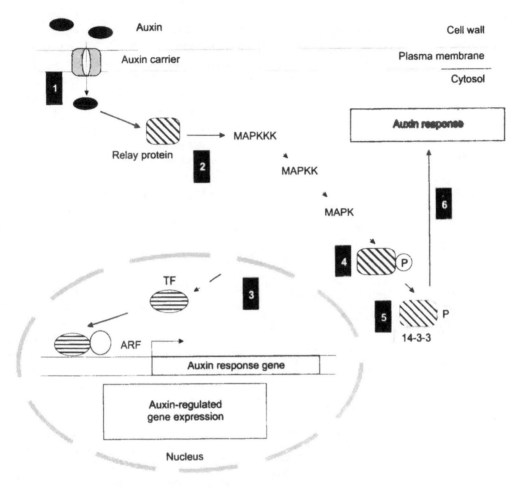

Fig. 14.13 MAPK transduction using auxin signaling as hypothetical example. 1) Auxin enters cell via protein carrier on plasma membrane, then activates relay protein. 2) Relay protein activates MAPKKK (MAPK kinase kinase), which activates MAPKK (MAPK kinase). In turn, MAPKK activates MAPK. 3) Activated MAPK may trigger responses that induce transcription factor (TF) interaction with auxin-response factor (ARF), thus increasing expression of auxin-regulated genes. 4) MAPK transduction cascade may also phosphorylate target protein. 5) A 14-3-3 protein binds to target protein, altering its conformation to its active form. 6) The activated target protein continues signal transduction steps that regulate auxin-induced cytoplasmic events.

leaf epidermis (Bergmann *et al.*, 2004). Signaling components that regulate stomatal distribution were identified through studies of mutants exhibiting altered stomatal patterns. One mutant, *tmm* (too many mouths), produces numerous juxtaposed stomata (Fig. 14.14). TMM is thought to regulate

precise unequal cell division that is the initial step in stomatal development (Geisler *et al.*, 2003). In *tmm* mutants, cell division occurs in a random pattern, altering the spatial distribution of stomata in the leaf epidermis. The TMM protein is a plasma membrane leucine-rich repeat

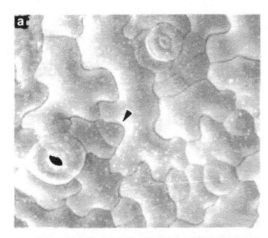

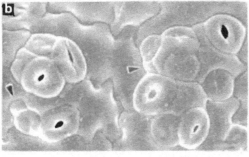

Fig. 14.14 Distribution of stomatal patterning in wild type and mutants of too many mouths. (a) Epidermal cells from wild-type leaf showing normal distribution of stomata and location of asymmetric division, which will form a new stoma (at arrowhead).
(b) Epidermal cells of a leaf of the *tmm* mutant showing clusters of stomata and additional cell divisions where new stomata will form (at arrowheads) Images copyrighted by the American Society of Plant Biologists and reprinted by kind permission. Images from Geisler *et al.* 2000. Images appeared on page 2078 as Figures 3A and C.

receptor-like protein that perceives a developmental or positional signal (Geisler *et al.*, 2003). The TMM protein may transmit information to an MAPKKK called YODA (Bergmann *et al.*, 2004). If *YODA* is overexpressed, no guard cells are produced

in the epidermis, whereas the *yoda* loss-of-function mutant undergoes too many early cell divisions, which affects other patterning events, causing an increased number of guard cells in the leaf epidermis and a surplus of stomata per unit area. Therefore, the YODA MAPKKK may act as a key regulator of epidermal cell development.

Small proteins known as 14-3-3 proteins are important in relaying phosphorylation events. They were first identified during cataloging of proteins in mammalian brain tissue, where they received the three number designation commonly used as the protein's name (Sehnke *et al.*, 2002). In *Arabidopsis*, 14-3-3 proteins have been found to be involved in stress, cold, and blue light signaling. The 14-3-3 proteins act by direct association of phosphorylated proteins such as CDPKs, β-amylase, other protein kinases, and transcription factors. The 14-3-3 proteins dimerize with the target protein, binding to a specific 15-amino acid sequence. It is thought that this binding alters conformation of the target protein to its active form. Therefore, 14-3-3 proteins are considered integral parts of protein kinase transduction cascades. Analysis of proteins for the specific 14-3-3 amino acid target sequence revealed over 500 proteins in *Arabidopsis* that contain two or more 14-3-3 binding sites. Further analysis is required to elucidate the specific roles of the 14-3-3 proteins and their target proteins.

SIGNAL TRANSDUCTION PATHWAYS

Transduction of specific signals involves activation of multiple targets as well as degradation of transduction components in regulation of the response. Genetic and molecular experimental approaches have elucidated many components of plant signal transduction components but the connections between receptors, transduction cascade components, and targets are still being determined. The sections below discuss three transduction pathways based on current experimental evidence, to serve as models of the emerging concepts of plant signaling.

GA Regulation of Amylase Gene Expression

GA is a key hormone in regulation of seed germination (induction of embryonic growth, resulting in root and shoot emergence from the seed coat) and release from dormancy (a physiological state wherein seeds do not germinate). One of the early events in GA-induced germination involves activation of transcription (and then translation) of hydrolytic enzymes such as α-amylase, cell wall degradative enzymes, and proteases. During germination α-amylase breaks down carbohydrate stores to provide energy in the form of glucose, which seedlings use for growth. In a simplified model of GA-induced a-amylase gene expression, GA binds to a receptor in the plasma membrane, activating a G protein transduction cascade (Fig. 14.15) (Gomi and Matsuoka, 2003). The transduction cascade then activates a protein kinase that phosphorylates a nuclear GAI protein (identified in GA insensitive mutants). GAI acts as a negative regulator of an MYB-type transcription factor when unphosphorylated, while phosphorylated GAI is targeted for proteasome degradation. Thus in this pathway, GA removes the suppressor of the GA-activated MYB (GAMYB) transcription factor. With degradation of GAI, transcription proceeds to produce GAMYB, which binds to the regulatory region of the a-amylase promoter, enhancing transcription of that gene.

Phytochrome Regulation of Transcription and Ion Channels

In phytochrome signaling, Pr absorbs red light, which transforms it to its active Pfr form. Yeast two-hybrid experiments revealed that active phytochrome interacts primarily with nuclear proteins such as the PIF transcription factors, and Aux/IAA (in cross talk with auxin signaling). Phytochrome kinase substrate 1 (PKS1) is the only phytochrome-interacting protein that is exclusively cytoplasmic, and is proposed to function by inactivating Pfr (Schäfer and Bowler, 2002). Active Pfr may act by stimulating Ca^{2+} influx (Spalding, 2000), and possibly heterotrimeric G protein transduction cascades, initiating red light-regulated cytoplasmic changes (Wang and Deng, 2002). In a hypothetical model, change from Pr to Pfr leads to activation of a plasma membrane Ca^{2+} channel, causing an increase in cytoplasmic Ca^{2+} level that might activate a signal transduction cascade that targets light-regulated cytoplasmic proteins (Fig. 14.16). Direct phytochrome regulation of gene expression involves movement of active Pfr to the nucleus where it interacts with transcription factors such as the homodimer of PIF3 proteins (Fig. 14.16) (Wang and Deng, 2002). The phytochrome-PIF3 complex promotes transcription of the primary target gene, which encodes an MYB-like transcription factor that regulates secondary light-regulated gene targets. The level of nuclear Pfr is also regulated by COP (constitutive photomorphogenic) proteins that complex to form the COP9 signalosome, a group of proteins involved in degradation and regulation of light-regulated transcription factors in the nucleus. It is proposed that the levels of Pfr, PIFs, and other intermediates in phytochrome signaling are controlled through

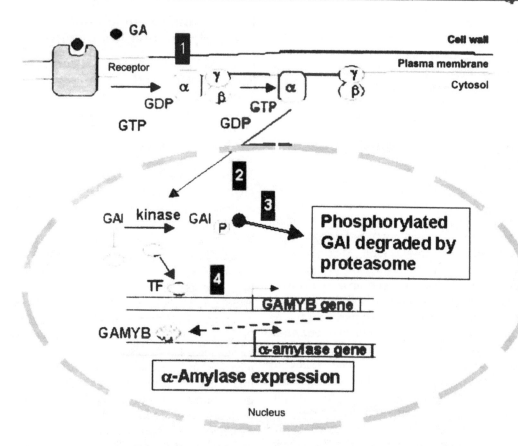

Fig. 14.15 Hypothetical model of GA-induced α-amylase gene expression. 1) GA binds to receptor in plasma membrane and activates G protein transduction cascade. 2) Activated Gα activates protein kinase that phosphorylates nuclear GAI protein. 3) Phosphorylated GAI protein targeted for proteasome degradation. 4) Removal of GAI protein may activate transcription factor (TF) and enhance transcription of GA-activated *MYB* (*GAMYB*) gene. After transcription, translation, and import of GAMYB into the nucleus (indicated by dash arrow), GAMYB binds to the regulatory region of a-amylase promoter, thus promoting its transcription.

proteolysis by the COP9 signalosome. Overall, Pfr level is regulated by multiple events: inactivation by PKS1 binding, proteolysis by the COP9 signalosome, and also by reversion to Pr in darkness.

Signaling in Nodule Initiation

In the complex signaling system between a plant host and its bacterial symbiont, plant-produced substances attract symbiotic bacteria, causing a chemotactic migration of bacteria to the root. Bacteria-produced Nod factors (lipochitin oligosaccharide) act as a chemical signal that triggers infection by the symbiont, root hair curling, and gene expression changes necessary for initiation of root nodule formation (Geurts and Bisseling, 2002). Nod factors induce rapid membrane depolarization with rapid uptake of Ca^{2+} into the cell (~1 min after Nod factor addition in experimental systems). In a current model of the Nod

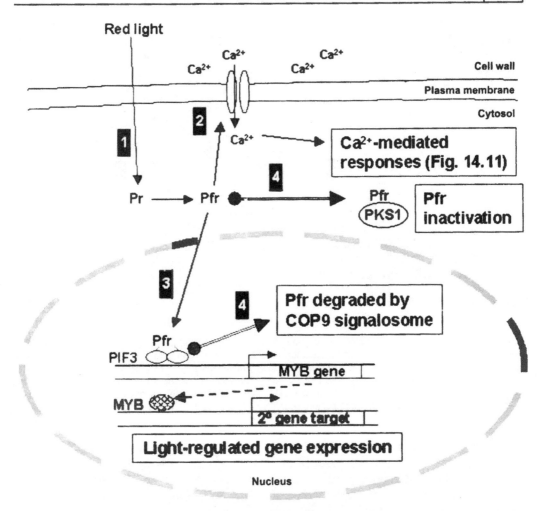

Fig. 14.16 Hypothetical model of phytochrome signaling. 1) Red light initiates change from Pr to Pfr. 2) Pfr leads to activation of plasma membrane Ca²⁺-channel, causing increase in cytoplasmic Ca²⁺ level, leading to signal transduction cascade that affects red-light regulated cytoplasmic events. 3) Pfr also moves to nucleus where it interacts with homodimer phytochrome-interacting factor 3 (PIF3), initiating transcription of primary gene target that encodes MYB-like transcription factor. After transcription, translation, and import of MYB into the nucleus (indicated by dash arrow), MYB regulates expression of secondary gene targets. (4) Pfr level regulated by inactivation both by PKS1 binding and proteolysis by COP9 signalosome.

factor interaction, the Nod factor signal binds to LysM receptor kinases at the plasma membrane, causing dimerization and activation of receptors (Radutolu *et al.*, 2003) (Fig. 14.17). Experimental evidence suggests that the receptor complex induces phospholipid-signaling pathways via PLC and PLD, to produce PA as a secondary messenger (Geurts and Bisseling, 2002). PA may activate key transduction components that in turn activate Ca²⁺ channels at the plasma membrane or internal membranes

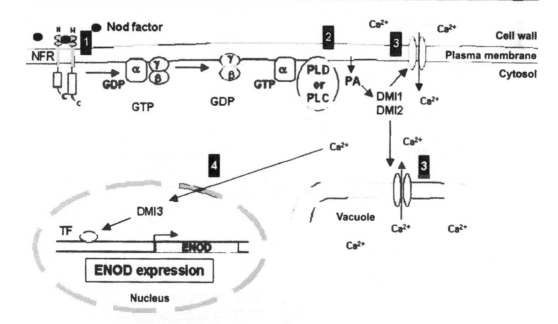

Fig. 14.17 Hypothetical model for Nod-factor regulation of Ca^{2+} influx and ENOD gene expression. 1) Nod factor receptors (NFR) on plasma membrane dimerize upon binding to Nod factor signal. 2) Activated receptor induces phospholipid signaling via G protein activation of PLD or PLC pathways to produce PA. 3) PA activates DMI1 and DMI2 proteins causing Ca^{2+} influx at plasma membrane or through internal membranes such as vacuole. (4) Increased Ca^{2+} initiates cascade to activate DMI3, which increases ENOD gene expression, possibly through activation of transcription factor (TF).

(such as the vacuole) causing an influx of Ca^{2+} to the cytosol. Genetic analyses of *dmi* (does not make infection) mutants reveal gene products that block early steps of nodulation. The *dmi1* and *dmi2* mutants lack Ca^{2+} pulses, indicating that DMI1 and DMI2 proteins are positioned earlier than Ca^{2+} influx in the transduction cascade, while *dmi3* mutants reveal that the DMI3 protein is positioned at a later step (Geurts and Bisseling, 2002). *DMI2* encodes an LRR-RLK, which may interact at the plasma membrane (Limpens and Bisseling, 2003). Ca^{2+} influx within the Nod-induced root hair occurs in pulses that move along the length of the root hair away from the point of perception. One hypothesis is that the wave begins with a rapid Ca^{2+} influx through the plasma membrane Ca^{2+} chan-

nels, followed by later influxes initiated at the Ca^{2+} channels in the internal membrane systems (White and Broadley, 2003). An early physiological response to the Nod factor is a characteristic curling at the root hair tip, an effect sometimes observed within hours. Changes in cytoskeletal arrangement within the cell lead to changes in directional growth that cause the tip of the root hair to curl. Nod factors also induce DMI3, which activates expression of specific *ENOD* (early nodulation) genes in both the root hair and inner cell layers. Many *ENOD* genes encode cell wall proteins.

SIGNALING NETWORKS

In reality, signal transduction cascades do not function independently, but share steps with other signaling pathways. As mentioned in Table 14.2, developmental events such as wounding, phototropism, and seed germination involve the interaction of environmental and chemical signals and signaling pathways. These types of interactions form intricate networks. Characterization of major changes in transcription events will provide more detail and greater understanding of the cross talk among the pathways. However, the major challenge becomes piecing together the information to understand how signaling networks function.

Phototropism is an example of a complex system involving light perception and hormone transport as part of the signaling mechanism. Blue light regulates phototropism by activating the primary photoreceptor phototropin. However, there is also evidence that the pigments cryptochrome and phytochrome act as secondary receptors. Phototropism also relies on auxin accumulation on the shaded side of the stem to produce an auxin gradient, which in turn stimulates auxin-induced cell elongation on the shaded portion of the stem. The PIN3 (pin-formed) auxin-efflux carrier protein, localized in the plasma membrane, regulates lateral auxin transport (Friml et al., 2002). Upon light induction, PIN3 relocates, causing a change in auxin transport toward the shaded side of the stem. Increased auxin in the shaded side then signals increased cellular elongation that causes characteristic curvature toward light. This auxin-transport regulation of unilateral growth can also be found in gravity-sensing tissue, indicating a possible role for PIN3-directed auxin gradient changes in gravitropism.

Recent experimental evidence also reveals a high level of cross talk in stress response pathways. Even though plants respond to a variety of signals, those pathways induced by discrete signals may share common targets, which result in the same cellular response (e.g. reduced growth). For example, significant signaling overlap is associated with UV damage, systemin-wounding, and oligosaccharide elicitor responses (Holley et al., 2003). The interaction of these pathways may be coordinated by a common MAPK pathway, resulting in a common stress response.

Changes that occur from seed dormancy to seed germination are under many regulatory mechanisms. Germination is characterized by the initial imbibition of water into the seed, activation of hydrolytic enzymes that release food stores, and growth of the embryonic root and shoot. In *Arabidopsis*, environmental signals such as cold and light are required for seed germination. Internal signaling via ethylene, GA, brassinosteroids, and ABA interact in regulation of seed germination (Fig. 14.18) (Bentsink and Koornneef, 2002). In *Arabidopsis* seed germination, light activates phytochrome, which is responsible for increased GA biosynthesis via activation of GA 3β-hydroxylase, the enzyme that catalyzes active GA production (Bentsink and Koornneef, 2002). Responses to GA include activation of α-amylase to break down starch in order to provide energy for cellular growth within the embryo. GA, BR, and ethylene all act as positive regulators of seed germination in *Arabidopsis*, whereas ABA directly inhibits growth of the embryo. Studies of the interactions of ethylene, ABA, and sugar response mutants revealed complex cross talk among these molecules. Ethylene acts as a negative regulator of ABA, thus inducing germination in dormant seeds. The role of sugar as a signal in seed germination is complex. It usually acts as a negative regulator by stimulating ABA biosynthesis; however, under high ABA conditions sugar can stimulate germination,

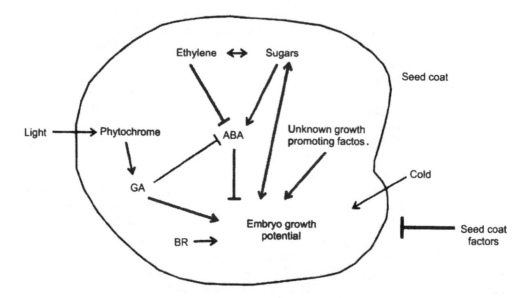

Fig. 14.18 Overview of network of environmental and hormonal signaling events that regulate seed germination in *Arabidopsis*. *Arabidopsis* seeds require both cold and light treatments prior to germination, shown as positive regulators induce germination. GA, ethylene, and brassinosteroid (BR) also act as positive regulators, whereas ABA and seed coat factors negatively regulate (⊥) germination. In this model, light activates phytochrome, which is responsible for increased GA biosynthesis and induction of embryo growth. Ethylene acts as negative regulator of ABA, thus inducing germination in dormant seeds. Sugar is shown as a negative regulator by its stimulation of ABA biosynthesis. However, under high ABA conditions sugar can stimulate germination. In addition, there is experimental evidence for involvement of seed coat factors, unknown growth factors, and cross-talk between ethylene and sugar in the complex regulation of seed germination in *Arabidopsis*. Image copyrighted by American Society of Plant Biologists and reprinted by kind permission. Images from Bentsink and Koornneef 2002. Image appeared on page 10 as Figure 2.

possibly by providing an energy source for the seedlings. In addition, there is experimental evidence that ethylene and sugar cross talk, chemical factors produced in the seed coat, and unknown growth factors contribute to the regulation of gene germination. Thus seed germination depends upon a set of complex interactions of environmental factors and hormones. These complex transduction scenarios involving cross talk among multiple pathways continue to be examined and experimentally resolved.

FUTURE CONSIDERATIONS

With the advent of genomic approaches, the components involved in signal transduction networks are rapidly being identified. Sequence homology provides a powerful tool for preliminary identification of potential signal-transduction proteins. However, it has become clear that the complexity of these networks involves hundreds of interacting signaling components with significant cross talk among pathways. The models will change as we reach the goal of the National Science Foundation's 2010 Project to "support research to determine the function of all genes in the model plant *Arabidopsis thaliana.*" While understanding the function of all genes will provide a huge body of information that will contribute to new models of signal transduction networks for plants, a major challenge will be to identify which gene products act as key regulatory switches that provide the basic mechanisms that regulate complex physiological changes.

References

Arteca, R.N. 1996. *Plant Growth Substances—Principles and Applications.* Chapman & Hall, New York, NY, pp. 1-27.

Assmann, S.M. 2002. Heterotrimeric and unconventional GTP binding proteins in plant cell signaling. Plant Cell 14: S355-S373.

Bayliss, W.M. and Starling, E. 1904. The chemical regulation of the secretory process. Proc. Roy. Soc. (Series B) 73: 310-322.

Bentsink, L. and Koornneef, M. 2002. Seed dormancy and germination. In: *The Arabidopsis Book.* C.R. Somerville and E.M. Meyerowitz, (eds.). Amer. Soc. Plant Biologists, Rockville, MD, USA, doi/10.1199/tab.0050. http://www.aspb.org/publications/ arabidopsis/.

Bergmann, D., Lukowitz, W., and Somerville, C. 2004. Stomatal development and pattern controlled by MAPKK kinase. Science 304: 1494-1497.

Bishop, G.J. and Koncz, C. 2002. Brassinosteroids and plant steroid hormone signaling. Plant Cell 14: S97-S110.

Braam, J., Sistrunk, M.L., Polisensky, D.H., Xu, W., Purugganan, M.M., Antosiewicz, D.M., Campbell, P., and Johnson, K.A. 1997. Plant responses to environmental stress: regulation and functions of the *Arabidopsis TCH* genes. Planta 203: S35-S41.

Cheng, S.-H., Willmann, M.R., Chen, H.-C., and Sheen, J. 2002. Calcium signaling through protein kinases. The *Arabidopsis* calcium-dependent protein kinase gene family. Plant Physiol. 129: 469-485.

Council of Biology Editors, Style Manual Committee. 1994. Scientific Style and Format: The CBE Manual for Authors. Cambridge Univ. Press, New York, NY, (eds.) pp. 369 (6th ed.).

De Long, A., Mockaitis, K., and Christensen, S. 2002. Protein phosphorylation in the delivery of and response to auxin signals. Plant Molec. Biol. 49: 285-303.

Diévart, A. and Clark, S.E. 2003. Using mutant alleles to determine the structure and function of leucine-rich repeat receptor-like kinases. Curr. Opin. Plant Biol. 6: 507-516.

Fields, S. and Song, O. 1989. A novel genetic system to detect protein-protein interactions. Nature 340: 245-246.

Friml, J., Wiśniewska, J., Benková, E., Mendgen, K., and Palme, K. 2002. Lateral relocation of auxin efflux regulator PIN3 mediates tropism in *Arabidopsis.* Nature 415: 806-809.

Fu, Y., Li, H., and Yang, Z. 2002. The ROP2 GTPase controls the formation of cortical fine F-actin and the early phase of directional cell expansion during *Arabidopsis* organogenesis. Plant Cell 14: 777-794.

Geisler, M.J., Nadeaw, J.A., and Sack, F.D. 2000. Orientated asymmetric divisions that generate the stomatal spacing pattern in *Arabidopsis* are dissupted by the too many months mutations. Plant Cell 12: 2075-2096.

Geisler, M., Jablonska, B., and Springer, P.S. 2002. Enhancer trap expression patterns provide a novel teaching resource. Plant Physiol. 130: 1747-1753.

Geisler, M.J., Deppong, D.O., Nadeau, J.A., and Sack, F.D. 2003. Stomatal neighbor cell polarity and division in *Arabidopsis.* Planta. 216: 571-579.

Geurts, R. and Bisseling, T. 2002. Rhizobium Nod factor perception and signaling. Plant Cell 14: S239-S249.

Gomi, K. and Matsuoka, M. 2003. Gibberellin signalling pathway. Curr. Opin. Plant Biol. 6: 489-493.

Hellmann, H. and Estelle, M. 2002. Plant development: Regulation by protein degradation. Science 297: 793-798.

Holk, A., Rietz, S., Zohn, M., Quader, H., and Scherer, G.F.E. 2002. Molecular identification of cytosolic, polatin-related phospholipases A from *Arabidopsis* with potential functions in plant signal transduction. Plant Physiol. 130:90-101.

Holley, S.R., Yalamanchili, R.P., Moura, D.S., Ryan, C.A., and Stratmann, J.W. 2003. Convergence of signaling pathways induced by systemin, oligosaccharide elicitors, and ultraviolet-B radiation at the level of mitogen-activated protein kinases in *Lycopersicon peruvianum* suspension-cultured cells. Plant Physiol. 132: 1728-1738.

Hutchinson, C.E. and Kieber, J.J. 2002. Cytokinin signaling in *Arabidopsis*. Plant Cell 14: S47-S59.

Hwang, I., Chen, H-C., and Sheen, J. 2002. Two-component signal transduction pathways in *Arabidopsis*. Plant Physiol. 129: 500-515.

Jones, A.M. 2002. G-protein-coupled signaling in *Arabidopsis*. Curr. Opin. Plant Biol. 5: 402-407.

Jones, M.A., Shen, J.-J., Fu, Y., Li, H., Yang, Z., and Grierson, C.S. 2002. The *Arabidopsis* Rop2 GTPase is a positive regulator of both root hair initiation and tip growth. *Plant Cell* 14: 763-776.

Kepinski, S. and Leyser, O. 2002. Ubiquitination and auxin signaling: A degrading story. Plant Cell 14: S81-S95.

Khan, A.R., Johnson, K.A., Braam, J., and James, M.N. 1997. Comparative modeling of the three-dimensional structure of the calmodulin-related TCH2 protein from *Arabidopsis*. Proteins 27: 144-153.

Li, J. 2003. Brassinosteroids signal through two receptor-like kinases. Curr. Opin. Plant Biol. 6: 494-499.

Limpens, E. and Bisseling, T. 2003. Signaling in symbiosis. Curr. Opin. Plant Biol. 6: 343-350.

McCourt, P. 1999. Genetic analysis of hormone signaling. Annu. Rev. Plant Physiol. Plant Molec. Biol. 50: 219-243.

Molendijk, A.J., Bischoll, F., Rajendrakumar, C.S.V., Friml, J., Braun, M., Giloy, S., and Palme, K. 2001. *Arabidopsis thaliana* Rop GTPases are localized to tips of root hairs and control polar growth. EMBO J. 20: 2779-2788.

Munnik, T. and Musgrave, A. 2001. Phospholipid signaling in plants: Holding on to phospholipase D. *Science STKE* DOI: 10.1126/stke.2001.111.pe42.

Peck, S.C. 2003. Early phosphorylation events in biotic stress. Curr. Opin. Plant Biol. 6: 334-338.

Radutolu, S., Madsen, L.H., Madsen, E.B., Felle, H.H., Umehara, Y., Grønlund, M. *et al.* 2003. Plant recognition of symbiotic bacteria requires two LysM receptor-like kinases. Nature 425: 585-592.

Rolland, F., Moore, B., and Sheen, J. 2002. Sugar sensing and signaling in plants. Plant Cell 14: S185-205.

Roman, G., Lubarsky, B., Kieber, J.J., Rothenberg, M., and Ecker, J.R. 1995. Genetic analysis of ethylene signal transduction in *Arabidopsis thaliana*: five novel mutant loci integrated into a stress response pathway. Genetics 139: 1393-1409.

Ryan, C.A., Pearce, G., Scheer, J., and Moura, D.S. 2002. Polypeptide hormones. Plant Cell 14: S251-S264.

Schäfer, E. and Bowler, C. 2002. Phytochrome-mediated photoperception and signal transduction in higher plants. EMBO Repts. 3: 1042-1048.

Sehnke, P.C., DeLille, J.M., and Ferl, R.J. 2002. Consummating signal transduction: The role of 14-3-3 proteins in the completion of signal-induced transitions in protein activity. Plant Cell 14: S339-S354.

Spalding, E.P. 2000. Ion channels and the transduction of light signals. Plant Cell Envir. 23: 655-674.

Suharsono, U., Fujisawa, Y., Kawasaki, T., Iwasaki, Y., Satoh, H., and Shimamoto, K. 2002. The heterotrimeric G protein α subunit acts upstream of the small GTPase Rac in disease resistance in rice. Proc. Natl. Acad. Sci. USA 99: 13307-13312.

Taiz, L. and Zeiger, E. 2002. *Plant Physiology*. Sinauer Asso., Inc., Sunderland, MA, USA (3rd ed.).

Turner, J.G., Ellis, C., and Devoto, A. 2002. The jasmonate signal pathway. Plant Cell 14: S153-S164.

Wang, H. and Deng, X.W. 2002. Phytochrome signaling mechanism. In: *The Arabidopsis Book*. Somerville, C.R. and Meyerowitz, E.M. (eds.). Amer. Soc. Plant Biologists, Rockville, MD, doi/10.1199/tab.0074, http://www.aspb.org/publications/arabidopsis/.

Wang, K.L.-C., Hai, L., and Ecker, J.R. 2002. Ethylene biosynthesis and signaling networks. Plant Cell 14: S131-S151.

White, P.J. and Broadley, M.R. 2003. Calcium in plants. Ann. Bot. 92: 487-511.

Yang, Z. 2002. Small GTPases: Versatile signaling switches in plants. Plant Cell 14: S375-S388.

Zhao, J., Reng, P., Schmitz, R.J., Decker, A.D., Tax, F.E., and Li, J. 2002. Two putative BIN2 substrates are nuclear components of brassinosteroid signaling. Plant Physiol. 130: 1221-1229.

Zhu, T. 2003. Global analysis of gene expression using GeneChip microarrays. Curr. Opin. Plant Biol. 6: 415-425.

Index

T - #0619 - 101024 - C0 - 229/179/25 - PB - 9781578083763 - Gloss Lamination